APPLIED CALCULUS FOR SCIENTISTS AND ENGINEERS

A Journey in Dialogues

Volume 1

Frank Blume

With Contributions from

Calvin Piston

Cover credit: © Chaoss | Dreamstime.com

ISBN-10: 150102065X
ISBN-13: 978-1501020650

For
Ferdi and Babette

Acknowledgments

First and foremost, I would like to express my gratitude to Calvin Piston, not only for his various contributions to this text, but also for his support during my sabbatical leave in the fall of 2001. I also am thankful to my colleague Peter Pohle for his skillful editing of spirals on pinecones, and to Kathy Mayer for scanning three hand-drawn pigs.

Preface and Guidelines for the Reader

This book is intended as an introduction to higher mathematics for undergraduates in science, engineering, and mathematics. The material covered is sufficient for a four semester course sequence that includes one semester on differential equations, but it is equally well suited for a more traditional course consisting of two semesters of single variable calculus and perhaps a third semester on vector calculus. The presentation of vector calculus in Part IX is almost entirely independent of the preceding parts on differential equations and linear algebra, and to skip these parts therefore causes no problems at all.

To successfully study the text, the reader needs a working knowledge of elementary concepts in arithmetic and algebra. Familiarity with trigonometry and analytic geometry is helpful, but not required.

Centered on the idea that effective teaching must bring knowledge to life, the book is designed to present calculus as a living discipline with a multiplicity of applications in science and engineering, and with close ties to the history and philosophy of the Western World. Indeed, almost every subject is developed in the context of a meaningful application, and it is almost always the application that comes first. We have taken great care to select examples that are not only relevant from a mathematical point of view, but also stimulating and interesting for the average reader. The variety of topics is far-ranging, from mechanics and the laws of motion to electric circuits and even a rather careful introduction to elementary probabilistic phenomena.

To enhance the quality of exposition we have inserted discussions in dialogue form at crucial junctions throughout the text. A dialogue is more likely to catch the reader's attention than ordinary mathematical prose and allows for the insertion of typical student questions that are ordinarily raised only in the classroom. We hope that this approach will increase the text's accessibility and render the relevant problem-solving strategies more transparent.

The dialogues involve three participants: the Teacher, Sophie—a very bright student—and Simplicio—a student of more average abilities. The name "Simplicio" has been adopted from Galileo's dialogues concerning *The Two New Sciences* and his *Dialogues on the Two Chief Systems of the World.*

The development of calculus was part of a historical process that profoundly changed the course of Western Civilization. To raise the reader's awareness of this larger cultural context we have included historical chapters on the lives of famous mathematicians and various philosophical discussions concerning the nature of science and mathematics.

Regarding the text's organization, we wish to emphasize that differential equations are covered before vector calculus because most students in science and engineering will encounter examples of differential equations far more frequently and far earlier than applications of vector calculus. This arrangement of the subjects does not cause any conceptual problems and is pedagogically beneficial. Furthermore, throughout the text there are remarks, exercises, sections, and entire chapters marked with a ★. Typically, a ★ indicates a higher degree of difficulty or identifies content that is not essential to the overall development. For instance, in Chapter 3 we introduce a rigorous definition of limits of functions that is interesting from a theoretical point of view, but not essential for understanding any of the subsequent applications. In general, the reader is well advised not to consider passages marked with a ★ as a source of frustration. If something doesn't make sense it is best to leave it aside and to go on with the main part of the text.

FRANK BLUME

Contents

Introduction

If asked to name one characteristic of the world we live in that does not appear to undergo change we might be well advised to point to the constancy of change itself. Change in endless variations occurs everywhere and at all times. Not surprisingly then, any quantitative inquiry into the nature of the physical universe requires an understanding of the mathematics of change, also known as *calculus*. To be more precise, the theme of change is central to *differential calculus*, while the mathematical description of accumulation of change lies at the heart of *integral calculus*. Common to both of these, and in this sense most fundamental, is the theme of approximation.

Historically the calculus was developed in response to problems posed by investigations into the laws of motion. Following the pioneering work of scientists such as Galileo and Kepler, Isaac Newton introduced in the late 17th century his "method of fluxions" as the mathematical foundation for the formulation of the fundamental laws of mechanics. At about the same time, Gottfried Wilhelm Leibniz independently arrived at his own version of the calculus which was in some ways slightly different from Newton's and in some ways slightly better. The invention of the calculus stands out as a towering achievement in the history of human thought. Unrivaled in its power and scope, it has played a dominant role in the development of science for more than three hundred years.

Accepting history as our guide, we will now illustrate some of the most basic concepts of differential and integral calculus with examples relating to the laws of motion. Here the theme of change will be encountered in changes of position and time.

Finding Velocity from Position and Time

Teacher: Let us begin with a seemingly simple question: how do we measure velocity?

Simplicio: As distance per time, I suppose.

Teacher: What exactly do you mean by that?

Simplicio: Well, as we observe, for example, two objects moving at different velocities, we notice that the faster object covers a greater distance in the same amount of time than the slower one. In other words, the ratio of distance and time is greater for the faster object than for the slower one. So it appears plausible to take this ratio as a measure for velocity.

Teacher: It does seem plausible, but to explore this idea a little further let me ask you a more specific question: how would you determine the velocity of an object that passes through the positions s_1 and s_2 (on a given coordinate axis) at the respective times t_1 and t_2?

Simplicio: I would evaluate the quotient $(s_2 - s_1)/(t_2 - t_1)$. For under the assumptions you stated, $s_2 - s_1$ is the distance traveled and $t_2 - t_1$ the length of time.

Teacher: Your answer regarding the velocity is correct, but I disagree with your assertion that $s_2 - s_1$ is the distance traveled. Consider, for example, an object that is moving back and forth until it comes to a halt at its original position. In this case the total change in position is zero because $s_1 = s_2$, but the total distance traveled is certainly different from zero.

Simplicio: That makes sense. I suppose it is therefore more accurate to define velocity as change in position over change in time rather than distance over time.

Teacher: Exactly.

Sophie: I must admit that I am not entirely comfortable with this definition, because I don't see how we are justified in characterizing the velocity by a single number—the ratio of $s_2 - s_1$ and $t_2 - t_1$—given that the velocity of an object is not necessarily constant over the time span from t_1 to t_2.

Teacher: That is an excellent point, and I should have been more careful in my use of language. Instead of "velocity" I should have said "average velocity," for it is the average velocity that we determine via the ratio $(s_2 - s_1)/(t_2 - t_1)$. By contrast, the velocity of an object at one particular moment in time is referred to as the *instantaneous velocity.* In our notation we express this distinction by writing $\bar{v}$ for the average velocity and v for the instantaneous velocity. Thus, we have

$$\bar{v} = \frac{s_2 - s_1}{t_2 - t_1}. \tag{1}$$

0.1 Example. In Figure 1 the path of a car on its way from Siloam Springs to Tulsa is given with some of the points in time and position indicated. Since in the interval from 13 *min* to 61 *min* the change in

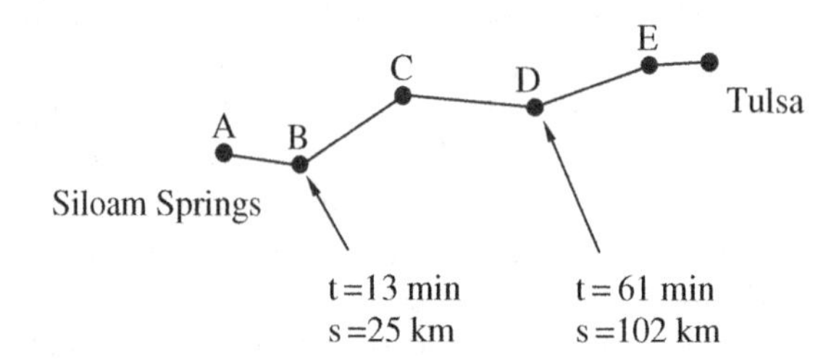

Figure 1: path of a car.

position is 102 − 25 *km*, formula (1) allows us to determine the average velocity between the points B and D as follows:

$$\bar{v}_{BD} = \frac{102 - 25}{61 - 13}\frac{km}{min} = \frac{77}{48}\frac{km}{min}.$$

0.2 Exercise. On the map in Figure 1, find the coordinates t and s for the points C and E and determine the average velocity $\bar{v}_{CE}$. You may assume the distances from B to C, C to D, and D to E to be equal. Furthermore, the time it takes to travel from B to C is seven eighths of the time needed to get from D to E, which in turn is equal to half the time spent on the segment from B to D.

Sophie: I understand how we can find the average velocity of an object given its position at certain points in time, but what about the instantaneous velocity?

Simplicio: It's simple: we only need to determine the change in position at a particular moment in time and divide by the change in t.

Teacher: That sounds simple indeed, but how long is one moment in time? Is it one minute, one second, or one tenth of a second?

Simplicio: It's as small as you like. The smallest interval of time that you can imagine.

Sophie: But there can be no such thing as a "smallest interval of time." For we can always cut a given interval in half to get an even smaller one—unless, of course, the length of the interval is zero.

Teacher: This is exactly where we have a problem. For if the time span $t_2 - t_1$ is zero, the fraction $(s_2 - s_1)/(t_2 - t_1)$ is undefined. To illustrate how this difficulty can be overcome we will now go back in history to Galileo's discovery of the *law of falling bodies* in the early 17th century.

Let us consider an object that is dropped from a certain height close to the surface of the earth. Neglecting all forces other than gravity, the law that describes the distance fallen in dependence on time was originally stated by Galileo in the following form:

The distances fallen in successive time intervals increase in proportion to the odd integers.

So suppose that in the first interval of time, an object has fallen one unit of distance. (Note: here the units are not necessarily feet, meters, seconds or any other common units of measurement, but simply units of time and length that may have been chosen for one particular experiment.) Then in the second interval of time it will fall an additional three units, followed by five units in the third. We record these values in the following table:

time	distance fallen
0	0
1	1
2	1+3=4
3	1+3+5=9
4	1+3+5+7=16
5	1+3+5+7+9=25
6	1+3+5+7+9+11=36
7	1+3+5+7+9+11+13=49

Given this table, we notice that the relation between distance and time is accurately described by the equation $s = t^2$ ($s = 1^2 = 1$ for $t = 1$, $s = 2^2 = 4$ for $t = 2$ and so forth). The algebraic relation $s = t^2$ can also be represented by a graph that shows the value of s in dependence on t. For every t on the horizontal axis we mark the value $s = t^2$ on the vertical axis and plot a point at position (t, t^2). The diagram in Figure 2 shows the resulting curve.

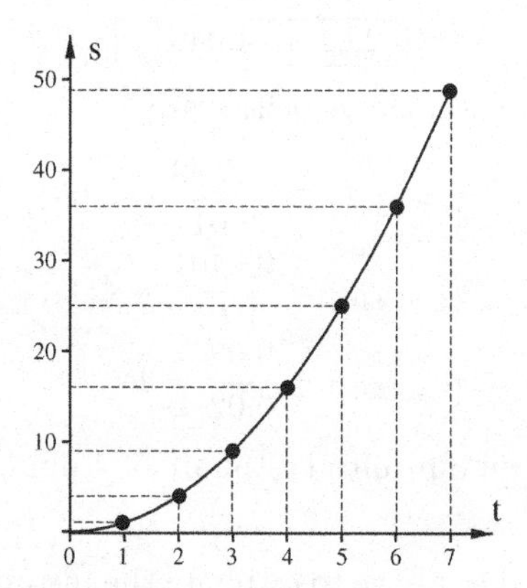

Figure 2: time/distance graph for a falling body.

Using the values in the table above, we see that the average velocity over the time interval from $t_1 = 1$ to $t_2 = 2$ is

$$\bar{v}_{[1,2]} = \frac{4-1}{2-1} = 3 \quad \text{(measured in units of distance per units of time).}$$

Here we use the interval notation $[1, 2]$ to denote the time span from $t_1 = 1$ to $t_2 = 2$. In general, such a time span will be denoted by $[t_1, t_2]$ (see also Chapter 1).

0.3 Exercise. Using the table above or the equation $s = t^2$, find the average velocities over the time intervals $[0, 2]$, $[1, 4]$, and $[2, 6]$.

Teacher: Let us now return to the problem of finding the instantaneous velocity. We already saw that the length of the interval $[t_1, t_2]$ cannot be reduced to zero as this would cause the ratio in formula (1) to be undefined. However, to make this interval very small is obviously possible.

Sophie: Are you suggesting that we ought to *approximate* the instantaneous velocity by calculating average velocities over successively smaller time intervals?

Teacher: Exactly. To illustrate this method I would like to show you how to estimate the instantaneous velocity of a falling body at time $t = 2$. To begin with I will use the equation $s = t^2$ to construct a new table of values for s with values for t that are close to 2.

time	distance
1.9	3.61
2.0	4.00
2.1	4.41

According to this table, the average velocities over the intervals $[1.9, 2.0]$, $[2.0, 2.1]$, and $[1.9, 2.1]$ are

$$\bar{v}_{[1.9,2.0]} = \frac{0.39}{0.1} = 3.9,$$
$$\bar{v}_{[2.0,2.1]} = \frac{0.41}{0.1} = 4.1,$$
$$\bar{v}_{[1.9,2.1]} = \frac{0.8}{0.2} = 4.$$

Thus, it appears that the instantaneous velocity at time $t = 2$ is approximately equal to 4. To strengthen our confidence in the accuracy of this result, it is helpful to generate improved estimates with even smaller intervals of time. Choosing values for t that are only 0.01 units apart, we obtain:

time	distance
1.99	3.9601
2.00	4.0000
2.01	4.0401

The corresponding values for the average velocities are

$$\bar{v}_{[1.99,2.00]} = \frac{0.0399}{0.01} = 3.99,$$
$$\bar{v}_{[2.00,2.01]} = \frac{0.0401}{0.01} = 4.01,$$
$$\bar{v}_{[1.99,2.01]} = \frac{0.08}{0.02} = 4.$$

These results convincingly confirm our initial estimate of 4 units of distance per unit of time for the instantaneous velocity at $t = 2$.

0.4 Exercise. Use the same method as above to estimate the instantaneous velocity at $t = 4$ and $t = 6$.

Simplicio: This is all very interesting, but we still have not found the *precise* value of the instantaneous velocity at $t = 2$. How are we going to come up with an exact solution if all we ever do is to calculate average velocities?

Teacher: Simplicio is right, the crucial step is still missing: we need to find the *limiting value* of the average velocities over successively smaller time intervals close to $t = 2$. This idea of finding a *limit* is one of the most important concepts in calculus, and we will study it very carefully in Chapters 2 and 3. For the moment, though, we need to be content with only approximate values for the instantaneous velocity.

Sophie: Teacher, you showed us how the equation $s = t^2$ can be represented graphically. So I am wondering whether it might not be possible to somehow read the value of the instantaneous velocity at $t = 2$ from the graph in Figure 2.

Teacher: The concept of instantaneous velocity is indeed easily given a geometric interpretation. To explain this, let me connect two points (t_1, s_1) and (t_2, s_2) on the graph described by the equation $s = t^2$ by a straight line (see Figure 3). Such a line is called a *secant line*, and its *slope* is by definition the ratio of the differences in the s- and t-values. In other words:

The slope of the secant line is equal to the average velocity $\bar{v} = \dfrac{s_2 - s_1}{t_2 - t_1}$.

Consequently, the calculation of average velocities over successively smaller intervals of time that we used to estimate the instantaneous velocity has a geometric analogue in the computation of secant

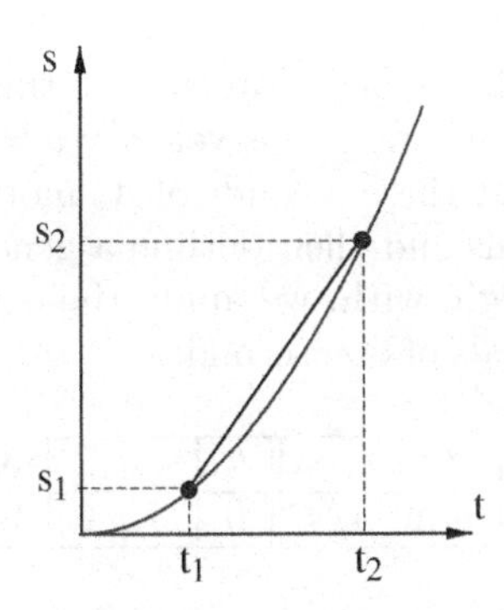

Figure 3: graph with secant line.

slopes through points on the graph that are successively closer to each other. To appreciate the significance of this observation let us take a more detailed look at the point $(2, 4)$ on the graph in Figure 2. As we magnify the region near this point, the graph appears to be almost a straight line (see Figure 4).

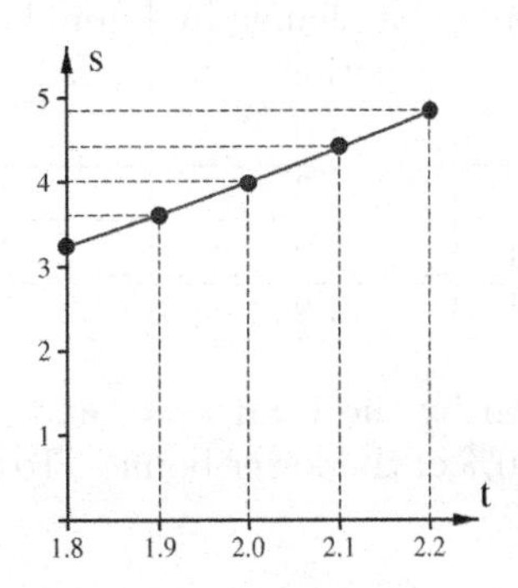

Figure 4: zooming in on the time/distance graph.

Sophie: Let me see whether I understand this: in calculating slopes of secant lines that connect the point $(2, 4)$ with other points close by, we essentially determine average velocities over small intervals of time close to $t = 2$. Since these average velocities approach the instantaneous velocity at $t = 2$, it follows that the secant slopes approach the instantaneous velocity as well. Furthermore, the fact that the graph close to $(2, 4)$ is almost a straight line shows that the secant slopes also approach the slope of the *tangent line* to the graph at $(2, 4)$. Thus, we arrive at the following conclusion:

The instantaneous velocity is equal to the tangent slope of the graph that shows s in dependence on t.

Simplicio: I am afraid this is beyond me.

Teacher: Don't worry about your lack of understanding at this point, early in the course. You will be given ample opportunity to revisit this argument in the lessons to come.

Finding Position from Velocity and Time

The theme of describing accumulation of change is related to the problem of finding the change in an object's position given its instantaneous velocity at any point in time. In the simple case of a constant velocity, the problem is easily solved by multiplying the velocity with the change in time. For instance, if an object travels at the constant velocity $v = 3\,m/s$ (meters per second), then, over a time span of 2 seconds, the change in its position is $3 \cdot 2 = 6\,m$.

To discuss what happens when the velocity varies over time, we will consider the example of an object that at time t travels with the instantaneous velocity $v = 3t$. Our goal will be to find the change in the object's position during the first three seconds of its motion. As in the previous section, we will at first determine only approximations and then outline a general method for finding exact solutions in the form of *limiting values.* To begin with, we apply the equation $v = 3t$ to generate a table that records the object's velocity at intervals of one second:

t in *sec*	0	1	2	3
v in m/s	0	3	6	9

For the first second, from $t = 0$ to $t = 1$, the least value for the object's velocity is $0\,m/s$ and the greatest is $3\,m/s$. Thus, during the first second, the object travels no less than 0 and no more than 3 meters. For the second second, the least velocity is $3\,m/s$ and the greatest is $6\,m/s$. Consequently, the change in position has to be between $3\,m$ and $6\,m$. We repeat this calculation for the third second and summarize our results:

time interval	lower bound for change in position in m	upper bound for change in position in m
$[0, 1]$	0	3
$[1, 2]$	3	6
$[2, 3]$	6	9
total	9	18

The resulting approximation, as given by the total lower and upper bounds, is rather crude, with a margin of error of $18 - 9 = 9\,m$ or 100% of the lower bound. To improve this estimate we set up a new table with time intervals of length 0.25:

time interval	min. velocity in m/s	max. velocity in m/s	lower bound for change in position in m	upper bound for change in position in m
$[0.00, 0.25]$	0.00	0.75	0.0000	0.1875
$[0.25, 0.50]$	0.75	1.50	0.1875	0.3750
$[0.50, 0.75]$	1.50	2.25	0.3750	0.5625
$[0.75, 1.00]$	2.25	3.00	0.5625	0.7500
$[1.00, 1.25]$	3.00	3.75	0.7500	0.9375
$[1.25, 1.50]$	3.75	4.50	0.9375	1.1250
$[1.50, 1.75]$	4.50	5.25	1.1250	1.3125
$[1.75, 2.00]$	5.25	6.00	1.3125	1.5000
$[2.00, 2.25]$	6.00	6.75	1.5000	1.6875
$[2.25, 2.50]$	6.75	7.50	1.6875	1.8750
$[2.50, 2.75]$	7.50	8.25	1.8750	2.0625
$[2.75, 3.00]$	8.25	9.00	2.0625	2.2500
total			12.3750	14.6250

To demonstrate how this table was generated we calculate explicitly the values for the second interval $[0.25, 0.50]$. Using the equation $v = 3t$, we find the minimal and maximal velocities to be $3 \cdot 0.25 = 0.75$ and $3 \cdot 0.50 = 1.50$ respectively. Since the length of the time interval $[0.25, 0.50]$ is 0.25, the lower bound for the change in position is $0.75 \cdot 0.25 = 0.1875$, while the upper bound is $1.50 \cdot 0.25 = 0.3750$.

Given the lower and upper estimates for the total change in position in the second table, we see that the margin of error has been narrowed down to $14.625 - 12.375 = 2.25\,m$, and it could be narrowed down further by reducing again the length of the time intervals.

0.5 Exercise. Using the method explained above and taking time intervals of length 0.2 seconds each, find lower and upper bounds for the total change in position between $t = 0$ and $t = 3$.

In both tables above, the average of the total lower and upper bounds is $13.5\,m$. We will see that in our particular example this average is actually exactly equal to the change in position over the interval from $t = 0$ to $t = 3$, but the reader is warned against any generalizations. In "most" cases the average of upper and lower bounds will not give us the correct answer.

From a geometric point of view, our estimates for the total change in position represent approximations for the area under the graph given by the equation $v = 3t$. To justify this claim we argue as follows: In order to find a lower estimate for the change in position over an interval $[t_1, t_2]$, we multiply the least velocity $v = 3t_1$ by the length of the time interval, which is $t_2 - t_1$. The product $3t_1(t_2 - t_1)$ is equal to the area of a rectangle of height $3t_1$ and width $t_2 - t_1$. Therefore, in adding up the lower estimates for the changes in position over each of the subintervals between $t = 0$ and $t = 3$, we obtain a lower estimate for the area under the graph given by the equation $v = 3t$. Similarly, an upper estimate for this area is represented by the sum of the upper estimates for the changes in position (see Figure 5). The precision of these estimates increases as the width of the rectangles decreases, and in the *limit*, as

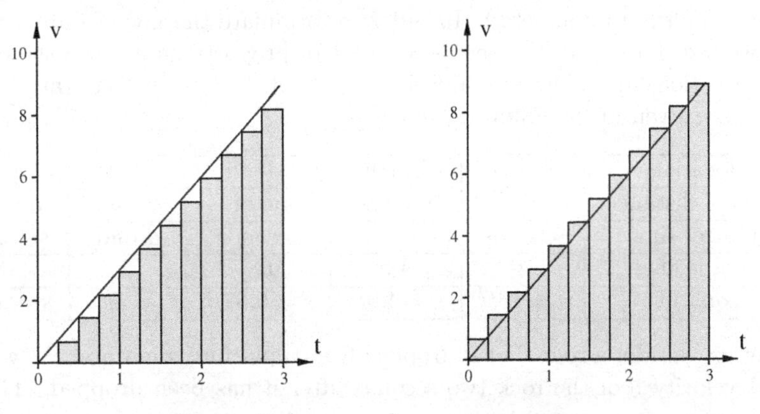

Figure 5: lower and upper estimates.

the width approaches zero, the total area of the rectangles (that is, the estimate for the total change in position) approaches the area under the graph. Thus, we arrive at the following conclusion:

> The change in the position of an object is equal to the area between the t-axis and the graph that shows its velocity in dependence on t. (2)

Remark. If a moving object reverses its direction of travel, then the velocity changes from positive to negative or vice versa. In such a case the graph of v is partly above and partly below the t-axis, and the total change in position is to be determined by subtracting the areas below the t-axis from those above the t-axis. We will discuss this case in more detail in Chapter 17.

Furthermore, in the example above, the area under the graph of $v = 3t$ over the interval $[0, 3]$ is simply the area of a right triangle of height 9 and width 3. Consequently, according to the principle stated in (2), the total change in position over the time interval $[0, 3]$ is $9 \cdot 3/2 = 13.5$.

★ *Remark.* For the skeptical reader we would like to add a general note of caution regarding the relation between calculus concepts such as "instantaneous velocity" and the physical reality that these concepts are supposed to convey. The fact that certain observable phenomena at a given level of resolution can be efficiently described in the language of calculus does not at all imply that the abstract world of mathematics is a true mirror image of the physical world around us. There are numerous objections that can be brought to bear against such a naive identification of a theoretical model with the objective reality that it is intended to represent. For instance, to accept as real the concept of instantaneous velocity requires that we believe in a *continuous* flow of time. For otherwise an infinite succession of approximations with average velocities taken over *arbitrarily small* intervals of time would be inconceivable. But what if time is discrete? What if the notion of continuous time is just as illusory as the seamlessly flowing motion on a movie screen that results from a rapid display of discrete static images? In this case, the concept of instantaneous velocity would evaporate into ontological thin air—without diminishing, of course, its usefulness in calculations. The reader with an interest in questions of this sort is encouraged to consult the philosophical literature in order to deepen his or her confusion (see, for example, [Ste]). ★

Additional Exercises

0.6. In our discussion of Galileo's law of falling bodies we did not specify any particular units of measurement for time and distance. In practice, though, we want units that are properly adjusted to a given situation. Changing from one system of units to another requires a multiplication by a certain scaling factor (for example: feet to inches, multiply by 12; centimeters to meters, multiply by 0.01; minutes to seconds, multiply by 60; etc.). In order to formulate the law of falling bodies in a specific system of units, we therefore need to insert a constant of proportionality into the equation describing the distance/time relationship. Thus, our equation will be of the form $s = ct^2$ rather than $s = t^2$. Some specific values of c are given in the following table:

location	Earth	Earth	Moon	Mars	Jupiter
unit of distance	feet	meters	meters	meters	meters
unit of time	seconds	seconds	seconds	seconds	seconds
value of c	16	4.9	0.81	2	13
equation	$s = 16t^2$	$s = 4.9t^2$	$s = 0.81t^2$	$s = 2t^2$	$s = 13t^2$

As an example, let us consider a rock that is dropped from a platform on Mars. We wish to approximate the instantaneous velocity v of the rock two seconds after it has been dropped. The average velocity over the interval from $t = 2\,s$ to $t = 2.1\,s$ is

$$\bar{v}_{[2,2.1]} = \frac{2 \cdot 2.1^2 - 2 \cdot 2^2}{0.1} = 8.2\,\frac{m}{s} \approx v.$$

Following this example, use the law of falling bodies to estimate the instantaneous velocity of an object dropped...

a) on Jupiter, after 1.5 seconds (in m/s).

b) on Earth, after 3 seconds (in ft/s).

c) on Earth, after 3 seconds (in m/s).

d) on the moon, after 3 seconds (in m/s).

0.7. An alternative formulation of the law of falling bodies asserts that velocity and time are related via the equation $v = 2ct$. Given this law, estimate the distance an object has fallen after two seconds by approximating the area under the graph given by the equation $v = 2ct$. (In this problem you can also determine the exact value for the distance as the area of a triangle.)

0.8. Assume that a rock drops $16t^2$ feet in t seconds.

a) How far does the rock fall during...

the first 0.25 seconds?

the first second?

the second second?

b) How far does the rock fall in the interval from...

$t = 1$ to $t = 1.1$ seconds?
$t = 1$ to $t = 1.01$ seconds?
$t = 1$ to $t = 1.001$ seconds?

c) Use b) to find three different estimates for the instantaneous velocity of the rock at time $t = 1$.

0.9. What are the upper and lower estimates for the change in an object's position if, during a 5 second interval, the object never travels faster than 10 meters per second and never slower than 8 meters per second?

0.10. Assume that a rocket's velocity after t seconds is t^2 feet per second. Find lower and upper estimates for the change in the rocket's position in the first 3 seconds using 9 subintervals.

0.11. Assume an object is moving along a straight line in such a way that its position after t seconds is $s = t^3/3$. Show that for $t > t_0$ we have $\bar{v}_{[t_0,t]} = (t^2 + t_0 t + t_0^2)/3$.

0.12. Assume that an object is moving along a straight line such that its position after t seconds is $s = 3t^2 + 1$ (measured in meters).

a) Find the change in the object's position between the times $t = 1$ and $t = 3$.

b) Find the average velocity of the object between the times $t = 1$ and $t = 3$.

c) Find a general formula for the object's average velocity between the times t and $t + 2$.

0.13. The *velocity* graphs of two cars (A and B) are given in Figure 6. We assume the cars start from the same position at time $t = 0$ and are heading in the same direction. On which of the following intervals is the distance between them increasing and on which is it decreasing: $[0, 0.25]$, $[0.75, 1]$, and $[1, 1.25]$? Explain your answer.

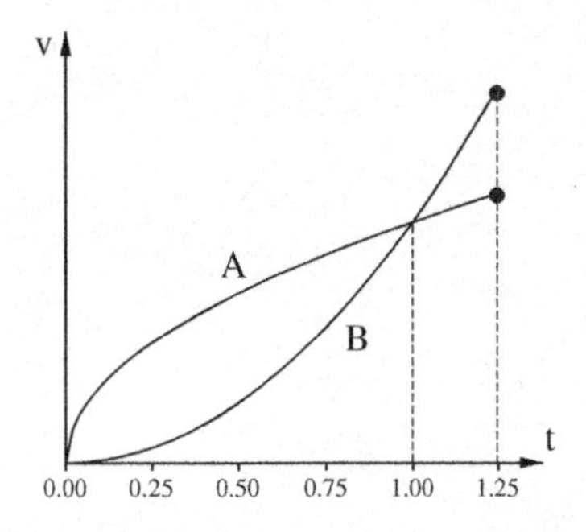

Figure 6: velocity graphs of two cars.

0.14. **a)** A car travels 30 miles with an average velocity of 40 *mph* and then another 30 miles with an average velocity of 60 *mph*. What is the car's average velocity for the entire 60 mile trip?

b) A car travels for 30 minutes with an average velocity of 40 *mph* and then for another 30 minutes with an average velocity of 60 *mph*. What is the car's average velocity for the entire 1 hour trip?

c) A car is to travel two miles. For the first mile the average velocity is 30 *mph*. Given these conditions, is it possible to average 60 *mph* for the entire trip? Explain your answer.

0.15. An object travels t^3 feet in t seconds.

a) How far does the object travel during the time span from $t = 1$ to $t = 1.2$ seconds?

b) What is the object's average velocity for this time span?

c) Answer the questions in a) and b) for the time span from $t = 2$ seconds to $t = 2.05$ seconds.

0.16. The table below shows the speedometer readings of a car for a time span when its velocity was increasing.

time	12:00	12:20	12:40	1:00	1:20	1:40	2:00
velocity in *mph*	30	36	42	47	50	54	57

Estimate the distance that the car traveled between 12:00 and 2:00.

PART I

DIFFERENTIATION

Chapter 1

Functions

The purpose of this chapter is introduce some fundamental ideas relating to functions and sets. Due to the nature of the subject, our discussion will appear more abstract than in most of the rest of the text, but we ask the reader to bear with us, for a certain degree of abstraction is necessary if we are to express ourselves clearly, using adequate language and notation.

A Brief Prologue on Sets

Not attempting to give a precise definition, we may regard a *set* to be an arbitrary collection of objects of our awareness. Sets may consist of apples and oranges as well as of cars and chairs or just about anything else that we can somehow conceive of. In mathematics, of course, apples and oranges are not commonly encountered, and at least in the first part of this text, sets will generally be composed of numbers. For example, $\{1, 2, 3\}$ is the set containing the numbers $1, 2$, and 3. The curly brackets are standard notation indicating that we are dealing with a set. We also refer to the numbers $1, 2$, and 3 as the *elements* of the set $\{1, 2, 3\}$. In general, if an object a is an element of a set A, we express this relation by writing $a \in A$ (e.g. $2 \in \{1, 2, 3\}$, but *not* $0 \in \{1, 2, 3\}$). Furthermore, a set A is said to be a *subset* of a set B, written as $A \subset B$, if every element of A is also an element of B (e.g. $\{1, 2\} \subset \{1, 2, 3\}$ but *not* $\{0, 2\} \subset \{1, 2, 3\}$).

1.1 Exercise. Which of the three sets $\{1, 2, 3, 4\}$, $\{1, 2, 5\}$, and $\{1, 2\}$ are subsets of $\{1, 2, 3, 4\}$?

The following list of definitions establishes the most important operations with sets and provides simple illustrations in each case:

a) The *union* $A \cup B$ of two sets A and B is the set of all elements that are contained in A *or* B.
Examples: $\{1, 5\} \cup \{2, 3, 4\} = \{1, 2, 3, 4, 5\}$ and $\{1, 5\} \cup \{1, 2\} = \{1, 2, 5\}$.

b) The *intersection* $A \cap B$ consists of all elements that are contained in both A *and* B.
Examples: $\{1, 3, 5\} \cap \{1, 2, 3\} = \{1, 3\}$, $\{1, 2\} \cap \{1, 2, 3\} = \{1, 2\}$, and $\{1, 2\} \cap \{3, 4\} = \emptyset$, where we denote by $\emptyset$ the empty set that contains no elements at all.

c) The *difference* $A \setminus B$ is the set of all elements that are in A but not in B.
Examples: $\{1, 2, 3\} \setminus \{1, 4\} = \{2, 3\}$ and $\{1, 2, 3\} \setminus \{4, 5\} = \{1, 2, 3\}$.

1.2 Exercise. Identify the elements in each of the following sets: $\{1, 2\} \cup \{2, 5, 6\}$, $\{1, 2, 3, 4\} \cap \{2, 4, 7\}$, and $\{1, 2, 3, 4\} \setminus \{2, 4, 7\}$.

Most frequently encountered in calculus is the set $\mathbb{R}$ of all *real numbers.* Intuitively, we think of a real number as a point on a coordinate axis, for example the x-axis in an xy-coordinate system (to give a more precise definition of real numbers is beyond the scope of this text). Some "typical" real numbers are 1, $-1/3$ and 0.356 but also $\sqrt{2}$ or π. A commonly occurring type of subset of $\mathbb{R}$ is the *interval:* for $a, b \in \mathbb{R}$ with $a < b$, the interval from a to b is the set of all real numbers between a and b. What we always need to be a little careful about is the inclusion or exclusion of the *boundary points* a and b. Inclusion is indicated by a bracket and exclusion by a parenthesis. Thus, there are four possible types of intervals:

$$[a,b],\ [a,b),\ (a,b],\ \text{and } (a,b).$$

The first in the list is said to be a *closed* interval, the next two are called *half-open* and the last is referred to as *open*. For instance, $[-3,5)$ is the half-open interval that contains all real numbers greater than or equal to -3 and strictly less than 5. We also allow $\pm\infty$ as values for a and/or b. The interval $[0,\infty)$, for example, is the set of all real numbers greater than or equal to 0, $(-\infty,1)$ is the set of all real numbers strictly less than 1, and $(-\infty,\infty)$ is equal to $\mathbb{R}$ itself.

1.3 Exercise. Which of the numbers -2, -1, $1/2$, 1, and $3/2$ are contained in the interval $[-1,1)$?

Definition of a Function

There are numerous instances in mathematics when we consider quantities or objects in dependence on or correspondence to other quantities or objects. Certain types of such correspondences are referred to as *functions*. The concept of a function is extremely important and indeed more fundamental to mathematics than even the concept of a number (although not as fundamental as the notion of a set). Two examples of functions that we already encountered in the Introduction are the position and velocity of a falling body in dependence on time. To give a purely mathematical example, we may consider the equation $y = x^2$. An equation of this form can be interpreted as a *rule of correspondence* or *rule of assignment* in the sense that it specifies for a given input value x the output value y to be x^2. However, for a complete definition of a function we need to specify not only a functional relation between input and output values (possibly represented by an equation), but also the sets that these input and output values are taken from.

1.4 Definition. Let A and B be nonempty sets. A *function* f from A to B is a rule of correspondence that assigns to each element in A *exactly one* element in B. The standard notation that symbolizes the dependence of an *output* $f(x)$ in B on an *input* x in A is

$$\begin{aligned} f : A &\to B \\ x &\mapsto f(x). \end{aligned}$$

Simplicio: I find this rather confusing. What, for example, are these sets A and B?

Teacher: Your confusion, Simplicio, is understandable, but once you have gotten used to the notation in Definition 1.4, you will find it helpful in reminding you that a function is more than just an equation like, for example, $y = x^2$. A full description of a function requires three things: a set A of possible input values, called the *domain*, a set B that contains the output values, and a rule of assignment symbolized by the notation $x \mapsto f(x)$ (in words: "x is assigned the value $f(x)$" or "x is mapped to $f(x)$").

Simplicio: That doesn't answer my question. You are saying that A contains the input values and B the output values, but what exactly are these sets A and B—that is, what are these input and output values?

Sophie: As I understand it, the letters A and B are variables representing sets just as x and y are variables representing numbers in an equation of the form $y = x^2$. We can only say what A and B actually are when we consider a specific example.

Teacher: Sophie is right. There are so many different types of functions that it would be entirely impossible to compile a complete list of all available choices for A and B. In the first part of this course A and B will usually be sets composed of real numbers like, for example, intervals. Eventually, though, as we progress to more advanced topics, there will be many variations in the choices of A and B, and we will have to be very careful not to confuse these different types of functions (see also Example 1.7 below).

Simplicio: Okay, I accept that, but I still would like to know what the arrows '$\to$' and '$\mapsto$' are about.

Teacher: The arrow in the first line $f : A \to B$ indicates that elements in A are assigned elements in B, while the arrow in the second row $x \mapsto f(x)$ symbolizes the actual rule of assignment. We say that an element x in A is assigned the element $f(x)$ in B. Here it is important to point out that not all values in B are necessarily values of the function f. In other words, the set of output values of f, which is referred to as the *range of* f and denoted by $R(f)$, forms a subset of B but is not necessarily equal to B. In order to understand the distinction between B and $R(f)$ it is best to consider an example.

1.5 Example. The equation $y = x^2$ can be interpreted as a rule of assignment in the sense that every value x is assigned the value x^2. So we may write

$$x \mapsto x^2.$$

In order to define a function, a rule of assignment alone is not sufficient—we also need to say exactly which values are permitted as input for x. In other words, we need to specify the domain A. There are many possible definitions for A, and which of these we pick may depend on the larger context, but for simplicity we will in this example choose A to be the set $\mathbb{R}$ of all real numbers. Then the output values x^2 are real numbers as well, and we may set $B := \mathbb{R}$ (the colon in front of the equal sign indicates that B is *defined* to be $\mathbb{R}$). Having thus specified A and B, we introduce the following function:

$$\begin{aligned} f : \mathbb{R} &\to \mathbb{R} \\ x &\mapsto x^2. \end{aligned} \tag{1.1}$$

The range $R(f)$ is in this case *not* equal to $B = \mathbb{R}$, because the square of any number $x \in \mathbb{R}$ is always greater than or equal to zero. In other words, the only possible output values of f are the positive real numbers including zero, i.e.,

$$R(f) = [0, \infty).$$

It would not have been wrong to replace $B = \mathbb{R}$ in the definition of f with $[0, \infty)$ so that (1.1) would read

$$\begin{aligned} f : \mathbb{R} &\to [0, \infty) \\ x &\mapsto x^2. \end{aligned}$$

However, in working with functions it is usually sufficient to be given only the general type rather than the exact range of output values. So the less precise notation in (1.1) is not only acceptable but may even be preferable. Moreover, for simplicity we will frequently replace the second row $x \mapsto x^2$ with the more familiar looking equation $f(x) := x^2$ (again the colon in front of the equal sign indicates that $f(x)$ is *defined* to be x^2). So instead of (1.1) we will also write "$f : \mathbb{R} \to \mathbb{R}$ is a function defined by the equation $f(x) := x^2$."

1.6 Exercise. Determine the range of the function $f : [-1, 2] \to \mathbb{R}$, $x \mapsto 2x - 1$.

Whenever we want to explicitly allow for the possibility that f is only defined on a subset D of $\mathbb{R}$ (rather than all of $\mathbb{R}$) without being really specific with regard to the choice of D, we will use the notation $f : D \subset \mathbb{R} \to \mathbb{R}$.

1.7 Example. To give an example of a function whose domain and range are not subsets of $\mathbb{R}$ we define A to be the set of all days of the year 1997 represented by their date pair. The pair for January 1st is $(1, 1)$ and for December 31st it is $(12, 31)$. So we set

$$A := \{(1, 1), \ldots, (1, 31), \ldots, (12, 1), \ldots, (12, 31)\}.$$

Furthermore, by B we denote the set of the seven days of the week, i.e.,

$$B := \{\textit{Mo, Tu, Wd, Th, Fr, Sa, Su}\}.$$

Since every day of the year 1997 corresponds to exactly one day of the week we may define a function $f : A \to B$ via the following rule of assignment:

$$(month, day) \mapsto \text{day of the week of } month/day/1997.$$

Given this definition, it follows that $(4, 21) \mapsto Mo$ or equivalently $f(4, 21) = Mo$, because the 21st of April 1997 was a Monday.

1.8 Example. Most important in defining a function is to make sure that the rule of correspondence assigns *exactly one* element in the set of output values to every element in the domain (see Definition 1.4). In order to give an example of a rule of correspondence that violates this condition we will now consider a modification of Example 1.7. We define B as above, but denote by A the set of integers from 1 to 31, i.e.,

$$A := \{1, \ldots, 31\}.$$

We say that a number $n \in A$ corresponds to a day of the week $d \in B$ if in 1997 there was at least one month for which the nth day of that month was the day of the week d. According to this rule, the number $21 \in A$ corresponds to $Mo \in B$ because, as mentioned above, the 21st of April 1997 was a Monday. However, 21 also corresponds to each of the other days of the week except Saturday, because in 1997 every day of the week except Saturday was the 21st of at least one month of that year. This shows that our modified rule of correspondence does not satisfy the uniqueness requirement for the assignment of values in B to values in A, and it therefore *does not define a function.* (As a matter of fact, *each* element in A is assigned more than one element in B. Think about it!)

In the context of the preceding examples it needs to be emphasized that it is quite permissible for a function to assign the same element in the range to two or more elements in the domain. For instance, in Example 1.7 we have $f(4, 21) = f(4, 28) = Mo$, because the 21st and the 28th of April 1997 were both Mondays. So the uniqueness requirement regarding the rule of assignment applies to the direction from the domain to the range but not vice versa.

Operations with Functions

Whenever we are dealing with functions that take values in $\mathbb{R}$ (i.e., the range is a subset of $\mathbb{R}$), it is possible to define sums, scalar multiples, products, and quotients of functions.

1.9 Definition. Let A be a nonempty set, $\lambda \in \mathbb{R}$, and let $f : A \to \mathbb{R}$ and $g : A \to \mathbb{R}$ be two real-valued functions with domain A. Then we define the functions $f + g : A \to \mathbb{R}$, $\lambda f : A \to \mathbb{R}$, and $fg : A \to \mathbb{R}$ via the equations

$$\begin{aligned}(f+g)(x) &:= f(x) + g(x);\\ (\lambda f)(x) &:= \lambda f(x);\\ (fg)(x) &:= f(x)g(x).\end{aligned}$$

If $g(x) \neq 0$ for all $x \in A$, then we can also define the *quotient* $f/g : A \to \mathbb{R}$ via the equation

$$\left(\frac{f}{g}\right)(x) := \frac{f(x)}{g(x)}.$$

Remark. in mathematical language numbers are often referred to as *scalars.* This is why the function λf is said to be a *scalar multiple of* f.

Simplicio: It seems there is a pattern emerging, for each time we introduce a new definition I get lost. What, for example, is the significance in writing $(f + g)(x) := f(x) + g(x)$? Aren't we here simply stating the obvious—$f + g = f + g$? How are we to regard such a trivial equation as a meaningful definition?

Teacher: You have a talent for pointing your finger at the right spot, Simplicio. There are some subtleties hidden in the defining equation $(f+g)(x) := f(x)+g(x)$ that are worth explaining. The term $f+g$ on the left-hand side as well as the individual letters f and g on the right-hand side each represent the *name of a function.* In order to define the function that goes by the name $f+g$, we need a rule of assignment in the form of a defining equation. So in the equation $(f+g)(x) := f(x)+g(x)$ we define the *value of the function $f+g$ at x* to be equal to the sum of the values of f and g at x.

Sophie: I must admit, I also feel we are splitting hairs.

Teacher: For beginners it is often difficult to see why it is necessary to be so precise, but we study mathematics not only to be able to perform certain calculations according to certain rules but, more importantly, to develop our minds—to learn how to think. It is therefore crucial that we express ourselves clearly, and form an accurate understanding of the concepts at hand.

1.10 Example. Let the functions $f : \mathbb{R} \to \mathbb{R}$ and $g : \mathbb{R} \to \mathbb{R}$ be defined via the equations $f(x) := x^2$ and $g(x) := x+1$. Then the value of the sum $f+g$ at x is

$$(f+g)(x) = f(x) + g(x) = x^2 + x + 1,$$

and the value of the product of f and g at x is

$$(fg)(x) = f(x)g(x) = x^2(x+1) = x^3 + x^2.$$

1.11 Exercise. Let f and g be functions defined on $\mathbb{R}$ with $f(x) := 1/(x^2+1)$ and $g(x) := x^2+1$. Write a defining equation for the quotient f/g and determine its range.

Graphs of Functions

The graph of a function is an important tool that frequently provides useful information concerning a function's essential properties. To be sure, though, the sketching of a graph is not so much a mathematical necessity as it is a concession to the remarkable frailty of the human mind when confronted with matters of abstraction. It may guide our intuition where an algebraic formula alone appears too barren. For many important types of functions it is actually not possible to sketch a graph (we will encounter examples of this sort in Part IX on vector calculus), but for functions whose domain and range are subsets of $\mathbb{R}$ it always works—at least in principle.

1.12 Example. Let us again consider the function $f : \mathbb{R} \to \mathbb{R}$, $x \mapsto x^2$, i.e., $f(x) := x^2$. In order to sketch the graph of f we draw a coordinate system with a horizontal axis representing the domain of f and a vertical axis containing the range. The graph of f is defined to be the set of all points of the

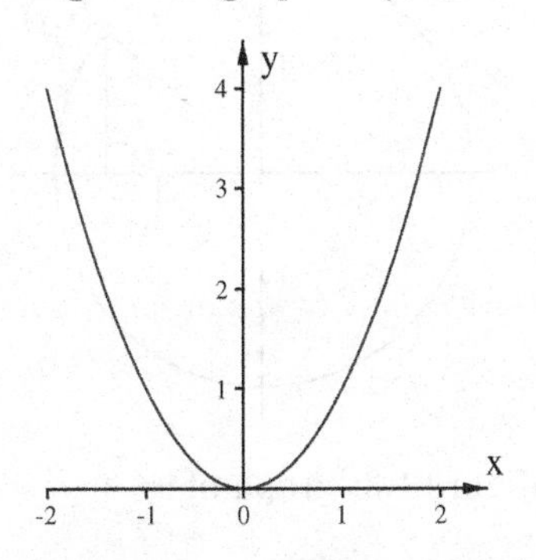

Figure 1.1: graph of $f(x) = x^2$.

form $(x, f(x))$ with $x \in \mathbb{R}$ (see Figure 1.1). According to this definition, $(-1/2, 1/4)$, $(1,1)$, and $(2,4)$ are, for example, points on the graph of f because $(-1/2)^2 = 1/4$, $1^2 = 1$ and $2^2 = 4$.

1.13 Exercise. Sketch the graph of the function $f : [0,4] \to \mathbb{R}$, $x \mapsto \sqrt{x}$.

★ *Remark.* From a more abstract point of view we can always define the graph of a function $f : A \to B$ to be the set of all points $(x, f(x))$ with $x \in A$. However, in this sense the word "graph" is slightly misleading, because it may not be descriptive of a nice geometric object, such as a curve in an xy-coordinate system. The sets A and B, or the rule of assignment between them, may be so complicated that any attempt at visualizing the graph of f would be futile. ★

Now we wish to discuss a criterion that allows us to recognize whether a given curve in an xy-coordinate system is the graph of a function. The defining property of a function is that every element in the domain is assigned *exactly one* element in the range. Consequently, if $f : D \subset \mathbb{R} \to \mathbb{R}$ is a function and $x \in D$, then the line that runs parallel to the y-axis and passes through $(x, 0)$ intersects the graph of f exactly once—in the point $(x, f(x))$ (see Figure 1.2).

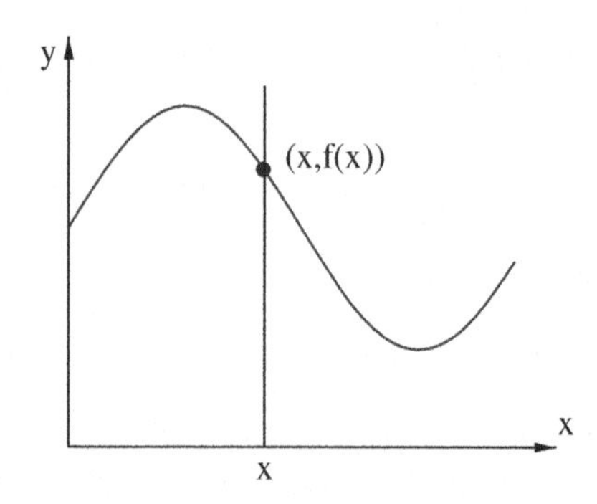

Figure 1.2: the vertical line test.

Conversely, if every vertical line intersects a given curve at most once, then this curve is the graph of a function depending on x. We summarize these observations in the following rule, which is commonly referred to as the *vertical line test:*

> A curve in an xy-coordinate system represents the graph of a function if, and only if, every line parallel to the y-axis intersects the curve at most once.

1.14 Example. Let us consider the equation $x^2 + y^2 = 4$. The set of all points (x, y) that satisfy this equation is a circle of radius 2 (see Figure 1.3). To demonstrate this we observe that, according to the

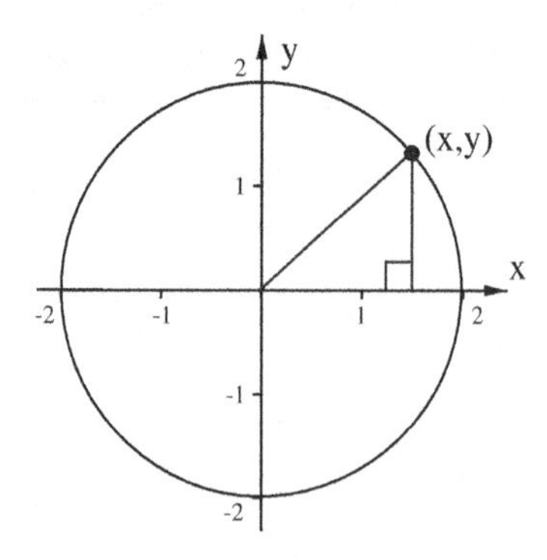

Figure 1.3: graph of $x^2 + y^2 = 4$.

theorem of Pythagoras (see Appendix A), the square of the distance from a point (x, y) to the origin $(0, 0)$ is $x^2 + y^2$. Therefore, the coordinates x and y satisfy the equation $x^2 + y^2 = 4$ if and only if the square of the distance from (x, y) to the origin is equal to 4, or equivalently, if and only if the distance from (x, y) to the origin is equal to 2. But what is the set of all points at distance 2 from the origin? It is a circle of radius 2 with its center at the origin. Given this observation, we may infer that the equation $x^2 + y^2 = 4$ does *not* define y as a function of x, because every vertical line that passes through the x-axis at some point in the interval $(-2,2)$ intersects the circle shown in Figure 1.3 exactly twice thereby violating the criterion stated in the vertical line test above.

Additional Exercises

1.15. Write each of the following sets as a single interval: $[1,2) \cup [2,3]$, $[1,2] \setminus \{2\}$, $[1,2] \setminus \{1,2\}$, $[0,3] \setminus [1,3]$, $\mathbb{R} \setminus (-\infty,0)$, $[1,2] \cap (3/2,3)$, $[1,2] \cap [2,3]$, and $(a,b) \cup (b,a)$ for $a,b \in \mathbb{R}$ with $a \neq b$.

1.16. Use function notation to describe how the second variable depends on the first in each of the examples listed below. Then determine the largest possible domain of each function (as a subset of $\mathbb{R}$) and the corresponding range. Note: only those values that are meaningful in the context of the given example should be included in the domain.

a) 1st variable: one leg L of a right triangle whose second leg is 2 units in length.
2nd variable: the hypotenuse H of the right triangle.

b) 1st variable: the hypotenuse H of a right triangle.
2nd variable: one leg L of the right triangle given that the other leg is 5 units in length.

c) 1st variable: the temperature C in degrees Celsius.
2nd variable: the temperature F in degrees Fahrenheit.

1.17. Let a function f be defined by the equation $f(x) := \dfrac{x-3}{(2-x)\sqrt{x^2+3}}$.

a) What is the largest possible domain of f (as a subset of $\mathbb{R}$)?

b) Find the values $f(3)$ and $f(-3)$.

c) Find the value $f(f(1))$.

1.18. What is the largest possible domain of the function $f(x) := \sqrt{x^2-4}$ (as a subset of $\mathbb{R}$)?

1.19. Is it possible to use the equation $y^2 = x^2$ as a rule of correspondence that defines y as a function of x with domain $\mathbb{R}$? Explain your answer.

1.20. Find values m and b such that for $f(x) := mx + b$ we have $f(0) = 3$ and $f(2) = -1$.

Remark. Functions of the form $f(x) = mx + b$ are said to be *affine*, and they are called *linear* if $b = 0$. The graph of an affine function is a straight line passing through the point $(0,b)$ with slope m.

1.21. Find defining equations for two affine functions f and g (see the remark above) with domain $[0,2]$ and range $[1,9]$ such that f is not equal to g. Sketch the graph of each.

1.22. Find a defining equation for an affine function f (see the remark above) that satisfies the equation $f(x+2) = f(x) + 6$ for all $x \in \mathbb{R}$.

1.23. For $k \neq 0$, the equation $2x + ky = -k$ implicitly defines y as an affine function of x. Solve the equation for y and find in each of the cases below a value for k (if there is one) such that the straight line representing the graph of y has the stated property.

a) The slope of the line is equal to 3.

b) The line passes through the point $(0,1)$.

c) The line is horizontal.

1.24. Let $g(x) := 1/x$ for all $x \in \mathbb{R} \setminus \{0\}$. Evaluate or simplify as far as possible each of the following expressions:

$$g(3-2),\ g(3)-g(2),\ \frac{g(2.25)-g(2)}{2.25-2},\ \frac{g(2+t)-g(2)}{(2+t)-2}.$$

1.25. Sketch the graph of a function $f : [0, 3] \to \mathbb{R}$ that satisfies the following conditions: $f(0) = 1$; $f(3) = -1$; and the equation $f(x) = 0$ has...

a) exactly one solution in the interval $[0, 3]$.

b) exactly two solutions in the interval $[0, 3]$.

c) exactly three solutions in the interval $[0, 3]$.

d) has infinitely many solutions in the interval $[0, 3]$.

Note: you need to draw four different graphs corresponding to a), b), c), and d).

1.26. For the functions $f(x) := \sqrt{x+1}$ and $g(x) := x^2 - 3$, find the defining equations for fg and f/g, and determine in each case the largest possible domain (as a subset of $\mathbb{R}$).

1.27. For the function shown in Figure 1.4 answer the following questions:

a) What are the domain and range of f?

b) What is the value of f at $x = 2$?

c) For which value(s) of x is $f(x)$ equal to 3?

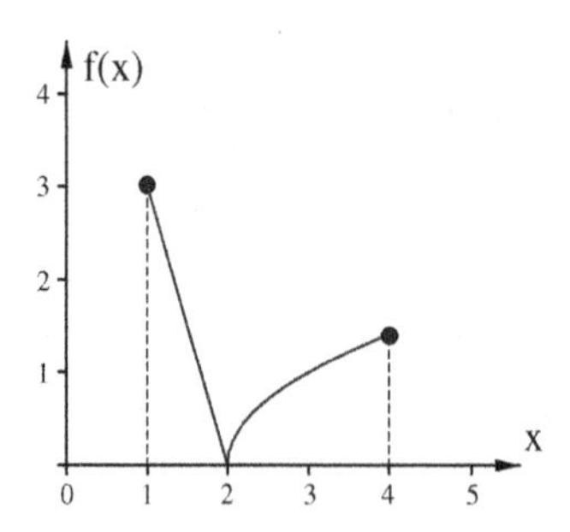

Figure 1.4: graph of f.

Chapter 2

Limits of Functions

Definition of a Limit

Teacher: Following up on our discussion in the Introduction, where we encountered the idea of limiting values in the context of analyzing the relation between position and velocity, the goal of today's lesson is to make more precise (but not entirely precise) the notion of the *limit of a function.* So let us consider a function $f : \mathbb{R} \to \mathbb{R}$. In order to find the limit of f at a certain point $x_0 \in \mathbb{R}$, we need to examine the values $f(x)$ for values x that are different from x_0, but very close to x_0. In other words, $|x - x_0|$ should be strictly greater than 0, but very small. We then say that f has the limit L at x_0 if the values $f(x)$ approach L as x approaches x_0.

Simplicio: The way you phrase it—"x approaches x_0"—one might think that x has grown legs.

Teacher: The image you suggest is a bit unorthodox, but not altogether inadequate. In saying that x approaches x_0 we imagine a sequence of values that are chosen successively closer to x_0. Then, along this sequence of values for x, the values $f(x)$ approach the limit L. To give an example, let us determine the limit of the function $f(x) := x^2$ at $x_0 = 2$.

Simplicio: Why again do we write $f(x) := x^2$ instead of simply $f(x) = x^2$?

Sophie: The colon in front of the equal sign indicates that we are dealing with a definition, and thus serves to remind us that the equation is not derived from any prior information.

Simplicio: All these new symbols and notations strike me as rather tedious. Do we really need to be that precise?

Teacher: Let me tell you, Simplicio, the pattern is always the same: a student will start out being sloppy about his notation, then his algebra goes down the drain, six months later he has grown long hair and pretty soon you find him selling drugs in the basement of the math department. Just yesterday again the papers reported on the arrest of two guys who had been known as violent notational offenders.

Simplicio: The prospect is sobering indeed: I am going to see the barber right after class.

Teacher: Good. Let us return then to the problem of identifying the limit of $f(x) := x^2$ at $x_0 = 2$. The following table illustrates that, as x approaches 2, the values $f(x)$ approach $2^2 = 4$:

x	1	1.5	1.9	1.99	2	2.01	2.1	2.5	3
$f(x)$	1	2.25	3.61	3.9601	4	4.0401	4.41	6.25	9

Thus, we say that the limit of f at 2 equals 4, and we write

$$\lim_{x \to 2} f(x) = 4.$$

Simplicio: It hardly seems surprising that the limit is 4, because $4 = 2^2$ is the value of f at 2. So it appears, the taking of the limit of a function f at a point x_0 is simply a matter of elementary arithmetic—all we need to do is to evaluate $f(x_0)$.

Teacher: I am afraid your conclusion is premature. How, for example, would we find the limit of $f(x) := (x^2 - 1)/(x - 1)$ at $x_0 = 1$? In this case f is *not even defined* at x_0, because a direct evaluation of f at 1 yields the indeterminate expression $0/0$. We can, of course, artificially define f to have any arbitrary value at $x_0 = 1$, but even that wouldn't help us in finding the limit $\lim_{x \to 1} f(x)$, because the values x that approach $x_0 = 1$ are assumed to be different from x_0.

Sophie: So what do we do in a situation like this?

Teacher: Unfortunately, there are no general rules. Each case may require a different approach. However, for the function $f(x) = (x^2 - 1)/(x - 1)$, a simple algebraic manipulation will do the trick. Since

$$x^2 - 1 = (x - 1)(x + 1),$$

the fraction $(x^2 - 1)/(x - 1)$ simplifies to $x + 1$. But now it is easy to see that, as x approaches 1, the value of $x + 1$ approaches $1 + 1 = 2$. Thus, we may write

$$\lim_{x \to 1} f(x) = \lim_{x \to 1} \frac{x^2 - 1}{x - 1} = \lim_{x \to 1} (x + 1) = 2.$$

Simplicio: I still don't get it. Why are we not content to confine ourselves to the study of limits that are easily determined by a direct evaluation of a function f at a point x_0? Why do we make life more difficult than it needs to be by considering examples of functions that are undefined at x_0?

Teacher: Do you remember that in the Introduction we talked about finding the instantaneous velocity of an object as the limit of the average velocities over successively smaller time intervals?

Simplicio: Kind of.

Teacher: Well, Simplicio, as we make the lengths of the time intervals smaller and smaller, the distance that the object travels in each of these intervals also approaches zero, and in the limit the change in position over the change in time is of the form $0/0$ (as in the example above). So even some of the most elementary examples of limits that we encounter in calculus already are not as trivial as you would like them to be. We will discuss this issue in more detail in Chapter 4.

Sophie: Teacher, I have a question: you said that there is no general method for determining the value of a limit...

Teacher: Quite so.

Sophie: ...but can we at least be certain that a limit does always exist?

Teacher: This is an excellent question, and the answer is "no." Limits do not always exist. Let's say, for instance, that a function $f : \mathbb{R} \to \mathbb{R}$ is defined by the equation

$$f(x) := \begin{cases} 1 & \text{if } x < 1 \\ 2 & \text{if } x \geq 1 \end{cases}. \tag{2.1}$$

In this case the limit at $x_0 = 1$ does not exist, because, as we approach $x_0 = 1$ from the left, the values of f approach 1 (in fact they are equal to 1), and as we approach $x_0 = 1$ from the right, the values of f approach 2. So there is no uniform approach to a single value (see also the graph of f as shown in Figure 2.1).

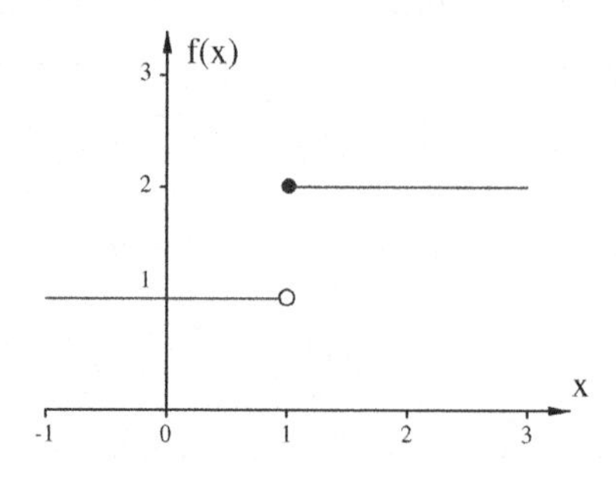

Figure 2.1: graph of f.

Sophie: I notice, though, that in this example the limit of f exists at all points except $x_0 = 1$. So I am wondering whether perhaps limits always exist almost everywhere except possibly at a few isolated points?

Teacher: I am sorry to disappoint you again, but there really is nothing we can say about the existence of limits in general. I encourage you take a look at the ⋆remark⋆ below, where a function is described that is defined on all of $\mathbb{R}$ but does not have an existing limit at a single point.

★ *Remark.* In order to come up with a function that is defined on all of $\mathbb{R}$ but does not have a limit at any point $x_0 \in \mathbb{R}$, we make use of the distinction between *rational* and *irrational* numbers: a number is said to be rational if it can be written as a fraction of two integers, and a number is said to be irrational if it is not rational. Since every real number has a finite or infinite decimal expansion, it is clear that every real number can be arbitrarily closely approximated by rational numbers. For example, the number π, which can be shown to be irrational, is approximately equal to $3.1416 = 31416/10000$. Surprisingly, though, it can also be shown that in a certain sense "most" real numbers are not rational (see also the ⋆remark⋆ on p.437) and that every interval in $\mathbb{R}$, no matter how small, contains infinitely many rational numbers as well as infinitely many irrational numbers. Using these facts, it is not very difficult to prove that the "popcorn"-function

$$f(x) := \begin{cases} 0 & \text{if } x \text{ is a rational number,} \\ 1 & \text{if } x \text{ is an irrational number} \end{cases}$$

does not have a limit at any point $x_0 \in \mathbb{R}$! ★

2.1 Definition. Assume that D is a subset of $\mathbb{R}$ and x_0 is a real number such that x_0 can be arbitrarily closely approximated by values $x \in D$ with $x \neq x_0$. (This, for example, is *not* the case for $D = [0, 1]$ and $x_0 = 2$, but it is the case for $D = [0, 1)$ and $x_0 = 1/2$ or $x_0 = 1$.) If $f : D \to \mathbb{R}$ is a function and $L \in \mathbb{R}$ a constant, then we say that the *limit* of f at x_0 is equal to L if the values $f(x)$ approach L as x approaches x_0 with $x \neq x_0$. In this case we write

$$\lim_{x \to x_0} f(x) = L.$$

For a more rigorous definition of the limit of a function, the reader is referred to Chapter 3.

2.2 Exercise. Let D be the set of all real numbers different from 2, i.e., $D = \mathbb{R} \setminus \{2\}$. Find $\lim_{x \to 2} f(x)$ for the function $f : D \to \mathbb{R}$ defined by the equation $f(x) := (x^2 - 4)/(x - 2)$.

2.3 Exercise. Given the following table of values of a function f, what would you expect to be the value of $\lim_{x \to 1} f(x)$?

x	0	0.5	0.9	0.99	1.01	1.1	1.5	2
$f(x)$	1.7	2.1	2.41	2.4961	2.5011	2.62	2.92	3.3

One-Sided Limits

As we saw above, the function defined in (2.1) does not have a limit at $x_0 = 1$. The problem here is that in the definition of the limit of a function we did not specify whether we approach x_0 with values $x > x_0$ or $x < x_0$. For the limit of f to exist, it must not make a difference whether we approach x_0 from the left ($x < x_0$) or from the right ($x > x_0$)—in both cases $f(x)$ should approach the same value L. This observation leads us to the concept of *one-sided limits:* even if $\lim_{x \to x_0} f(x)$ does not exist, it may still be of interest to know whether the values $f(x)$ approach a certain value L if for the approach of x to x_0 we allow only values greater than x_0 or only values less than x_0.

2.4 Definition. Assume that D is a subset of $\mathbb{R}$, and let x_0 be a real number such that x_0 can be arbitrarily closely approximated with values $x \in D$ that are greater than x_0 ($x > x_0$). For a function

$f : D \to \mathbb{R}$ and a real number L, we say that the *right-hand limit* of f at x_0 is equal to L if the values $f(x)$ approach L as x approaches x_0 from the right ($x > x_0$). The corresponding notation is

$$\lim_{x \to x_0^+} f(x) = L.$$

Similarly, if x_0 can be arbitrarily closely approximated with values $x \in D$ that are less than x_0 ($x < x_0$), we say that the *left-hand limit* of f at x_0 is equal to L if the values $f(x)$ approach L as x approaches x_0 from the left ($x < x_0$). In this case we write

$$\lim_{x \to x_0^-} f(x) = L.$$

2.5 Example. For the function defined in (2.1) we have

$$\lim_{x \to 1^-} f(x) = 1 \quad \text{and} \quad \lim_{x \to 1^+} f(x) = 2.$$

2.6 Exercise. Find $\lim_{x \to 0^+} f(x)$ and $\lim_{x \to 0^-} f(x)$ for the function $f : \mathbb{R} \to \mathbb{R}$ defined by the equation

$$f(x) := \begin{cases} x^2 - 1 & \text{for } x < 0, \\ x + 1 & \text{for } x \geq 0. \end{cases}$$

Intuitively it makes perfect sense to think that the limit of a function at a point x_0 should exist if the left- and right-hand limits exist and are equal, because in that case the values $f(x)$ approach the same value regardless of whether x approaches x_0 from the left or from the right. It can be shown that this is indeed true, although an exact proof would require the use of a more precise definition for the limit of a function than the one given in Definition 2.1 (see Chapter 3).

2.7 Theorem. *Assume that D is a subset of $\mathbb{R}$, and let x_0 be a real number such that x_0 can be arbitrarily closely approximated with values $x \in D$ that are less than x_0 ($x < x_0$) as well as with values $x \in D$ that are greater than x_0 ($x > x_0$). Then, for a function $f : D \to \mathbb{R}$ the limit $\lim_{x \to x_0} f(x)$ exists if and only if the one-sided limits $\lim_{x \to x_0^+} f(x)$ and $\lim_{x \to x_0^-} f(x)$ exist and are equal. In this case we have*

$$\lim_{x \to x_0} f(x) = \lim_{x \to x_0^+} f(x) = \lim_{x \to x_0^-} f(x).$$

2.8 Exercise. Given the statement of Theorem 2.7, explain why the function $f : \mathbb{R} \to \mathbb{R}$ defined in Exercise 2.6 does not have a limit at $x_0 = 0$.

Rules for Limits

We wish to discuss limits of sums, scalar multiples, products, and quotients of functions (see Definition 1.8). Considering first the case of the product of two functions $f : D \subset \mathbb{R} \to \mathbb{R}$ and $g : D \subset \mathbb{R} \to \mathbb{R}$, we assume that for some $x_0, L, M \in \mathbb{R}$ we have

$$\lim_{x \to x_0} f(x) = L \quad \text{and} \quad \lim_{x \to x_0} g(x) = M.$$

Then the limit $\lim_{x \to x_0}(f(x)g(x))$ is simply the product of the limit of $f(x)$ and the limit of $g(x)$ at x_0, because, as $f(x)$ approaches L and $g(x)$ approaches M, the value of $f(x)g(x)$ approaches LM. In other words, $\lim_{x \to x_0}(f(x)g(x)) = LM$. Of course, this equation can be shown to be true with a precise proof, but again we would have to rely on the more rigorous definition of the limit that will be given in Chapter 3. For sums, scalar multiples, and quotients of functions, matters are no more complicated than for products, and we may therefore state the following theorem:

2.9 Theorem. *Suppose we are given two functions $f : D \subset \mathbb{R} \to \mathbb{R}$ and $g : D \subset \mathbb{R} \to \mathbb{R}$ such that*

$$\lim_{x \to x_0} f(x) = L \quad \text{and} \quad \lim_{x \to x_0} g(x) = M.$$

for some $x_0, L, M \in \mathbb{R}$. Then

$$\lim_{x \to x_0} (f(x) + g(x)) = L + M,$$
$$\lim_{x \to x_0} (f(x)g(x)) = LM,$$

and for all $\lambda \in \mathbb{R}$ we have

$$\lim_{x \to x_0} (\lambda f(x)) = \lambda L.$$

If in addition we assume that $M \neq 0$, then it also follows that

$$\lim_{x \to x_0} \frac{f(x)}{g(x)} = \frac{L}{M}.$$

Completely analogous rules also hold for left- and right-hand limits respectively.

Teacher: To see whether you have properly grasped the meaning of Theorems 2.7 and 2.9, I would like you to consider two functions f and g defined on $\mathbb{R}$ that satisfy the following equations:

$$\begin{aligned} \lim_{x \to 1} f(x) &= 0, \\ \lim_{x \to 1^+} g(x) &= 1, \\ \lim_{x \to 1^-} g(x) &= -1. \end{aligned} \tag{2.2}$$

Do these conditions allow us to decide whether the limit of the product fg exists at $x_0 = 1$?

Simplicio: I think so. First of all, the statement of Theorem 2.7 clearly allows us to infer that the limit of g does not exist at $x_0 = 1$. For if it did, then contrary to the assumptions in (2.2), the left- and right-hand limits of g would have to be equal. Therefore, the limit of fg at $x_0 = 1$ does not exist either, because Theorem 2.9 says that in order for the limit of fg to exist, the limits of both f and g must exist as well.

Teacher: I agree with the first part of your statement: the limit of g does indeed not exist at $x_0 = 1$. However, I am a bit troubled with regard to the second part. What do you think, Sophie?

Sophie: I am not sure, but in Theorem 2.9 it only says that the limit of fg exists if the limits of f and g exist. Simplicio concluded that the limit of fg does *not* exist, because the limits of f and g do *not* both exist. So it appears he somehow got the logic reversed.

Teacher: You are right. It's a problem of logic. Let's say we denote by A the statement "the limits of f and g exist at $x_0 = 1$" and by B the statement "the limit of fg exists at $x_0 = 1$." Then Theorem 2.9 asserts that A implies B. However, to say that A implies B is not the same as saying that the negation of A implies the negation of B. In other words, it does not follow that the nonexistence of either the limit of f or the limit of g at $x_0 = 1$ implies the nonexistence of the limit of fg at $x_0 = 1$.

Simplicio: I'm afraid I am lost.

Teacher: Okay, let me try to explain it with a simple example. Let us denote by A the statement "Socrates is a man and all men are mortal" and by B the statement "Socrates is mortal." Then it certainly is correct to say that A implies B, because if it is true that all men are mortal and that Socrates is a man, then Socrates must be mortal. Do you agree?

Simplicio: Of course.

Teacher: Good. Now let us examine whether the negation of A implies the negation of B. The negation of A is the statement "Socrates is not a man or there is a man who is not mortal" and the negation of B is "Socrates is not mortal." Is it true then that the negation of A implies the negation of B?

Simplicio: No, not really, because if Socrates were, for example, a dog, then the negation of A would be true, because Socrates were in this case not a man, but the statement "Socrates is not mortal" would obviously be false.

Teacher: Very good, you got it.
Simplicio: That makes sense now, and I do see my mistake.
Teacher: I'm glad to hear that, but we still haven't settled the question regarding the existence of the limit of fg at $x_0 = 1$. To address this issue again, let me point out that according to the first equation in (2.2), the limit of f at $x_0 = 1$ exists and is equal to zero. Therefore, Theorem 2.7 implies that the left- and right-hand limits of f at $x_0 = 1$ exist and are both equal to zero as well. Consequently Theorem 2.9, which is valid also for one-sided limits, in conjunction with the third equation in (2.2) allows us to infer that

$$\lim_{x\to 1^-} f(x)g(x) = \lim_{x\to 1^-} f(x) \lim_{x\to 1^-} g(x) = 0\cdot(-1) = 0,$$
$$\lim_{x\to 1^+} f(x)g(x) = \lim_{x\to 1^+} f(x) \lim_{x\to 1^+} g(x) = 0\cdot 1 = 0.$$

Given these results, we may again apply Theorem 2.7 to conclude that

$$\lim_{x\to 1} f(x)g(x) = 0.$$

Sophie: It's interesting how a little bit of logic can clear things up!
Teacher: Proper use of logic indeed is vital to the study of calculus, because, as I said earlier, mathematics is not about memorizing patterns of calculation—it's about learning how to think.

Additional Exercises

2.10. For the functions f and g shown in Figure 2.2, determine whether the following limits exist, and find the values of those that do exist.

a) $\lim_{x\to -1} f(x)$ **b)** $\lim_{x\to 1} f(x)$ **c)** $\lim_{x\to -1} g(x)$ **d)** $\lim_{x\to 1} g(x)$ **e)** $\lim_{x\to 1}(2f(x)+3g(x))$
f) $\lim_{x\to 2} f(x)g(x)$ **g)** $\lim_{x\to 0} \dfrac{f(x)}{g(x)}$ **h)** $\lim_{x\to 0} \dfrac{g(x)}{f(x)}$ **i)** $\lim_{x\to 2} g(f(x))$ **j)** $\lim_{x\to -1}(f(x)+g(x))$

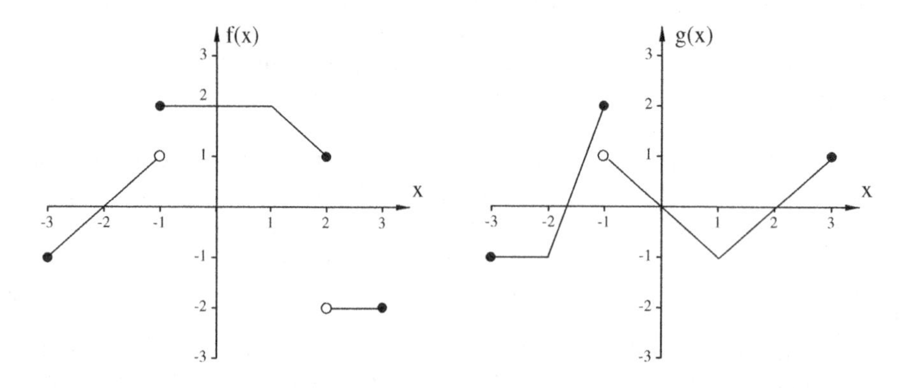

Figure 2.2: graphs of f and g.

2.11. Using the graphs given in Figure 2.2, determine the values of the following one-sided limits:

a) $\lim_{x\to -1^-} f(x)$ **b)** $\lim_{x\to -1^+} f(x)$ **c)** $\lim_{x\to -1^-} g(x)$
d) $\lim_{x\to -1^+} g(x)$ **e)** $\lim_{x\to 0^-} f(x+2)$ **f)** $\lim_{x\to -1^-} f(x^2)$

2.12. Use algebraic simplification or a table of values to evaluate the following limits:

a) $\lim_{x\to 4} \frac{x^2-16}{x-4}$ b) $\lim_{x\to -2} \frac{x^2+3x+2}{x+2}$ c) $\lim_{x\to -3} \frac{x^2-9}{x+3}$

d) $\lim_{x\to -1+} -\frac{1+x}{\sqrt{1-x^2}}$ e) $\lim_{x\to 1} \frac{1-x}{1-\sqrt{x}}$ f) $\lim_{x\to 1} \frac{x^3-1}{x^2-1}$

2.13. Let a function $f : [1/2, 4] \setminus \{2\} \to \mathbb{R}$ be defined by the equation $f(x) := \dfrac{x^3-4x^2+7x-6}{x-2}$.

a) Sketch the graph of f. *Hint.* Use long division to simplify the defining equation for f.

b) Find the range of f.

c) Find the value of $\lim_{x\to 2} f(x)$ from the graph you sketched in a).

2.14. For $f(x) := \dfrac{x^2-1}{x-1}$ evaluate $\lim_{x\to 1} f(x)^2$ and $(\lim_{x\to 1} f(x))^2$.

2.15. Assume that $f : \mathbb{R} \to \mathbb{R}$ is a function and $x_0 \in \mathbb{R}$ such that $\lim_{x\to x_0} f(x)^2$ exists. Is it necessarily true that $\lim_{x\to x_0} f(x)$ exists as well? Explain your answer.

2.16. For $f(x) := 2x^2 - 1$, evaluate the following limits:

a) $\lim_{h\to 0} \frac{f(2+h)-f(2)}{h}$ b) $\lim_{h\to 0} \frac{f(x+h)-f(x)}{h}$ (Your answer will depend on x.)

2.17. For $g(t) := t^2 + t - 2$, evaluate the following limits:

a) $\lim_{h\to 0} \frac{g(1+h)-g(1)}{h}$ b) $\lim_{h\to 0} \frac{g(t+h)-g(t)}{h}$ (Your answer will depend on t.)

Chapter 3

★ Limits Made More Precise

Limits of Sequences

In our definition of the limit of a function in 2.1 we relied on an intuitive understanding of what it means for x to "approach" x_0 or for $f(x)$ to "approach" L. From a mathematical point of view, such an intuition-based definition is, of course, not very satisfactory, and the purpose of the present chapter is therefore to outline a more rigorous approach. We begin our discussion by introducing the notion of a *sequence*: a sequence is an infinite succession of real numbers a_n indexed by positive integers n. For example, the sequence corresponding to the defining equation $a_n := 1/n$ is

$$\frac{1}{1}, \frac{1}{2}, \frac{1}{3}, \frac{1}{4}, \ldots$$

In indexing the values a_n by n we are essentially describing a functional relationship between n and a_n. In other words, if we denote by $\mathbb{N}$ the set of all positive integers, then the rule of assignment $n \mapsto a_n$ defines a function with domain $\mathbb{N}$ and values $a_n \in \mathbb{R}$. This observation motivates the following definition:

3.1 Definition. A *sequence* is a function

$$\begin{aligned} a : \mathbb{N} &\to \mathbb{R} \\ n &\mapsto a_n. \end{aligned}$$

The standard notation for sequences is $(a_n)_{n=1}^{\infty}$.

To discuss exactly how the limit of a sequence is to be defined, we consider again the sequence $(1/n)_{n=1}^{\infty}$. As n increases, the values $a_n = 1/n$ decrease to zero. In order to understand what really is meant by "decreasing to zero," we pick a small positive number like, for example, 0.1. Eventually, that is for large values of n, all the terms in the sequence $(1/n)_{n=1}^{\infty}$ will be less than 0.1. This observation follows from the simple fact that for all $n \in \mathbb{N}$ with $n > 10$ we have $a_n = 1/n < 1/10 = 0.1$. Of course, the validity of this statement does not depend on the particular choice of the value 0.1. If instead of 0.1 we choose an arbitrary number $\varepsilon > 0$, then $a_n = 1/n < \varepsilon$ for all $n \in \mathbb{N}$ that are greater than $1/\varepsilon$. This shows that no matter how small a value we choose for ε, we will always be able to conclude that a_n is smaller than ε if only n is sufficiently large, that is, if only n is larger than $N := 1/\varepsilon$. In other words, for a given $\varepsilon > 0$ we can find an $N \in \mathbb{R}$ such that for $n > N$ the distance $|a_n - 0|$ between a_n and the limiting value 0 is less than ε.

As a second example we consider the sequence $(a_n)_{n=1}^{\infty}$ defined by the equation

$$a_n := 1 + \frac{(-1)^n}{n} \quad \text{for all } n \in \mathbb{N}.$$

In listing the first few terms of this sequence,

$$a_1 = 0,\ a_2 = \frac{3}{2},\ a_3 = \frac{2}{3},\ a_4 = \frac{5}{4},\ a_5 = \frac{4}{5}, \dots$$

we observe an apparent approach to the limiting value 1. To rigorously justify this assertion we choose, as in the previous example, an arbitrary number $\varepsilon > 0$. Then, for $n > 1/\varepsilon =: N$ we have $1/n < \varepsilon$ and therefore,

$$-\varepsilon < \frac{(-1)^n}{n} < \varepsilon.$$

Hence

$$|a_n - 1| = \left|\frac{(-1)^n}{n}\right| < \varepsilon \text{ for all } n > N.$$

In other words, for any number $\varepsilon > 0$ we can find an $N \in \mathbb{R}$ such that for $n > N$ the distance $|a_n - 1|$ between a_n and the limiting value 1 is less than ε.

3.2 Definition. A sequence $(a_n)_{n=1}^{\infty}$ is said to *converge* to a number $L \in \mathbb{R}$ if for every $\varepsilon > 0$ there exists an $N \in \mathbb{R}$ such that

$$|a_n - L| < \varepsilon \text{ for all } n > N.$$

In this case L is referred to as the *limit* of the sequence $(a_n)_{n=1}^{\infty}$, and we write

$$L = \lim_{n\to\infty} a_n.$$

3.3 Example. We wish to show that the sequence $(1/\sqrt{n})_{n=1}^{\infty}$ converges to 0. With reference to Definition 3.2 we begin by selecting an arbitrary $\varepsilon > 0$. Next we need to find an $N \in \mathbb{R}$ such that $|1/\sqrt{n} - 0| < \varepsilon$ for all $n > N$. Since the inequality $|1/\sqrt{n} - 0| < \varepsilon$ is obviously equivalent to the inequality $1/\varepsilon^2 < n$, we set $N := 1/\varepsilon^2$. This yields

$$\left|\frac{1}{\sqrt{n}} - 0\right| < \varepsilon \text{ for all } n > N,$$

and it therefore follows that $\lim_{n\to\infty} 1/\sqrt{n} = 0$ as desired.

3.4 Exercise. Show that $\lim_{n\to\infty} \left(1 + \frac{1}{n^2}\right) = 1$.

Limits of Functions

We stated in Chapter 2 that a function $f : D \subset \mathbb{R} \to \mathbb{R}$ has the limit L at x_0 if the values $f(x)$ approach L as x approaches x_0. To clarify this definition, let us assume that we are given a *sequence* $(x_n)_{n=1}^{\infty}$ in the domain D of f such that $x_n \neq x_0$ for all $n \in \mathbb{N}$ and $\lim_{n\to\infty} x_n = x_0$. Then the values x_n approach x_0 (in the sense of Definition 3.2), and if the limit of f at x_0 is L, then the values $f(x_n)$ approach L. In other words, the *sequence* $(f(x_n))_{n=1}^{\infty}$ converges to L. (Note: the restriction $x_n \neq x_0$ is necessary, because we want to allow for the possibility that the limit of f at x_0 is different from the value of f at x_0 or that f is not even defined at x_0 (see Chapter 2)). In order for the limit of f at x_0 to exist, we need to require the convergence behavior of $(f(x_n))_{n=1}^{\infty}$ to be dependent only on the assumption $\lim_{n\to\infty} x_n = x_0$ rather than the particular values in the sequence $(x_n)_{n\in\mathbb{N}}$. More precisely, the sequence $(f(x_n))_{n=1}^{\infty}$ ought to converge to the same limit L for *all* sequences $(x_n)_{n=1}^{\infty}$ in the domain D for which $x_n \neq x_0$ and $\lim_{n\to\infty} x_n = x_0$. Since this approach to the problem of defining the limit of f obviously presupposes the existence of at least one such sequence $(x_n)_{n=1}^{\infty}$ in D, we are led to introduce the following definitions:

3.5 Definition. Let $D \subset \mathbb{R}$ be an arbitrary set. A point $x_0 \in \mathbb{R}$ is said to be an *accumulation point of* D if there is a sequence $(x_n)_{n=1}^{\infty}$ such that $x_n \in D \setminus \{x_0\}$ for all $n \in \mathbb{N}$ and $\lim\limits_{n\to\infty} x_n = x_0$.

3.6 Definition. Let $f : D \subset \mathbb{R} \to \mathbb{R}$ be a function and assume that $x_0 \in \mathbb{R}$ is an accumulation point of D. Then for $L \in \mathbb{R}$ we say that f has the *limit* L *at* x_0 if $\lim_{n\to\infty} f(x_n) = L$ for every sequence $(x_n)_{n=1}^{\infty}$ that satisfies the following conditions: $x_n \in D \setminus \{x_0\}$ for all $n \in \mathbb{N}$ and $\lim_{n\to\infty} x_n = x_0$. In this case we write $\lim_{x\to x_0} f(x) = L$.

3.7 Exercise. Use Definition 3.6 to show that $\lim_{x\to 2} x^2 = 4$.

The following theorem provides an alternative characterization for the limit of a function which is frequently easier to apply than the one in Definition 3.6.

3.8 Theorem. *If $f : D \subset \mathbb{R} \to \mathbb{R}$ is a function and $x_0 \in \mathbb{R}$ is an accumulation point of D, then $\lim_{x\to x_0} f(x) = L$ if and only if for every $\varepsilon > 0$ there is a $\delta > 0$ such that*

$$|f(x) - L| < \varepsilon \text{ for all } x \in D \text{ that satisfy the inequality } 0 < |x - x_0| < \delta.$$

For a proof of this theorem based on Definition 3.6 see [Rud1].

Additional Exercises

3.9. Use Definition 3.5 to explain why a point $x_0 \in \mathbb{R}$ is an accumulation point of a set $D \subset \mathbb{R}$ if and only if for all $\varepsilon > 0$ there exists an $x \in D$ with $0 < |x - x_0| < \varepsilon$.

3.10. Use Definition 3.6 to determine the following limits:

$$\text{a) } \lim_{x\to 1} x^3 \qquad \text{b) } \lim_{x\to 2} \frac{x^2-4}{x-2} \qquad \text{c) } \lim_{x\to 2} \sqrt{x+1}$$

3.11. Use the alternative characterization in Theorem 3.8 to determine the limits in Exercise 3.10.

Chapter 4

The Derivative

Definition of the Derivative

Teacher: In the Introduction we discussed the law of falling bodies in relation to the concept of instantaneous velocity. We saw that with an appropriate choice of the units of measurement, the distance s an object has fallen after t units of time is given by the equation $s = t^2$.

Sophie: So we ought to regard s as a function of t?

Teacher: Yes indeed, and it is therefore better to write $s(t)$ instead of s in order to make the functional dependence of s on t more apparent. In the same way we will denote the instantaneous velocity of the falling body by $v(t)$, because it is a function of time as well.

Simplicio: It really doesn't seem relevant to me whether we write v or $v(t)$. What I would like to know is how we can possibly calculate $v(t)$ if all we have is the equation $s(t) = t^2$.

Teacher: You should not underestimate the value of adequate notation, Simplicio, but you are right in directing our attention to the calculation of $v(t)$, because this is the central problem that we need to address.

Sophie: I assume we will proceed as in our first lesson and approximate the instantaneous velocity by average velocities over successively smaller intervals of time. What I still don't understand, though, is how this method will ever generate an exact answer. It seems so fuzzy.

Teacher: It seems fuzzy, but it actually is not. Do you remember our discussion of limits?

Sophie: Yes I do.

Teacher: Is it not true that a limit is a single completely precise value despite the fact that it results from a "fuzzy" process of approximation?

Sophie: I see. So what we need to do in order to determine $v(t)$ is to take the limit of the average velocities over intervals of time that are successively smaller and closer to t.

Teacher: Quite so, and for this purpose we usually consider time intervals of the form $[t, t + \Delta t]$, where we think of Δt as a small number different from zero.

Simplicio: What does the Δ stand for?

Teacher: It is really not important whether we write Δt or some other symbol like, for example, the letter h, but traditionally, the Greek letter Δ stands for "difference" and is supposed to convey the fact that Δt is the difference between $t + \Delta t$ and t. But let us return now to the problem of calculating $v(t)$. Since the object's average velocity over the time interval $[t, t + \Delta t]$ is equal to the change in distance $s(t + \Delta t) - s(t)$ over the change in time $(t + \Delta t) - t = \Delta t$, it follows that

$$\bar{v}_{[t,t+\Delta t]} = \frac{s(t + \Delta t) - s(t)}{\Delta t}. \tag{4.1}$$

Sophie: There is a little problem here that confuses me. You said that we assume Δt to be different from zero...

Teacher: That is correct. If Δt were equal to zero, then the expression on the right-hand side of equation (4.1) would be undefined. Furthermore, in taking a limit we always consider only values that are different from the value that is being approached. Since in our case Δt approaches zero, we choose Δt to be different from zero just as in Definition 2.1 where we required x to be different from x_0.

Sophie: ... That I understand, but my problem is that in considering the interval $[t, t+\Delta t]$, we assumed not only $\Delta t \neq 0$, but implicitly also $\Delta t > 0$. For if Δt were less than zero, we would have $t + \Delta t < t$, and the average velocity would have to be calculated over the interval $[t+\Delta t, t]$ instead of $[t, t+\Delta t]$.

Teacher: Your point is well taken, but I can assure you that there is no serious difficulty. If Δt is less than zero then, as you said, the average velocity has to be determined over the interval $[t+\Delta t, t]$. The change in position over this interval is $s(t) - s(t+\Delta t)$, and the change in time is $t - (t + \Delta t) = -\Delta t$. Thus, we obtain

$$\bar{v}_{[t+\Delta t, t]} = \frac{s(t) - s(t+\Delta t)}{-\Delta t} = \frac{s(t+\Delta t) - s(t)}{\Delta t},$$

which is exactly the same expression as in (4.1).

Sophie: That answers my question.

Teacher: Good, let us proceed then to determine $v(t)$ by taking the limit of $(s(t+\Delta t) - s(t))/\Delta t$ as Δt approaches zero:

$$\begin{aligned} v(t) &= \lim_{\Delta t \to 0} \frac{s(t+\Delta t) - s(t)}{\Delta t} \\ &= \lim_{\Delta t \to 0} \frac{(t+\Delta t)^2 - t^2}{\Delta t} \qquad \text{(because } s(t) = t^2\text{)} \\ &= \lim_{\Delta t \to 0} \frac{2t\Delta t + (\Delta t)^2}{\Delta t} \\ &= \lim_{\Delta t \to 0} (2t + \Delta t) \\ &= 2t. \end{aligned}$$

This calculation shows that the instantaneous velocity corresponding to the position function $s(t) = t^2$ is given by the equation $v(t) = 2t$.

4.1 Exercise. Find the instantaneous velocity in the more general case that $s(t) = ct^2$ for some constant $c \in \mathbb{R}$ (see also Exercise 0.6).

To capture the essence of the preceding discussion—the taking of the limit of the average velocities—we introduce the following general definition:

4.2 Definition. Let $f : I \to \mathbb{R}$ be a function defined on an open interval I. We say that f is *differentiable at a point* $x_0 \in I$ if the limit

$$\lim_{\Delta x \to 0} \frac{f(x_0 + \Delta x) - f(x_0)}{\Delta x}$$

exists. The value of this limit is then referred to as the *derivative of f at x_0* and is denoted by $f'(x_0)$. Furthermore, f is said to be *differentiable* if it is differentiable at every point $x_0 \in I$, and the function $f' : I \to \mathbb{R}$, defined by the equation

$$f'(x) := \lim_{\Delta x \to 0} \frac{f(x + \Delta x) - f(x)}{\Delta x},$$

is in this case referred to as the *derivative function* of f.

The derivative function f' is frequently also denoted by

$$\frac{df}{dx} \quad \text{or} \quad \frac{d}{dx} f,$$

and for the derivative of f at a point x_0 we may write

$$\frac{d}{dx} f(x_0) \quad \text{or} \quad \left.\frac{df}{dx}\right|_{x=x_0}$$

in place of $f'(x_0)$. The motivation for this notation is that the *difference quotient*

$$\frac{f(x+\Delta x)-f(x)}{\Delta x},$$

which appears in the definition of f', is commonly abbreviated as $\Delta f/\Delta x$, and the taking of the limit as Δx approaches 0 is symbolized by replacing Δ with d. Thus, the process of differentiation turns $\Delta f/\Delta x$ into df/dx.

★ *Remark.* The df and dx are often thought of as "infinitely small numbers" or "infinitesimals." However, the concept of infinitesimals within the real number system is rather obscure, because there is in fact only one real number that may be regarded as infinitely small (in absolute value), and that number is zero. Thus, in the fraction df/dx we run into a division by zero which is undefined. For more than a hundred years the absence of a precise definition for the limit of a function, and therefore the derivative of a function, posed a serious problem for the development of mathematics, and the notation df/dx, suggesting the use of infinitely small numbers, has to be seen in this historical context. In modern mathematics a way has actually been found to make precise the definition of derivatives via infinitesimals, but a detailed discussion of this subject is far beyond the scope of this text (see [Gol]). ★

4.3 Example. Let us consider the function $f:\mathbb{R}\to\mathbb{R}$ given by the equation $f(x):=x^3$. We wish to find the value of the derivative of f at $x_0=1$. According to the definition of the derivative, we need to find the limit of the difference quotient

$$\frac{\Delta f}{\Delta x}=\frac{f(1+\Delta x)-f(1)}{\Delta x}=\frac{(1+\Delta x)^3-1}{\Delta x}$$

as Δx approaches zero. The following table shows some values of this difference quotient for values of Δx that are progressively closer to zero:

Δx	-1	-0.5	-0.1	-0.01	0.01	0.1	0.5	1
$\Delta f/\Delta x$	1	1.75	2.71	2.9701	3.0301	3.31	4.75	7

Given this table, we expect that $f'(1)=3$. To verify this claim we perform the following calculation:

$$\begin{aligned}f'(1)&=\lim_{\Delta x\to 0}\frac{(1+\Delta x)^3-1}{\Delta x}=\lim_{\Delta x\to 0}\frac{1+3\Delta x+3\Delta x^2+\Delta x^3-1}{\Delta x}=\lim_{\Delta x\to 0}\frac{3\Delta x+3\Delta x^2+\Delta x^3}{\Delta x}\\&=\lim_{\Delta x\to 0}\frac{\Delta x(3+3\Delta x+\Delta x^2)}{\Delta x}=\lim_{\Delta x\to 0}(3+3\Delta x+\Delta x^2)=\lim_{\Delta x\to 0}3+3\lim_{\Delta x\to 0}\Delta x+\lim_{\Delta x\to 0}\Delta x^2\\&=3+0+0=3.\end{aligned}$$

This proves that indeed $f'(1)=3$.

4.4 Example. For the function $f:\mathbb{R}\setminus\{0\}\to\mathbb{R}$, defined by the equation $f(x):=1/x$, we wish to determine the derivative function f':

$$\begin{aligned}f'(x)&=\lim_{\Delta x\to 0}\frac{f(x+\Delta x)-f(x)}{\Delta x}=\lim_{\Delta x\to 0}\frac{1/(x+\Delta x)-1/x}{\Delta x}=\lim_{\Delta x\to 0}\frac{x-(x+\Delta x)}{(x+\Delta x)x\Delta x}\\&=\lim_{\Delta x\to 0}\frac{-1}{(x+\Delta x)x}=-\frac{1}{x^2}.\end{aligned}$$

4.5 Exercise. For the function $f:\mathbb{R}\to\mathbb{R}$, defined by the equation $f(x):=x^3$, show that

$$f'(x)=3x^2.$$

Hint. Your calculation will differ from the calculation in Example 4.3 only in that 1 needs to be replaced with x.

Remark. There is an alternative way to calculate the derivative of a function that is also frequently used. According to Definition 4.2, the derivative of a differentiable function $f : I \to \mathbb{R}$ at a point x_0 is equal to $\lim_{\Delta x \to 0}(f(x_0 + \Delta x) - f(x_0))/\Delta x$. Setting $x := x_0 + \Delta x$, we see that x approaches x_0 as Δx approaches zero and therefore,

$$\boxed{f'(x_0) = \lim_{\Delta x \to 0} \frac{f(x_0 + \Delta x) - f(x_0)}{\Delta x} = \lim_{x \to x_0} \frac{f(x) - f(x_0)}{x - x_0}.}$$

4.6 Example. For the function $f(x) := x^2$, the above characterization of the derivative yields

$$f'(x_0) = \lim_{x \to x_0} \frac{f(x) - f(x_0)}{x - x_0} = \lim_{x \to x_0} \frac{x^2 - x_0^2}{x - x_0} = \lim_{x \to x_0} \frac{(x - x_0)(x + x_0)}{x - x_0} = \lim_{x \to x_0} (x + x_0) = 2x_0$$

for all $x_0 \in \mathbb{R}$. A similar factorization also works for $g(x) := \sqrt{x}$:

$$g'(x_0) = \lim_{x \to x_0} \frac{\sqrt{x} - \sqrt{x_0}}{x - x_0} = \lim_{x \to x_0} \frac{\sqrt{x} - \sqrt{x_0}}{(\sqrt{x} - \sqrt{x_0})(\sqrt{x} + \sqrt{x_0})} = \lim_{x \to x_0} \frac{1}{\sqrt{x} + \sqrt{x_0}} = \frac{1}{2\sqrt{x_0}}$$

for all $x_0 \in (0, \infty)$. For the cube root function $h(x) := \sqrt[3]{x}$ matters are already slightly more complicated as the following calculation shows:

$$\begin{aligned} h'(x_0) &= \lim_{x \to x_0} \frac{\sqrt[3]{x} - \sqrt[3]{x_0}}{x - x_0} = \lim_{x \to x_0} \frac{\sqrt[3]{x} - \sqrt[3]{x_0}}{\sqrt[3]{x}^3 - \sqrt[3]{x_0}^3} = \lim_{x \to x_0} \frac{\sqrt[3]{x} - \sqrt[3]{x_0}}{(\sqrt[3]{x} - \sqrt[3]{x_0})(\sqrt[3]{x}^2 + \sqrt[3]{x}\sqrt[3]{x_0} + \sqrt[3]{x_0}^2)} \\ &= \lim_{x \to x_0} \frac{1}{\sqrt[3]{x}^2 + \sqrt[3]{x}\sqrt[3]{x_0} + \sqrt[3]{x_0}^2} = \frac{1}{3\sqrt[3]{x_0}^2} \end{aligned}$$

for all $x_0 \in \mathbb{R}^3$.

4.7 Exercise. Find the derivative of $f(x) := \sqrt[4]{x}$ at a given point $x_0 \in \mathbb{R}$ and then generalize your result to determine the derivative of the nth root of x for an arbitrary positive integer n.

Instantaneous Velocity and Acceleration

Let us denote by $s(t)$ the position in dependence on time of an object that is moving along a straight line. From our discussion in the previous section we have learned that the object's instantaneous velocity $v(t)$ is equal to the limit of the average velocities $(s(t + \Delta t) - s(t))/\Delta t$ as Δt approaches zero. In other words, we have

$$v(t) = s'(t).$$

If the velocity changes over time, the object is said to be *accelerating.* Just as the average velocity of an object is defined as change in position over change in time, so the *average acceleration* is defined as change in velocity over change in time. In taking the analogy one step further, the *instantaneous acceleration* $a(t)$ is then obtained by taking the limit of the average accelerations over successively smaller intervals of time close to t. Hence

$$a(t) = \lim_{\Delta t \to 0} \frac{v(t + \Delta t) - v(t)}{\Delta t} = v'(t).$$

Remark. The instantaneous acceleration is positive if the velocity is increasing and negative if the velocity is decreasing. Similarly, the instantaneous velocity is positive if $s(t)$ is increasing and negative if $s(t)$ is decreasing (see also Theorem 4.11 below).

4.8 Exercise. Use the defining equation for the instantaneous acceleration to find $a(t)$ for $v(t) = 2t$.

Geometric Interpretation of the Derivative

Let $f : \mathbb{R} \to \mathbb{R}$ be a differentiable function. To understand the geometric meaning of the derivative $f'(x)$, we begin by drawing a *secant line* through two points $(x, f(x))$ and $(x + \Delta x, f(x + \Delta x))$ on the graph of f as shown in Figure 4.1 The slope m of the secant line is the quotient of the lengths of the vertical and horizontal sides of the slope triangle in Figure 4.1. Since these lengths are $f(x+\Delta x) - f(x)$

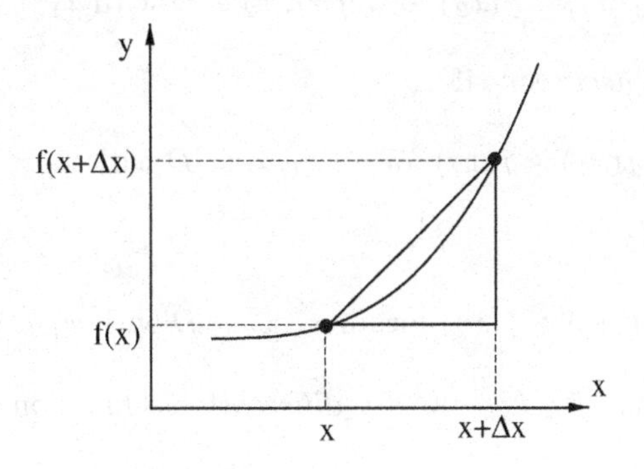

Figure 4.1: secant slope of f.

and Δx respectively, we have

$$m = \frac{f(x + \Delta x) - f(x)}{\Delta x}.$$

In other words, the difference quotient of f gives us the slope of the secant line passing through the points $(x, f(x))$ and $(x+\Delta x, f(x+\Delta x))$. As Δx approaches zero, the slope of the secant lines approaches the slope of the tangent line to the graph of f at the point $(x, f(x))$ (see Figure 4.2), and the value of the difference quotient of f approaches $f'(x)$. This observation leads us to the *geometric interpretation*

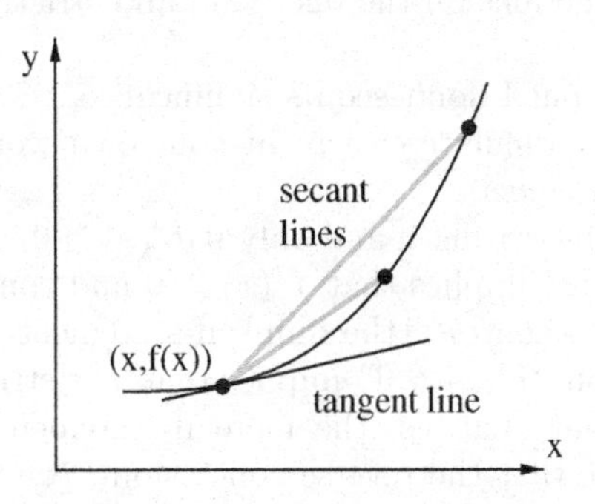

Figure 4.2: secant lines approaching the tangent line.

of the derivative:

The value $f'(x)$ is equal to the *slope of the tangent line* to the graph of f at the point $(x, f(x))$.

Therefore, the *equation of the tangent line* to the graph of f at a point $(x_0, f(x_0))$ is

$$y(x) = f'(x_0)(x - x_0) + f(x_0).$$

4.9 Exercise. Verify the equation above and then find the equation of the tangent line to the graph of the function $f(x) := x^2$ at the point $(2, 4)$.

The geometric interpretation of the derivative allows us to generate valuable information concerning the shape of the graph of a function. For example, if $f'(x) > 0$ for all x in a certain interval I, then f has to be increasing over this interval, because the slope of the tangent lines is always positive. Similarly, f has to be decreasing if $f'(x) < 0$.

4.10 Definition. A function $f : D \subset \mathbb{R} \to \mathbb{R}$ is said to be *increasing* if

$$f(x_1) \leq f(x_2) \text{ for all } x_1, x_2 \in D \text{ with } x_1 < x_2.$$

It is called *strictly increasing* if

$$f(x_1) < f(x_2) \text{ for all } x_1, x_2 \in D \text{ with } x_1 < x_2.$$

Furthermore, f is said to be *decreasing* if

$$f(x_1) \geq f(x_2) \text{ for all } x_1, x_2 \in D \text{ with } x_1 < x_2$$

and *strictly decreasing* if

$$f(x_1) > f(x_2) \text{ for all } x_1, x_2 \in D \text{ with } x_1 < x_2.$$

4.11 Theorem. *Assume that $f : I \to \mathbb{R}$ is a differentiable function defined on an open interval I. Then*

f is increasing if and only if $f'(x) \geq 0$ for all $x \in I$,
f is strictly increasing if $f'(x) > 0$ for all $x \in I$,
f is decreasing if and only if $f'(x) \leq 0$ for all $x \in I$, and
f is strictly decreasing if $f'(x) < 0$ for all $x \in I$.

Teacher: Did you notice that in Theorem 4.11 there is a difference in logic between the statements about increasing and decreasing functions on the one hand and strictly increasing and strictly decreasing functions on the other hand?

Sophie: I noticed the difference, but I don't see its significance.

Teacher: Perhaps you can first formulate clearly in your own words what constitutes the difference between the two types of statements.

Sophie: Well, if I say that "f is increasing if and only if $f'(x) \geq 0$," then I have a two-way implication: the statement "f is increasing" implies that $f'(x) \geq 0$ and conversely, the statement "$f'(x) \geq 0$" implies that f is increasing. By contrast, the simple if-statement "f is strictly increasing if $f'(x) > 0$" only says that the assumption "$f'(x) > 0$" implies that f is strictly increasing.

Teacher: Very good. Given your analysis, the mere if-statement in the case of strictly increasing functions appears to indicate that the reverse conclusion "f strictly increasing implies $f'(x) > 0$" is false.

Simplicio: Yes, but how is that possible? How could a function that is strictly increasing have a derivative that is not always positive? If the derivative of a function is negative at a certain point x_0 then, according to the geometric interpretation of the derivative, the tangent line has negative slope, and close to x_0 the function would therefore have to be decreasing.

Teacher: I agree, but your mistake is in assuming that the derivative has to be negative somewhere if it is not strictly greater than zero everywhere. You failed to consider the possibility that the derivative is *equal* to zero.

Sophie: I see. What you are suggesting is that we ought to be looking for a strictly increasing function the derivative of which is equal to zero at one or perhaps several points.

Teacher: Exactly. Let's consider, for example, the function $f : \mathbb{R} \to \mathbb{R}$ defined by the equation $f(x) := x^3$. We saw in Exercise 4.5 that the derivative of f is equal to $3x^2$. Consequently, $f'(0) = 0$, and in particular, it is not true that $f'(x) > 0$ for all $x \in \mathbb{R}$. However, f is strictly increasing according to Definition 4.10, because if $x_1 < x_2$ then $f(x_1) = x_1^3 < x_2^3 = f(x_2)$. In accordance with these observations, the graph of f in Figure 4.3 shows a strictly increasing function with a "flat spot" at $x = 0$ where the derivative vanishes.

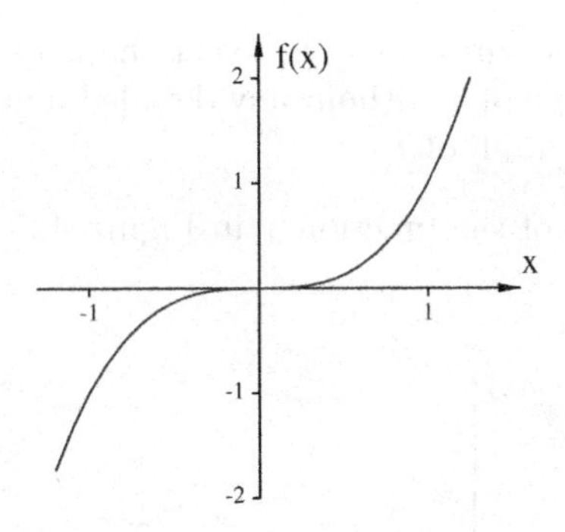

Figure 4.3: graph of $f(x) = x^3$.

4.12 Example. As an application of Theorem 4.11, we wish to discuss how we can produce an *approximate* sketch of the graph of the derivative of a function given only the graph of the function itself. Consider, for instance, the graph in Figure 4.4. Here we are shown a function f that is increasing on the intervals $[-3/2, 0]$ and $[2, 3]$ and decreasing on $[0, 2]$. Consequently, we have $f'(x) \geq 0$ for all $x \in [-3/2, 0] \cup [2, 3]$ and $f'(x) \leq 0$ for all $x \in [0, 2]$. We also see that $f'(x)$ is different from 0 everywhere except at $x = 0$ and $x = 2$. Furthermore, f' has its largest value at $x = -3/2$ and its smallest value at approximately $x = 1$.

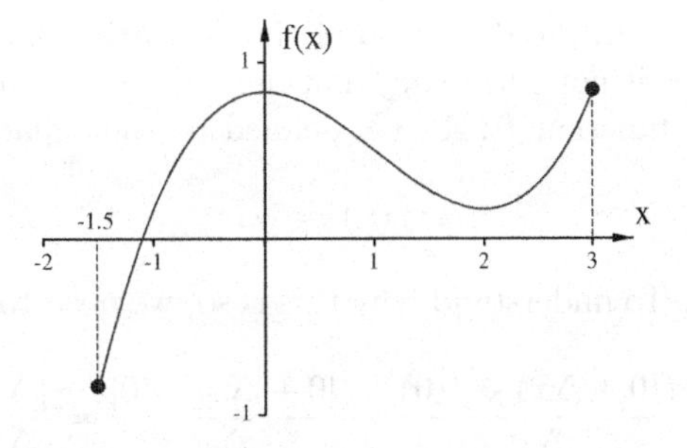

Figure 4.4: graph of f.

Consistent with this information, a possible graph of f' is shown in the diagram on the left-hand side in Figure 4.5:

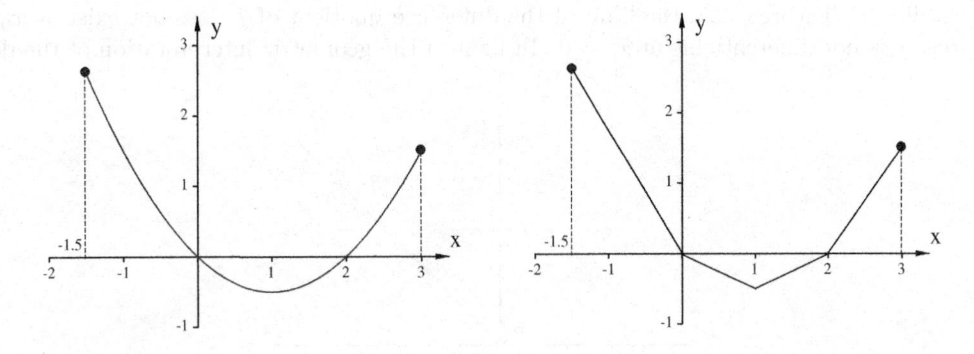

Figure 4.5: graph of f' and alternative graph of f'.

To illustrate that the information generated from the graph of f is incomplete, the diagram on the right-hand side in Figure 4.5 shows another possible graph of f' that also matches all the properties listed above. There is no way to be sure about the exact shape of the graph of f', unless we are given a precise definition for f. The graph of f without any detailed numerical or graphical analysis can only give us an approximation for the graph of f'.

4.13 Exercise. Given the graph of the function f in Figure 4.6, make an approximate sketch of the graph of f'.

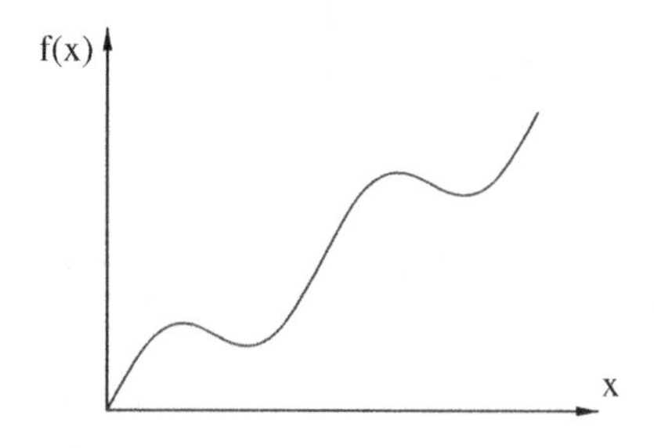

Figure 4.6: graph of f.

A Nondifferentiable Function

Since limits feature prominently in the definition of the derivative, and since limits do not necessarily always exist (see Chapter 2), it is not surprising that derivatives of functions do not necessarily always exist either. For instance, the function $f : \mathbb{R} \to \mathbb{R}$ defined by the equation

$$f(x) := |x|$$

is not differentiable at $x_0 = 0$. To understand why this is so, we need to examine the difference quotient of f at zero:

$$\frac{f(0 + \Delta x) - f(0)}{\Delta x} = \frac{|0 + \Delta x| - |0|}{\Delta x} = \frac{|\Delta x|}{\Delta x}.$$

If $\Delta x < 0$ then $|\Delta x|/\Delta x = -1$, and if $\Delta x > 0$ then $|\Delta x|/\Delta x = 1$. (For example, for $\Delta x = 0.1$ we have $|0.1|/0.1 = 0.1/0.1 = 1$ and for $\Delta x = -0.1$ we have $|-0.1|/(-0.1) = 0.1/(-0.1) = -1$.) This shows that

$$\lim_{\Delta x \to 0^-} \frac{|\Delta x|}{\Delta x} = -1 \neq 1 = \lim_{\Delta x \to 0^+} \frac{|\Delta x|}{\Delta x}.$$

Thus, according to Theorem 2.7, the limit of the difference quotient of f does not exist at $x_0 = 0$. In other words, f is not differentiable at $x_0 = 0$. In light of the geometric interpretation of the derivative

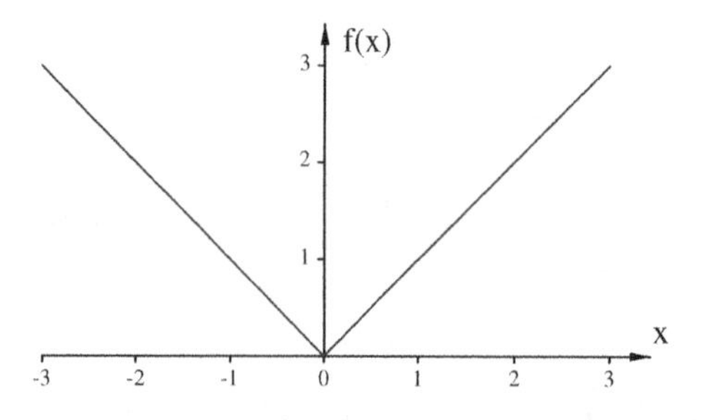

Figure 4.7: graph of $f(x) = |x|$.

it is also intuitively easy to see that f cannot be differentiable at $x_0 = 0$, because at $(0, 0)$ the V-shaped graph of f (see Figure 4.7) does not have a well defined tangent slope. In general, the graph of a differentiable function must be "smooth"—it cannot have any sharp edges.

Additional Exercises

4.14. Using the definition of the derivative, find the value of $f'(2)$ for each of the following functions:

a) $f(x) = x$ **b)** $f(x) = x^2 - x$ **c)** $f(x) = x^4$

d) $f(x) = x^2 - 2x + 3$ **e)** $f(x) = \dfrac{(x-1)(x+1)}{x^2-1}$ **f)** $f(x) = \dfrac{1}{x^2}$

4.15. The position $s(t)$ of an object moving along a straight line is given by the graph in Figure 4.8.

a) Find the average velocity over the time intervals $[0, 4]$ and $[1, 4]$.

b) Find the equation of the secant line that connects the points $(1, s(1))$ and $(4, s(4))$.

c) Find the equation of the tangent line to the graph of s at $(1, s(1))$.

d) For what times t is the object's instantaneous velocity positive? For what times is it negative? When is it zero?

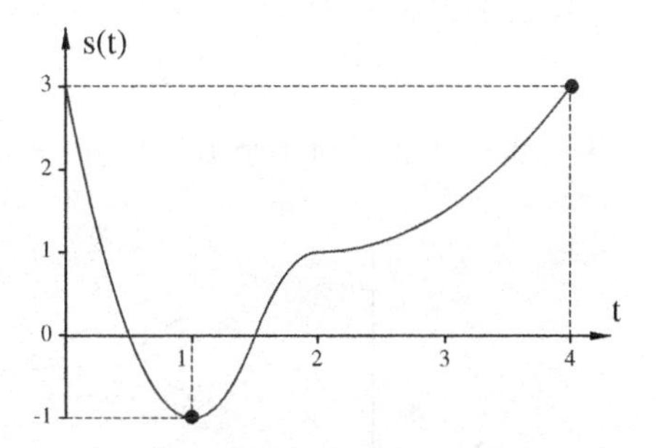

Figure 4.8: graph of s.

4.16. Let $f(t)$ be a function that satisfies the equation $f(1+h) - f(1) = 3h + 4h^2 - 5h^3$ for all $h \in \mathbb{R}$. Show that f is differentiable at 1 and find the value of f' at 1.

4.17. Using the result of Exercise 4.14d, find the equation of the tangent line to the graph of $f(x) := x^2 - 2x + 3$ at $(2, f(2))$. Then sketch the graph of f and the tangent line on the same set of axes.

4.18. Suppose that $f(1) = 2$ and $f'(1) = 3$. Estimate the values of $f(1.1)$ and $f(0.9)$.

4.19. Suppose that $f(2) = -1$ and $f'(2) = 1.5$. Estimate the values of $f(2.1)$ and $f(1.9)$.

4.20. How do we interpret the expression $\dfrac{f(x_2) - f(x_1)}{x_2 - x_1}$, if...

a) $f(x)$ is the speed of an airplane x seconds after takeoff.

b) $f(x)$ is the number of individuals in a population at time x.

c) $f(x)$ is the y-coordinate of the points on the graph of f.

d) $f(x)$ is the area under a velocity graph over the time interval $[0, x]$ (you may assume the velocity to be positive).

4.21. Using a table of values, estimate the value of $f'(7)$ for $f(x) := \sqrt[3]{x^2 + 15}$.

4.22. Given the graph of the function u in Figure 4.9, estimate the following values:

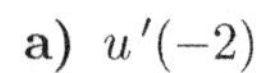

a) $u'(-2)$

b) $u'(-1)$

c) $u'(0)$

d) $u'(1)$

e) $u'(3.5)$

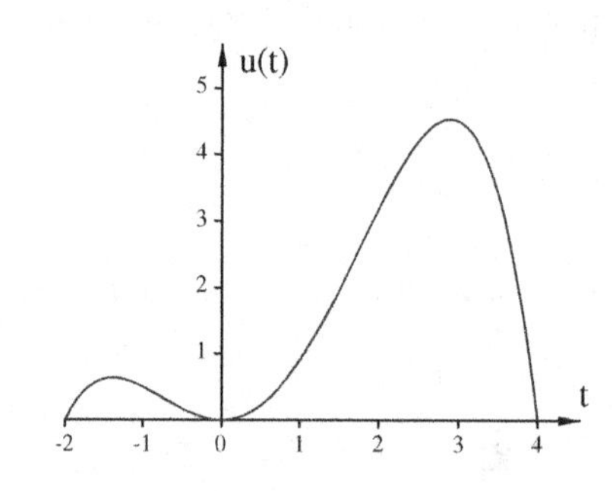

Figure 4.9: graph of u.

4.23. Suppose that $f : \mathbb{R} \to \mathbb{R}$ is a function for which $\lim_{x\to 2}(f(x) - f(2))/(x - 2) = 0$. Label each of the statements given below as TRUE, FALSE, or POSSIBLY TRUE/POSSIBLY FALSE. Explain your reasoning.

a) $f'(2) = 0$

b) $f(2) = 0$

c) $\lim_{x\to 2} f(x) = f(2)$

4.24. In Figure 4.10 you are shown the graphs of two functions f and g. Sketch the graph of the derivative of $f - g$.

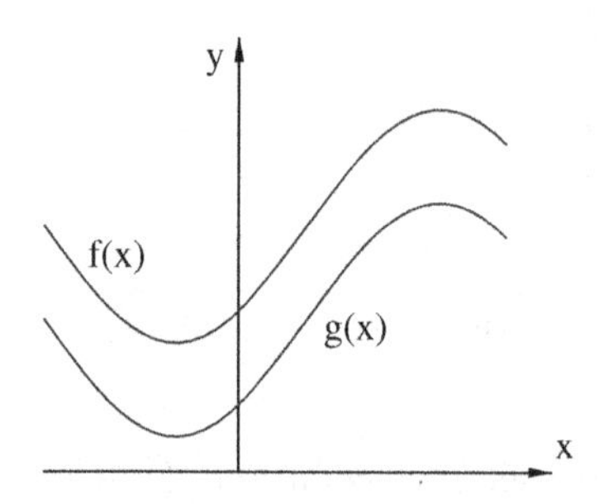

Figure 4.10: graphs of f and g.

4.25. For each of the functions f, g, and h shown in Figure 4.11, make an approximate sketch of the graph of the derivative.

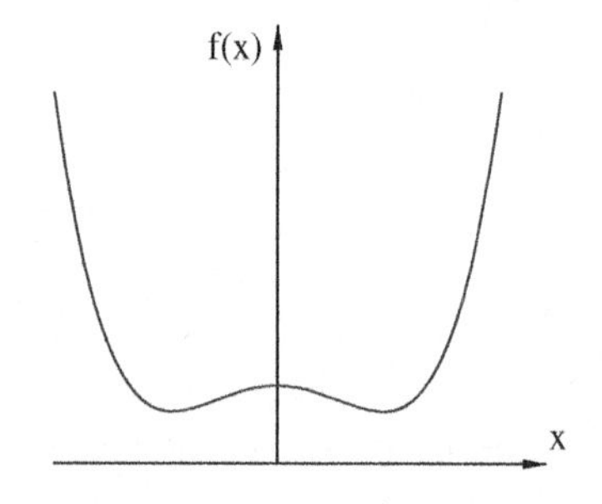

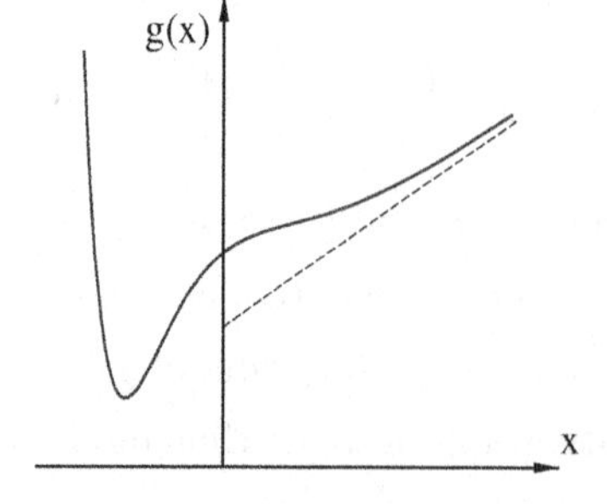

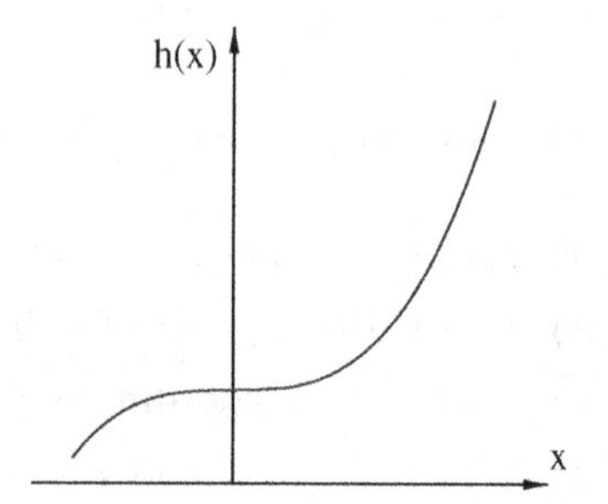

Figure 4.11: graphs of f, g, and h.

4.26. An alternative definition for the derivative of a function $f : I \to \mathbb{R}$ defined on an open interval I is given by the equation

$$f'(x) := \lim_{\Delta x \to 0} \frac{f(x + \Delta x) - f(x - \Delta x)}{2\Delta x}.$$

a) What is an appropriate graphical interpretation of this definition?

b) Use this definition to find $f'(2)$ for $f(x) := 3x^2$.

4.27. If the function w records a person's weight as a function of the person's age, then what does the derivative of w represent?

4.28. If the function v records the trade-in-value of a car as a function of the car's age, then what does the derivative of v represent?

Chapter 5

The Origins of Modern Mathematics

To discern the causes that brought about the great transition from the Middle Ages to the modern world is no easy task. In our time of postmodern deconstruction, the traditional answers, pointing for example to the rediscovery of ancient learning in the Renaissance, to economic shifts, or rapid advances in science and technology as catalysts for change, are seriously challenged. The very notion, in fact, that a Scientific Revolution did indeed occur has fallen out of fashion.[??] And it is fair enough: we never can set up truly objective criteria for the designation of distinct historical periods, and no feat of scholarship, no matter how broad in scope, can ever encompass the whole of human life in all its vast complexity in any bygone era. However, there also can be no doubt whatsoever that dramatic change in Western Civilization has somehow been effected. We no longer live under the direct cultural and political control of the Catholic Church, and the *Summa Theologica* of Saint Thomas Aquinas, this great medieval synthesis of Aristotelian philosophy and Christian dogma, no longer sets the standard in our intellectual discourse.

What happened in the 15th, 16th, and 17th centuries is perhaps best likened to the psychological developments that mark the passage from childhood to adolescence. Here as there, a new consciousness is born, but a precise timeline or chain of causation is difficult to ascertain.

With the advent of the Italian Renaissance, the human spirit emerged from the shadow of the Church's parental authority to assert its autonomy and test its strength. Inasmuch as this quest for intellectual emancipation was characterized by a greater appreciation for the worth and dignity of the individual or by aspirations for increased political freedom and religious tolerance, our assessment of the legacy of modernity will be unambiguously positive. With scientific methods of disease control at our command, the bubonic plague no longer ravages the nations of Europe, and the Age of Enlightenment has put an end to the witch hunts and the Inquisition. On the darker side, however, a note of caution is in order, for throughout the modern era men and women of sincere conviction have expressed their sense of unease at the parting with tradition and the uncontested rule of reason. In the days of the Copernican Revolution, for instance, John Donne voiced his discomfort at the Earth's exile from its familiar place at the center of the universe in his poem *An Anatomy of the World:*[1]

> [The] new philosophy calls all in doubt,
> The element of fire is quite put out;
> The sun is lost, and th'earth, and no man's wit
> Can well direct him where to look for it.
> And freely men confess that this world's spent,
> When in the planets, and the firmament
> They seek so many new; then see that this
> Is crumbled out again to his atomies.
> 'Tis all in pieces, all coherence gone;
> All just supply, and all relation:
> Prince, subject, father, son, are things forgot,

For every man alone thinks he hath got
To be a phoenix, and that then can be
None of that kind, of which he is, but he.

Francisco Goya apparently was motivated by similar sentiments when he etched a somber vision of monsters produced by "the dream of reason."

Figure 5.1: *The Dream of Reason Produces Monsters* by Goya (1799).

On this less favorable view, it may indeed appear that "all coherence" was lost when the modern analytic mind set out to demonstrate that reality in its entirety is but an aggregate of "atomies." Driven by the underlying presumption of the comprehensive reach of the human intellect, modern thought has spread before us barren lands of fragmentary knowledge in which the human soul can find no place of rest. As our conceptual scientific understanding advanced, the world receded ever further from the immediacy of direct experience toward the realm of abstraction. Images that once were tied to deeply felt beliefs have faded into pale blue rational reflections. A mountain range no longer is the place where legends tell of dragons slain by heroes of the past. For we have learned to look at it more truthfully, or so we think, as we record the elevation of its peaks and measure its extent in degrees of latitude. Under the austere light of reason, reality is stripped bare of its metaphysical attachments, and, sensing the void, we feel unsettled.

Given this modern predisposition of ours to cast aside anything that reason cannot fathom, it comes as no surprise that *René Descartes*, one of the great founding figures of modern philosophy, began his intellectual journey from a position of systematic, unexcepting doubt. Born in 1596 into an old distinguished family at La Haye, he was left to the care of his father—a successful lawyer and councillor of the parliament in the province of Touraine—when his mother died a few days after giving birth. Due to his fragile health, young René was forced to stay at home until, at the age of eight, he entered the Jesuit college of La Flèche. Possessed of "an extreme desire to acquire instruction," he strove to obtain "a clear and certain knowledge" of "all that is useful in life."[2] Yet by the time of graduation, he had grown disillusioned with the obscurity of traditional scholastic learning* and found himself "embarrassed with so many doubts and errors that it seemed to [him] that the effort to instruct [himself] had no effect other than the increasing discovery of [his] own ignorance."[3] Throughout his education, only mathematics could satisfy his longing for clarity and logical necessity. In his famous *Discourse on Method* he later wrote:

> Most of all I was delighted with Mathematics because of the certainty of its demonstrations and the evidence of its reasoning; but I did not yet understand its true use, and, believing

*The curriculum at La Flèche was aimed at integrating medieval scholasticism with the classical learning of the Renaissance.

> that it was of service only in the mechanical arts, I was astonished that, seeing how firm and solid was its basis, no loftier edifice had been reared thereupon.[4]

Upon leaving La Flèche, Descartes briefly immersed himself in the pleasures of social life, but with boredom setting in he soon retired for another two years of study. At the end of this time, having completed a degree in law, he determined that a more thorough reading of "the great book of the world"[5] was needed to advance his understanding of the practical affairs of men. In 1618, at the outbreak of the Thirty Years' War in Germany, he therefore enlisted in the army of Prince Maurice of Nassau and a year later transferred his allegiance to the Duke of Bavaria. As an unpaid volunteer he was not exposed to the full rigors of military life and was able to set aside some time for study and reflection. Thus, it happened on a night in November, when his regiment was detained by the onset of winter in the small town of Ulm, that a nightly vision announced to him his destiny. In his diary we read:

> 10 Nov. 1619: I was filled with enthusiasm, discovered the foundation of a marvelous science, and at the same time my vocation was revealed to me.[6]

His calling it turned out was to erect a comprehensive rationalistic philosophy that was to exert a decisive influence on the development of modern thought for more than three hundred years. At the outset of his intellectual tour de force Descartes resolved to strictly adhere to four fundamental principles which he deemed sufficient to ensure his mind would not stray from the path toward the "clear and certain knowledge"[7] that he so desired:

(i) Accept nothing as true except that which reason can confirm with indubitable clarity.

(ii) Divide up any given problem into as many parts as possible.

(iii) Proceed in an orderly fashion from the simplest objects to those of greatest complexity.

(iv) Conduct enumerations and reviews so general and comprehensive that nothing is omitted and all facts are accounted for.

Having thus established the methodological groundwork, Descartes now withdrew to a position of radical doubt in order to seek out the fundamental truths on which to build his edifice of knowledge. He went so far as to assert that no sensory experience or mental representation could ever be trusted to be more than a dream or possibly a deceptive projection induced by a "malignant demon."[8] Yet at the very instance when he had become convinced "that there was nothing in all the world, that there was no heaven, no earth, that there were no minds, nor any bodies"[9] he found himself unable to deny that the very act of doubting affirmed his own existence: *cogito ergo sum*—I think therefore I am. It was this proclamation of self-awareness of the thinking subject that Descartes chose as his point of departure. From here he ventured on to resurrect a coherent vision of reality, offering proofs for the existence of God and physical substance in the process. The end result was a fully developed metaphysical dualism that reduced the entire physical universe, including animals and human bodies, to a mechanical clockwork and allowed for a spiritual component to exist only in the interaction of God with the minds of men.

What is remarkable in the unfolding of this rationalistic philosophy is the insistence on mathematical rigor in deducing propositions of higher order from self-evident first principles. Indeed, Descartes not only looked upon mathematics as an ideal of perfection for the acquisition of true knowledge, but also contributed to its progress. With his work *La Géométrie*, which he attached as one of three appendices to his *Discourse on Method*, he became known as the author of coordinate geometry (or, as we would say, *analytic geometry*). The fundamental insight on which he based his theory is described in the opening sentence of *La Géométrie:*

> Any problem in geometry can easily be reduced to such terms that a knowledge of the lengths of certain straight lines is sufficient for its construction.[10]

In other words, geometric problems can be solved by algebraic means. To illustrate this idea, Descartes goes on to explain how the arithmetic operations of multiplication and division can be represented by constructions involving similar triangles. For example, if we wish to form the product xy of two given numbers x and y, we may proceed as follows: First we draw a coordinate axis† on which two points A and B are marked at the respective distances of 1 and x units from the origin at P (see Figure 5.2). On a line L that passes through the origin at an arbitrary angle we mark a third point C at a distance

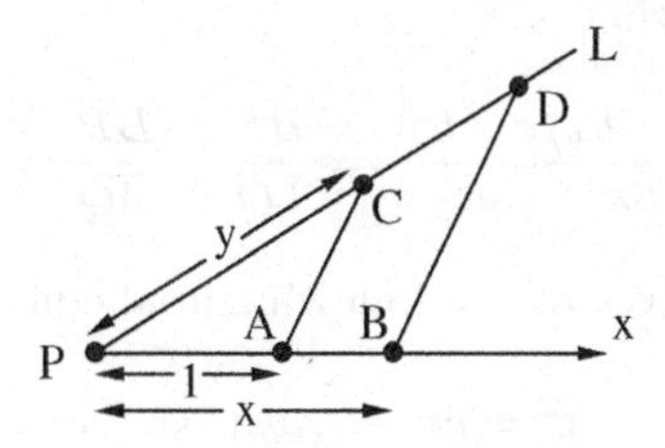

Figure 5.2: constructing the product xy.

of y units from P. To complete the construction we insert a line segment parallel to $\overline{AC}$ that connects B with a point D on L. Since the triangles $\triangle ACP$ and $\triangle BDP$ are similar, we may infer that

$$x = \frac{x}{1} = \frac{\overline{BP}}{\overline{AP}} = \frac{\overline{DP}}{\overline{CP}} = \frac{\overline{DP}}{y}.$$

Consequently, the product xy is equal to the length of the line segment from D to P.

5.1 Exercise. Explain how the value of the quotient x/y can be constructed geometrically.

After discussing a number of similar constructions concerning the extraction of square roots and in particular also roots of quadratic equations, Descartes turns his attention to the important problem of representing a curve by an algebraic equation. One of the examples he gives is shown in Figure 5.3, where we imagine the triangle $\triangle KLN$ to be sliding up and down along the y-axis.‡ For each of the

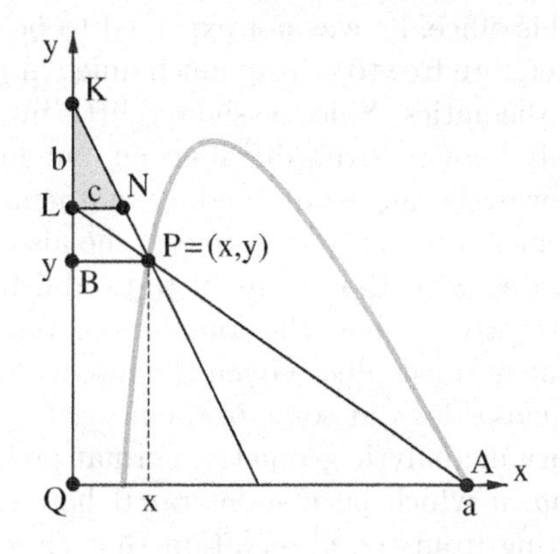

Figure 5.3: constructing a hyperbola.

triangle's successive positions we determine a point $P = (x, y)$ as the intersection of the extension of the line segment $\overline{KN}$ with the line passing through L and the given point $A = (a, 0)$. In this manner of construction P traces out a curve, which in fact turns out to be a hyperbola. In order to find the corresponding xy-equation, Descartes first considers the similar triangles $\triangle KLN$ and $\triangle KBP$ to infer that

$$\frac{\overline{BK}}{x} = \frac{\overline{BK}}{\overline{BP}} = \frac{\overline{KL}}{\overline{LN}} = \frac{b}{c},$$

†Descartes actually used a line segment on which a unit length was marked instead of a regular coordinate axis.

‡The coordinate axes are here introduced for better recognizability—they do not appear in the original manuscript.

or equivalently,

$$\overline{BK} = \frac{bx}{c}.$$

To proceed he observes $\overline{BL}$ to be equal to $\overline{BK} - b$, and upon examining the similar triangles $\triangle LBP$ and $\triangle LQA$ he is able to conclude that

$$\frac{bx/c - b}{bx/c - b + y} = \frac{\overline{BL}}{\overline{LQ}} = \frac{\overline{BP}}{\overline{AQ}} = \frac{x}{a}.$$

After rearranging the terms he arrives at the following final equation:

$$x^2 = cx - \frac{c}{b}xy + ax - ac. \tag{5.1}$$

5.2 Exercise. Plot the curve described by equation (5.1) for $a = 4$, $b = 1$ and $c = 1/2$.

With the concept of describing a curve by an algebraic equation now at his disposal, Descartes provides further evidence for the power of his method by demonstrating how the points of intersection of two curves can be determined as simultaneous solutions of the corresponding equations, and in another section of his work he also offers an elegant solution of the famous four-lines problem of Pappus which involves the general equation of a conic section.[§]

In the light of such remarkable accomplishments, it is no small surprise to learn that René Descartes, the foremost philosopher of his age, was surpassed in brilliance by a mathematical amateur from the French countryside: *Pierre de Fermat.* By 1637, the year when *La Géométrie* was published, Fermat had already independently developed a competing version of analytic geometry, which in several respects proved superior. Not only did he make better use of coordinate axes, but he also focused more consistently on the analysis of geometric curves, and his exposition was in general possessed of greater clarity.

Born in 1601, as the son of a leather-merchant, Pierre de Fermat was a man of retiring disposition who spent his entire professional life as a provincial judge and councillor of the parliament in the town of Toulouse. Due to the nature of his office, he was not expected to be very actively engaged in the social life of his community and was therefore free to devote much quiet time to the study of ancient languages and, above all, his research in mathematics. Since he showed little interest in publication, his astonishing discoveries did not become widely known during his lifetime, but in retrospect there can be no doubt that, next to Isaac Newton, he was the most outstanding mathematician of the 17th century. Apart from the aforementioned achievements in analytic geometry, he also made seminal contributions to the theory of numbers,[¶] he collaborated with Blaise Pascal to establish the basic principles of the theory of probability, and, most importantly, he laid the foundations on which Newton and Leibniz would later build the differential and integral calculus. Given the momentous significance of his insights, it is appropriate that we now take a closer look at some of them.

In 1637, expanding on his work in analytic geometry, Fermat produced a manuscript on *The Method of Finding Maxima and Minima* in which he demonstrated how certain optimization problems (see Chapter 14) can be solved starting from the observation that close to a maximum or minimum of a continuous curve, the change in y is very small compared to the change in x. As a first example, he considered the problem of maximizing the area of a rectangle given its circumference. In denoting the circumference by $2a$ and the side lengths by x and $a - x$ respectively (see Figure 5.4), Fermat essentially

---

[§] The problem posed by Pappus was to describe the curve consisting of all points P for which the sum of the squares of the distances from four given lines in the plane is constant. Appolonius had already shown the curve to be a conic section, but his proof had been laborious and cumbersome compared to that given by Descartes.

[¶] In the theory of numbers a conjecture known as Fermat's last theorem has gained a certain notoriety for its stubborn refusal to yield to the concerted efforts of mathematicians at establishing its validity. The conjecture asserts that the equation $x^n + y^n = z^n$ has no solutions in the positive integers for any integer exponent n greater than 2. Fermat had written a note in the margin of one of his books claiming to "have discovered a truly marvelous proof of this, which, however the margin [was] not large enough to contain." It took more than 300 hundred years until finally, in 1996, the British mathematician Andrew Wiles announced a proof of Fermat's last theorem that, after careful scrutiny, was declared free of error.

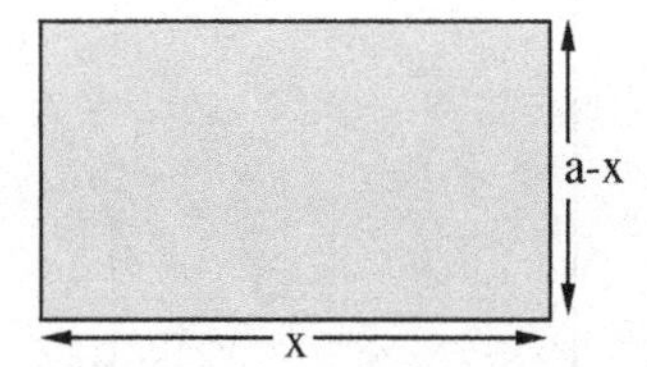

Figure 5.4: a rectangle of circumference $2a$.

faced the problem of maximizing the function $f(x) := x(a-x)$, which represents the rectangle's area. As we plot the graph of f we notice that, close to the point $(x_0, f(x_0))$ where the maximum is assumed (see Figure 5.5), the graph is almost flat. Consequently, if $x_0 + e \approx x_0$ (i.e., $e \approx 0$), then

$$x_0(a-x_0) = f(x_0) \approx f(x_0+e) = (x_0+e)(a-x_0-e),$$

or equivalently,

$$0 \approx e(a-2x_0) - e^2.$$

At this point Fermat first divided both sides by e, and then set e equal to zero to infer that $x_0 = a/2$. In other words, relying on the concepts introduced in Chapters 2 and 4, we may say that he took the limit of the difference quotient

$$\frac{f(x_0+e)-f(x_0)}{e} = a - 2x_0 - e$$

as e approaches zero, and postulating the limit $a - 2x_0$ to be zero at the point where the maximum occurs, he deduced that $x_0 = a/2$. Upon drawing the desired conclusion that the largest rectangle of a given circumference is a square, Fermat remarked with great satisfaction that "we can hardly expect to find a more general method."[11]

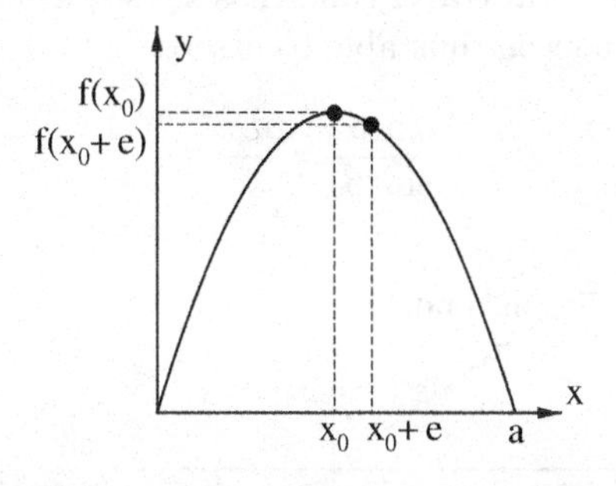

Figure 5.5: approaching the maximum of f.

5.3 Exercise. Apply the method described above to locate the minimum of the function $f(x) := 2x^2 - 2$.

At the next stage in his development Fermat turned his attention to the more general problem of identifying the slope of the tangent line at an arbitrary point on a curve (rather than at a maximum or minimum where the slope is zero). In considering a "generalized parabola"[12] described by the equation $y = x^n$, he was able to show that the tangent slope at any of its points is nx^{n-1}, or as we might say, he proved that

$$\frac{dy}{dx} = nx^{n-1}.$$

(Note: this result is known as the *power rule* for differentiation and will be established in Chapter 10.) Furthermore, he demonstrated that the area between a "generalized parabola" and the x-axis over an

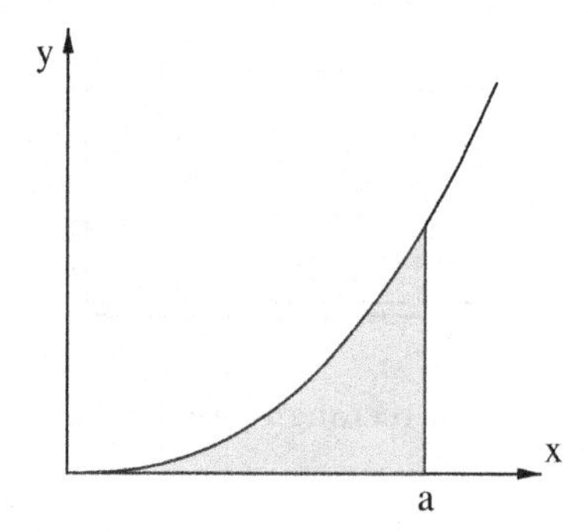

Figure 5.6: area under the graph of $y = x^n$.

interval $[0, a]$ (see Figure 5.6) is

$$\frac{a^{n+1}}{n+1}.$$

In Chapters 17 and 19 we will classify this problem as belonging to the realm of integral calculus and learn to write Fermat's result in the form $\int_0^a x^n\,dx = a^{n+1}/(n+1)$.

In the most famous application of his methods, Fermat analyzed the refraction of light at the interface of two media of different optical densities such as air and water. The quantitative description of this phenomenon had been given by Snell: if α is the angle of incidence, β the angle of refraction (see Figure 5.7) and k the refractive index (for, say, the transition from air to water), then

$$\frac{\sin(\alpha)}{\sin(\beta)} = k.$$

Fermat was able to theoretically derive this law (see also our discussion in Chapter 12) relying only on the following two assumptions: the speed of light in air (denoted by c_a) is greater than the speed of light in water (denoted by c_w), and the path by which a light ray travels from one point to another (P and Q in Figure 5.7) minimizes the total travel time. His subsequent calculation showed k to be equal to the quotient of c_a over c_w, and he was thus able to assert that

$$\frac{\sin(\alpha)}{\sin(\beta)} = \frac{c_a}{c_w}.$$

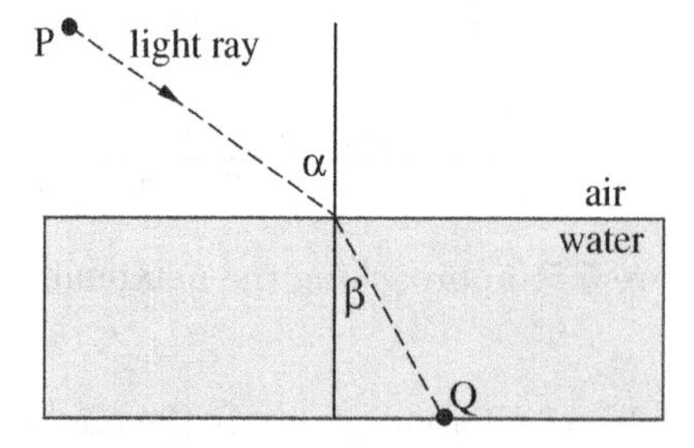

Figure 5.7: refracted light ray.

Given this brief survey of Fermat's mathematical legacy, it may not surprise us that Pierre Simon Laplace referred to him as "the true inventor of the differential calculus."[13] Fermat continued to pursue his research and to fulfill his official duties until his death in 1665. Fifteen years earlier, Descartes had succumbed to the arctic rigors of the Swedish winter on the occasion of a visit to the court of Queen Christina. If it must be admitted that as a mathematician Pierre de Fermat was the greater of the two, it still remains for us to marvel at the seminal achievements in mathematics, science, and philosophy, that both—Fermat and Descartes—contributed to the progress of Western Civilization. It was they who set the stage for the astonishing developments in modern mathematics that were to come, and it was Descartes in particular who paved the way for the Age of Enlightenment. Of him it has been said that "he lived by thought and for thought alone... never was an existence more noble than his."[14]

Chapter 6

Designing a Radar Antenna (Part 1)

A radar antenna is a device for receiving low frequency electromagnetic waves, also known as radio waves. Since in typical applications (such as flight navigation or radio astronomy) the intensity of incoming waves is very low, radar antennas are designed to amplify signals by projecting them from a bowl-shaped reflector toward a common focal point at which the receiver is placed (see Figure 6.1). To determine a mathematical equation that describes the exact shape the reflector must have in order to achieve such a focusing effect is the purpose of the present chapter.

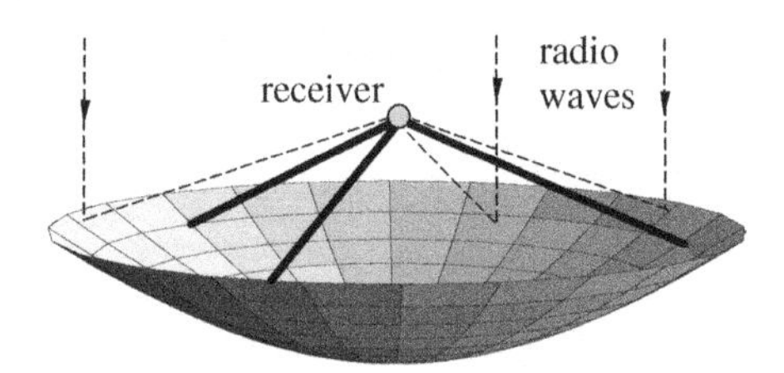

Figure 6.1: radar antenna.

Given that the signal source is usually located at a great distance, we may assume that the incoming signals travel in a direction nearly parallel to the direction from the receiver to the signal source. This assumption is crucial, for it allows us to construct a radar antenna with circular symmetry relative to the axis from the receiver to the signal source. In other words, the shape of a radar antenna is completely determined by any of its cross-sections through the receiver and parallel to the axis of symmetry. Placing one of these cross-sections in an xy-coordinate system with the receiver at the origin (see Figure 6.2), we are thus left with the task of finding a function $f : \mathbb{R} \to \mathbb{R}$ such that every line parallel to the y-axis is "reflected off the graph of f" in such a way that the reflected line passes through the origin.

In order to make sense of the notion of reflecting a line off the graph of a function we need to

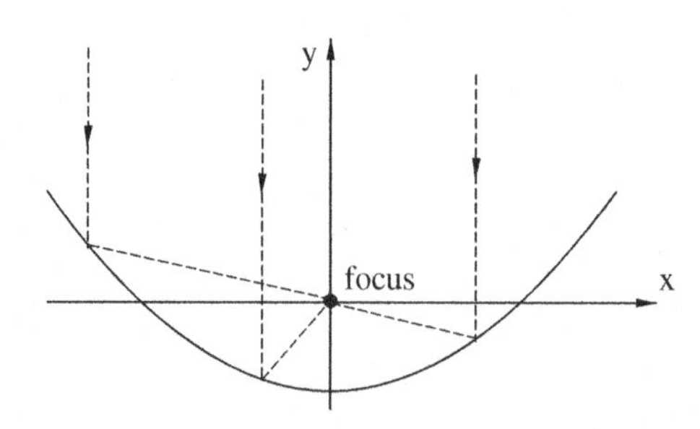

Figure 6.2: cross-section of a radar antenna.

understand the *law of reflection*: if light (which is also an electromagnetic wave and therefore obeys the same physical laws as radio waves) is reflected off a plane surface (such as a plane mirror), then the angle of each light ray relative to the surface will be the same before and after the reflection (see Figure 6.3). In order to generalize this law to the case of curved surfaces we need to address the problem

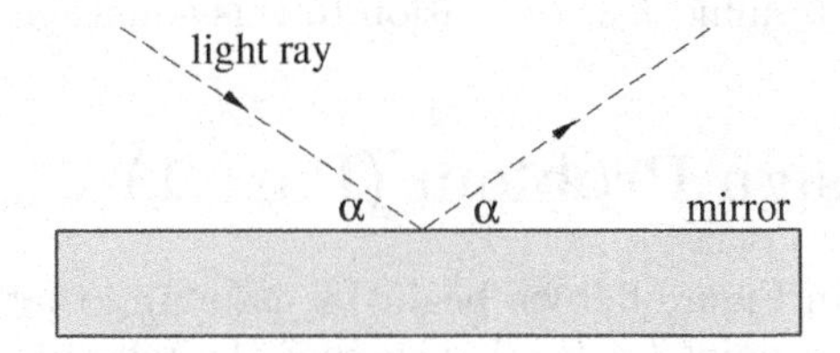

Figure 6.3: the law of reflection.

of measuring angles relative to a curved surface. The correct approach here is to measure angles relative to tangent lines at the surface. More precisely, if the shape of the intersection of a (curved) surface with the plane spanned by a light ray and its reflection off the surface is described by the graph of a function f, then the *general law of reflection* can be stated as follows:

> When a light ray is reflected off the graph of f at a point $(x, f(x))$, then the angles of the light ray and the reflected light ray relative to the tangent line to the graph of f at $(x, f(x))$ are equal.

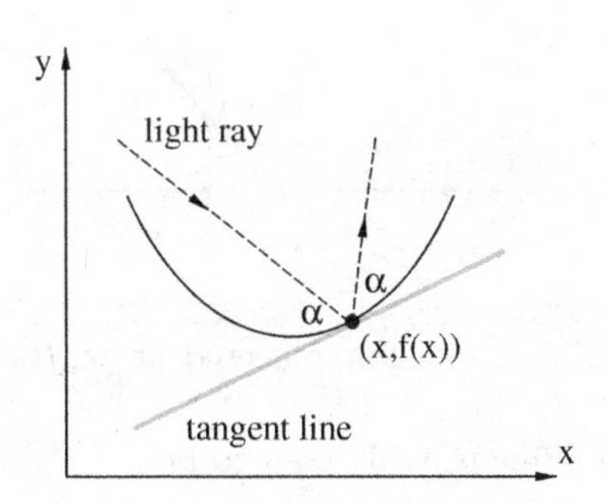

Figure 6.4: the general law of reflection.

Simplicio: I understand the general law of reflection, but how can we use it to find a defining equation for a function f with a focusing property as shown in Figure 6.2? This seems to be a really difficult problem, and all we have to work with is a simple reflection property for light rays.

Teacher: When we are given a problem like this, Simplicio, it is very important that we have the right mind-set. For example, it is usually advisable to think first in terms of general concepts rather than specific methods of calculation. In our case, this means that we need to find a link between the fundamental ideas that we introduced in the first four chapters and the general law of reflection, which is, as you said correctly, the only additional piece of information that we have to work with.

Simplicio: That sounds a bit too vague. I don't see how such a general method of problem-solving translates into any practical steps.

Teacher: It is natural for you to still be confused at this stage, early in the course, because problem-solving thinking requires a lot of experience. However, one of our most important goals is to develop a capacity for rational thought, and it is therefore crucial that we tackle nontrivial problems as frequently and as early as possible.

Sophie: I have an idea: according to the general law of reflection, the angles of the incoming and the reflected light rays are to be measured relative to tangent lines, and the geometric interpretation of the derivative also involves the concept of tangent lines to a graph. Could this be the link we are looking for?

Teacher: I think so. What about you, Simplicio?

Simplicio: I can see where Sophie is headed, but the trouble is that we don't have a formula for either f or f'. It seems impossible to make use of the geometric interpretation of the derivative, if the derivative itself is unknown.

Teacher: That's a good point, but there is a way out: we will use Sophie's idea to find an equation that relates f to f' and then determine f as a solution to this equation.

Solution of the Design Problem (Part 1)

With reference to the set-up in Figure 6.2 we begin by assuming that a line parallel to the y-axis is reflected off the graph of f at a point $(x, f(x))$ such that the reflected line passes through the origin. Denoting by α the angle between the vertical line and the tangent line to the graph of f at $(x, f(x))$, we may apply the general law of reflection to conclude that the angle between the reflected line and the tangent line is equal to α as well. Trivially, the angle α appears a third time in our diagram as the angle between the tangent line and the continuation of the vertical line beyond the point $(x, f(x))$ in the negative direction (see Figure 6.5). Considering the slope triangle with vertices at $(0,0)$, $(x,0)$, and

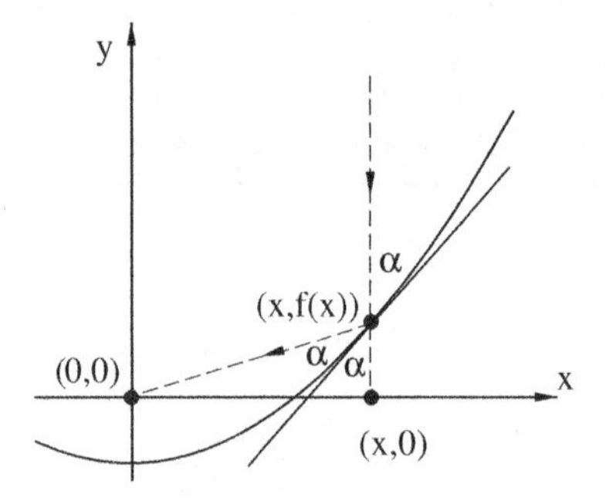

Figure 6.5: line reflected at $(x, f(x))$.

$(x, f(x))$, the slope of the reflected line is easily seen to be

$$m_r = \frac{f(x)}{x}. \tag{6.1}$$

Concerning the slope m_t of the tangent line, we may apply the geometric interpretation of the derivative to infer that

$$m_t = f'(x). \tag{6.2}$$

In order to express m_r and m_t in terms of α, we need to review some elementary trigonometry: given a right triangle as shown in Figure 6.6, with hypotenuse c and side lengths a and b, we define

$$\sin(\alpha) := \frac{a}{c}, \quad \cos(\alpha) := \frac{b}{c},$$
$$\tan(\alpha) := \frac{a}{b}, \quad \cot(\alpha) := \frac{b}{a}.$$

For the purpose of solving our design problem we will actually need only $\cot(\alpha)$, but for later reference we have also listed here the definitions of $\sin(\alpha)$, $\cos(\alpha)$, and $\tan(\alpha)$. Regarding the question of whether these trigonometric functions are well-defined in the sense that their values depend only on α, we wish to point out that the ratio of any two sides in a right triangle does not depend on the size of the triangle, but only on its shape, which in turn is completely determined by the value of α. (Remember that the ratios of corresponding sides in any pair of similar triangles are equal.)

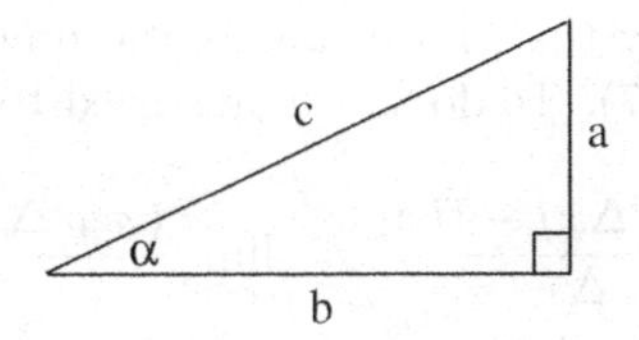

Figure 6.6: right triangle.

6.1 Exercise. Use the definitions above to prove that

$$\tan(\alpha) = \frac{\sin(\alpha)}{\cos(\alpha)} \quad \text{and} \quad \cot(\alpha) = \frac{\cos(\alpha)}{\sin(\alpha)}.$$

6.2 Exercise. Show that $\sin(\pi/6) = \cos(\pi/3) = 1/2$ and $\sin(\pi/3) = \cos(\pi/6) = \sqrt{3}/2$. *Hint.* Consider a right triangle that is obtained by cutting an equilateral triangle in half.

6.3 Exercise. Show that $\sin(\pi/4) = \cos(\pi/4) = \sqrt{2}/2$. *Hint.* Consider a right triangle with two equal sides.

Referring again to the slope triangle with vertices at $(0,0)$, $(x,0)$, and $(x, f(x))$ in Figure 6.5, we may apply the definition of cotangent to conclude that

$$m_r = \cot(2\alpha). \tag{6.3}$$

Similarly, in considering the slope triangle underneath the tangent line with vertices at $(x,0)$, $(x, f(x))$, and the point of intersection of the tangent line with the x-axis, we notice that

$$m_t = \cot(\alpha). \tag{6.4}$$

Unfortunately, it is not possible to work with equations (6.3) and (6.4) as they are, because the arguments of cotangent (i.e., the terms inside the parentheses) are not equal—in (6.3) the argument is 2α, and in (6.4) it is α. To overcome this difficulty, we need to express $\cot(2\alpha)$ in terms of $\cot(\alpha)$. A trigonometric formula that allows us to do just that is the double angle formula for cotangent:

$$\cot(2\alpha) = \frac{\cot^2(\alpha) - 1}{2\cot(\alpha)}. \tag{6.5}$$

(A proof of this identity can be found in Appendix A.) Combining now the equations (6.1), (6.2), (6.3), (6.4), and (6.5), we obtain

$$\frac{f(x)}{x} = m_r = \cot(2\alpha) = \frac{\cot^2(\alpha) - 1}{2\cot(\alpha)} = \frac{m_t^2 - 1}{2m_t} = \frac{f'(x)^2 - 1}{2f'(x)}.$$

Hence

$$0 = f'(x)^2 - \frac{2f(x)f'(x)}{x} - 1. \tag{6.6}$$

Using the quadratic formula (see Appendix D) to solve this equation for $f'(x)$ yields

$$\boxed{f'(x) = \frac{f(x)}{x} \pm \sqrt{\frac{f(x)^2}{x^2} + 1}.} \tag{6.7}$$

This is our defining equation for the cross-sectional shape of a radar antenna. In Chapter 31 we will solve this equation for $f(x)$ using integration (which is the subject of Parts II and IV). At this point, though, we are already able to verify that, for example, the function $f(x) := x^2 - 1/4\,(= x^2 - 0.25)$ is *one possible solution* of equation (6.7). To do so we first need to determine the derivative of f:

$$\begin{aligned} f'(x) &= \lim_{\Delta x \to 0} \frac{f(x+\Delta x) - f(x)}{\Delta x} = \lim_{\Delta x \to 0} \frac{(x+\Delta x)^2 - 1/4 - (x^2 - 1/4)}{\Delta x} \\ &= \lim_{\Delta x \to 0} \frac{2x\Delta x + \Delta x^2}{\Delta x} = \lim_{\Delta x \to 0} (2x + \Delta x) = 2x. \end{aligned}$$

Replacing $f(x)$ and $f'(x)$ in (6.7) with $x^2 - 1/4$ and $2x$ respectively, it remains to be shown that

$$2x = \frac{x^2 - 1/4}{x} \pm \sqrt{\frac{(x^2 - 1/4)^2}{x^2} + 1}.$$

Rewriting this equation in the equivalent form of (6.6) yields

$$0 = (2x)^2 - \frac{2(x^2 - 1/4)2x}{x} - 1.$$

Since this identity is easily seen to be correct using elementary algebra, we may conclude that a radar antenna, whose cross-sections (through its center and parallel to the direction of the incoming signals) have the same shape as the *parabolic* graph of the function $f(x) = x^2 - 1/4$, will have the desired focusing property.

6.4 Exercise. Show that for all $p > 0$ the function $f_p : \mathbb{R} \to \mathbb{R}$ defined by the equation $f_p(x) := (px^2 - 1/p)/2$ is a solution of equation (6.7), and verify that $f_2(x) = x^2 - 1/4$.

We will prove in Chapter 31 that the cross-sectional shape of any radar antenna is given by the graph of a function f_p as defined in Exercise 6.4. In other words, for every radar antenna, there exists a value $p > 0$ such that the cross-sectional shape of that particular radar antenna is given by the graph of f_p.

Additional Exercises

6.5. Use the results of Exercises 6.2 and 6.3 to find the values of $\tan(\pi/6)$, $\cot(\pi/3)$, $\tan(\pi/3)$, $\cot(\pi/6)$, $\tan(\pi/4)$, $\cot(\pi/4)$, and $\cot(5\pi/6)$.

6.6. Find the equation of the line shown in Figure 6.7 under the assumption that $\alpha = \pi/6$.

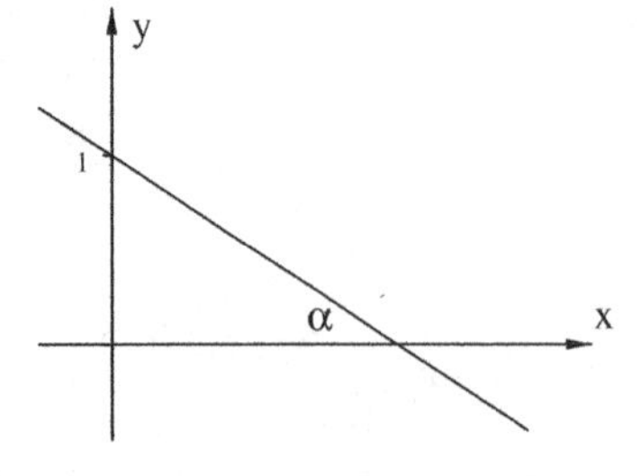

Figure 6.7: graph of a line.

6.7. Find the equation of the line shown in Figure 6.7 under the assumption that $\cot(2\alpha) = 1/2$.

6.8. For each of the functions below write the equation of the line connecting the origin with the point $(x, f(x))$.

a) $f(x) = 4 - x^2$ **b)** $f(x) = \sqrt{2x + x^2}$ **c)** $f(x) = x^4 - 2x^2$

6.9. Find the equation of a line...

a) passing through $(1, 2)$ and $(-2, 3)$.

b) passing through $(0, 5)$ and parallel to the line given by the equation $y - 2x = 3$.

c) perpendicular to the line given by the equation $2x + 3y = 4$ and passing through the point $(2, -1)$.

6.10. Assume that a light ray in an xy-coordinate system is traveling parallel to the y-axis (in the downward direction) and is reflected off the graph of the function $f(x) = x^3$ at the point $(1/2, 1/8)$. Find the equation of the line that represents the path of the reflected light ray. *Hint.* The slope of the reflected line can be determined by combining in this order the equations (6.3), (6.5), (6.4), and (6.2).

6.11. Assume that a light ray is traveling *upward* along the line $x = 3/2$ toward the graph of the function $f(x) := x^2 - 1/4$ (see Figure 6.8). Find the equation of the line that describes the path of the reflected light ray, and explain your reasoning.

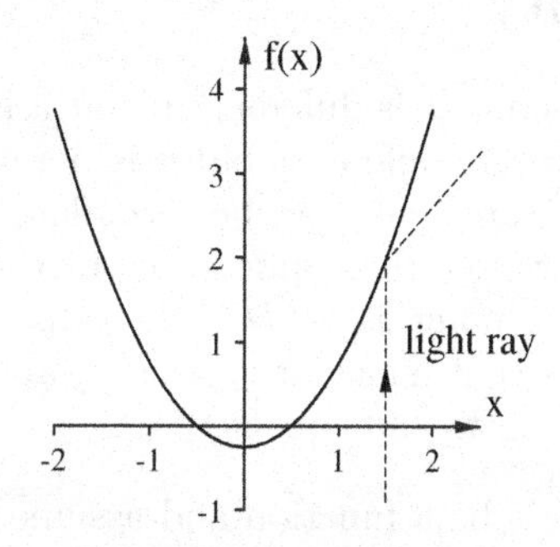

Figure 6.8: a light ray traveling upward.

Chapter 7

A Theoretical Matter: Continuity

Definition of Continuity

In Chapter 4 we learned that a function f is differentiable at a point x_0 if its graph has a well-defined tangent slope at $(x_0, f(x_0))$. However, differentiability is a rather strong condition to impose on a function, and many mathematical statements can be formulated without requiring quite so much. A frequently encountered condition weaker than differentiability is *continuity:* we say that a function $f : D \subset \mathbb{R} \to \mathbb{R}$ is continuous at a point $x_0 \in D$ if for values of x close to x_0 the values $f(x)$ are close to $f(x_0)$. In other words, f is continuous at x_0 if $f(x)$ approaches $f(x_0)$ as x approaches x_0, or equivalently, if the limit of f at x_0 exists and equals $f(x_0)$.

7.1 Definition. Let $f : D \subset \mathbb{R} \to \mathbb{R}$ be a function and assume that $x_0 \in D$ is a point in the domain of f that can be arbitrarily closely approximated by values $x \in D$ that are different from x_0 ($x \neq x_0$). Then f is said to be *continuous at* $x_0 \in D$ if

$$\lim_{x \to x_0} f(x) = f(x_0).$$

If x_0 is an *isolated point* of D in the sense that x_0 cannot be arbitrarily closely approximated by values $x \in D$ with $x \neq x_0$, then, by convention, f is said to be continuous at x_0 as well. Furthermore, we say that f is *continuous* if f is continuous at every point $x_0 \in D$.

Remark. The concept of isolated points is best illustrated with some examples. For $D := [0,1] \cup \{2,3\}$ the points 2 and 3 are isolated points of D, whereas the sets $[0,1]$ and $[0,1] \cup (2,3)$ do not have any isolated points. To include the special case of isolated points in Definition 7.1 is necessary in order to establish a notion of continuity that is consistent with the generally accepted use of this term in mathematics. However, in practical situations we will only very rarely encounter functions that are defined at isolated points.

★ *Remark.* A rigorous definition of the notion of isolated points can be stated as follows: given a subset D of $\mathbb{R}$, we say that a point $x_0 \in D$ is an *isolated point* of D if there exists an $\varepsilon > 0$ such that $(x_0 - \varepsilon, x_0 + \varepsilon) \cap D = \{x_0\}$. ★

7.2 Example. For the function $f : \mathbb{R} \to \mathbb{R}$ defined by the equation $f(x) := |x|$ we have

$$\lim_{x \to x_0} f(x) = \lim_{x \to x_0} |x| = |x_0| = f(x_0)$$

for all $x_0 \in \mathbb{R}$. This shows that f is continuous, and it also illustrates that continuity is a condition weaker than differentiability because, from our discussion in Chapter 4, we know that f is not differentiable at $x_0 = 0$.

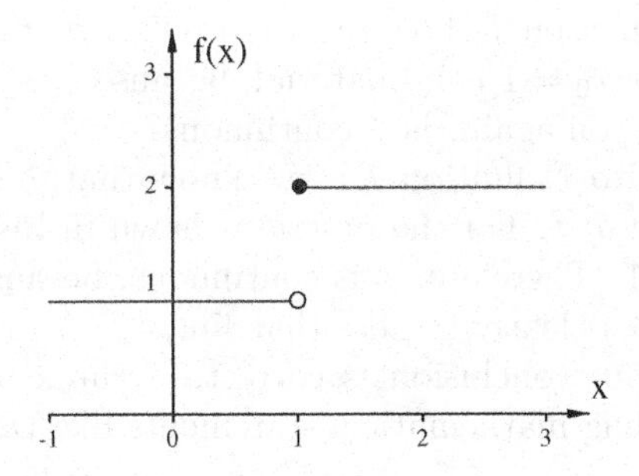

Figure 7.1: a function with a discontinuity at $x_0 = 1$.

7.3 Example. As another example, let us consider the function $f : \mathbb{R} \to \mathbb{R}$ (see Figure 7.1) defined by the equation

$$f(x) := \begin{cases} 1 & \text{if } x < 1, \\ 2 & \text{if } x \geq 1. \end{cases}$$

This function is not continuous at $x_0 = 1$ because the left- and right-hand limits of f are not equal at $x_0 = 1$ and, by implication, the limit of f at $x_0 = 1$ does not exist (and is, in particular, not equal to $f(1)$).

Teacher: Let us define a function $f : \mathbb{R} \setminus \{1\} \to \mathbb{R}$ (i.e., the domain of f is the set of all real numbers different from 1) by the equation

$$f(x) := \begin{cases} 1 & \text{if } x < 1, \\ 2 & \text{if } x > 1. \end{cases}$$

The graph of f is almost the same as the graph in Figure 7.1:

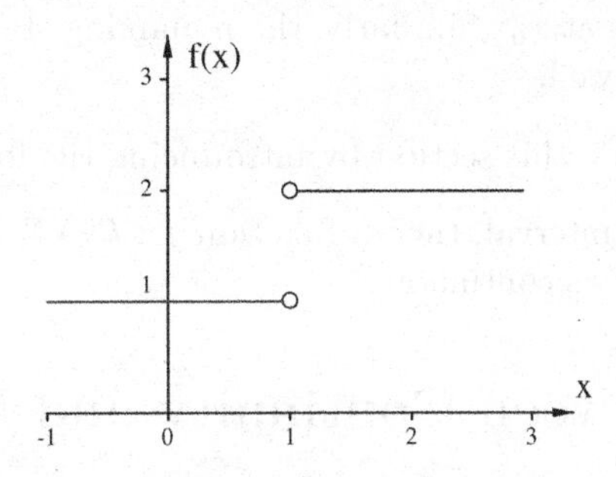

Figure 7.2: graph of f.

What do you think—is this function continuous or not?
Simplicio: Clearly, the answer is "no," because f jumps from 1 to 2 at $x_0 = 1$.
Teacher: What do you think, Sophie?
Sophie: I agree with Simplicio, and I don't even understand why you are asking this question, because the answer seems so entirely obvious.
Teacher: This time both of you got fooled. To help you identify your mistake I would like you to look back at Definition 7.1: we say that a function f is continuous at x_0 if...
Simplicio: ...the limit of f at x_0 is equal to $f(x_0)$.
Teacher: Correct. But this requires that the function f actually be defined at x_0, because otherwise the expression $f(x_0)$ cannot be evaluated. For this reason we find in Definition 7.1 the condition "$x_0 \in D$," where D is the domain of f. In other words, it is meaningful to ask whether f is continuous at a point x_0 only if x_0 is contained in the domain of f.

Sophie: So with regard to the function f shown in Figure 7.2, we would have to say that the question of whether f is continuous at $x_0 = 1$ is immaterial, because f is not defined at 1.
Teacher: Exactly. So let me ask you again: is f continuous?
Sophie: Yes, because, according to Definition 7.1, we know that f is continuous if f is continuous at every point x_0 *in the domain of* f. For the function shown in Figure 7.2, the domain consists of all real numbers different from 1. Therefore, f is continuous, because to the left and to the right of 1 the function is constant, which clearly implies that $\lim_{x\to x_0} f(x) = f(x_0)$ for all $x_0 \neq 1$.
Teacher: Very good, Sophie: your conclusion is correct. I think our discussion illustrates well how careful we need to be in reading mathematical statements like Definition 7.1—every detail matters.

7.4 Exercise. Which of the following two functions is continuous?

a) $f : \mathbb{R} \to \mathbb{R}$, $f(x) := \begin{cases} 1/x & \text{if } x \neq 0, \\ 0 & \text{if } x = 0. \end{cases}$

b) $f : \mathbb{R} \setminus \{0\} \to \mathbb{R}$, $f(x) := 1/x$.

As in the case of limits of functions, there are some simple rules concerning sums, scalar multiples, products, and quotients of continuous functions:

7.5 Theorem. *If $f : D \subset \mathbb{R} \to \mathbb{R}$ and $g : D \subset \mathbb{R} \to \mathbb{R}$ are continuous functions and $\lambda \in \mathbb{R}$, then $f + g$, λf, and fg are continuous as well. Furthermore, if $g(x) \neq 0$ for all $x \in D$, then f/g is continuous as well.*

Proof. In order to prove that $f + g$ is continuous, it is sufficient to show that $f + g$ is continuous at every point $x_0 \in D$ that is not an isolated point of D. Since f and g are assumed to be continuous, it follows that $\lim_{x\to x_0} f(x) = f(x_0)$ and $\lim_{x\to x_0} g(x) = g(x_0)$ for all $x_0 \in D$ that are not isolated points of D. Consequently, for any such point x_0 Theorem 2.9 implies that

$$\lim_{x\to x_0} (f(x) + g(x)) = \lim_{x\to x_0} f(x) + \lim_{x\to x_0} g(x) = f(x_0) + g(x_0).$$

Thus, $f + g$ is indeed continuous at x_0. Similarly, the remaining statements in Theorem 7.5 are direct consequences of Theorem 2.9 as well. □

For later reference we conclude this section by introducing the following definition:

7.6 Definition. If I is an open interval, then a function $f : I \to \mathbb{R}$ is said to be *continuously differentiable* if f is differentiable and f' is continuous.

On the Relation between Continuity and Differentiability

The example of the function $f(x) := |x|$, which we discussed in Chapter 4 and Example 7.2, shows that it is possible for a function to be continuous but not differentiable. In other words, continuity does not imply differentiability. Conversely, though, the inference of continuity from differentiability is always valid. To see this, we appeal to the alternative definition of the derivative given in Chapter 4, p.34: a function $f : \mathbb{R} \to \mathbb{R}$ is differentiable at a point $x_0 \in \mathbb{R}$ if the limit

$$\lim_{x\to x_0} \frac{f(x) - f(x_0)}{x - x_0}$$

exists. Given this statement, the assumption of differentiability at a point x_0 allows us to conclude that

$$f(x) - f(x_0) = \left(\frac{f(x) - f(x_0)}{x - x_0}\right)(x - x_0) \approx f'(x_0)(x - x_0)$$

whenever x is "sufficiently close" to x_0. Since $f'(x_0)$ is a constant (it doesn't depend on x), $f'(x_0)(x - x_0)$ approaches zero as x approaches x_0 and, by implication, $f(x) - f(x_0)$ approaches zero as well.

Consequently, the limit of f at x_0 must be equal to $f(x_0)$, and this proves that f is continuous at x_0. The following sequence of equations makes this argument precise:

$$\begin{aligned}\lim_{x\to x_0} f(x) &= f(x_0) + \lim_{x\to x_0}(f(x) - f(x_0)) = f(x_0) + \lim_{x\to x_0}\left(\frac{f(x)-f(x_0)}{x-x_0}\right)(x-x_0)\\ &= f(x_0) + f'(x_0)\cdot 0 = f(x_0).\end{aligned}$$

Thus, we have established the following theorem:

7.7 Theorem. *Let $f : I \to \mathbb{R}$ be a function defined on an open interval I and assume that f is differentiable at a point $x_0 \in I$. Then f is continuous at x_0. Furthermore, if f is differentiable, then f is continuous.*

★ An Alternative Characterization of Continuity

As explained above, a function is continuous if small changes in x result in small changes in the values $f(x)$. To further explore this idea, we pick an arbitrary positive number ε (typically small), say $\varepsilon = 0.1$. If a given function f is continuous at a point x_0, then it should be possible to conclude that the values $f(x)$ do not differ by more than $\varepsilon = 0.1$ from $f(x_0)$ if only x is "sufficiently close" to x_0. So there should be a (small) number, say $\delta > 0$, such that the distance between $f(x)$ and $f(x_0)$ is smaller than ε whenever the distance between x and x_0 is less than δ. It is clear that the validity of this statement should not depend on the particular choice of 0.1 as our value for ε. We should be able to draw the same conclusion for any number $\varepsilon > 0$ no matter how small.

7.8 Theorem. *A function $f : D \subset \mathbb{R} \to \mathbb{R}$ is continuous at a point $x_0 \in D$ if and only if for every $\varepsilon > 0$ we can find a $\delta > 0$ such that for all $x \in D$ we have*

$$|f(x) - f(x_0)| < \varepsilon \quad \textit{whenever} \quad |x - x_0| < \delta.$$

For a proof of this theorem (which is not difficult using the concepts introduced in Chapter 3) the reader is referred to [Rud1].

Additional Exercises

7.9. Given a real number $x_0 \neq 0$, we define

$$f(x) := \begin{cases} \dfrac{x^2 - x_0^2}{x - x_0} & \text{if } x \neq x_0,\\ 0 & \text{if } x = x_0. \end{cases}$$

Answer the following questions, and explain your reasoning in each case.

a) Is f defined at x_0?

b) Does the limit of f at x_0 exist? If so, what is its value?

c) Is f continuous at x_0?

d) Is f differentiable at x_0?

7.10. Let $x_0 \in \mathbb{R}$, and assume that $f : D \subset \mathbb{R} \to \mathbb{R}$ is a function such that $L := \lim_{x\to x_0} f(x)$ exists. Label each of the following statements as TRUE, FALSE, or POSSIBLY TRUE/POSSIBLY FALSE. Explain your reasoning.

a) f is defined at x_0.

b) $f(x_0) = L$.

c) f is continuous at x_0.

7.11. Assume that $f : D \subset \mathbb{R} \to \mathbb{R}$ is continuous at a point $x_0 \in \mathbb{R}$, which is not an isolated point of D. Label each of the following statements as TRUE, FALSE, or POSSIBLY TRUE/POSSIBLY FALSE. Explain your reasoning.

a) f is defined at x_0.

b) $f(x_0) = \lim_{x \to x_0} f(x)$.

c) $\lim_{x \to x_0} f(x)$ exists.

d) f is differentiable at x_0.

7.12. Let $f(x) := \begin{cases} 2 & \text{if } x < 0, \\ 3 - x & \text{if } 0 \leq x \leq 1, \\ x^2 + 1 & \text{if } x > 1. \end{cases}$

a) Sketch the graph of f.

b) Find $f(-1)$, $f(0)$, $f(1)$, and $f(2)$.

c) Is f continuous at $x_0 = 0$? Justify your answer.

d) Is f continuous at $x_0 = 1$? Justify your answer.

7.13. Which of the functions f, g, and h shown in Figure 7.3 is/are continuous?

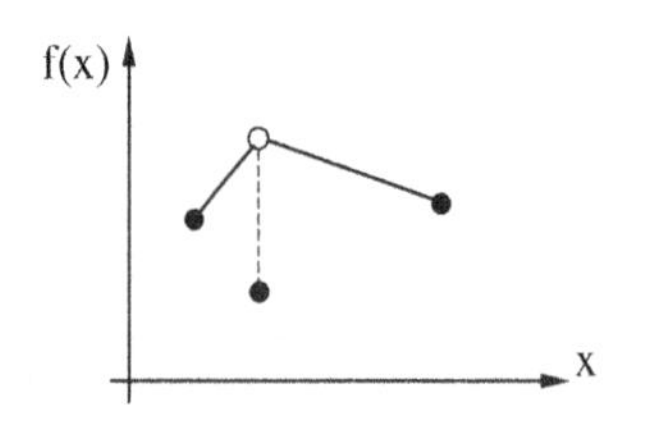

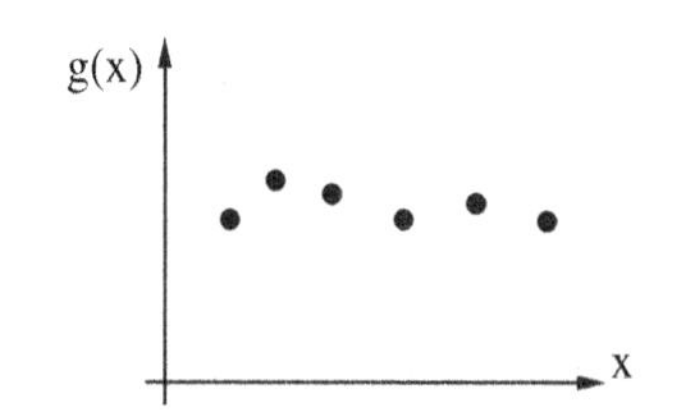

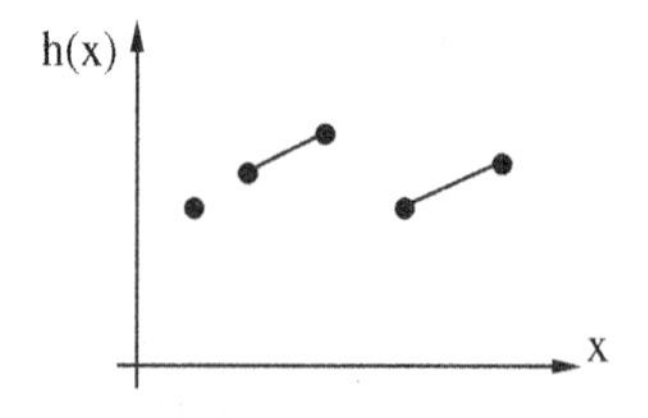

Figure 7.3: graphs of f, g, and h.

7.14. For given $a, b \in \mathbb{R}$, let $f_{a,b}(t) := \begin{cases} at^2 & \text{if } t \leq 1, \\ t + 2b & \text{if } t > 1. \end{cases}$

a) Draw the graph of $f_{a,b}$ for $a = b = 1$.

b) Find all values for a and b for which $f_{a,b}$ is continuous.

c) Find all values for a and b for which $f_{a,b}$ is differentiable.

d) Draw the graph of $f_{a,b}$ for the values of a and b that you found in c).

7.15. Which of the following statements are true for all functions $f : [1, 2] \to \mathbb{R}$? Explain your reasoning.

a) If $f(1) < 0$ and $f(2) > 0$, then there is a point $z \in (1, 2)$ such that $f(z) = 0$.

b) If f is continuous and $f(1) < 0$ and $f(2) > 0$, then there is a point $z \in (1, 2)$ such that $f(z) = 0$.

c) If there is a point $z \in (1, 2)$ such that $f(z) = 0$, then $f(1)$ and $f(2)$ have different signs (i.e., either $f(1) > 0$ and $f(2) < 0$ or $f(1) < 0$ and $f(2) > 0$).

d) If f is continuous and $f(x) \neq 0$ for all $x \in [1, 2]$, then $f(1)$ and $f(2)$ have the same sign.

7.16. Given an arbitrary function $f : \mathbb{R} \to \mathbb{R}$, label each of the following statements as TRUE, FALSE, or POSSIBLY TRUE/POSSIBLY FALSE. Explain your reasoning.

a) $\lim_{x\to 1} f(x) = f(1)$.

b) If $\lim_{x\to 0} f(x)/x = 1$, then $f(0) = 0$.

c) If $\lim_{x\to 0} f(x)/x = 1$, then $\lim_{x\to 0} f(x) = 0$.

d) If $\lim_{x\to 0}(f(x) - f(0))/x = 3$, then $f'(0) = 3$.

e) If $\lim_{x\to 0}(f(x) - f(0))/x = 3$, then $\lim_{x\to 0} f(x) = f(0)$.

f) If $\lim_{x\to 1^-} f(x) = 1$ and $\lim_{x\to 1^+} f(x) = 3$, then $\lim_{x\to 1} f(x) = 2$.

7.17. Is it true that a function $f : \mathbb{R} \to \mathbb{R}$ is not differentiable at a point $x_0 \in \mathbb{R}$ if it is not continuous at x_0?

7.18. A calculus student, shopping at WAL-MART, suddenly realizes that he cannot recall the value of f at $x = 1$. Fortunately, he notices two sales clerks at the cash register involved in an angry discussion concerning just this very question. One of the clerks shouts at the other that the function f is continuous at 1, and to lend his assertion more credibility he follows through by punching the other clerk in the nose. Visibly shaken by the weight of the argument, and just before falling unconscious, the unfortunate clerk stammers in reply that the function f satisfies the equation $f(x) = 5x^2$ for all $x \neq 1$. Given this information, what should the student conclude concerning the value of f at $x = 1$?

7.19. Draw a graph of a function that is continuous on $[0,6]$ but not differentiable at $x = 2$.

7.20. Draw a graph of a function $f : [0, 4] \to \mathbb{R}$ that is not continuous at $x_0 = 2$ but for which $\lim_{x\to x_0} f(x)$ exists for every $x_0 \in [0, 4]$.

Chapter 8

Newton's Laws and Rocket Motion (Part 1)

Newton's Fundamental Laws of Mechanics

Imagine that you are tranquilly floating inside a spacecraft when suddenly a little green guy tries to get in at the rear. Courageously you open the window to throw a stone at the intruder as a deterrent. In such a situation it would be interesting to find out what the alien might think at the moment of impact, but we will be content to consider the following, more prosaic question: by how much does the velocity of the spacecraft change when the stone is sent flying into space?

To find an answer we begin by looking at some examples from everyday life. What happens, for instance, when you sit in a rocking chair and flick a finger at a marble in your hand to send it flying through the air? The answer is "nothing much"—the marble will quickly fall to the ground, and that's it. By contrast, if you push a very heavy object away from you instead of a small marble, you will not only see the object flying through the air, but also feel yourself rocking backwards in the chair. The reason nothing happens in the case of the marble is that the mass of the marble is very small compared to the combined mass of the chair and your body. There may actually be a very slight backward rocking motion, but it will hardly be perceptible.

One of the differences between your experience in the rocking chair and the situation in the spaceship is the absence of frictional forces such as air resistance in the near perfect vacuum in which the spaceship moves. This is important because, unhindered by frictional forces, even a very small stone thrown at low velocity will cause a very slight increase in the velocity of the spaceship in the forward direction (which in our example is the direction opposite to the direction in which the stone is thrown). If a big stone is thrown at high velocity, the increase in velocity is of course greater. Experiments suggest that in general the increase in velocity depends on the product of the mass and the velocity of the object that is pushed away. In other words, it depends on the object's *momentum*. Given an object of mass m moving at velocity v (along a straight line), the product

$$p := mv$$

is referred to as the momentum of the object. Typically, the momentum will, of course, vary over time (and we will indicate this by writing $p(t)$ instead of p), but in *absence of external forces*, Newton's *law of the preservation of momentum* says that

$$\boxed{\text{the total momentum of a physical system remains constant.}}$$

This law is also known as the *first fundamental law of mechanics*. Let us examine what it tells us about the stone/spaceship system that we discussed above. To begin with, we need to assume that there are no external forces. This is a very realistic assumption for a spaceship in interstellar space where frictional forces are essentially nonexistent and gravitational attractions from distant stars are negligible. In denoting by m_1 and m_2 the masses of the spaceship and the stone respectively, we observe that the total mass of the stone/spaceship system is

$$m = m_1 + m_2.$$

If the initial velocity of the spaceship (before the stone has been thrown) is v_0, then the initial momentum of the system is

$$p = mv_0 = m_1 v_0 + m_2 v_0. \tag{8.1}$$

If we further assume that the stone is thrown at speed u in the direction opposite to the spaceship's direction of motion, then the resulting velocity of the stone is $v_0 - u$. To find the resulting velocity v

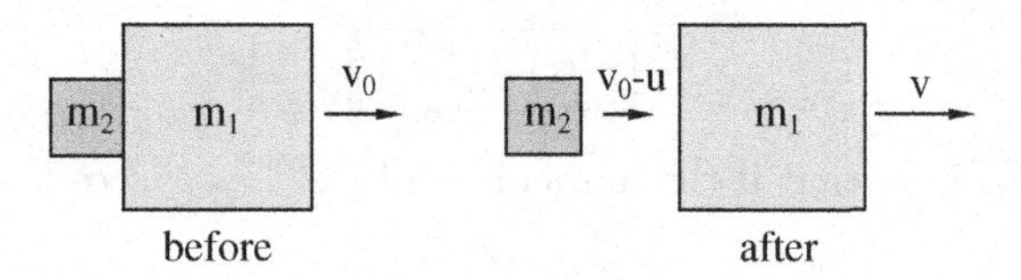

Figure 8.1: preservation of momentum.

of the spaceship, we apply the law of the preservation of momentum to infer that the total momentum of the stone/spaceship system remains constant (i.e., the total momentum is the same before and after the stone has been thrown). Since the momentum after throwing the stone is $m_1 v + m_2(v_0 - u)$, we may use (8.1) to deduce that

$$m_1 v_0 + m_2 v_0 = m_1 v + m_2(v_0 - u).$$

Solving for v yields

$$v = v_0 + \frac{m_2}{m_1} u. \tag{8.2}$$

Consequently, the increase in velocity is $v - v_0 = m_2 u / m_1$.

8.1 Example. Assume that the mass of the spaceship is $m_1 = 10000\ kg$, the mass of the stone is $m_2 = 0.25\ kg$, and the speed of the stone relative to the spaceship is $u = 30\,mi/h$. In this case the velocity of the spaceship increases only by

$$\frac{m_2}{m_1} u = \frac{0.25 \cdot 30}{10000} \frac{mi}{h} = 0.00075 \frac{mi}{h},$$

but if on the other hand the spaceship is rather small with $m_1 = 1000\ kg$, and a big rock is thrown at high velocity, with say $m_2 = 100\ kg$ and $u = 100\ mi/h$, then the velocity increases by

$$\frac{m_2}{m_1} u = \frac{100 \cdot 100}{1000} \frac{mi}{h} = 10 \frac{mi}{h},$$

which may already cause a noticeable effect.

8.2 Exercise. Find the value for u (in terms of m_1, m_2, and v_0) for which the velocity of the spaceship increases from v_0 to $2v_0$.

There is a very elegant way to express the law of the preservation of momentum in the language of calculus: If we denote by $p(t)$ the total momentum of a physical system, then, in absence of external forces, $p(t)$ is equal to a constant, say p_0, and therefore,

$$p'(t) = \lim_{\Delta t \to 0} \frac{p(t + \Delta t) - p(t)}{\Delta t} = \lim_{\Delta t \to 0} \frac{p_0 - p_0}{\Delta t} = 0.$$

So the law of the preservation of momentum assumes the form

$$\boxed{p'(t) = 0.}$$

Remark. What we have just shown is essentially the following:

The derivative of a constant function is equal to zero.

This statement makes intuitively perfect sense, because the graph of a constant function in an xy-coordinate system is a horizontal line of slope zero. The reverse statement, that any differentiable function (defined on an open interval) with a derivative identically equal to zero is constant, is also true, but a rigorous proof of this fact will be postponed until Chapter 40, p.306.

In the case that external forces are present, Newton's *second fundamental law of mechanics* says that the change in momentum of a system is equal to the total external force applied to the system. In other words, if we denote by $F(t)$ the external force, then (for a straight line motion) we have

$$\boxed{p'(t) = F(t).}$$

For later reference, we also note that if the mass of an object has the *constant* value m, then

$$p'(t) = \frac{d}{dt}mv(t) = mv'(t) = ma(t) \quad \text{(see Chapter 4, p.34).}$$

So in this case, Newton's law says that *force is equal to mass times acceleration*:

$$\boxed{ma(t) = F(t).}$$

Rocket Motion

Teacher: Let us consider a rocket in interstellar space that is burning propellant gas through its boosters at a constant relative speed u. How can we approach the problem of finding a formula for the rocket's velocity $v(t)$?

Sophie: I remember that whenever we approach a given problem, we begin by trying to devise some general strategy. In our case this general strategy certainly has to involve the law of the preservation of momentum, but I am not exactly sure how.

Teacher: Well, we have already discussed one application of the law of the preservation of momentum—the stone thrown at the alien intruder. So we may want to explore whether there are any parallels between this and the current example.

Sophie: In both cases mass is expelled at a certain relative speed u—in the example of the alien intruder a stone was thrown away from the spaceship, and in the present example propellant gas is burned and ejected through the rocket's boosters.

Simplicio: So are we just going to replace the mass of the stone m_2 in equation (8.2) with the mass of the fuel on board the rocket?

Teacher: It's not quite that simple, because you have failed to take into account the following important difference between the two cases: while the stone is thrown at one particular moment in time, the rocket burns its fuel in a continuous process. Consequently, the velocity of the rocket does not increase abruptly but in small increments over an extended period of time.

Simplicio: Then what are we going to do?

Teacher: Since change in the rocket's velocity $v(t)$ is continuous, we probably want to consider a sufficiently small time interval $[t, t + \Delta t]$ over which the velocity is nearly constant. To see how this will solve our problem, we first need to determine the momentum of the rocket at time t.

Sophie: If we denote by m the mass of the rocket, then I think the rocket's momentum at time t is $p(t) = mv(t)$.

Teacher: That is almost correct, but remember that the burning of fuel causes the mass of the rocket to decrease over time.

Sophie: I see. Then we'd better write $m(t)$ instead of m in order to express the dependence of the mass of the rocket on t. This yields $p(t) = m(t)v(t)$.

Teacher: Correct.
Simplicio: Am I right in assuming that the next step is to apply the law of the preservation of momentum to conclude that $p'(t) = 0$?
Teacher: Unfortunately, that would be a mistake.
Simplicio: Why is that?
Teacher: In the present example it is not the momentum of the rocket, which Sophie chose to denote by $p(t)$, that remains constant but rather the *combined momentum* of the rocket and the propellant gas ejected into space. So the propellant/rocket system plays the same role as the stone/spaceship system in our introductory example.
Simplicio: I understand what you are saying, but I would never have been able to figure it out myself, and I find it thoroughly confusing.
Teacher: The confusion really is due to a poor choice of notation. It would have been better to denote the momentum of the rocket by a different letter like, for example, q and to reserve the letter p for the total momentum of the propellant/rocket system. With this change in notation the momentum of the rocket is given by the equation

$$q(t) = m(t)v(t), \tag{8.3}$$

and the constant total momentum p of the propellant/rocket system does now indeed satisfy the equation $p'(t) = 0$. To continue from here, we need to find an estimate for the momentum of the propellant gas that is ejected into space between the times t and $t + \Delta t$. In this time span the mass of the rocket is reduced from $m(t)$ to $m(t + \Delta t)$ and, by implication, the mass ejected is $m(t) - m(t + \Delta t)$. Since the relative speed of the propellant gas in the direction opposite to the rocket's direction of motion is u, the momentum of the ejected propellant gas is approximately

$$(m(t) - m(t + \Delta t))(v(t) - u). \tag{8.4}$$

Simplicio: Why "approximately"? As far as I can tell, the expression in (8.4) is exactly equal to the momentum of the ejected propellant gas.
Teacher: No, it is not. For the rocket's velocity changes continuously from $v(t)$ to $v(t + \Delta t)$—it does not retain the constant value $v(t)$ for the entire time span from t to $t + \Delta t$ as we implicitly assumed in our derivation of (8.4).
Simplicio: To represent the rocket's velocity by $v(t)$ seems a bit arbitrary, though. Couldn't we just as well have chosen $v(t + \Delta t)$ or perhaps the average of $v(t)$ and $v(t + \Delta t)$?
Teacher: We could have indeed, but it wouldn't make any difference, because if Δt is sufficiently small, then $v(t) \approx v(t + \Delta t)$, and in the limit, as Δt approaches zero, we would arrive at the same resulting equation.
Simplicio: Thank you, that answers my question.
Teacher: Well, then we are ready to invoke the law of the preservation of momentum in its correct form: over the interval from t to $t + \Delta t$ the total momentum of the propellant/rocket system remains constant. Therefore, the rocket's momentum at time t is equal to...
Sophie: ...the sum of the momenta of the rocket at time $t + \Delta t$ and the propellant gas burned between the times t and $t + \Delta t$.
Teacher: Furthermore, since the rocket's momentum at time $t + \Delta t$ is $m(t + \Delta t)v(t + \Delta t)$, the estimate in (8.4) allows us to infer that the total approximate momentum of the propellant/rocket system at time $t + \Delta t$ is

$$m(t + \Delta t)v(t + \Delta t) + (m(t) - m(t + \Delta t))(v(t) - u). \tag{8.5}$$

Thus, the law of the preservation of momentum asserts the approximate equality of the expressions in (8.3) and (8.5):

$$m(t)v(t) \approx m(t + \Delta t)v(t + \Delta t) + (m(t) - m(t + \Delta t))(v(t) - u).$$

Rearranging the terms and dividing both sides by Δt yields

$$\frac{m(t + \Delta t)v(t + \Delta t) - m(t)v(t)}{\Delta t} \approx \frac{m(t + \Delta t) - m(t)}{\Delta t}(v(t) - u). \tag{8.6}$$

The quotient on the left-hand side is simply the difference quotient of $q(t) = m(t)v(t)$, and as Δt approaches zero, this difference quotient approaches the derivative $q'(t)$. Furthermore, in taking the limit, the error in our estimate of the momentum of the ejected propellant gas converges to zero, and the approximate identity (8.6) becomes exact:

$$q'(t) = \lim_{\Delta t \to 0} \frac{m(t+\Delta t) - m(t)}{\Delta t}(v(t) - u) = m'(t)(v(t) - u). \tag{8.7}$$

To further simplify this equation, we will need to express $q'(t)$ in terms of $m(t), m'(t), v(t)$, and $v'(t)$, and this requires that we find a formula for the derivative of the product of $m(t)$ and $v(t)$.

The Product Rule

In order to determine the derivative of $q(t) = m(t)v(t)$, we will first explore the following more general question: given two differentiable functions $f : I \to \mathbb{R}$ and $g : I \to \mathbb{R}$ defined on an open interval I, what is the derivative of the product function fg? (The tempting answer $f'g'$ is **wrong!**) The best intuitive approach to this problem is to interpret the product $f(x)g(x)$ as the area of a rectangle with side lengths $f(x)$ and $g(x)$ (at least if $f(x)$ and $g(x)$ are both positive). If we assume that with a small change Δx in x the functions f and g *increase* from $f(x)$ to $f(x + \Delta x)$ and from $g(x)$ to $g(x + \Delta x)$ respectively, then, as illustrated in Figure 8.2, the area of the rectangle increases from A by the amount

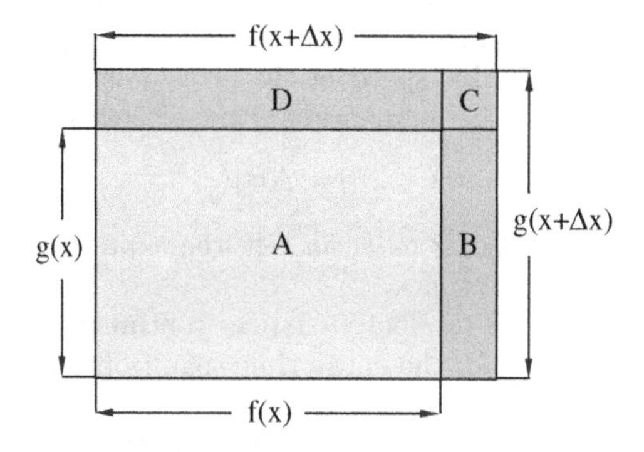

Figure 8.2: illustration of the product rule.

$B + C + D$ to $A + B + C + D$. Since

$$\begin{aligned} B + C + D &= \underbrace{f(x+\Delta x)g(x+\Delta x)}_{A+B+C+D} - \underbrace{f(x)g(x)}_{A} \\ &= \underbrace{(f(x+\Delta x) - f(x))g(x+\Delta x)}_{B+C} + \underbrace{(g(x+\Delta x) - g(x))f(x)}_{D}, \end{aligned}$$

it follows that

$$\begin{aligned} (fg)'(x) &= \lim_{\Delta x \to 0} \frac{f(x+\Delta x)g(x+\Delta x) - f(x)g(x)}{\Delta x} \\ &= \lim_{\Delta x \to 0} \frac{(f(x+\Delta x) - f(x))g(x+\Delta x) + (g(x+\Delta x) - g(x))f(x)}{\Delta x} \\ &= \lim_{\Delta x \to 0} \frac{f(x+\Delta x) - f(x)}{\Delta x} g(x+\Delta x) + \lim_{\Delta x \to 0} \frac{g(x+\Delta x) - g(x)}{\Delta x} f(x) \\ &= f'(x)g(x) + f(x)g'(x). \end{aligned}$$

To justify the last step in this calculation in which we replaced $\lim_{\Delta x \to 0} g(x+\Delta x)$ with $g(x)$, we observe that, according to Theorem 7.7, the differentiability of g implies that g is also continuous, and this in turn allows us to conclude that $\lim_{\Delta x \to 0} g(x + \Delta x) = g(x)$. Thus we have established the following theorem, commonly referred to as the *product rule:*

8.3 Theorem. *If $f : I \to \mathbb{R}$ and $g : I \to \mathbb{R}$ are differentiable functions defined on an open interval I, then fg is differentiable as well, and*

$$(fg)'(x) = f'(x)g(x) + f(x)g'(x)$$

for all $t \in I$.

In applying this theorem to equation (8.3), we obtain

$$q'(t) = m'(t)v(t) + m(t)v'(t).$$

(Several other examples of applications of the product rule will be disussed in Chapter 10.) Combining the equation above with (8.7) yields

$$m'(t)v(t) + m(t)v'(t) = m'(t)(v(t) - u)$$

or equivalently,

$$\boxed{m(t)v'(t) = -m'(t)u.} \tag{8.8}$$

This equation is known as the *fundamental equation of rocket motion in absence of external forces.*

8.4 Example. Let us assume that a rocket in interstellar space ejects propellant gas through its boosters at a constant rate beginning at time $t = 0$. By M_F we denote the mass of the fuel on board at time $t = 0$, by M_R the mass of the rocket without fuel, and by T the total time it takes to burn the fuel. Since the fuel is assumed to be burned at a constant rate, the graph of $m(t)$ for $0 \leq t \leq T$ is a

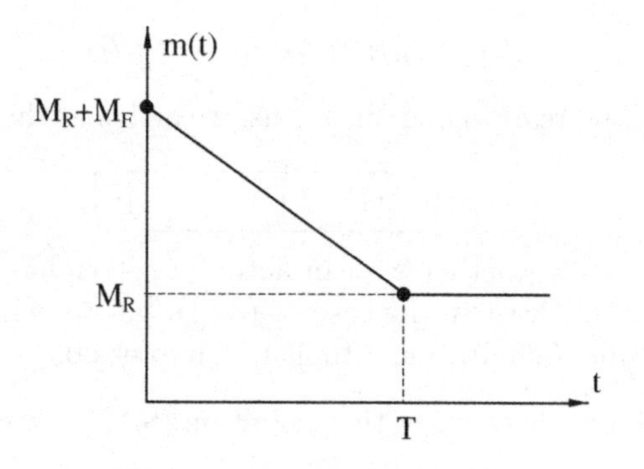

Figure 8.3: graph of $m(t)$.

straight line of slope $-M_F/T$ (see Figure 8.3). Given that $m(0) = M_R + M_F$, it therefore follows that

$$m(t) = \begin{cases} -M_F t/T + M_R + M_F & \text{for } 0 \leq t \leq T, \\ M_R & \text{for } t > T, \end{cases} \tag{8.9}$$

and the geometric interpretation of the derivative, as stated in Chapter 4, implies that

$$m'(t) = -\frac{M_F}{T} \quad \text{for } 0 \leq t < T. \tag{8.10}$$

This result can of course also be deduced directly from the definition of the derivative by evaluating the limit of the difference quotient of m (see Exercise 8.8). Combining (8.8) and (8.9) with (8.10) yields

$$\left(-\frac{M_F t}{T} + M_R + M_F\right) v'(t) = \frac{M_F u}{T},$$

or equivalently,

$$v'(t) = \frac{M_F u}{(M_R + M_F)T - M_F t}$$

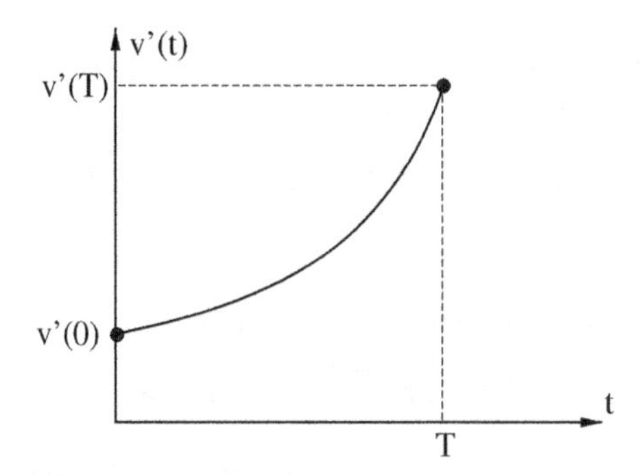

Figure 8.4: graph of $v'(t)$.

for $0 \leq t < T$. Since $m(t)$ assumes the constant value M_R for $t > T$, it follows that $m'(t) = 0$ for $t > T$, and by implication $v'(t)$ must equal zero as well. Consistent with these observations, the graph of v' is shown in Figure 8.4. It is not surprising that the acceleration $a(t) = v'(t)$ increases until all fuel has been burned, because as long as the rocket is burning fuel its mass is decreasing, and the forward thrust of its boosters therefore generates the greatest acceleration toward the end of the burning cycle when t is close to T.

If a rocket is subject to external forces, such as the gravitational pull close to the surface of the earth, we need to invoke Newton's second fundamental law of mechanics which asserts that the change in total momentum is equal to the external force $F(t)$ (see p.64). In this case (8.7) is to be replaced with the equation

$$q'(t) - m'(t)(v(t) - u) = F(t). \tag{8.11}$$

Again using the product rule and rearranging the terms, we arrive at the *fundamental equation of rocket motion*:

$$\boxed{m(t)v'(t) = F(t) - m'(t)u} \tag{8.12}$$

Naturally, the question now arises whether we can actually determine $v(t)$ from either equation (8.8) or (8.12). The answer is "yes" (at least in the case of (8.8)), but we will postpone this discussion until we have introduced the relevant computational tools in Chapter 30.

8.5 Exercise. Use the product rule to show that equation (8.12) is a consequence of (8.11).

Additional Exercises

8.6. Using the law of the preservation of momentum, fill in the missing value in each of the rows of the following table:

m_1	m_2	v_0	u	v
1000	50	150	10	
750		200	50	280
1000	500	100		50

8.7. Given the diagram in Figure 8.2, identify the products that represent each of the following areas:

a) A b) B c) C d) D e) $A+B$ f) $B+C$

8.8. Use a limit of difference quotients to verify equation (8.10).

8.9. Suppose that a rocket has a mass of 350 *kg* not including an initial mass of 85 *kg* of fuel on board. Furthermore, assume that fuel is burned at the rate of 0.24 *kg/s* and that $u = 50\,km/s$.

a) Find the equation for $m(t)$.

b) Sketch the graph of $m(t)$.

c) Use the definition of the derivative to find the equation for $m'(t)$.

d) Find the equation for $v'(t)$.

e) Sketch the graph of $v'(t)$ and use the geometric interpretation of the derivative to make an approximate sketch of the graph of $v(t)$ under the additional assumption that $v(0) = 0$.

8.10. Given two differentiable functions f and g, apply the product rule to decide which of the following expressions is/are equal to $(fg)'f + f'fg$.

a) $2f'fg + f^2g'$

b) $f'fg' + f'g^2$

c) $(f^2g)'$

Chapter 9

Miraculous Insights

> I seem to have been only a boy playing on the seashore, and diverting myself in now and then finding a smoother pebble or prettier shell than ordinary, whilst the great ocean of truth lay all undiscovered before me.[1]

These are the words of a man who assumes a place of singular, almost unequaled importance in the history of ideas. To himself *Isaac Newton* may have appeared a mere boy searching for little gems along the way, but to the world he appeared a prophet in whose work the scientific revolution reached its climax.

The beginnings of his life were humble indeed. The women who assisted at his birth on Christmas Day 1642 entertained little hope to see him live. He had been born prematurely and was so tiny as to fit "inside a quart pot."[2] Adding to the odds against survival was the poor health of his mother who still suffered from the shock of her husband's death three months earlier. Further uncertainties loomed as the year 1642 had witnessed the outbreak of the English civil war. Some of the early battles between the armies of Charles I and the Roundheads under Oliver Cromwell were fought within a few miles of Woolsthorpe where the Newtons had their home. In the face of such adversities the little infant clung to life with a tenacity that bespoke a physical constitution on par with the power of his intellect.

When Isaac was four years old his mother Hannah moved to the nearby town of Grantham where she married the Reverend Barnabas Smith. Her son was left to the care of his grandmother, Margery Ayscough, but in the marriage agreement he was assured of a parcel of land that together with the inheritance from the paternal estate yielded an annual income of about £80. Upon the Reverend's death in 1653 Hannah returned to Woolsthorpe, taking with her the three children she had borne her second husband.

In 1655 Isaac was sent off to attend King's School at Grantham where Henry Stokes was the headmaster. At first he showed little interest in his academic work and ranked second to last in his class. Instead of studying Latin and Greek, as required by his teachers, he preferred to amuse himself with mechanical contraptions of his own design. Dr. William Stukeley, a friend of Newton's in his later years, is quoted as saying:

> Every one that knew Sir Isaac, or have heard of him, recount the pregnancy of his parts when a boy, his strange inventions and extraordinary inclination for mechanics. That instead of playing among the other boys, when from school, he always busyed himself in making knicknacks and models of wood in many kinds: for which purpose he had got little saws, hatchets, hammers and a whole shop of tools, which he would use with great dexterity.[3]

The list of his inventions is long: "kites with lanterns to scare the credulous villagers,"[4] a water clock, a sundial, work boxes and toys, and most prominently a mill in which he placed a mouse to turn the wheels. "To pay a visit to Isaac's mouse miller"[5] quickly became popular among the townsfolk and the peasants from the surrounding countryside.

As for the boy's lamentable study habits, it may be said that it took a kick in the stomach to cure him of his laziness. For when once a classmate attacked him thus he did not content himself with physical revenge, which he inflicted skillfully, but resolved to also overtake the bully in his academic rank. As it turned out, the feat was easily accomplished—not only did he surpass the bully but he quickly rose to first student of the school. It was at this time also that Henry Stokes took note and recognized the lad to be endowed with talents far beyond the ordinary.

At about the age of fifteen his mother called him back to Woolsthorpe to manage her estate but, absentminded as he was, Isaac proved ill suited for the task. Stukeley notes:

> When at home if his mother ordered him into the fields to look after the sheep, the corn, or upon any rural employment, it went on very heavily through his manage. His chief delight was to sit under a tree, with a book in his hands, or to busy himself with his knife in cutting wood for models of somewhat or other that struck his fancy, or he would go to a running stream, and make little millwheels to put into the water.... The dams, sluices and other hydrostatic experiments were his care without regarding the sheep, corn, or such matters under his charge, or even remembering dinnertime.[6]

Despite these obvious failures, his mother at first refused to listen when her brother, the Reverend William Ayscough, and Henry Stokes entreated her to send Isaac back to school in preparation for the university. Eventually, though, when it became undeniable that her son was not called to live as a yeoman in Lincolnshire, she consented, and in June 1662, after some additional work at Grantham, Isaac Newton set out for Cambridge.

During his first year at Trinity College he was tutored by Benjamin Pulleyn, a Regius Professor of Greek. In the tradition of medieval scholasticism, the curriculum was largely aimed at the study of classical literature and peripatetic philosophy. However, driven by an unquenchable thirst for knowledge, Newton soon began to independently explore the more recent works of such pioneers of science as Copernicus, Brahe, Kepler, Galileo, and Descartes. In 1663 he stopped taking notes on Aristotle's *Organon* in his commonplace book and opened a new section entitled *Quaestiones quaedam Philosophicae (Certain Questions Concerning Philosophy).* On the top of the first page we find the momentous entry:

> Amicus Plato amicus Aritoteles magis amica veritas. (I am a friend of Plato, I am a friend of Aristotle, but truth is my greater friend.)[7]

His notebook shows that by 1664 he had internalized the essence of the scientific method—the synthesis of Cartesian rationalism (see Chapter 5) and Baconian empiricism (as expounded in the *New Organon* of Francis Bacon). Reason and observation, so he realized, must be closely joined if the laws of nature are to be revealed.

It was at this time also that Newton began to immerse himself more deeply in the study of mathematics, which he deemed crucial for the progress of his thought. His forays into the field, though, were not as effortless as the magnitude of his genius might lead us to suspect. Starting with a book on trigonometry, he failed to understand the demonstrations and turned instead to Euclidean geometry. He covered the whole of Euclid's *Elements*, but it took a second reading before he really grasped the text's intent. Next on his list was Oughtred's *Clavis Mathematica*, "which he understood, though not entirely."[8] Upon taking up Descartes' *La Géométrie* (see Chapter 5) he only progressed in stages while repeatedly rereading the initial sections. Eventually, though, "he made himself master of the whole,"[9] and in 1665, when the outbreak of the plague had forced him to return to Woolsthorpe, his ceaseless efforts resulted in the creation of the *method of fluxions*, that is, the *calculus.* The plague years 1665–66, in fact, are known as Newton's *anni mirabiles.* In this short span of time, the better part of which he spent in isolation at his home in Lincolnshire,* he laid the foundations for all his major discoveries in optics, mathematics, physics, and celestial mechanics. Some fifty years later he recalled this period of frenzied mental activity in a memoir:

> In the beginning of the year 1665 I found the Method of approximating series and the Rule for reducing any dignity of any Binomial into such a series. The same year in May I found

*Newton apparently did not spend the full two years in Lincolnshire, but returned to Cambridge between March 20 and late June 1666.

> the method of Tangents of Gregory and Slusius, and in November had the direct method of fluxions and the next year in January had the Theory of Colours and in May following I had entrance to the inverse method of fluxions. And the same year I began to think of gravity extending to the orb of the Moon and having found out how to estimate the force with which [a] globe revolving within a sphere presses the surface of the sphere from Kepler's rule of the periodical times of the Planets being in sesquialterate proportion of their distances from the centres of their Orbs, I deduced that the forces which keep the Planets in their Orbs must [be] reciprocally as the squares of their distances from the centres about which they revolve: and thereby compared the force requisite to keep the Moon in her Orb with the force of gravity at the surface of the earth, and found them answer pretty nearly. All this was in the two plague years of 1665 and 1666. For in those days I was in the prime of my age of invention and minded Mathematics and Philosophy more than at any time since.[10]

The "direct method of fluxions" that Newton refers to here is the differential calculus, and the "inverse method" is the integral calculus. As for the remaining discoveries, we may first want to take a closer look at Newton's work on infinite series by examining binomials of the form $(1+x)^n$. Using elementary algebra, it is not difficult to verify that

$$(1+x)^2 = 1+2x+x^2 = 1+\frac{2}{1}x+\frac{2\cdot 1}{1\cdot 2}x^2,$$
$$(1+x)^3 = 1+3x+3x^2+x^3 = 1+\frac{3}{1}x+\frac{3\cdot 2}{1\cdot 2}x^2+\frac{3\cdot 2\cdot 1}{1\cdot 2\cdot 3}x^3,$$

and in general

$$(1+x)^n = 1+\frac{n}{1}x+\frac{n(n-1)}{1\cdot 2}x^2+\frac{n(n-1)(n-2)}{1\cdot 2\cdot 3}x^3+\cdots+\frac{n(n-1)\cdots 2\cdot 1}{1\cdot 2\cdots(n-1)n}x^n$$

for any nonnegative integer n (see also Appendix E and the ⋆remark⋆ in Chapter 34, p.262). What Newton was able to show is that for an arbitrary real number α (in place of the nonnegative integer n) the binomial $(1+x)^\alpha$ can be represented by an infinite sum, or *series* (see Chapters 41 and 42, and in particular also Exercise 42.31), in which the coefficients are built according to the same pattern as above:

$$(1+x)^\alpha = 1+\frac{\alpha}{1}x+\frac{\alpha(\alpha-1)}{1\cdot 2}x^2+\cdots+\frac{\alpha(\alpha-1)\cdots(\alpha-k+1)}{1\cdot 2\cdots k}x^k+\ldots \text{ (ad infinitum).}$$

This result is commonly referred to as the *binomial theorem.*

Turning our attention now to Newton's mention of the "Theory of Colours" it is instructive to read the opening paragraph of a paper he published in the *Transactions* of the Royal Society in 1672:

> To perform my late promise to you, I shall without further ceremony acquaint you, that in the beginning of the year 1666 (at which time I applied myself to the grinding of optic glasses of other figures than spherical), I produced a triangular glass prism, to try therewith the celebrated phenomenon of colours. And for that purpose having darkened my chamber, and made a small hole in my window shuts, to let in a convenient quantity of the sun's light, I placed my prism at his entrance, that it might be thereby refracted to the opposite wall. It was at first a very pleasing diversion to view the vivid and intense colours produced thereby; but after a while applying myself to consider them more circumspectly, I was surprised to see them in an oblong form; which according to the received laws of refraction, I expected would have been circular.[11]

In the process of his investigations Newton was led to conclude that a light ray's color is directly correlated with its refractivity (as described by the refractive index)—a discovery that may be said to form the basis of the science of spectroscopy. Of great significance was also his invention of the reflecting telescope which paved the way for dramatic improvements in observational astronomy.

Simplicio: I must admit, Newton's account of his discoveries in the years 1665 and 1666 left me somewhat puzzled. That he invented the calculus and made observations concerning the spectral decomposition of light I can accept, but as regards his claim to have "deduced that the forces which keep the Planets in their Orbs must [be] reciprocally as the squares of their distances from the centres about which they revolve" I feel entirely confused.

Teacher: To properly present the matter we need to begin with the fall of an apple.

Simplicio: An apple?

Teacher: Yes, indeed. Close to the end of his life Newton recalled in a conversation with Stukeley that his first thoughts concerning the universal law of gravitation were stirred when at his mother's home he watched an apple fall from a tree.

Sophie: It dropped on his head, didn't it.

Teacher: That may be a myth, but the rest of the story seems well supported.

Simplicio: But what is the point?

Teacher: The point, Simplicio, is that Newton realized the force that pulled the apple to the ground to be of the same nature as the force that keeps the moon on its orbit round the earth. In either case the motion is caused by an attractive force in the direction from the object toward the center of the earth.

Simplicio: Wait a minute, the periodic motion of the moon seems very different from the linear descent of a falling apple. Why, then, would I conclude the corresponding motive forces to be pointing in the same direction?

Teacher: We need to be careful to distinguish between the direction of motion and the direction of acceleration. For the apple the two directions are identical, and the resulting motion therefore proceeds on a straight line, but for the moon they are perpendicular. Consequently, the law of the preservation of momentum—the first of Newton's fundamental laws of mechanics (see Chapters 8 and 68—the latter in Volume 2)—implies that at any point in time the moon would drift off into space on a tangent to its orbit (the projected path in Figure 9.1) if the central force from the earth did not restrain it. In other words, the same gravitational force that causes the apple to accelerate

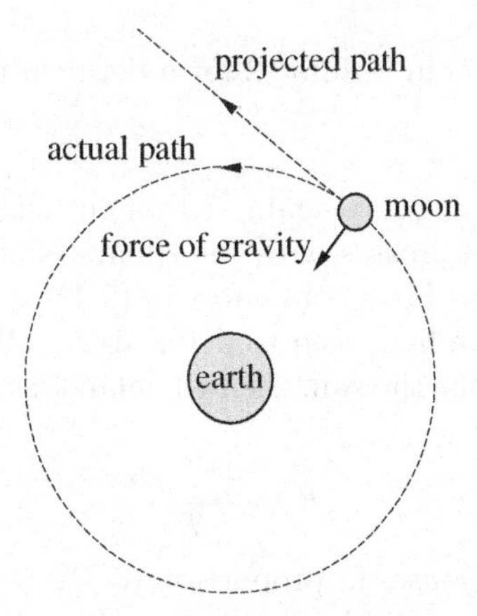

Figure 9.1: actual and projected paths of the moon.

downward effects, when applied to the moon, only a change in the direction of motion but not in the speed. As a matter of course, the same phenomenon can also be observed for the almost circular motion of the earth about the sun. Here, as there, the gravitational force only serves to change the direction of motion while the speed remains constant.†

Simplicio: So there is no force needed to propel a planet (or the moon) forward on its orbit?

Teacher: Indeed, the continual forward movement is guaranteed solely by the law of the preservation of momentum. Johannes Kepler, the discoverer of the three fundamental laws of planetary motion,

†The earth actually does experience minor fluctuations in its speed, because its path is not perfectly circular, but rather slightly elliptic. For a more detailed discussion the reader is referred to Chapter 70 in Volume 2.

still believed that there was a Holy Spirit force pushing the planets from behind, but all that is really needed is a gravitational center of attraction. If you find the concept difficult to grasp, a simple thought experiment perhaps will help: have you ever attached a ball to a string and let it spin about in a circle at a constant speed?

Simplicio: Certainly.

Teacher: Then what was the force you felt?

Simplicio: The ball was trying to pull away toward the outside so that I had to hold on tight.

Teacher: In other words, you experienced a centrifugal force and counterbalanced it by exerting on the ball an opposing centripetal force of equal magnitude (see Figure 9.2).

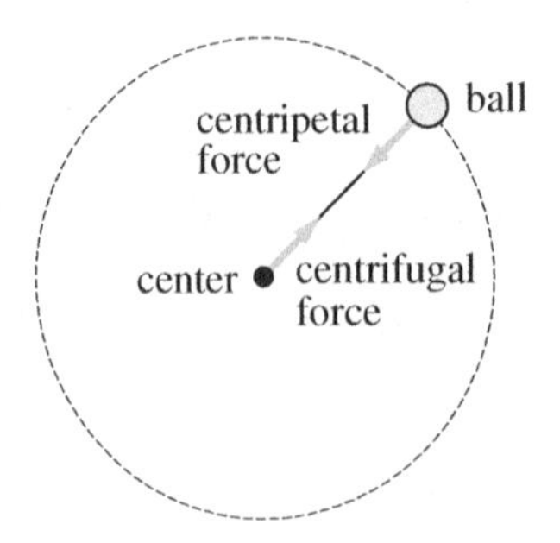

Figure 9.2: centrifugal and centripetal forces.

Sophie: Do I understand correctly that you are trying to suggest an analogy between the centripetal force exerted on the ball and the gravitational force exerted on a planet?

Teacher: That indeed is my intent. But there is more to be said, because for a ball of mass m revolving at velocity v and distance R about a given center, the magnitude F of the centripetal force is easily measured to be

$$\frac{mv^2}{R} \quad \text{(see Chapter 68 in Volume 2 for a detailed theoretical explanation),} \tag{9.1}$$

and therefore...

Simplicio: Just a second, I am getting lost again. I thought the purpose of your deliberations was to prove that the force of gravity decreases with the square of the distance. If that is the case, why is there an R rather than an R^2 in the denominator in (9.1)?

Teacher: To answer this question we first need to determine v. If T is the time it takes for a planet to complete one full revolution about the sun, then the planet's constant speed is $2\pi R/T$. Substituting this term for v in (9.1) yields

$$F = \frac{4\pi^2 mR}{T^2}.$$

Simplicio: So the force actually *increases* in proportion to R?

Teacher: We are not done yet. For we still have not taken into account Newton's reference to "Kepler's rule of the periodical times of the Planets being in sesquialterate proportion of their distances from the centres of their Orbs." The rule in question is the last of three fundamental laws of planetary motion that Johannes Kepler discovered in the first two decades of the 17th Century (see also Chapter 70 in Volume 2):

(i) The path of a planet is elliptic in shape and the position of the sun coincides with the location of one of the foci of the ellipse.

(ii) The area passed over by the line connecting the planet with the sun is the same for equal time intervals regardless of the position of the planet on its elliptic orbit.

(iii) The ratio of the square of the time T that a planet needs for one full revolution about the sun and the cube of the major radius a of the planet's elliptic path is the same for all planets in the solar system.[‡]

Put differently, the third law asserts that there exists a constant C, common to all planets in the solar system, such that $T^2 = Ca^3$. For the special case that a planet's orbit is a circle of radius R, it follows that $T^2 = CR^3$ (i.e., $T = \sqrt{C}R^{3/2}$ is in "sesquialterate proportion" of the distance from the center), and therefore

$$F = \frac{4\pi^2 mR}{CR^3} = \frac{4\pi^2 m}{CR^2}. \tag{9.2}$$

It is this equation that finally shows the gravitational force to decrease with the square of the distance. One more detail, though, needs to be added: according to Newton's third fundamental law of mechanics (actio = reactio), the gravitational attraction exerted by the sun on a planet must be equal in magnitude to the attraction exerted in the opposite direction from the planet on the sun (in perfect analogy to the centrifugal and centripetal forces in Figure 9.2). Consequently, there must exist a constant D such that

$$F = \frac{4\pi^2 M}{DR^2},$$

where we denote by M the mass of the sun. Since F has thus been shown to be proportional to both m and M, we may infer that there exists a constant γ—the so-called *gravitational constant*—such that

$$\boxed{F = \frac{\gamma m M}{R^2}.} \tag{9.3}$$

Given the fairly limited complexity of the computations leading up to the *universal law of gravitation*, as stated in (9.3), the skeptical reader perhaps is doubtful whether Newton really deserves to occupy that place of singular importance in the history of ideas that is usually assigned to him. After all, not only are the computations simple, but they also heavily depend on outside contributions such as Kepler's third law or the law of the preservation of momentum, which Galileo was the first to formulate[§] (see Chapter 15). To such a charge we must reply that anyone familiar with the breadth of Newton's vision will agree that it is almost impossible to overestimate the magnitude of his achievements.

The core of his comprehensive system is composed of the three fundamental laws of mechanics (i.e., preservation of momentum, $F = p' = ma$, and actio = reactio) and the universal law of gravitation. Any moving object anywhere in the universe (so we suppose) is subject to these laws. Whether we analyze the trajectories of stars in distant galaxies, the motion of a falling body, or the oscillations of a pendulum, it is always these laws to which we must appeal. Such a claim to universal validity stands in stark contrast to, for example, the limited scope of Kepler's work which specifically addresses only the motion of the planets. In fact, in what may well be his greatest accomplishment, Newton skillfully applied his method of fluxions (i.e., the calculus) to demonstrate that the laws of mechanics and the law of gravitation imply Kepler's laws as a necessary consequence (see Chapters 70 and 75 in Volume 2). Where Kepler had labored over table after table of observational data for more than twenty years, never certain where his next guess might lead him, Newton succeeded elegantly and brilliantly with the compelling force of logical deduction. He had found the golden key to unlock the secrets of the skies.

In 1687 Newton's scientific career culminated in the publication of his masterpiece the *Philosophiae Naturalis Principia Mathematica*, also known simply as the *Principia*, in which he gave a detailed account of his discoveries in mathematics, physics, and astronomy. The years that followed were anticlimactic, but upon occasion Newton still displayed the formidable powers of his intellect as, for instance, in 1696 when he met a challenge by Jean Bernoulli and quickly solved the brachistochrone problem concerning the path of fastest descent between two given points. On the whole, however, he dedicated the latter part of his life to administrative duties and official functions. He was appointed Warden of the Mint in April 1696 and then Master of the Mint in 1699. In 1703 he ascended to the presidency of the Royal Society, and was knighted by Queen Anne in 1705 at Trinity College. Regrettably, he also

[‡] A slightly more precise formulation of this law will be given at the end of Chapter 70 in Volume 2.

[§] Galileo's formulation of this law is usually referred to as the law of inertia.

spent considerable time engaged in a "shameful squabble"[12] with Leibniz over priority regarding the discovery of the calculus, a dispute that only served to discredit both contestants.

No account of Newton's life is complete without mention of the spiritual dimension of his personality. For strange as it may seem in retrospect, Newton himself considered his voluminous writings on Church history and theology more important than his scientific work. He also was profoundly interested in alchemy and obsessively searched "for the philosopher's stone and the elixir of life."[13] When he moved to London in 1696 as Warden of the Mint, he stored many of his writings on such matters in a chest that was passed on to his niece after his death. In 1936, the renowned economist John Maynard Keynes reassembled about half of the original contents and, having carefully studied Newton's papers, arrived at the following remarkable conclusion:

> In the eighteenth century and since, Newton came to be thought of as the first and greatest of the modern age of scientists, a rationalist, one who taught us to think on the lines of cold and untinctured reason.
>
> I do not see him in this light, I do not think that anyone who has pored over the contents of that box which he packed up when he finally left Cambridge... can see him like that. Newton was not the first of the age of reason. He was the last of the magicians, the last of the Babylonians and Sumerians, the last great mind which looked out on the visible and intellectual world with the same eyes as those who began to build our intellectual inheritance rather less than 10,000 years ago. Isaac Newton, a posthumous child born with no father on Christmas Day, 1642, was the last wonder-child to whom the Magi could do sincere and appropriate homage.[14]

In the early morning hours of March 20, 1727, the last of the magicians and foremost of modern scientists, the man who had lived a life of monastic abstinence committed to thought and duty alone, died at his home in Kensington at the age of eighty-four. On a tablet over the fireplace in the room where he was born, his miraculous insights into the workings of the cosmos are commemorated in the following words of Alexander Pope:[15]

Nature and Nature's laws lay hid in Night
God said: Let Newton be! And all was Light.

for all integers $n \geq 0$. In a case where n is less than zero, we apply this rule to $-n$ in place of n (because $-n > 0$ if $n < 0$) to conclude that the derivative of x^{-n} is $-nx^{-n-1}$. Using now the quotient rule (or the special rule for the differentiation of $1/g$ as given in Example 10.2), we obtain

$$\frac{d}{dx}x^n = \frac{d}{dx}\frac{1}{x^{-n}} = -\frac{(-n)x^{-n-1}}{(x^{-n})^2} = nx^{n-1}.$$

Having thus established the same rule for positive and negative values of n, we may formulate the *power rule* for differentiation as follows:

$$\boxed{\frac{d}{dx}x^n = nx^{n-1} \quad \text{for all integers } n.}$$

10.4 Example. Let a function f be defined by the equation $f(x) := x^5 - 2x^3 + 1$. We wish to find $f'(x)$. According to the sum rule and the rule for the differentiation of scalar multiples, we have

$$f'(x) = \frac{d}{dx}x^5 + \frac{d}{dx}(-2x^3) + \frac{d}{dx}1 = \frac{d}{dx}x^5 - 2\frac{d}{dx}x^3 + \frac{d}{dx}1.$$

Applying the power rule successively with 5, 3, and 0 in place of n yields

$$f'(x) = 5x^4 - 2 \cdot 3x^2 + 0 = 5x^4 - 6x^2.$$

10.5 Example. Let us compute the derivative of the function

$$f(x) := \frac{x^2}{x+1}.$$

Using the quotient rule, the sum rule, and the power rule, it follows that

$$\begin{aligned} f'(x) &= \frac{\left(\frac{d}{dx}x^2\right)(x+1) - x^2\frac{d}{dx}(x+1)}{(x+1)^2} = \frac{2x(x+1) - x^2\left(\frac{d}{dx}x + \frac{d}{dx}1\right)}{(x+1)^2} \\ &= \frac{2x(x+1) - x^2 \cdot 1}{(x+1)^2} = \frac{x^2+2x}{(x+1)^2}. \end{aligned}$$

10.6 Exercise. Find the following derivatives: $\frac{d}{dx}(x^4 - 3x^2 + 1)$ and $\frac{d}{dx}\left(\frac{x-2}{x^3+1}\right)$.

At this point of our development we are still very limited in our ability to find derivatives of specific functions, and the most important rule of differentiation has yet to be discussed. How, for example, would we take the derivative of a function of the form $(x^2+1)^n$? It certainly is possible to multiply (x^2+1) by itself n times so that the sum and power rules could be applied to the resulting expression, but for large n this method is not very convenient. It turns out to be much easier to consider $(x^2+1)^n$ as the "composition" of the functions $f(x) = x^n$ and $g(x) = x^2+1$ and then to apply a general rule for the differentiation of the composition of two functions known as the *chain rule*. How this all works and what it means to "compose" two functions will be explained in the next two sections.

Composition of Functions

Teacher: To introduce the notion of *composition of functions* we consider two functions $f, g : \mathbb{R} \to \mathbb{R}$ and define a new function $f \circ g : \mathbb{R} \to \mathbb{R}$ (that is, "f composed g") via the equation

$$(f \circ g)(x) := f(g(x)).$$

Simplicio: I am once again a bit confused. Are we multiplying f and g, or what exactly is the notation $f \circ g$ supposed to convey?

Teacher: It looks more complicated than it really is. Here is an example: let f and g be defined as $f(x) := x^5$ and $g(x) := x^2+1$. Then the definition above says that in order to find the value of $f \circ g$ at a point x, all we need to do is to find the value of f at the point $g(x)$, which in our case is x^2+1.

Simplicio: So am I right in concluding that $(f \circ g)(x) = f(g(x)) = f(x^2+1) = (x^5)^2+1 = x^{10}+1$?

Teacher: The first part, $(f \circ g)(x) = f(x^2+1)$, is correct, but the second part of your computation is wrong.

Sophie: It seems to me that Simplicio got the order reversed. Since $f(x) = x^5$, it follows that $f(x^2+1) = (x^2+1)^5$ rather than $(x^5)^2+1$.

Teacher: I agree. Simplicio was too careless in his use of function notation. In writing $f(x) = x^5$ we indicate that the function f produces its output values by raising the input values to the fifth power. So if the input value is x^2+1, then this entire term has to be raised to the fifth power to produce the output value $f(x^2+1) = (x^2+1)^5$.

Simplicio: Okay, I accept that, but why do we write $f \circ g$? Why do we not just always write $f(g(x))$ if evaluating f at the point $g(x)$ is what we really do in a given example?

Teacher: You already asked a similar question in one of our earlier lessons (see Chapter 1, p.16), and now as then the answer follows from an accurate understanding of function notation: $f \circ g$ is the *name* of the function whose *value* at a given point x is $f(g(x))$. Other than that, the symbol "$\circ$" has no intrinsic meaning attached to it. Just as with the more familiar symbols "+" and "−," it is nothing but a convention.

10.7 Exercise. Let $f : \mathbb{R} \to \mathbb{R}$ and $g : [0, \infty) \to \mathbb{R}$ be defined by the equations $f(x) := x^3 + x + 1$ and $g(x) := \sqrt{x}$. Find the value of the function $f \circ g$ at $x = 4$.

There is one difficulty that we have not discussed so far: in order for the expression $f(g(x))$ to be well defined, it is necessary for $g(x)$ to be contained in the domain of f. To illustrate this, let us consider the functions $f : [0, \infty] \to \mathbb{R}$ with $f(x) := \sqrt{x}$ and $g : \mathbb{R} \to \mathbb{R}$ with $g(x) := x^3$. In this case it is impossible to form the composition of f and g, because if $x < 0$, then also $x^3 < 0$, and the value $(f \circ g)(x) = f(g(x)) = f(x^3) = \sqrt{x^3}$ is undefined. However, if we restrict the domain of g to the set of positive real numbers, then the range of g also contains only positive real numbers and, by implication, $f(g(x))$ will be well defined for every x in the domain of g. Motivated by these observations, we introduce the following general definition:

10.8 Definition. Let A, B, C, and D be sets and let $g : A \to B$ and $f : C \to D$ be functions such that the range of g is a subset of C (i.e., $R(g) \subset C$). Then the composition $f \circ g : A \to D$ is defined via the equation

$$(f \circ g)(x) := f(g(x)) \quad \text{for all } x \in A.$$

10.9 Exercise. Let $f : [0, 1] \to \mathbb{R}$ be a given function, and assume that g satisfies the defining equation $g(x) := x^2$. Which of the intervals $[-2, 0], [0, 1]$, $[-1/2, 1/2]$, and $[2, 3]$ are possible choices for the domain of g given the requirement that the composition $f \circ g$ should be well defined? *Hint.* For which choices for the domain of g is the range of g a subset of the domain of f?

For later reference we conclude this section by recording the following fact concerning the continuity of compositions of continuous functions:

10.10 Theorem. If $f : D \subset \mathbb{R} \to \mathbb{R}$ and $g : E \subset \mathbb{R} \to \mathbb{R}$ are continuous functions such that the range of g is contained in the domain of f (i.e., $R(g) \subset D$), then $f \circ g$ is continuous on E.

10.11 Exercise. Prove Theorem 10.10.

The Chain Rule

In this section we will derive a general rule for the differentiation of the composition of two functions. This rule is known as the *chain rule* and can be regarded as the most important rule of differentiation in calculus. To get started we assume that $f : J \to \mathbb{R}$ and $g : I \to \mathbb{R}$ are differentiable functions defined on open intervals I and J such that the range of g is a subset of J. Then, according to Definition 4.2, we have

$$\begin{aligned}(f\circ g)'(x) &= \lim_{\Delta x\to 0}\frac{f(g(x+\Delta x))-f(g(x))}{\Delta x} = \lim_{\Delta x\to 0}\left(\frac{f(g(x+\Delta x))-f(g(x))}{g(x+\Delta x)-g(x)}\right)\left(\frac{g(x+\Delta x)-g(x)}{\Delta x}\right)\\ &= \lim_{\Delta x\to 0}\frac{f(g(x+\Delta x))-f(g(x))}{g(x+\Delta x)-g(x)}\lim_{\Delta x\to 0}\frac{g(x+\Delta x)-g(x)}{\Delta x}\\ &= \lim_{\Delta x\to 0}\frac{f(g(x+\Delta x))-f(g(x))}{g(x+\Delta x)-g(x)}\cdot g'(x).\end{aligned}$$

whenever $g(x+\Delta x)-g(x)\neq 0$. (Note: for a fully rigorous derivation of the chain rule, the case where $g(x+\Delta x)-g(x)=0$ would have to be considered as well.) Since $g(x+\Delta x)$ approaches $g(x)$ as Δx approaches zero (here again we use the fact that differentiability implies continuity as stated in Theorem 7.7), it follows that $\Delta y := g(x+\Delta x)-g(x)$ approaches zero as well. Furthermore,

$$\frac{f(g(x+\Delta x))-f(g(x))}{g(x+\Delta x)-g(x)} = \frac{f(g(x)+\Delta y)-f(g(x))}{\Delta y},$$

and as Δy approaches zero the term on the right-hand side of this equation approaches the derivative of f at $g(x)$. Consequently,

$$(f\circ g)'(x) = \lim_{\Delta y\to 0}\frac{f(g(x)+\Delta y)-f(g(x))}{\Delta y}\cdot g'(x) = f'(g(x))g'(x).$$

A concise formulation of this result is given in the following theorem:

10.12 Theorem. *If $g: I\to\mathbb{R}$ and $f: J\to\mathbb{R}$ are differentiable functions defined on two open intervals, I and J respectively, such that the range of g is a subset of J, then $f\circ g: I\to\mathbb{R}$ is differentiable as well and*

$$(f\circ g)'(x) = f'(g(x))g'(x).$$

Equivalently, we may also write

$$\frac{d}{dx}f(g(x)) = f'(g(x))g'(x)$$

or, in abbreviated form,

$$(f\circ g)' = (f'\circ g)g'.$$

10.13 Example. Let us determine the derivative of the function $h:\mathbb{R}\to\mathbb{R}$ defined by the equation $h(x) := (x^2+1)^5$. As explained in the dialogue in the previous section, h is the composition of the functions $f(x) := x^5$ and $g(x) := x^2+1$ (i.e., $h = f\circ g$ and $h(x) = f(g(x))$ for all $x\in\mathbb{R}$). Since

$$g'(x) = 2x \quad\text{and}\quad f'(x) = 5x^4,$$

the chain rule, as stated in Theorem 10.12, implies that

$$h'(x) = f'(g(x))g'(x) = 5g(x)^4 g'(x) = 5(x^2+1)^4(2x) = 10x(x^2+1)^4.$$

10.14 Example. To give a somewhat more complicated example, let us consider the function $h:\mathbb{R}\to\mathbb{R}$ defined by the equation $h(x) := ((x^2+2x-3)^{10}+x)^4$. In order to determine the derivative of h, we will actually have to apply the chain rule twice. To begin with, we set $f_1(x) := x^4$ and $g_1(x) := (x^2+2x-3)^{10}+x$ so that $h = f_1\circ g_1$. Since $f_1'(x) = 4x^3$, the chain rule implies that

$$h'(x) = f_1'(g_1(x))g_1'(x) = 4g_1(x)^3 g_1'(x) = 4((x^2+2x-3)^{10}+x)^3 g_1'(x). \tag{10.2}$$

To find the derivative of g_1, we first use the sum rule to conclude that

$$g_1'(x) = \frac{d}{dx}(x^2 + 2x - 3)^{10} + 1. \tag{10.3}$$

Next we define $f_2(x) := x^{10}$ and $g_2(x) := x^2 + 2x - 3$ and observe that $f_2(g_2(x)) = (x^2 + 2x - 3)^{10}$. Again using the chain rule, we may infer that

$$\frac{d}{dx}(x^2 + 2x - 3)^{10} = f_2'(g_2(x))g_2'(x) = 10g_2(x)^9(2x + 2) = 10(x^2 + 2x - 3)^9(2x + 2), \tag{10.4}$$

and combining (10.2) and (10.3) with (10.4), we obtain

$$h'(x) = 4((x^2 + 2x - 3)^{10} + x)^3(10(x^2 + 2x - 3)^9(2x + 2) + 1).$$

Additional Exercises

10.15. Compute the derivatives of the following functions:

a) $f(x) = x^4 - 2x^3 + 3$ **b)** $f(x) = (3x - x^3 + 2)x^2$ **c)** $f(x) = (3x^3 - 4)^{600}$

d) $f(x) = (3x - 4)^{1000}$ **e)** $f(x) = \left(x^2 + (x^2 + 1)^{100}\right)^{100}$ **f)** $f(x) = (x^2 + 1)^{10}(x - 2)^{10}$

g) $f(x) = \dfrac{x - 2x^2 + 7}{3 - x}$ **h)** $f(x) = \dfrac{2x^2 - 5}{x + 3}$ **i)** $f(x) = \dfrac{3}{x^4}$

j) $f(x) = \dfrac{x^5}{x^7 - 2}$ **k)** $f(x) = \left(\dfrac{x - 3}{4 - x^2}\right)^{-2}$ **l)** $f(x) = x^3 - 5x^2 + \dfrac{3}{x}$

10.16. Use the results of Example 4.6 to find the derivatives of the following functions:

a) $f(x) = \sqrt{x^2 - 2x + 3}$ **b)** $f(x) = \sqrt{x}^{-3}$ **c)** $f(x) = \sqrt[3]{x^2 - \sqrt{x}}$

10.17. Find the equation of the tangent line to the graph of each of the following functions at the indicated point:

a) $f(x) = 2x^3 - 2x^2 + 3x - 1$, at $x_0 = 1$.

b) $f(x) = \dfrac{2x^2 + 1}{x - 5}$, at $x_0 = 2$.

c) $f(x) = (x - 4x^2)^3$, at $x_0 = 0$.

d) $f(x) = \sqrt{5x - 4}$, at $x_0 = 4$ (use Example 4.6).

10.18. For each of the functions listed below, find a function $F(x)$ such that $F'(x)$ is equal to the given function.

a) $f(x) = x + 3$ **b)** $f(x) = x^2 - x + 1$ **c)** $f(x) = \dfrac{3}{x^2}$ **d)** $f(x) = 3x^2(x^3 - 1)^4$

10.19. Given that $f'(2) = 4$, $g'(2) = -3$, $f(2) = -1$, and $g(2) = 1$ do the following:

a) Find the values of $(f + g)'(2)$, $(fg)'(2)$, and $(f/g)'(2)$.

b) Find the equation of the tangent line to the graph of fg at $x_0 = 2$.

10.20. The width of a rectangle is increasing at a rate of 2 *cm/s*, and its length is increasing at a rate of 3 *cm/s*. At what rate is the area of the rectangle increasing when its width is 4 *cm* and its length is 5 *cm*?

10.21. Assume that the functions f and g are differentiable at $x_0 = 0$ and that $f(x)g(x) = x$ for all $x \in \mathbb{R}$. Show that exactly one of the two values $f(0)$ and $g(0)$ is equal to zero, and exactly one of them is different from zero. *Hint.* Differentiate both sides of the equation $f(x)g(x) = x$.

10.22. Given that $f'(2) = 4$, $g'(2) = -3$, $f(2) = -1$, is it possible to determine the value of $(f \circ g)'(2)$? If not, give one additional condition that would make it possible to determine this value.

10.23. Let $f : \mathbb{R} \to \mathbb{R}$ be a differentiable function and define $f_1 := f$, $f_2 := f \circ f_1$, $f_3 := f \circ f_2$, $f_4 := f \circ f_3$, etc. Find formulae for $f_2'(x)$, $f_3'(x)$, and $f_4'(x)$ in terms of f and f'. Do you see a pattern? Try to find a formula for $f_n'(x)$ in terms of f and f' for any integer $n \geq 1$.

10.24. Assume that $f, g : \mathbb{R} \to \mathbb{R}$ are two differentiable functions that satisfy the conditions $f' = g$ and $g' = -f$. Show that for $h(x) := f(x)^2 + g(x)^2$ we have $h'(x) = 0$ for all $x \in \mathbb{R}$.

10.25. Assume that $y : \mathbb{R} \to \mathbb{R}$ is a differentiable function such that $R(y) = (0, \infty)$. For which value of y (if any) is the rate of change of y^3 with respect to x equal to twelve times the rate of change of y with respect to x?

10.26. Given two differentiable functions f and g defined on $\mathbb{R}$, determine the derivative of

$$h(x) := \frac{f(g(x)^2 - x^3)}{f(x + f(x))}.$$

Chapter 11

Graphs of Functions

Maxima and Minima

In order to make an approximate sketch of the graph of a function, it is often sufficient to determine only a few points that are of particular significance for the shape of the graph. Most important in this context are the *extrema* of a function, that is, the points where a function assumes either a *maximum* or a *minimum*. We commonly distinguish two types of extrema: *global* extrema and *local* extrema. As an example, let us consider the graph of the function $f : [a, b] \to \mathbb{R}$ shown in Figure 11.1. Since the function

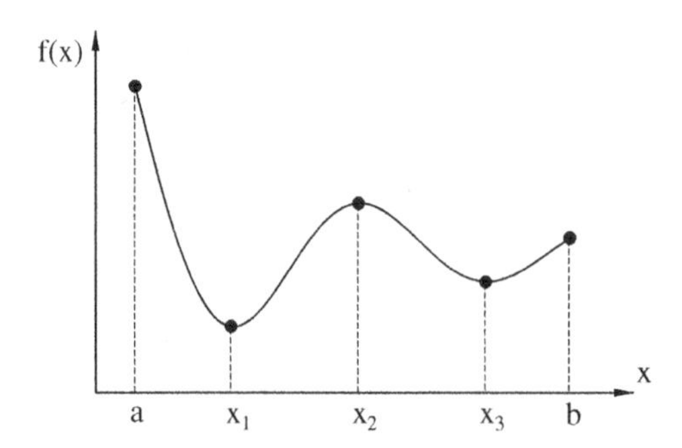

Figure 11.1: global and local extrema.

f has its largest value at a and its smallest value at x_1, we say that f has a global maximum at a and a global minimum at x_1. Furthermore, f has local maxima at x_2 and b and a local minimum at x_3. This means that for all values $x \in [a, b]$ that are "sufficiently close" to x_2 (or b) we have $f(x) \leq f(x_2)$ (or $f(x) \leq f(b)$) and, similarly, for all values $x \in [a, b]$ that are "sufficiently close" to x_3 we have $f(x) \geq f(x_3)$ (see also the ⋆remark⋆ below).

11.1 Definition. Let $f : D \subset \mathbb{R} \to \mathbb{R}$ be a function and let $x_0 \in D$. Then we say that f assumes a...

- **a)** *global maximum* at x_0 if $f(x) \leq f(x_0)$ for all $x \in D$.
- **b)** *global minimum* at x_0 if $f(x) \geq f(x_0)$ for all $x \in D$.
- **c)** *local maximum* at x_0 if $f(x) \leq f(x_0)$ for all $x \in D$ that are sufficiently close to x_0.
- **d)** *local minimum* at x_0 if $f(x) \geq f(x_0)$ for all $x \in D$ that are sufficiently close to x_0.
- **e)** *global/local extremum* at x_0 if f assumes either a global/local minimum or a global/local maximum at x_0.

★ *Remark.* More precise formulations of c) and d) are as follows:

c′) f assumes a *local maximum* at $x_0 \in D$ if there is an $\varepsilon > 0$ such that $f(x) \leq f(x_0)$ for all $x \in D$ with $|x - x_0| < \varepsilon$.

d′) f assumes a *local minimum* at $x_0 \in D$ if there is an $\varepsilon > 0$ such that $f(x) \geq f(x_0)$ for all $x \in D$ with $|x - x_0| < \varepsilon$.

In the sense of these definitions, the set of all points that are "sufficiently close" to x_0 simply consists of those $x \in I$ that satisfy the inequality $|x - x_0| < \varepsilon$. ★

11.2 Exercise. Is it true that every global extremum is also a local extremum? Explain your answer.

11.3 Exercise. Use Definition 11.1 to determine the local and global extrema of the function f shown in Figure 11.2.

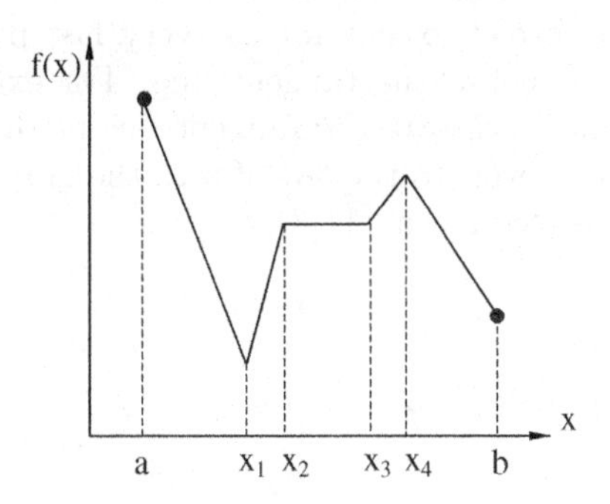

Figure 11.2: graph of f.

To provide a rationale for the introduction of the concept of *critical points* in Definition 11.4 below, we wish to point out that in Figure 11.1 the slope of the tangent lines to the graph of f at x_1, x_2, and x_3 is equal to zero. By implication, the derivative of f at any of these points is equal to zero as well.

11.4 Definition. Let $f : I \to \mathbb{R}$ be a differentiable function defined on an open interval I. We say that $x_0 \in I$ is a *critical point* of f if $f'(x_0) = 0$.

11.5 Exercise. Find the critical points of the function $f : \mathbb{R} \to \mathbb{R}$ defined by the equation $f(x) := x^3 - 3x^2 - 9x + 3$. *Hint.* Determine $f'(x)$ and then solve for x the equation $f'(x) = 0$.

11.6 Theorem. *Let $f : I \to \mathbb{R}$ be a differentiable function defined on an open interval I, and suppose that f has a local extremum at $x_0 \in I$. Then x_0 is a critical point of f, i.e., $f'(x_0) = 0$.*

Proof. ★ Without loss of generality we may assume that f has a local minimum at x_0, because the proof for the case that f has a local maximum at x_0 is completely analogous. If f has a local minimum at x_0 then, according to Definition 11.1d′, there is an $\varepsilon > 0$ such that $f(x) \geq f(x_0)$ for all $x \in I$ with $|x - x_0| < \varepsilon$. Therefore,

$$\frac{f(x) - f(x_0)}{x - x_0} \geq 0 \text{ if } x_0 < x < x_0 + \varepsilon \text{ and}$$
$$\frac{f(x) - f(x_0)}{x - x_0} \leq 0 \text{ if } x_0 - \varepsilon < x < x_0.$$

Using the alternative definition of the derivative as discussed on p.34, it follows that

$$0 \leq \lim_{x \to x_0+} \frac{f(x) - f(x_0)}{x - x_0} = f'(x_0) = \lim_{x \to x_0-} \frac{f(x) - f(x_0)}{x - x_0} \leq 0.$$

Consequently, $f'(x_0) = 0$, as desired. ★ □

Teacher: Let us determine the critical points of the function $f : \mathbb{R} \to \mathbb{R}$ defined by the equation $f(x) := x^3$. Since $f'(x) = 3x^2$, the only critical point of f is $x_0 = 0$, because $x_0 = 0$ is the only solution of the equation $3x^2 = 0$ (see Definition 11.4). Is it permissible, in light of Theorem 11.6, to infer that f has a local extremum at $x_0 = 0$?

Sophie: This is another one of those logic questions, and I think the answer is "no" because the statement of Theorem 11.6 only asserts that every local extremum is a critical point, but it doesn't say that every critical point is necessarily also a local extremum.

Simplicio: I see that Theorem 11.6 is logically a one-way statement, but I wonder whether it might not be possible to make it work the other way as well. After all, if the derivative of f is zero at x_0, that is, if we have a critical point at x_0, then the tangent line to the graph of f at $(x_0, f(x_0))$ is horizontal, and therefore we should expect to find a local extremum at x_0, because close to x_0 the graph of f is either above or below the tangent line.

Teacher: Everything you said is correct except for the very last part: close to x_0, the graph of f is not necessarily either above or below the tangent line. For example, in the case of the function $f(x) = x^3$ with a critical point at $x_0 = 0$, the tangent line to the graph of f at $(x_0, f(x_0)) = (0, 0)$ is identical with the x-axis. However, to the left of zero the graph of f is below the x-axis, while to the right it is above the x-axis (see Figure 11.3).

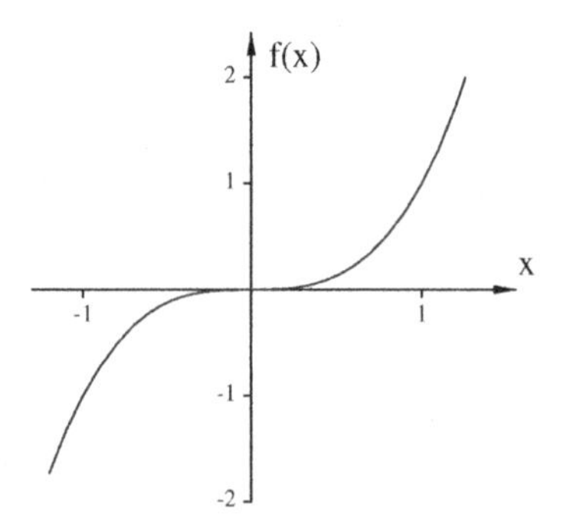

Figure 11.3: graph of $f(x) = x^3$.

Simplicio: Does that mean that Theorem 11.6 is not of much use since we can never be sure whether a critical point is actually a local extremum?

Teacher: Look at it this way: Let's say we want to determine the local extrema of a differentiable function f. Then, instead of examining all points in the domain of f, we can restrict our search to the critical points of f because, according to Theorem 11.6, every local extremum of a differentiable function is necessarily also a critical point.

Simplicio: I understand that, but my point is that as long as we don't have a method of distinguishing between critical points that are local extrema and those that aren't, we are not getting anywhere.

Teacher: That is true, but fortunately, such a method is easily devised. To show you how, I would like to examine the functions in Figure 11.4 that respectively assume a local maximum and a local minimum at a given point x_0. In the diagram on the left, f is increasing to the left of x_0 and decreasing to the right. Conversely, at the local minimum shown in the diagram on the right, the function is decreasing to the left of x_0 and increasing to the right. This observation is very helpful, because in Theorem 4.11 we established a link between the characterization of a function f as increasing or decreasing and the values of its derivative f'.

Sophie: I remember that: a differentiable function is increasing on an open interval I if and only if $f'(x) \geq 0$ for all $x \in I$, and it is decreasing if and only if $f'(x) \leq 0$ for all $x \in I$.

Teacher: Just so, and this leads us to the following theorem, known as the *first derivative test* for local extrema:

11.7 Theorem. *Let $f : I \to \mathbb{R}$ be a differentiable function and let $x_0 \in I$.*

a) *f assumes a local maximum at x_0 if for all $x \in I$ that are sufficiently close to x_0 the following*

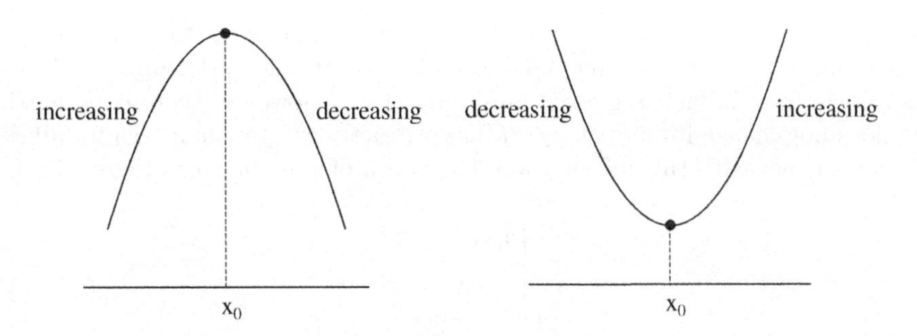

Figure 11.4: local maximum and local minimum.

conditions are satisfied:

$$f'(x) \geq 0 \text{ for } x < x_0 \text{ and } f'(x) \leq 0 \text{ for } x > x_0.$$

b) *f assumes a local minimum at x_0 if for all $x \in I$ that are sufficiently close to x_0 the following conditions are satisfied:*

$$f'(x) \leq 0 \text{ for } x < x_0 \text{ and } f'(x) \geq 0 \text{ for } x > x_0.$$

★ *Remark.* More precise formulations of a) and b) are as follows:

a′) f assumes a local maximum at x_0 if there is an $\varepsilon > 0$ such that for all $x \in I$ it is the case that

$$f'(x) \geq 0 \text{ whenever } x_0 - \varepsilon < x < x_0 \text{ and } f'(x) \leq 0 \text{ whenever } x_0 + \varepsilon > x > x_0.$$

b′) f assumes a local minimum at x_0 if there is an $\varepsilon > 0$ such that for all $x \in I$ it is the case that

$$f'(x) \leq 0 \text{ whenever } x_0 - \varepsilon < x < x_0 \text{ and } f'(x) \geq 0 \text{ whenever } x_0 + \varepsilon > x > x_0.$$

In analogy to Theorem 11.6, the statement of Theorem 11.7 does not assert any of the given conditions to be equivalent. It only says that a function has a local extremum at a point x_0 if certain conditions are met, but it does *not* claim these conditions to be satisfied whenever the function has a local extremum. In this context we encourage the reader to examine the function

$$f(x) := \begin{cases} x^2(1 + \sin(1/x)) & \text{if } x \neq 0, \\ 0 & \text{if } x = 0. \end{cases}$$

It can be shown that this function is differentiable and assumes a local minimum at $x_0 = 0$ despite the fact that it does not satisfy any of the conditions listed in Theorem 11.7. ★

11.8 Example. Let us consider the function $f : \mathbb{R} \to \mathbb{R}$ defined by the equation

$$f(x) := 2x(x-1)^3.$$

We wish to find the roots and local extrema of f. Solving the equation $0 = 2x(x-1)^3$ for x, we see that f has two roots at $x_1 = 0$ and $x_2 = 1$. In order to find the local extrema of f, we determine first the critical points of f as the roots of the derivative of f. Using the product rule, it is not difficult to verify that

$$f'(x) = 2(4x-1)(x-1)^2.$$

So in order to find the critical points of f, we need to solve for x the equation

$$0 = (4x-1)(x-1)^2.$$

The solutions are $x_3 = 1/4$ and $x_2 = 1$. Furthermore, it is easy to verify that $f'(x) \geq 0$ for all $x \in (1/4, \infty)$ and $f'(x) < 0$ for all $x \in (-\infty, 1/4)$. According to Theorem 11.7, it therefore follows that f has a local (and in fact, a global) minimum at x_3. However, there is no local extremum at x_2, because f' does not change its sign at x_2—$f'(x)$ is strictly greater than zero for all $x \in (1/4, 1)$ and all $x \in (1, \infty)$. Consistent with this information, the graph of f is shown in Figure 11.5.

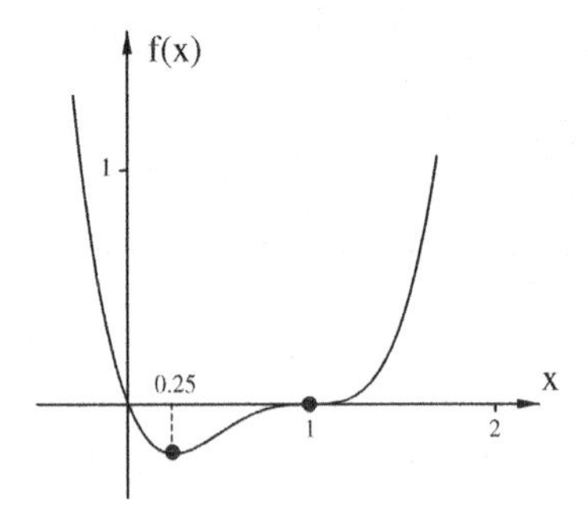

Figure 11.5: graph of $f(x) = 2x(x-1)^3$.

11.9 Exercise. Find all roots and all local extrema of the function $f : \mathbb{R} \to \mathbb{R}$ given by the equation $f(x) := (x-1)(x-3)(x-4)$, and make an approximate sketch of the graph of f.

Concavity and Points of Inflection

Let us consider the graphs in Figure 11.6 that show two functions defined on an interval $[a, b]$. Both

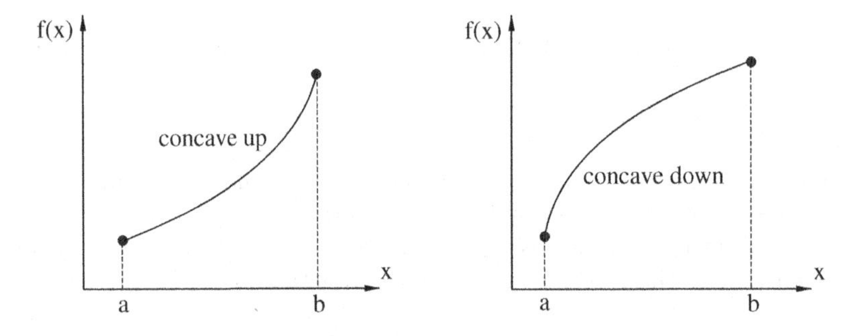

Figure 11.6: graphs that are concave up and down respectively.

functions have in common that they are increasing and assume a global minimum at a and a global maximum at b, but there is also a very obvious difference—the graph on the left in Figure 11.6 bends upward whereas the graph on the right bends downward. In standard terminology we say that the function on the left is *concave up* while the one on the right is *concave down*. A precise definition of these terms is given in the following ⋆remark⋆.

★ *Remark.* A function $f : I \to \mathbb{R}$ defined on an interval I is said to be *concave up* if for all $x, y \in I$ and all $\lambda \in [0, 1]$ we have

$$f(\lambda x + (1-\lambda)y) \leq \lambda f(x) + (1-\lambda)f(y),$$

and f is said to be *concave down* if

$$f(\lambda x + (1-\lambda)y) \geq \lambda f(x) + (1-\lambda)f(y).$$

The reader should verify that these definitions correctly express the intuitive notion of concavity as illustrated in Figure 11.6. Interestingly, it can be shown that a function that is concave up or down on a given interval is necessarily also continuous on that interval (for a proof see [Rud2]). ★

In order to find a convenient calculus criterion for determining the concavity type of a function, we observe that the slope of the graph on the left in Figure 11.6 is increasing while for the graph on the right it is decreasing. In other words, f' is increasing in the diagram on the left and decreasing in the diagram on the right. Given the statement of Theorem 4.11, we may therefore infer that the derivative of f' is greater than or equal to zero if and only if f is concave up, and less than or equal to zero if and only if f is concave down. Note: the derivative of f' is referred to as the *second derivative* of f and is denoted by f''.

11.10 Theorem. *Let $f : I \to \mathbb{R}$ be a function defined on an open interval I such that f is twice differentiable on I (i.e., f and f' are both differentiable on I). Then*

a) *f is concave up on I if and only if $f''(x) \geq 0$ for all $x \in I$ and*

b) *f is concave down on I if and only if $f''(x) \leq 0$ for all $x \in I$.*

Remark. At this point it is appropriate to say a few words about the general notation for *higher-order derivatives.* We have already mentioned that the second derivative of a function f is denoted by f'' and, not surprisingly, for the third derivative (i.e., the derivative of the second derivative) we write f'''. However, for derivatives of order greater than 3 we use the notation $f^{(n)}$ where n is an integer greater than 3. For example, the fifth derivative of f at a point x is denoted by $f^{(5)}(x)$. Alternatively, we may also denote the second and third derivatives by

$$\frac{d^2 f}{dx^2} \quad \text{and} \quad \frac{d^3 f}{dx^3},$$

and, in general, the nth derivative by

$$\frac{d^n f}{dx^n}.$$

11.11 Exercise. Find $\dfrac{d^3}{dx^3}(x+1)^4$ and determine $f''(x)$ and $f^{(4)}(x)$ for $f(x) := x^5 + x^6$.

With regard to the problem of sketching the graph of a function, it is of particular interest to identify the points where the concavity changes from concave down to concave up or vice versa. For example, the function $f(x) = x^3$, shown in Figure 11.3, changes from concave down to concave up at $x_0 = 0$.

11.12 Definition. Let $f : I \to \mathbb{R}$ be a function defined on an open interval I, and let $x_0 \in I$. Then x_0 is said to be a *point of inflection* if f is concave up to the immediate left of x_0 and concave down to the immediate right of x_0, or if f is concave down to the immediate left of x_0 and concave up to the immediate right of x_0.

★ *Remark.* More precisely, we say that x_0 is a *point of inflection* of f, if there is an $\varepsilon > 0$ such that either f is concave up on $(x_0 - \varepsilon, x_0)$ and concave down on $(x_0, x_0 + \varepsilon)$, or f is concave down on $(x_0 - \varepsilon, x_0)$ and concave up on $(x_0, x_0 + \varepsilon)$. ★

In order to develop a practical method for finding points of inflection, we observe that if f is twice differentiable and x_0 is a point of inflection of f then f' is either increasing to the immediate left of x_0 and decreasing to the immediate right of x_0, or f' is decreasing to the immediate left of x_0 and increasing to the immediate right of x_0. In other words:

A point of inflection of f is a local extremum of f'.

This statement, in conjunction with Theorem 11.6, implies that every point of inflection of f is a critical point of f' (assuming, of course, that f is twice differentiable). Therefore:

If x_0 is a point of inflection of f, then $f''(x_0) = 0$.

11.13 Example. Let us consider the function $f : \mathbb{R} \to \mathbb{R}$ defined by the equation

$$f(x) := \frac{1}{1+x^2}.$$

We wish to find the local extrema of f, the points of inflection, and the intervals on which f is concave up or down. Using the quotient rule to compute f', it is easy to see that

$$f'(x) = \frac{-2x}{(1+x^2)^2}$$

for all $x \in \mathbb{R}$. Consequently, the only critical point of f is $x_1 = 0$. Moreover, this critical point is a local (and, in fact, a global) maximum, because $f'(x) > 0$ for $x \in (-\infty, 0)$ and $f'(x) < 0$ for $x \in (0, \infty)$. In order to find the points of inflection of f, we need to determine the local extrema of f', and this in turn requires that we find the roots of f''. Applying the quotient rule to f', we obtain

$$f''(x) = \frac{2(3x^2-1)}{(1+x^2)^3}.$$

In solving for x the equation $f''(x) = 2(3x^2-1)/(1+x^2)^3 = 0$ we readily identify $x_2 = \sqrt{3}/3$ and $x_3 = -\sqrt{3}/3$ as the critical points of f'. Furthermore, since $f''(x) > 0$ whenever $x < -\sqrt{3}/3$ or $x > \sqrt{3}/3$ and $f''(x) < 0$ whenever $-\sqrt{3}/3 < x < \sqrt{3}/3$, we may infer that x_2 and x_3 are local extrema of f', because the sign of f'' changes at these two points. Consequently, x_2 and x_3 are the points of inflection of f. It also follows that f is concave up on the intervals $(-\infty, -\sqrt{3}/3)$ and $(\sqrt{3}/3, \infty)$ and concave down on $(-\sqrt{3}/3, \sqrt{3}/3)$. Consistent with this information, the graph of f is shown in Figure 11.7. Its shape suggests that $f(x)$ approaches zero as x tends to $\pm\infty$. To substantiate this claim,

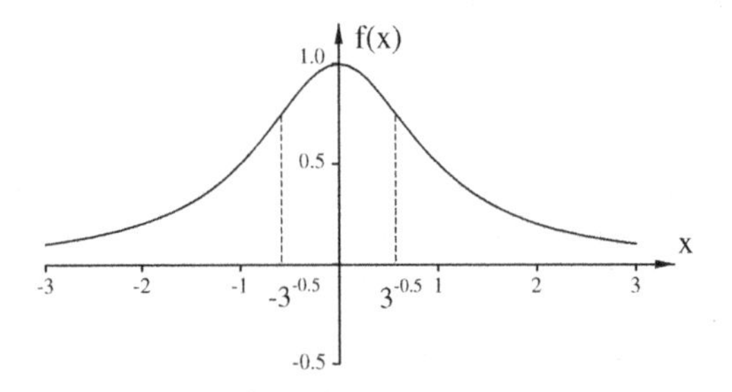

Figure 11.7: graph of $f(x) = 1/(1+x^2)$.

we will introduce in the next section the concept of *limits at infinity* (see in particular Example 11.19).

11.14 Exercise. Find the points of inflection of the function $f(x) = 2x(x-1)^3$ discussed in Example 11.8.

11.15 Exercise. Find the roots, local extrema, and points of inflection of the function $f : \mathbb{R} \to \mathbb{R}$ given by the equation $f(x) := x^3 - 3x^2 + 2x$.

To conclude the present section we wish to discuss how the second derivative of a function f—rather than the first derivative as in Theorem 11.7—can be used to determine whether a critical point of f is a local maximum or minimum. In Figure 11.8 we are given the graph of a function $f : [a, b] \to \mathbb{R}$ that is concave down and has a critical point at x_0. Intuitively, it is not difficult to understand that the downward concavity of f forces the critical point to be a local maximum. We can also compare this observation with the situation in Figure 11.3, where the critical point $x_0 = 0$ is not a local extremum, but a point of inflection. (Critical points that are also points of inflection are called *saddle points.*) Analogously, a function f has a local minimum at a critical point x_0 if close to x_0 the function is concave up. Given the statement of Theorem 11.6, we may infer that f assumes a local maximum at a

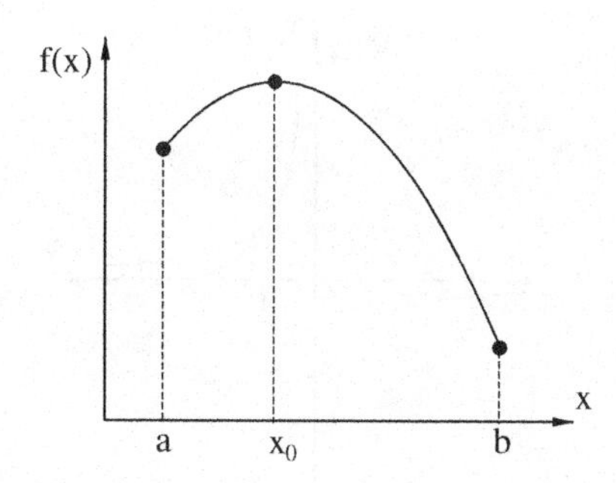

Figure 11.8: downward concavity at a local maximum.

critical point x_0 if close to x_0 we have $f''(x) \leq 0$, and that f has a local minimum at x_0 if close to x_0 we have $f''(x) \geq 0$. With some additional arguments it can be shown that the condition "$f''(x) \leq 0$ for all x close to x_0" is satisfied if $f''(x_0) < 0$, and similarly the condition "$f''(x) \geq 0$ for all x close to x_0" is satisfied if $f''(x_0) > 0$. So we only need to check the value of the second derivative at a single point! The following theorem, which is known as the *second derivative test* for local extrema, summarizes these observations (for a precise proof see [Rud1]):

11.16 Theorem. *Let $f : I \to \mathbb{R}$ be twice differentiable on an open interval I and assume that $x_0 \in I$ is a critical point of f such that f'' is continuous at x_0.*

a) *If $f''(x_0) < 0$, then f assumes a local maximum at x_0.*

b) *If $f''(x_0) > 0$, then f assumes a local minimum at x_0.*

11.17 Example. Let us consider again the function $f(x) = 1/(1+x^2)$. In Example 11.13, we used the first derivative test to show that f has a local minimum at $x_1 = 0$, but the same result also follows from the second derivative test, as stated in Theorem 11.16, simply by observing that $f''(x_1) = f''(0) = -2 < 0$.

11.18 Exercise. Use the second derivative test to decide whether the critical points of the function in Exercise 11.15 are local maxima or local minima.

Infinite Limits and Limits at Infinity

Let us consider the function $f : \mathbb{R} \setminus \{0\} \to \mathbb{R}$ (the domain of f is the set of all real numbers different from 0) defined by the equation

$$f(x) := \frac{1}{x}.$$

Taking a look at the graph of f as shown in Figure 11.9, we notice that as x increases (and in this sense we say "as x tends to ∞") the values $f(x)$ approach 0. The same is true when x tends to $-\infty$. Furthermore, as x approaches 0 from the right (left) the values $f(x)$ tend to ∞ ($-\infty$). In a case like this, we say that f has a *horizontal asymptote* at $y = 0$ and a *vertical asymptote* at $x = 0$. To formalize these observations we write

$$\lim_{x\to\infty} \frac{1}{x} = 0,$$
$$\lim_{x\to-\infty} \frac{1}{x} = 0,$$
$$\lim_{x\to 0^+} \frac{1}{x} = \infty,$$
$$\lim_{x\to 0^-} \frac{1}{x} = -\infty.$$

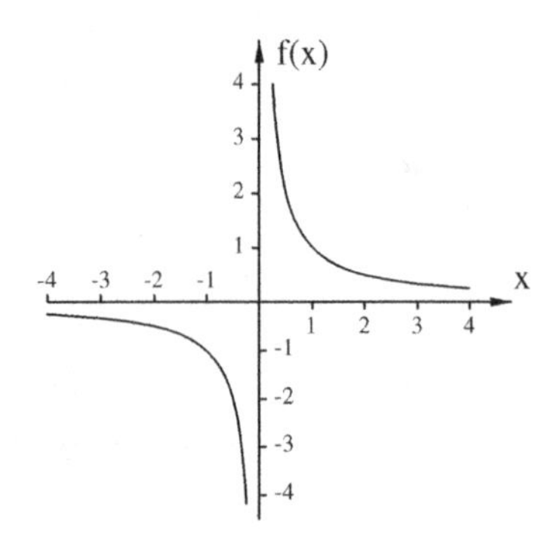

Figure 11.9: graph of $f(x) = 1/x$.

The only difference to the types of limits that we introduced in Definitions 2.1 and 2.4 is that now we also allow $\pm\infty$ as values for x_0 and L. The first two limits above are *limits at infinity* (positive or negative infinity), and the remaining two are what we call *infinite limits*. (Of course, it is also possible to have combinations of these two types. For instance, $\lim_{x\to\infty} x^2 = \infty$ is an infinite limit at infinity.)

11.19 Example. For the function $f(x) = 1/(1+x^2)$, discussed in Example 11.13, we find that

$$\lim_{x\to\infty} \frac{1}{1+x^2} = 0 = \lim_{x\to-\infty} \frac{1}{1+x^2}$$

because as x tends to $\pm\infty$, the value of $1 + x^2$ increases to ∞ and, by implication, the reciprocal $1/(1+x^2)$ approaches zero.

11.20 Example. We wish to determine the value of

$$\lim_{x\to\infty} \frac{x^2-1}{2x^2+x}.$$

Factoring out x^2 from the numerator and the denominator yields

$$\frac{x^2-1}{2x^2+x} = \frac{1-1/x^2}{2+1/x}.$$

As x tends to ∞, both $1/x$ and $1/x^2$ approach zero. Using the rules for limits as given in Theorem 2.9 (which can easily be shown to be valid also for finite limits at infinity), we thus obtain

$$\lim_{x\to\infty} \frac{x^2-1}{2x^2+x} = \frac{\lim_{x\to\infty}(1-1/x^2)}{\lim_{x\to\infty}(2+1/x)} = \frac{1-\lim_{x\to\infty} 1/x^2}{2+\lim_{x\to\infty} 1/x} = \frac{1-0}{2+0} = \frac{1}{2}.$$

11.21 Exercise. Find $\lim\limits_{x\to-\infty} \dfrac{2x^3+x-1}{3x^3-4}$. *Hint.* Factor out x^3 from the numerator and the denominator.

11.22 Example. Let us compute the value of

$$\lim_{x\to 2^-} \frac{x}{x^2-4}.$$

We notice that the denominator $x^2 - 4$ has a root at $x = 2$. Furthermore, if $x \in (0, 2)$, then $x > 0$ and $x^2 - 4 < 0$. This means that as we approach 2 from the left, the values of $x/(x^2 - 4)$ will be negative. Since $x^2 - 4$ approaches 0 as x approaches 2, the absolute value of $x/(x^2 - 4)$ will tend to ∞. These observations allow us to infer that

$$\lim_{x\to 2^-} \frac{x}{x^2-4} = -\infty.$$

11.23 Exercise. Determine $\lim\limits_{x\to 1^+} \dfrac{x+1}{x^3-1}$.

A Further Example

Let a function $f : \mathbb{R} \setminus \{1\} \to \mathbb{R}$ be defined by the equation

$$f(x) := \frac{x^3}{x-1}.$$

We wish to determine...

a) the roots of f,

b) the local extrema of f,

c) the points of inflection of f,

d) the intervals on which f is concave up or down,

e) the vertical asymptotes of f, and

f) the values of $\lim_{x\to\infty} f(x)$ and $\lim_{x\to-\infty} f(x)$.

It is easy to see that the only root of f is $x_1 = 0$. The first and second derivatives of f are

$$f'(x) = \frac{x^2(2x-3)}{(x-1)^2} \quad \text{and} \quad f''(x) = \frac{2x(x^2-3x+3)}{(x-1)^3}.$$

Therefore, the critical points of f are $x_1 = 0$ and $x_2 = 3/2$, and the only root of f'' is $x_1 = 0$ (this is so because the quadratic factor $x^2 - 3x + 3$ does not have any roots). Since $f'(x) \leq 0$ for all $x < 3/2$ with $x \neq 1$, it follows that f does not have a local extremum at x_1 (x_1 is a saddle point). However, f has a local minimum at x_2 because $f'(x) > 0$ for $x > 3/2$ (so the sign of $f'(x)$ changes from negative to positive at x_1). The point x_0 is a point of inflection of f because $f''(x) > 0$ for $x \in (-\infty, 0)$ and $f''(x) < 0$ for $x \in (0,1)$. This observation, in conjunction with the fact that f'' is positive on $(1,\infty)$, also shows that f is concave up on $(-\infty,0)$ and concave down on $(0,1)$. Consequently, the concavity of f changes at $x_3 = 1$, but since f is not defined at $x_3 = 1$ we do *not* refer to x_3 as a point of inflection. Next we wish to show that f has a vertical asymptote at x_3. To do so we need to determine

$$\lim_{x\to1^-} \frac{x^3}{x-1} \quad \text{and} \quad \lim_{x\to1^+} \frac{x^3}{x-1}.$$

As x approaches 1, the denominator $x-1$ approaches 0 and the numerator x^3 approaches 1. This implies that the absolute value of $x^3/(x-1)$ tends to ∞, and therefore f does indeed have a vertical asymptote at $x_3 = 1$. Moreover, since $x^3/(x-1)$ is negative for $x \in (0,1)$ and positive for $x \in (1,\infty)$, we find that

$$\lim_{x\to1^-} \frac{x^3}{x-1} = -\infty \quad \text{and} \quad \lim_{x\to1^+} \frac{x^3}{x-1} = \infty.$$

Finally, all we have left to do is to examine the limits at infinity in f) above. Intuitively, it is clear that the numerator x^3 in the definition of f will grow much faster than the denominator $x-1$ as x tends to ∞ so that we expect the limit of f at ∞ to be ∞. To prove this claim, we factor out x from the numerator and the denominator in the expression $x^3/(x-1)$. This yields

$$f(x) = \frac{x^3}{x-1} = \frac{x^2}{1-1/x}$$

and, therefore,

$$\lim_{x\to\infty} f(x) = \left(\lim_{x\to\infty} x^2\right)\left(\lim_{x\to\infty} \frac{1}{1-1/x}\right) = \infty \cdot \frac{1}{1-0} = \infty.$$

Similarly, it can be shown that $\lim_{x\to-\infty} f(x)$ equals ∞ as well. (Notice also that as x tends to $-\infty$, both x^3 and $x-1$ will be negative and $x^3/(x-1)$ will thus be positive.) Consistent with all this information, the graph of f is shown in Figure 11.10.

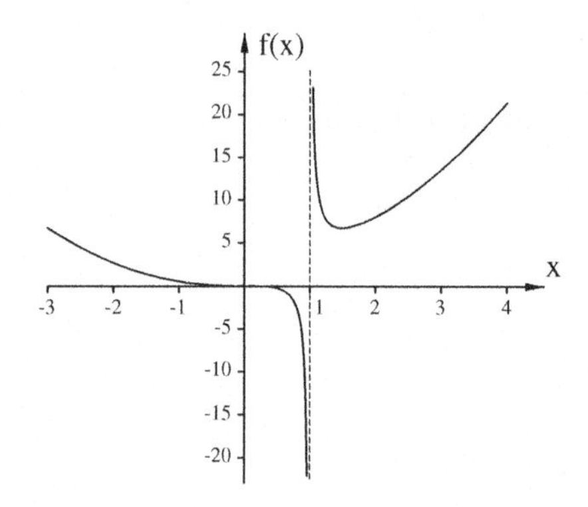

Figure 11.10: graph of $f(x) = x^3/(x-1)$.

Additional Exercises

11.24. Determine the intervals on which each of the following functions is increasing and those on which it is decreasing. Then find all local extrema.

a) $f(x) = 8x^9 - 9x^8$

b) $f(x) = 7x^9 - 18x^7$

c) $f(x) = 5x^6 + 6x^5 - 15x^4$

11.25. Find all points of inflection of the functions in Exercise 11.24 and list the intervals where the functions are concave up and where they are concave down.

11.26. Sketch a graph of a function f such that the derivative f' satisfies the properties in the following table:

x	$(-\infty, -1)$	-1	$(-1, 1)$	1	$(1, 3)$	3	$(3, \infty)$
$f'(x)$	positive	0	negative	0	negative	0	positive

11.27. Determine where the function $f(x) = (1000 - x)^2 + x^2$ is increasing and where it is decreasing.

11.28. Use the result of Problem 11.27 to decide whether 1000^2 is larger or smaller than $998^2 + 2^2$. Explain your answer.

11.29. For each of the functions given below, determine where f is increasing and where it is decreasing. Then sketch the graph of f and identify the local extrema.

a) $f(x) = |x - 1| + |x + 2|$

b) $f(x) = |x - 3| + |x| + |x + 1|$

11.30. Sketch the graph of a continuous function $f : \mathbb{R} \to \mathbb{R}$ that satisfies all of the following conditions:

a) $f'(x) < 0$ for all $x \neq 4$,

b) f is not differentiable at $x_0 = 4$, and

c) $f''(x) < 0$ for all $x \neq 4$.

11.31. Assume that $f : (-3, 3) \to \mathbb{R}$ is a twice differentiable function that satisfies the following conditions:

x	$(-3, -1)$	-1	$(-1, 0)$	0	$(0, 1)$	1	$(1, 3)$
$f'(x)$	positive	0	negative	negative	negative	0	negative
$f''(x)$	negative	negative	negative	0	positive	0	negative

a) Find the local extrema of f.

b) Find the points of inflection of f.

c) Sketch a graph of f that is consistent with the information given in the table.

11.32. For each of the sets of conditions given below, sketch a graph of a function f satisfying these conditions or explain why no such function exists.

a) $f'(x) > 0$, $f''(x) > 0$ and $f(x) < 0$ for all $x \in \mathbb{R}$.

b) $f'(x) < 0$ and $f''(x) > 0$ for all $x \in \mathbb{R}$.

c) $f''(x) > 0$ and $f(x) > 0$ for all $x \in \mathbb{R}$.

11.33. In Figure 11.11 you are given the graph of the *derivative* of a function $f : [0, 10] \to \mathbb{R}$. Mark...

a) the interval(s) on which f is increasing,

b) the interval(s) on which f is decreasing,

c) the interval(s) on which f is concave up,

d) the interval(s) on which f is concave down,

e) the local extrema of f,

f) the points of inflection of f, and

g) the global extrema of f.

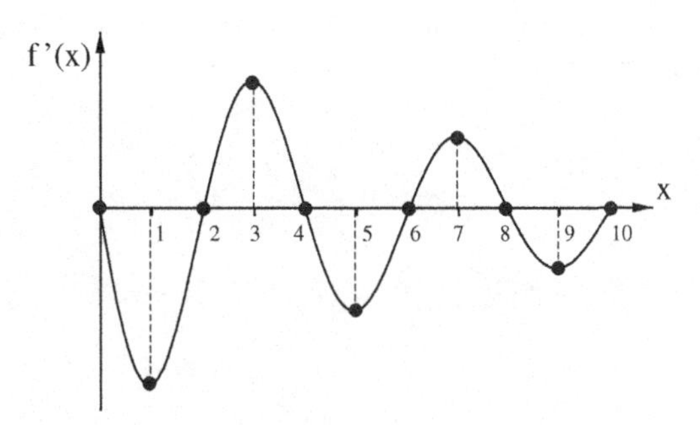

Figure 11.11: graph of $f'(x)$

11.34. Decide in each of the cases listed below whether the given conclusion necessarily follows from the given premise. Justify your answers.

a) Premise: $f : \mathbb{R} \to \mathbb{R}$ is increasing and differentiable.
Conclusion: $f'(x) \geq 0$ for all $x \in \mathbb{R}$.

b) Premise: $f : \mathbb{R} \to \mathbb{R}$ is continuous, $f'(x) > 0$ for all $x \in (a, b)$, and $f'(x) < 0$ for all $x \in (b, c)$ (where $a < b < c$).
Conclusion: f has a local maximum at $x = b$.

c) Premise: $f : \mathbb{R} \to \mathbb{R}$ is a function that satisfies the conditions $f'(x) > 0$ for all $x \in (a, b)$ and $f'(x) < 0$ for all $x \in (b, c)$ (where $a < b < c$).
Conclusion: f has a local maximum at $x = b$.

d) Premise: $f : \mathbb{R} \to \mathbb{R}$ is a function that has at least two local maxima.
Conclusion: f has at least one local minimum.

e) Premise: $f : \mathbb{R} \to \mathbb{R}$ has a global maximum at $x_0 \in \mathbb{R}$.
Conclusion: f has a local maximum at x_0.

f) Premise: $f : \mathbb{R} \to \mathbb{R}$ is differentiable and $f'(x) \neq 0$ for all $x \in \mathbb{R}$.
Conclusion: f has no local extremum.

11.35. Sketch the graphs of the functions given below, identifying all intervals where the function is increasing, decreasing, concave up, or concave down, and also all asymptotes.

$$\text{a) } f(x) := \frac{x - 14/13}{x^2 - 1} \quad \text{b) } f(x) := \frac{x}{x^2 + 1} \quad \text{c) } f(x) := \frac{x^2}{4 - x^2}$$

11.36. Assume that $f : \mathbb{R} \to \mathbb{R}$ satisfies the following conditions: f is twice differentiable, $f(0) = 1$, $f'(x) > 0$ for all $x \neq 0$, f is concave down on $(-\infty, 0)$, and f is concave up on $(0, \infty)$.

a) Sketch a possible graph of f.

b) Find the x-coordinate of all local minima of $g(x) := f(x^2)$.

c) On which intervals is the graph of g concave up?

d) Sketch a possible graph of g.

11.37. For which value(s) of x does the slope of the line tangent to the curve given by the equation $y = -x^3 + 3x^2 + 1$ have its largest value?

11.38. Sketch the graph of a function $f : [0, 1] \to \mathbb{R}$ that...

a) has exactly three local minima, but no local maximum.

b) neither has a global minimum nor a global maximum.

c) has exactly three local extrema, all of which are also global extrema.

d) satisfies the following conditions: f has local extrema at infinitely many points, and all local extrema are also global extrema.

Chapter 12

Optimization (Part 1)

To illustrate how the techniques developed in the previous chapter are relevant to applications of the calculus, we will study in this chapter examples of *optimization problems* from geometry, physics, and engineering.

The Distance from a Point to the Graph of a Function

Let us assume that $f : \mathbb{R} \to \mathbb{R}$ is a differentiable function and that $P = (x_0, y_0)$ is a point in an xy-coordinate system. We wish to determine the minimal distance between P and the graph of f (see Figure 12.1). Our strategy for solving this geometric problem is to express the distance from P to an

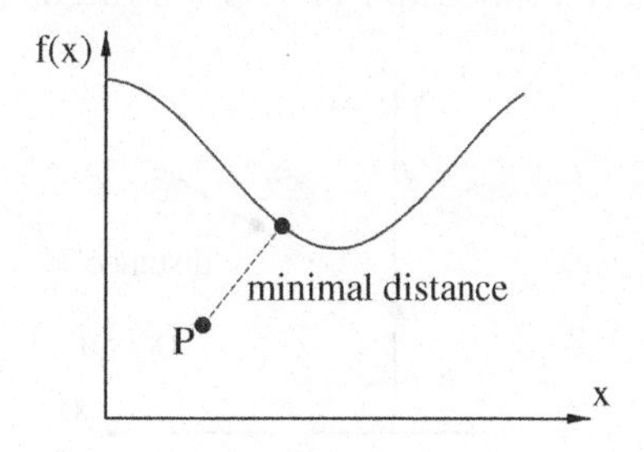

Figure 12.1: distance from a point to a graph.

arbitrary point $(x, f(x))$ on the graph of f as a function of x and then to determine the minimum of this distance function by setting its derivative equal to zero. For this purpose we first need to find a general formula for the distance d between any two points (x_1, y_1) and (x_2, y_2) in an xy-coordinate system (see Figure 12.2). Since the points (x_1, y_1), (x_2, y_2), and (x_2, y_1) form the vertices of a right triangle with

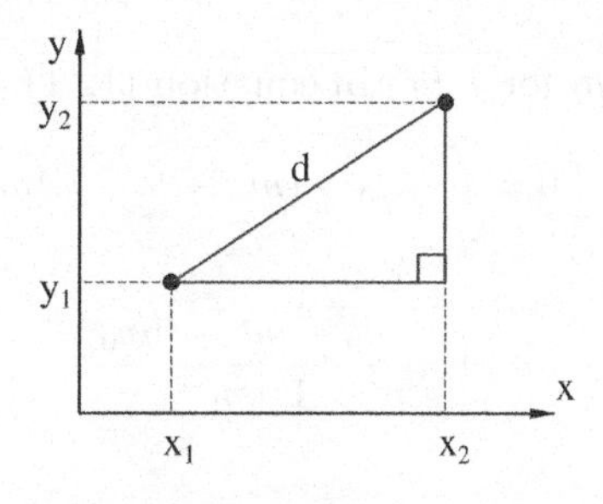

Figure 12.2: distance between two points.

hypotenuse d and side lengths $x_2 - x_1$ and $y_2 - y_1$, the theorem of Pythagoras implies that

$$d^2 = (x_2 - x_1)^2 + (y_2 - y_1)^2.$$

Using this formula with $P = (x_0, y_0)$ in place of (x_1, y_1) and $(x, f(x))$ in place of (x_2, y_2), the distance $d(x)$ from P to $(x, f(x))$ satisfies the equation

$$d(x)^2 = (x - x_0)^2 + (f(x) - y_0)^2,$$

or equivalently,

$$d(x) = \sqrt{(x - x_0)^2 + (f(x) - y_0)^2}.$$

Since the minimal value of d is the square root of the minimal value of the function

$$D(x) := d(x)^2,$$

it follows that D and d assume their respective minima at the same location on the x-axis. From a practical point of view, though, D is easier to work with than d, because the defining equation of D does not involve any square roots. In order to find the critical points of D, we need to set its derivative equal to zero. This yields

$$0 = D'(x) = 2(x - x_0) + 2(f(x) - y_0)f'(x),$$

or equivalently,

$$\boxed{0 = x - x_0 + (f(x) - y_0)f'(x).} \tag{12.1}$$

12.1 Example. Let us assume that the graph of f is a straight line with y-intercept b and slope m

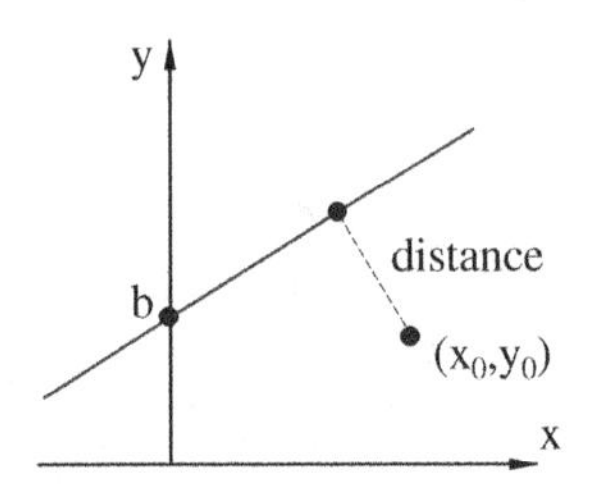

Figure 12.3: distance from a point to a line.

(see Figure 12.3). In this case we have

$$f(x) = mx + b$$

and, by implication,

$$D(x) = (x - x_0)^2 + (mx + b - y_0)^2.$$

Substituting $mx + b$ for $f(x)$ and m for $f'(x)$ in equation (12.1), we obtain

$$0 = x - x_0 + (mx + b - y_0)m$$

and solving for x yields

$$x = \frac{x_0 - mb + my_0}{1 + m^2}.$$

Thus, the minimal value of D is

$$D\left(\frac{x_0 - mb + my_0}{1 + m^2}\right) = \left(\frac{x_0 - mb + my_0}{1 + m^2} - x_0\right)^2 + \left(m\left(\frac{x_0 - mb + my_0}{1 + m^2}\right) + b - y_0\right)^2$$

$$= \frac{1}{(1+m^2)^2}\left((my_0 - mb - m^2x_0)^2 + (mx_0 + b - y_0)^2\right)$$
$$= \frac{(1+m^2)(mx_0 + b - y_0)^2}{(1+m^2)^2}$$
$$= \frac{(mx_0 + b - y_0)^2}{1+m^2}.$$

Taking the square root, we find the following formula for the minimal distance from the line to P:

$$\boxed{d_{min} = \frac{|mx_0 + b - y_0|}{\sqrt{1+m^2}}.}$$

12.2 Exercise. Find the minimal distance from the point $(2,3)$ to the line described by the equation $y = x/2 - 1$.

12.3 Example. We wish to determine the minimal distance from a point $P = (0, y_0)$ on the y-axis to

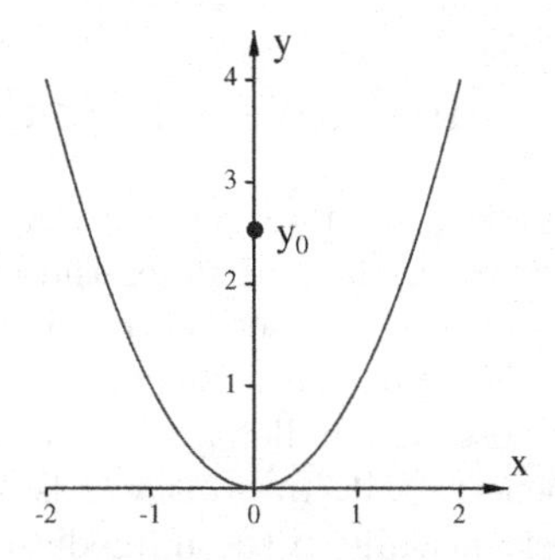

Figure 12.4: graph of $f(x) = x^2$ with a point marked on the y-axis.

the graph of the function

$$f(x) := x^2 \quad \text{(see Figure 12.4)}.$$

Since, in this case, x_0 is equal to zero, it follows that

$$D(x) = x^2 + (x^2 - y_0)^2,$$

and equation (12.1) assumes the form

$$0 = x + 2x(x^2 - y_0) = x(2x^2 - 2y_0 + 1).$$

Solving for x yields

$$x = 0 \quad \text{or} \quad x = \pm\sqrt{y_0 - \frac{1}{2}}.$$

Given these results, we need to distinguish two cases depending on the value of y_0. If $y_0 \leq 1/2$, then the only critical point is $x = 0$ and the minimal distance from the parabola to P is given by the equation $d_{min}^2 = D(0) = y_0^2$. Hence

$$\boxed{d_{min} = |y_0| \quad \text{if } y_0 \leq \frac{1}{2}.}$$

If on the other hand $y_0 > 1/2$, then D has the three distinct critical points

$$x_1 = 0,$$
$$x_2 = \sqrt{y_0 - \frac{1}{2}},$$
$$x_3 = -\sqrt{y_0 - \frac{1}{2}},$$

and the values of D at these three points are

$$D(x_1) = y_0^2,$$
$$D(x_2) = D(x_3) = y_0 - \frac{1}{2} + \left(y_0 - \frac{1}{2} - y_0\right)^2 = y_0 - \frac{1}{4}.$$

In order to find the minimal distance in this case, we need to determine the smaller of the two values y_0^2 and $y_0 - 1/4$. (Alternatively, we could also use the second derivative test to determine where D assumes a local minimum.) Since $y_0^2 - (y_0 - 1/4) = (y_0 - 1/2)^2 \geq 0$, it follows that $y_0^2 \geq y_0 - 1/4$, and therefore,

$$\boxed{d_{min} = \sqrt{y_0 - \frac{1}{4}} \quad \text{if } y_0 > \frac{1}{2}.}$$

12.4 Exercise. Find the distance from the points $(0, -2)$ and $(0, 5)$ to the parabola described by the equation $y = x^2$.

Fermat's Principle and the Laws of Reflection and Refraction

Teacher: Imagine yourself standing on a lake shore under a clear blue sky. As you look at some boats sailing in the distance, the surface of the lake assumes the appearance of a giant mirror reflecting light from the sun and far away objects. Closer to the shore, though, the picture changes—now there are pebbles and aquatic life visible beneath the surface with little interference from reflections.

Simplicio: That's not surprising because at smaller angles of observation, light is more likely to be reflected off the surface than to penetrate it. In other words, light from the surface at a far distance will typically reach us via a reflection while in the immediate vicinity of where we stand it is more likely to have passed from water into air before it enters our eyes.

Teacher: Your explanation is basically correct, but your lack of surprise is surprising indeed.

Simplicio: Why is that?

Teacher: Have you ever heard of the theory that light consists of individual particles?

Simplicio: Yes, they are called photons, if I'm not mistaken.

Teacher: Well then, how do you suppose individual photons make up their minds as to whether they should dive into the water or rather be reflected to stay in the air?

Simplicio: I guess some of them just don't like to get wet.

Sophie: Let's not be silly now, because the question really is intriguing. I never stopped to think what caused a process so commonly observed as light reflecting off a surface.

Teacher: The phenomenon of partial reflection is one of the many deep mysteries that the theory of quantum electrodynamics has revealed. According to this theory, physical reality at atomic scales is intrinsically probabilistic, and predictions concerning the behavior of certain particles such as photons are possible only at the level of statistical averages. In other words, there is no way to tell whether an individual photon will penetrate a surface or be reflected—what can be known is only the probability for such a reflection to occur.

Sophie: How do we calculate this probability?

Teacher: Given the limited scope of our present discussion, it is unfortunately not possible to answer this question, but I encourage you to take a look at the relevant literature such as the exposition in [Fe]. On a more elementary level, though, we have already come across some interesting problems that are more accessible. Consider, for instance, the well known *law of reflection* which asserts that the angle of a light ray with respect to a given surface will be the same before and after it is reflected off the surface (see Chapter 6). When Pierre de Fermat turned his attention to the study of light in 1657, he was able to show that the law of reflection as well as the *law of refraction* (which we are going to discuss below) can be derived from the following general law, known as *Fermat's principle*:

Light follows a path that minimizes the total travel time.

To see how he did it, let us consider a reflected light ray traveling from a point A to a point B as shown in Figure 12.5. Using the theorem of Pythagoras, it is easy to see that the distance between

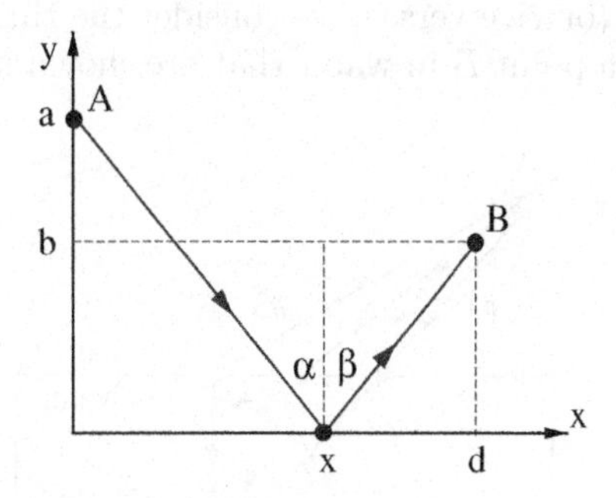

Figure 12.5: a reflected light ray.

A and $(x, 0)$ is $\sqrt{a^2 + x^2}$, and the distance between $(x, 0)$ and B is $\sqrt{b^2 + (d - x)^2}$. Therefore, the total distance traveled between the points A and B is

$$L(x) = \sqrt{a^2 + x^2} + \sqrt{b^2 + (d - x)^2}.$$

Since the speed of light in a given medium is constant, it is obvious that the problem of minimizing the travel time is equivalent to the problem of minimizing the traveled distance. So according to Fermat's principle, the reflection occurs at the point x where L assumes its minimal value. To find the location of this minimum we determine the critical points of L by setting the derivative $L'(x)$ equal to zero:

$$\begin{aligned} 0 = L'(x) &= \frac{d}{dx}\left(\sqrt{a^2 + x^2} + \sqrt{b^2 + (d - x)^2}\right) \\ &= \frac{2x}{2\sqrt{a^2 + x^2}} - \frac{2(d - x)}{2\sqrt{b^2 + (d - x)^2}} \quad \text{(see Example 4.6)} \qquad (12.2) \\ &= \frac{x}{\sqrt{a^2 + x^2}} - \frac{d - x}{\sqrt{b^2 + (d - x)^2}}. \end{aligned}$$

Hence,

$$\frac{x}{\sqrt{a^2 + x^2}} = \frac{d - x}{\sqrt{b^2 + (d - x)^2}}.$$

Referring again to Figure 12.5, we see that

$$\sin(\alpha) = \frac{x}{\sqrt{a^2 + x^2}} \quad \text{and} \quad \sin(\beta) = \frac{d - x}{\sqrt{b^2 + (d - x)^2}}. \qquad (12.3)$$

Consequently, $L'(x) = 0$ if and only if $\sin(\alpha) = \sin(\beta)$. Since α and β are angles between 0 and $\pi/2$, it follows that $\sin(\alpha) = \sin(\beta)$ if and only if $\alpha = \beta$. Thus we arrive at the familiar statement of the law of reflection:

The angle of incidence α is equal to the angle of reflection β.

12.5 Exercise. Verify equation (12.2) and use the second derivative test (Theorem 11.16) to show that the critical point of L is a local (and in fact, global) minimum.

Turning our attention now to the *law of refraction*, we need to recall that light travels at different velocities in media of different optical densities. Taking air and water as examples of two such media, we denote by c_a the speed of light in air and by c_w the speed of light in water. Experimental evidence shows that the ratio

$$k := \frac{c_a}{c_w},$$

the so-called *refractive index*, is approximately equal to 1.33 (in other words, c_a is significantly larger than c_w). In order to understand how the difference in magnitude between c_a and c_w affects light rays passing from air into water (or vice versa), we consider the three hypothetical paths of a light ray passing from a point A in air to a point B in water that are shown in Figure 12.6. Path #1 certainly is

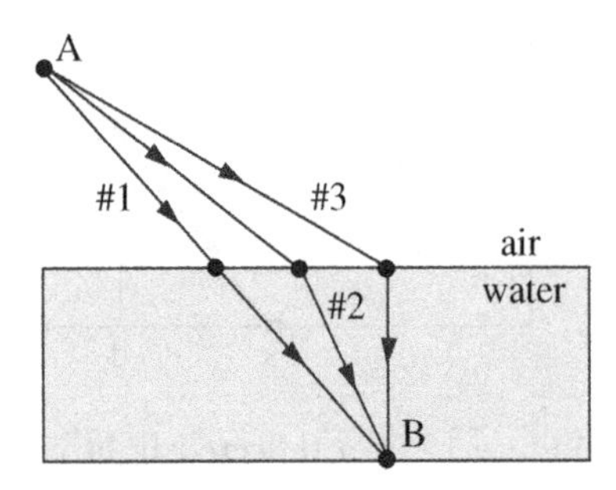

Figure 12.6: hypothetical paths of a light ray.

the shortest of the three paths, because it is simply a straight line. Consequently, we might be tempted to think that the travel *time* (see Fermat's principle) is shortest for path #1 as well. This conclusion, however, is false, because for #1 the distance traveled in water—the slow medium—is larger than for #2 and #3. In other words, in the case of path #1 a lot of time is wasted by traveling a long distance at a relatively low velocity. By contrast, for path #3 the distance traveled in water is minimal, but #3 is the longest of the three paths and the advantage gained by minimizing the distance traveled in water may be outweighed by a disproportionate increase in the total length of the path. The ideal path probably looks more like #2, where a moderate decrease of the distance traveled in water is balanced against a moderate increase in the total distance.

To identify the ideal path we consider Figure 12.7, where we are shown a light ray passing from air into water with an angle of incidence α and an *angle of refraction* β. Since time equals distance divided

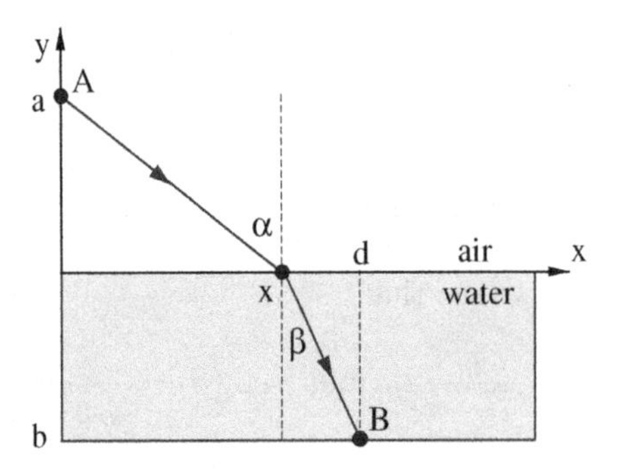

Figure 12.7: a refracted light ray.

by velocity, it follows that the light ray spends $\sqrt{a^2+x^2}/c_a$ units of time traveling from A to $(x, 0)$ and $\sqrt{b^2+(d-x)^2}/c_w$ units of time traveling from $(x, 0)$ to B. Thus, the total travel time is

$$T(x) = \frac{\sqrt{a^2+x^2}}{c_a} + \frac{\sqrt{b^2+(d-x)^2}}{c_w}.$$

To find the critical points of T, we determine the derivative of T and set it equal to zero:

$$0 = T'(x) = \frac{1}{c_a} \cdot \frac{x}{\sqrt{a^2+x^2}} - \frac{1}{c_w} \cdot \frac{d-x}{\sqrt{b^2+(d-x)^2}}.$$

Since the equations in (12.3) are easily seen to be valid for Figure 12.7 as well, we obtain

$$0 = \frac{\sin(\alpha)}{c_a} - \frac{\sin(\beta)}{c_w},$$

or equivalently,

$$\boxed{\frac{\sin(\alpha)}{\sin(\beta)} = \frac{c_a}{c_w} = k.} \tag{12.4}$$

This equation describes the *law of refraction* for a light ray at the interface between air and water.

12.6 Exercise. Use the second derivative test to show that the critical point of T is a local (and in fact global) minimum.

Remark. Our derivation of the law of refraction did not depend on the direction of travel of the light ray, and our result would therefore remain unchanged if the light source were at B instead of at A. Consequently, for light passing from water into air the angle of incidence is smaller than the angle of refraction, while the reverse is true for light traveling in the opposite direction from air into water.

The Storage Capacity of a CD

On a traditional phonograph record, sound is encoded in the surface structures of a thin spiral groove. A stylus in contact with the record's surface translates these structures into electric currents which in turn are converted into sound waves by means of a speaker. While undoubtedly ingenious, this method of analog recording is not fully satisfactory because any imperfections or any damage along the groove will diminish the quality of the reproduced sound.

By contrast, the digital encoding patterns on a compact disc allow for a near-perfect reproduction of sound that is also unaffected by repeated copying. In a digital recording session, electric currents generated by high frequency measurements of air pressure fluctuations in sound waves are sent to an analog-to-digital converter that transforms the readings into a string of integers. On the disc's surface, these integers are represented as ridges and valleys of differing lengths along a narrow spiral track imprinted on a very thin layer of aluminum. A CD player detects the information stored on this spiral track by measuring the degree to which a laser beam is reflected off the disc's surface. Whenever the laser traces a flat valley, a significant portion of its beam will be reflected, while a ridge largely serves to scatter the light thereby minimizing reflection. The varying intensities of the reflected light are picked up by the CD player's photo sensors, then transmitted as electric currents to a digital-to-analog converter, and eventually transformed into sound waves by a set of speakers.

After this brief description of the digital encoding and reproduction of sound (which certainly was not intended to be particularly accurate or even complete) we now turn our attention to the problem of determining the storage capacity of a CD. There are two design features that are of central importance in this context. The first of these is the laser wavelength. Both the width of the spiral track (*the track pitch*) on which data is encoded and the maximal amount of data on a track segment of unit length (*the linear density*) are functions of the wavelength. The following table shows this functional dependence for three different types of lasers:

laser color	wavelength in nanometers	track pitch in microns	linear density in bytes/mm
infrared	780	1.6	121
red	640	0.74	387
blue	410	0.32	800

The second design feature is the disc drive mechanics. Laser discs can be designed for either *constant angular velocity* (CAV) or *constant linear velocity* (CLV) disc drives (the corresponding patterns of data storage are shown in severely simplified form in Figure 12.8). The CLV drive must rotate faster when

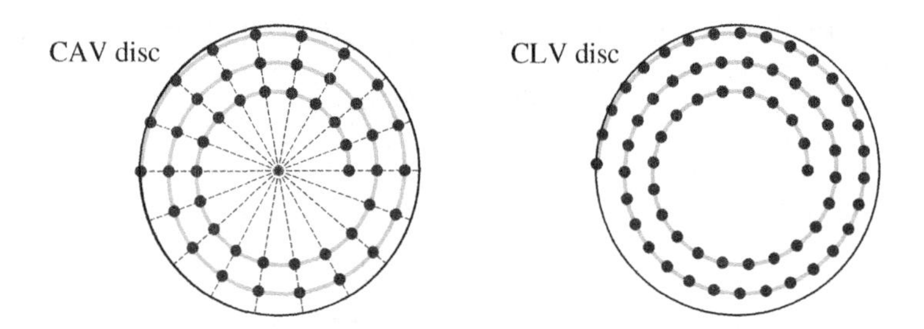

Figure 12.8: data storage on CAV and CLV discs.

reading the inner tracks, while the CAV drive rotates at the same speed at all times. The CLV has the advantage of greater storage capacity but the disadvantage of slower data access in random searches.

To determine the storage capacity of a CLV disc, we denote by δ the linear data density and by w the track pitch respectively (see the diagram on the left in Figure 12.9). In accordance with current industry standards (see the diagram on the right in Figure 12.9), we assume that the inner and outer radii of the storage region are $r = 22.5\,mm$ and $R = 58\,mm$ respectively.

A simple but fairly precise *estimate* for the length of the data spiral can be obtained as follows: we imagine that the spiral track in the annular region is unwound into a long, nearly rectangular region of width w. The area of this rectangular region is, in good approximation, equal to the area A of the annular storage region. Therefore, in denoting the length of the entire spiral by L, we find that

$$A = \pi R^2 - \pi r^2 \approx L \cdot w.$$

Hence

$$L \approx \frac{A}{w} = \frac{\pi(R^2 - r^2)}{w}. \tag{12.5}$$

Using the value for the track pitch of an infrared laser as well as the values for r and R given above (see also Figure 12.9), we find that

$$L \approx \frac{\pi(58^2 - 22.5^2)}{0.0016} \approx 5611179.00362\,mm \quad \text{(about 3.5 miles)}, \tag{12.6}$$

and the corresponding estimate for the storage capacity of a CLV disc is

$$\delta L = 121L \approx 678952659\ \textit{bytes}.$$

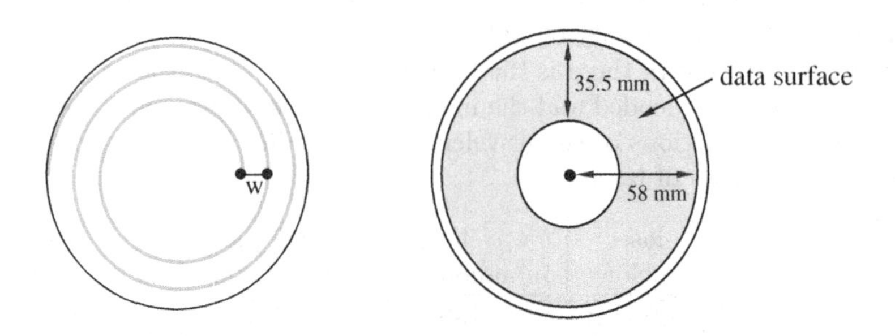

Figure 12.9: data spiral and storage region on a CD.

12.7 Exercise. Use formula (12.5) to estimate the length and the storage capacity of a CLV disc with $r = 22.5\,mm$ and $R = 58\,mm$ for a blue-light laser.

Teacher: Let me pose the following question: how far toward the center should the storage region on a CAV disc be extended if the storage capacity is to be maximized?

Simplicio: I am sure you wouldn't ask this question if the answer were as simple as I think it is, but my first response would be to make the region as large as possible. The larger the region, the longer the data spiral, and the larger the amount of data that can be stored. So ideally, the storage region should cover the entire CD.

Teacher: What do you think, Sophie?

Sophie: I agree with Simplicio on both counts. Given your penchant for asking trick questions, his answer is as implausible from a psychological viewpoint as it is plausible from a logical viewpoint.

Teacher: Since your psychological insight is obviously superior to your command of logic, I would like to direct your attention back to Figure 12.8. Here you can see that the storage density on a CAV disc is highest on the inner parts of the spiral and decreases toward the outside. This observation, combined with the fact that the laser wavelength imposes a physical limit on the linear data density, should cause you to reconsider your reasoning regarding the problem of maximizing the storage capacity.

Sophie: I think I am beginning to understand. All spiral segments on a CAV disc that complete a full 360° loop contain the same amount of data as the spiral's innermost loop (see Figure 12.8). Consequently, the storage density on the longest loops in the outer regions of the disc will be very low if the innermost loop is very small. For as you said, the linear density on the innermost loop cannot be increased beyond the physical limit imposed by the laser wavelength. Therefore, the apparent advantage of maximizing the storage area by extending the data spiral all the way to the center is likely to be outweighed by a corresponding loss in overall storage density.

Simplicio: Are we saying then that data should be stored only in the outer regions of the disc in order to guarantee that the storage density is high?

Teacher: It's a bit more complicated than that. We are facing here a typical optimization problem that requires us to find just the right dimensions for the storage region—not too big and not too small. To make this more precise, we denote by δ the linear density (which depends on the laser wavelength) and assume that data is stored in an annular region with inner radius r and outer radius R. Then, along the innermost loop of (approximate) length $2\pi r$ the maximal amount of data stored is

$$2\pi r\delta.$$

How can we use this result to determine the amount of data on the entire annular region between the inner and outer radii?

Simplicio: I would just multiply δ with the full length of the spiral.

Teacher: That would work for a CLV disc, but for a CAV disc we need to take into account that the storage density decreases toward the outside.

Sophie: I think the answer is approximately $2\pi r\delta$ times $(R-r)/w$, where w is the track pitch (see also Figure 12.9 below).

Simplicio: How did you figure that out?

Sophie: It's easy. As we move from r to R, the spiral winds around the center of the disc in $(R-r)/w$ full loops, and the amount of data stored on each of them is, as we already saw, equal to the amount of data on the innermost loop, which in turn we found to be $2\pi r\delta$.

Teacher: Sophie is right, and her observation shows that in order to determine the maximal storage capacity of a CAV disc we need to find the maximum of the function

$$f(r) := \frac{2\pi r\delta(R-r)}{w}. \tag{12.7}$$

Setting the derivative of f equal to zero yields

$$f'(r) = \frac{2\pi\delta(R-2r)}{w} = 0,$$

or equivalently,

$$r = \frac{R}{2}.$$

Since f is easily seen to have a maximum at this critical point, we may infer that the storage capacity of a CAV disc is maximal when the inner radius of the annular storage region is half the outer radius.

12.8 Exercise. Use the second derivative test to verify that f assumes a maximum at $r = R/2$.

12.9 Exercise. Find the maximal storage capacity of a CAV disc with $R = 58\,mm$ for an infrared and a blue-light laser.

Additional Exercises

12.10. A fence is to be built at minimal cost around a plot of land consisting of a rectangle joined with two semidiscs at the sides (see Figure 12.10). How would you choose the radius of the semicircular discs if the fence is to enclose a total area of $100\,m^2$, and if the price per meter of fence is three times higher for the semicircular sides than for the straight horizontal sides?

Figure 12.10: a rectangle joined with two semidiscs.

12.11. What is the maximal area of a rectangle inscribed in the region bounded by the lines $x = 0$ and $y = 1$ and the curve $y = x^3$ as shown in Figure 12.11?

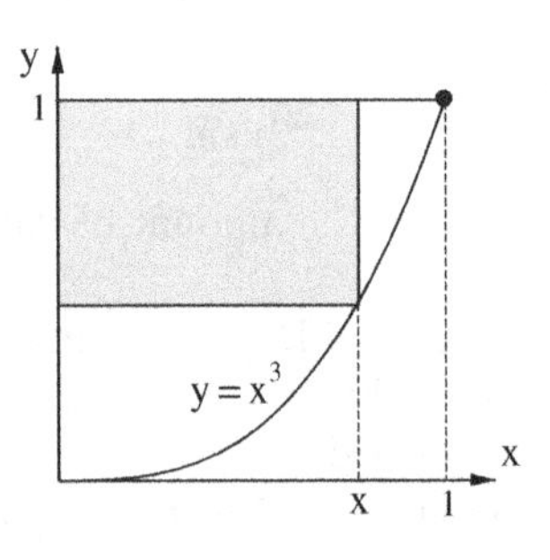

Figure 12.11: a rectangle inscribed in a parabolic region.

12.12. What is the maximal area of a triangle inscribed in a circle of radius 1 in the position shown in Figure 12.12?

12.13. Find the minimal distance from the line given by the equation $3x + 4y = 10$ to the origin $(0, 0)$, and determine the coordinates of the point P on the line that is closest to the origin.

12.14. Repeat Exercise 12.13 for the line given by the equation $x + 2y = 4$ and the point $(3, 3)$ (in place of $(0, 0)$).

12.15. Find the minimal distance from the graph of the function $f(x) := 2x^2 - 3x - 36/7$ to the origin $(0, 0)$, and determine the coordinates of the point P on the graph of f that is closest to the origin. *Hint.* You will encounter a cubic equation for which a simple integer solution can be guessed.

12.16. Repeat Exercise 12.15 for $f(x) := 5 - x^2$ and the point $(-3, 2)$ (in place of the origin $(0, 0)$).

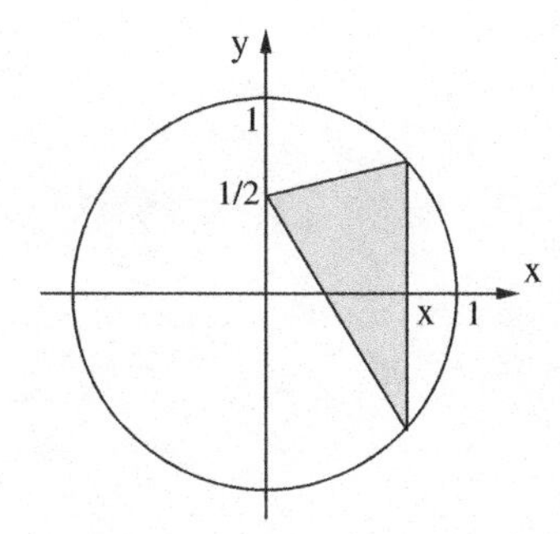

Figure 12.12: a triangle inscribed in a circle.

12.17. Find the point on the parabola $y = x^2 + 2$ that is closest to the line described by the equation $2x - y = 0$.

12.18. A billboard 54 feet wide is perpendicular to a straight road and is 18 feet east of the nearest point on the road. A car approaches the billboard from the south. From what point does a passenger in the car see the billboard at the widest angle?

12.19. How would you choose the height and radius of a $12\,oz$ $(= 355\,ml = 355\,ccm)$ cylindrical aluminum can (including top and bottom) if you wanted to minimize the amount of aluminum used.

12.20. A taxi driving at a constant speed of $60\,mph$ gets 30 miles per gallon of fuel, and the price of fuel is \$1.85 per gallon. Each increase in speed by 2 mph reduces the fuel efficiency by 0.5 miles per gallon. Other costs of running the taxi amount to \$2.35 per hour. Assuming the taxi is to travel a given distance D, you are to determine what constant speed will minimize the total cost of operating the taxi?

12.21. A man has parked his car 500 meters down stream on the opposite shore of a river that is 200 meters wide. Assuming that he can swim at a speed of $1.5\,km/h$ and walk at a pace of $6\,km/h$, where along the opposite shore should he leave the water if he wishes to minimize the total time needed to get back to his car, and what actually is the minimal time?

12.22. The hourly cost of fuel in operating a train is proportional to the square of its speed, and for a speed of $40\,km/h$ the cost is \$200. Other fixed expenses amount to \$800 per hour. Assuming the train is to travel a distance D, you are to find the speed at which it should run to minimize the total cost.

Chapter 13

Newton's Method of Approximation

Approximating Roots Using Tangent Lines

In continuation of our discussion in Chapter 12 we wish to determine the minimal distance from the point $P = (2, 3)$ to the graph of the function $h(x) := x^4$. Since the derivative of x^4 is $4x^3$, equation (12.1) assumes the form

$$0 = x - 2 + 4x^3(x^4 - 3),$$

or equivalently,

$$0 = 4x^7 - 12x^3 + x - 2. \tag{13.1}$$

At this point we are apparently stuck, because a formula (analogous to the quadratic formula) that allows us to compute the roots of a polynomial of degree seven does not exist. In fact, it can be shown that for degrees greater than four, a formula that expresses the roots of a polynomial in terms of its coefficients cannot exist for very precise theoretical reasons. It is, of course, possible that in the specific case of equation (13.1) an exact solution can be found with a bit of creative problem-solving thinking, but the prospect is uninspiring and, even if successful, our search would most likely be of little conceptual value. What we will concentrate on instead is to develop a method of approximation that will allow us to compute the solutions of (13.1) to an arbitrarily high degree of accuracy. In setting

$$f(x) := 4x^7 - 12x^3 + x - 2,$$

the problem of solving equation (13.1) for x is equivalent to the problem of finding the roots of f. Thus, it appears appropriate to broaden the scope of our discussion by considering the general problem of locating the roots of an arbitrary continuously differentiable function $f : \mathbb{R} \to \mathbb{R}$ (see Definition 7.6 and the ⋆remark⋆ below) within a given interval $[a, b]$. For convenience we usually try to choose a and b in such a way that f has exactly one root in $[a, b]$. An example of a simple and useful criterion that guarantees the existence and uniqueness of a root in $[a, b]$ is the following rule: if f is concave up on $[a, b]$ and $f(a) < 0 < f(b)$, then there exists exactly one $z \in [a, b]$ such that $f(z) = 0$ (for variations see Theorem 13.4). The validity of this rule is intuitively fairly obvious (see Figure 13.1), but for the skeptical reader the following ⋆remark⋆ provides some additional explanations.

★ *Remark.* If, as mentioned above, we assume f to be continuously differentiable, then f is in particular continuous (see Theorem 7.7) and cannot "jump" from negative to positive values or vice versa. Therefore, if $f(a) < 0 < f(b)$, then there has to exist at least one point z between a and b such that $f(z) = 0$. In rigorous introductions to the theory of calculus, this statement regarding the existence of a root of f is usually established as a consequence of a slightly more general statement known as the *intermediate value theorem* (see [Rud1] for a proof). Furthermore, the assumption of upward concavity implies that the slope of f is increasing and, by implication, the graph of f cannot intersect the x-axis twice. In other words, the root z must be unique. ★

In order to locate the root z of f in $[a, b]$, we employ the following stepwise process of approximation: In the first step we determine a point x_1 on the x-axis as the point of intersection of the tangent line to the graph of f at $(b, f(b))$ with the x-axis (see Figure 13.1). Then, given x_1, we draw a tangent line to the graph of f at the point $(x_1, f(x_1))$ and determine x_2 as the point of intersection of this second tangent line with the x-axis. Now we repeat the process by determining x_3 as the point of intersection of the tangent line to the graph of f at $(x_2, f(x_2))$ with the x-axis and, continuing in this manner, we generate a sequence of values x_n such that $\lim_{n\to\infty} x_n = z$.

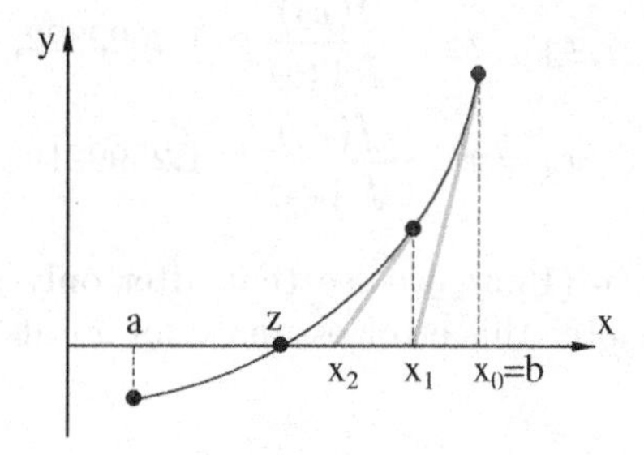

Figure 13.1: Newton's method.

In order to find a formula that allows us to calculate x_n, we start from the initial value $x_0 := b$. Since the equation of the tangent line to the graph of f at $(x_0, f(x_0))$ is

$$y = f'(x_0)(x - x_0) + f(x_0) \quad \text{(see Chapter 4, p.35)},$$

we can determine x_1 by setting y equal to zero and solving the resulting equation for x. This yields

$$x_1 = x_0 - \frac{f(x_0)}{f'(x_0)}.$$

Given this value for x_1, we observe that the equation for the tangent line to the graph of f at $(x_1, f(x_1))$ is

$$y = f'(x_1)(x - x_1) + f(x_1).$$

Again setting y equal to zero and solving for x, we find the value for x_2:

$$x_2 = x_1 - \frac{f(x_1)}{f'(x_1)}.$$

Continuing in this fashion, we readily observe that in general, the approximation x_{n+1} can be determined from x_n via the equation

$$\boxed{x_{n+1} = x_n - \frac{f(x_n)}{f'(x_n)}.} \tag{13.2}$$

The process of repeatedly calculating x_{n+1} from x_n is referred to as *Newton's method of approximation.*

13.1 Example. We wish to find an approximation for the root of the function

$$f(x) := x^3 - 2.$$

Given this definition, it is easy to verify that $f(0) < 0$ and $f(2) > 2$. Furthermore, f is concave up on the interval $[0, 2]$ because the second derivative $f''(x) = 6x$ is greater than or equal to zero on $[0, 2]$. Using elementary algebra, it is also easy to see that the root of f is

$$z = \sqrt[3]{2} \approx 1.2599211. \tag{13.3}$$

In order to demonstrate the effectiveness of Newton's method, we approximate z by repeatedly applying equation (13.2): setting $x_0 := 2$ and using the fact that $f'(x) = 3x^2$, we obtain

$$x_1 = x_0 - \frac{f(x_0)}{f'(x_0)} = 1.5,$$
$$x_2 = x_1 - \frac{f(x_1)}{f'(x_1)} \approx 1.2962963,$$
$$x_3 = x_2 - \frac{f(x_2)}{f'(x_2)} \approx 1.2609322,$$
$$x_4 = x_3 - \frac{f(x_3)}{f'(x_3)} \approx 1.2599219.$$

In comparing x_4 with the value in (13.3), we see that after only four successive approximations the error is already less than 10^{-6}. Note: this error estimate for x_4 also follows from the observation that $f(x_4) > 0$ and $f(x_4 - 10^{-6}) < 0$.

13.2 Exercise. Use Newton's method to find an approximation for $\sqrt[5]{3}$ with a margin of error less than 10^{-4}.

13.3 Example. Let us consider the function

$$f(x) := x(x-2)^3 + \frac{3}{2}.$$

The graph of f is shown in Figure 13.2. By inspection, we notice that $f(1.6) > 0$ and $f(0.5) < 0$. In

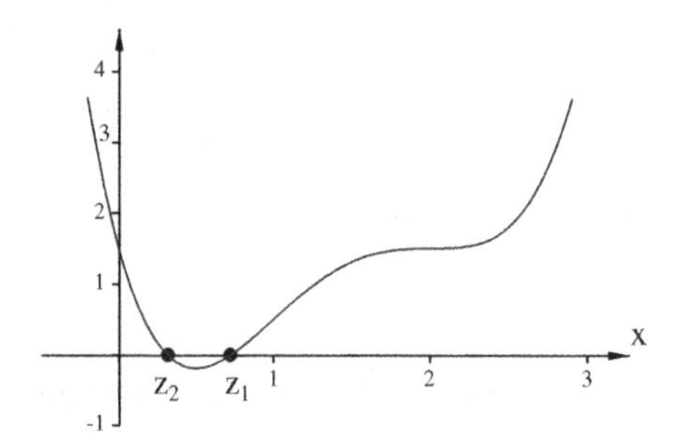

Figure 13.2: graph of $f(x) = x(x-2)^3 + 3/2$.

order to determine the root z_1 in the interval $[0.5, 1.6]$ we may be tempted to simply apply Newton's method with $x_0 := 1.6$ and $f'(x) = (x-2)^2(4x-2)$, but here is what happens:

$$x_1 = x_0 - \frac{f(x_0)}{f'(x_0)} \approx -0.385,$$
$$x_2 = x_1 - \frac{f(x_1)}{f'(x_1)} \approx -0.051,$$
$$x_3 = x_2 - \frac{f(x_2)}{f'(x_2)} \approx 0.158,$$
$$x_4 = x_3 - \frac{f(x_3)}{f'(x_3)} \approx 0.268.$$

Simplicio: It seems there is something wrong. According to Figure 13.2, the value of z_1 should be somewhere between 0.5 and 1, but none of the values above are even close to this range.

Teacher: You are right, something is wrong. To sort things out I would like to direct your attention to the fact that close to the point $x_0 = 1.6$ the graph of f in Figure 13.2 is concave down. By contrast, the function shown in Figure 13.1 is concave up over the entire interval from a to b. Can you see how this difference between the two graphs is relevant in regard to the applicability of Newton's method?

Sophie: I think so, because the fact that f is concave down close to x_0 means that the tangent line at the point $(x_0, f(x_0))$ is *above* the graph of f rather than below as in Figure 13.1. For this reason our first approximation is far out on the left at -0.385.

Teacher: Yes, and the value -0.385 is even to the left of z_2 which, in fact, causes the sequence of values generated by Newton's method to converge to z_2 instead of z_1.

Simplicio: So what are we going to do?

Teacher: We need to choose a different starting point x_0. Given the diagram in Figure 13.2, we notice that close to z_1 the function f appears to be concave up. To verify this claim we take the second derivative of f:

$$f''(x) = 12(x-2)(x-1).$$

Since $12(x-2)(x-1) > 0$ for all $x < 1$, it follows that f is concave up on the interval $(-\infty, 1)$. Therefore, in choosing for example $x_0 := 0.9$ (note that $f(0.9) > 0$), we can use Newton's method to generate a sequence of values rapidly converging to z_1:

$$x_1 = x_0 - \frac{f(x_0)}{f'(x_0)} = 0.743957,$$
$$x_2 = x_1 - \frac{f(x_1)}{f'(x_1)} = 0.727209,$$
$$x_3 = x_2 - \frac{f(x_2)}{f'(x_2)} = 0.726832.$$

With only three approximations, the error is less than 10^{-6}, because $f(x_3) > 0$ and $f(x_3 - 10^{-6}) < 0$.

Next we consider the case that f is concave down on an interval $[a, b]$ with $f(a) < 0$ and $f(b) > 0$. As illustrated in Figure 13.3, we can now generate a sequence of values x_n approaching z from the left

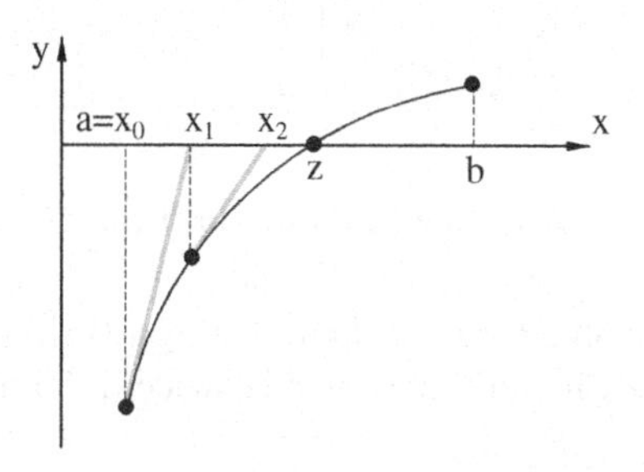

Figure 13.3: Newton's method in the case that f is concave down.

if we choose $x_0 := a$ as our starting point. The following theorem summarizes these observations and also provides rules for successful approximations in the case that $f(a) > 0$ and $f(b) < 0$.

13.4 Theorem. *Let $f : [a, b] \to \mathbb{R}$ be a differentiable function.*

a) *If $f(a) < 0 < f(b)$, then Newton's method will generate a sequence of values approaching a unique root (see the ⋆remark⋆ on p.108) of f in $[a, b]$ if one of the following conditions is satisfied:*

(i) f is concave up on $[a, b]$ and $x_0 := b$.

(ii) f is concave down on $[a, b]$ and $x_0 := a$.

b) *If $f(a) > 0 > f(b)$, then Newton's method will generate a sequence of values approaching a unique root of f in $[a, b]$ if one of the following conditions is satisfied:*

(i) f is concave up on $[a, b]$ and $x_0 := a$.

(ii) f is concave down on $[a, b]$ and $x_0 := b$.

Note: Theorem 13.4 is not a statement of equivalence, because it may very well happen that the sequence x_n converges to the desired root even if none of the conditions in the theorem are met.

13.5 Exercise. Find the approximate location of the roots of the function $f(x) := x^3 + 9x^2/2 - 12x + 5$ by using Newton's method. The error in the approximation should be less than 0.01. *Hint.* In order to make a good guess at the approximate positions of the roots, it may be helpful to determine first the local extrema of f.

★ *Remark.* Newton's method does not always generate a convergent sequence even if the function f under consideration has a root. For instance, the function $f : \mathbb{R} \to \mathbb{R}$ defined by the equation $f(x) := \sqrt[3]{x}$ has a root at $z = 0$, but there is no starting value $x_0 \neq 0$ for which the sequence of values x_n generated by Newton's method converges to a finite limit—in fact, the absolute values $|x_n|$ always diverge to ∞. To see why, we observe that $f'(x) = \sqrt[3]{x}^{-2}/3$ and

$$x_{n+1} = x_n - \frac{\sqrt[3]{x_n}}{\sqrt[3]{x_n}^{-2}/3} = -2x_n.$$

Given this equation it is easy to verify that $x_n = (-1)^n 2^n x_0$, and therefore, $\lim_{n\to\infty} |x_n| = \lim_{n\to\infty} 2^n |x_0| = \infty$. A graphical illustration of this result is given in Figure 13.4.

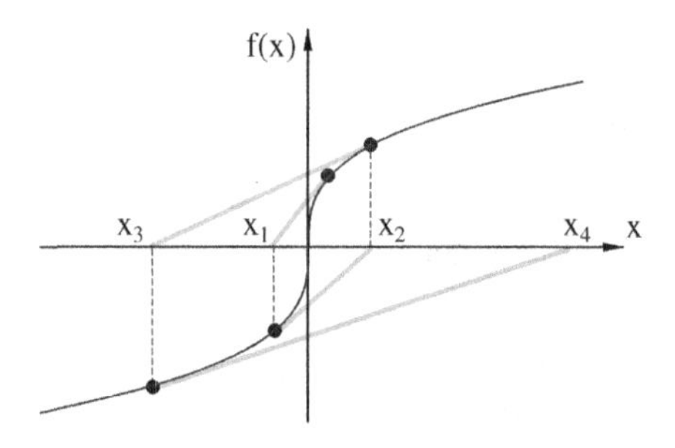

Figure 13.4: Newton's method for $f(x) = \sqrt[3]{x}$.

It can be shown (see Exercises 53.29 and 53.30 in Volume 2) that Newton's method always generates a convergent sequence of values x_n if the function f is twice differentiable and if for some $\lambda \in (0, 1)$ we have

$$|f(x)f''(x)| < \lambda f'(x)^2 \text{ for all } x \in \mathbb{R}.$$

The reader should verify that this condition is not satisfied for the function $f(x) = \sqrt[3]{x}$. ★

13.6 Example. Returning now to our introductory example, our objective is to determine an approximation for the minimal distance from the point $(2, 3)$ to the graph of the function $h(x) = x^4$. We already saw that in order to solve this problem we need to find the roots of the function $f(x) = 4x^7 - 12x^3 + x - 2$. Given the diagram in Figure 13.5, we readily observe that the minimal distance is assumed for values of x in the interval $[1, 2]$. (Note: in Figure 13.5, the scale on the y-axis is narrower than on the x-axis, and the minimal distance therefore appears to be assumed at a position slightly to the right of the actual one, but there can be no doubt that $[1, 2]$ is the right interval.) Since $f''(x) = 168x^5 - 72x = 24x(7x^4 - 3)$, it follows that $f''(x)$ is greater than zero whenever x is greater than $\sqrt[4]{3/7}$. In particular, f is concave up on the interval $[1, 2]$. Observing further that $f(2) = 416 > 0$ and $f(1) = -9 < 0$, we may be certain

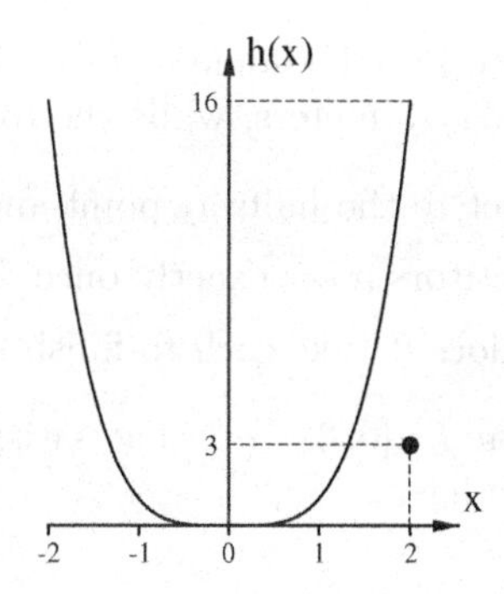

Figure 13.5: graph of h with the point $(2,3)$.

that Newton's method will generate a sequence of values converging to the unique root of f in $[1,2]$ if we set $x_0 := 2$ (see Theorem 13.4a, case (ii)). The first seven approximations are as follows:

$$x_1 = x_0 - \frac{f(x_0)}{f'(x_0)} \approx 1.747725894,$$

$$x_2 = x_1 - \frac{f(x_1)}{f'(x_1)} \approx 1.551907657,$$

$$x_3 = x_2 - \frac{f(x_2)}{f'(x_2)} \approx 1.416305181,$$

$$x_4 = x_3 - \frac{f(x_3)}{f'(x_3)} \approx 1.344915739,$$

$$x_5 = x_4 - \frac{f(x_4)}{f'(x_4)} \approx 1.325333675,$$

$$x_6 = x_5 - \frac{f(x_5)}{f'(x_5)} \approx 1.323994505,$$

$$x_7 = x_6 - \frac{f(x_6)}{f'(x_6)} \approx 1.323988538.$$

The error in the estimate x_7 is less than 10^{-9} because $f(x_7) \approx 3 \cdot 10^{-12} > 0$ and $f(x_7 - 10^{-9}) \approx -5.26 \cdot 10^{-8} < 0$. Since the function D (see Chapter 12) is, in our case, given by the equation $D(x) = (x-2)^2 + (x^4-3)^2$, it follows that the minimal distance from the point $(2,3)$ to the graph of h is

$$d_{min} \approx \sqrt{D(x_7)} = \sqrt{(x_7-2)^2 + (x_7^4-3)^2} \approx 0.67992.$$

13.7 Exercise. Find an approximation for the minimal distance from the graph of the function $h(x) := x^3 - 1$ to the point $(4,2)$. The margin of error in the x-coordinate of the point on the graph of h where the minimum is assumed should be less than 10^{-5}.

Additional Exercises

13.8. Explain why the equation $x^2 = 1 - x^6$ has exactly two solutions and approximate the solutions with a margin of error less than 0.01.

13.9. Find approximations for the root(s) of the function $f(x) := x^3 + x - 4$ with a margin of error less than 10^{-5}.

13.10. A hare and a tortoise compete in a 1 kilometer race. Remarkable as this may sound, after t minutes the hare has run $500(2\sqrt{t} + \sqrt[3]{t})/3$ meters, while the tortoise has run $100t + 250\sqrt{t}$ meters.

a) How long does it take each to get to the halfway point and who arrives at this point first?

b) After the race starts, the competitors meet exactly once. When and where does this happen?

c) Who wins the race? How long does it take each to finish the race?

13.11. Sketch the graph of a function $f : [0,3] \to \mathbb{R}$ that satisfies the following conditions: $f(0) = 1$, $f(3) = -1$, and the equation $f(x) = 0$ has...

a) exactly one solution on $[0,3]$.

b) exactly two solutions on $[0,3]$.

c) no solution on $[0,3]$.

13.12. Suppose that $f : \mathbb{R} \to \mathbb{R}$ is a function that satisfies the following conditions: $f'(x), f''(x) > 0$ for all $x \in \mathbb{R}$, and the equation $f(x) = 0$ has exactly one solution $z \in \mathbb{R}$. If x_1 is an estimate for z generated by Newton's method from a given starting value $x_0 < z$, then which of the following inequalities must be satisfied?

a) $x_1 > z$

b) $x_1 < z$

13.13. Explore what happens in applying Newton's method to the function $f(x) := x^2 + 1$ for the starting values $x_0 = 1$, 2, and 10.

13.14. Use Newton's method to find an approximation for the square root of 2 with a margin of error less than 10^{-10}.

13.15. Use Newton's method to determine the point of intersection of the curves described by the equations $y = x^2 + 1$ and $y = x^3$. The margin of error in the x-coordinate should be less that 10^{-5}.

PART II

INTEGRATION

Chapter 14

Introduction to Integration

Antiderivatives and the Law of Falling Bodies Revisited

Teacher: In the Introduction, we discussed the problem of finding the distance an object travels given its velocity as a function of time. In today's lesson we will explore this subject in more detail and generalize our results to a larger context. To begin with, we will take a closer look at Galileo's law of falling bodies. Initially, we formulated this law as a rule of proportionality relating the distance an object falls in successive time intervals to the odd integers (see p.2). Choosing appropriate units of measurement, we proceeded to derive the functional relation $s(t) = t^2$, where $s(t)$ denoted the distance an object had fallen after t units of time. It is surprising, though, that we never made any reference at all to the properties of the falling body—neither its mass nor its shape nor any other quality were mentioned. This suggests that the law of falling bodies is universally valid for all types of objects. Unfortunately, our everyday experience does not seem to support such a far-reaching claim—a falling feather obviously does not descend as speedily as a rock.

Sophie: I don't see that there is necessarily a contradiction. If we say that the distance an object falls in successive time intervals is proportional to the odd integers, then the difference in velocity between a feather and a rock could be due to a difference in the constant of proportionality. Let's say, for instance, that we have chosen our units of measurement in such a way that, as explained in the Introduction, the rock drops one unit of distance in the first unit of time, then three units of distance in the second unit of time, five units of distance in the third unit of time, and so forth. Then, using the same units of measurement, the proportionality of successive distances to the odd integers in the case of a falling feather would be guaranteed if there were a constant of proportionality c such that the feather drops c units of distance in the first unit of time, $3c$ units of distance in the second unit of time, $5c$ units of distance in the third unit of time, and so forth. So both the rock and the feather would obey the same fundamental law, but the velocity of the feather would be small compared to the velocity of the rock if the value of c were small compared to 1.

Teacher: This is a very keen observation, and from a logical point of view there is no mistake in your argument, but nature just doesn't work that way.

Simplicio: So the law of falling bodies, as we formulated it, applies only to a certain class of objects?

Teacher: No, it applies to all objects in exactly the same way. In particular, there really is no difference in velocity between a falling rock and a falling feather.

Simplicio: I am certainly always glad to learn something new, but aren't we going a bit too far this time? If I took a feather and a rock and dropped them right here in front of me, the rock would reach the ground long before the feather. There is absolutely no doubt about it. So your assertion is blatantly contradicting experience and common sense.

Teacher: I am glad to hear you referring to "common sense," because it is common sense that is most profitably employed in support of my assertion that all objects fall at the same rate.* Let me try

*In our exposition here, we follow closely Galileo's original argument in his *Dialogues Concerning Two New Sciences*.

to make my case: am I correct in assuming that the observed difference in the velocity of a falling feather and a falling rock is, in your view, a specific instance of a more general law which says that the velocity of a falling body increases with its mass—heavy objects fall faster than lighter ones?

Simplicio: Basically that is what I say, but I think that the shape and density of an object also play a role—not only its mass. (Note: density = mass/volume)

Teacher: Okay, but you are convinced that of two falling bodies with exactly the same shape and density, the velocity of the lighter one would be less than the velocity of the heavier one.

Simplicio: Certainly.

Teacher: So it would also follow that two objects with equal density and equal shape fall at the same rate if their masses are equal as well.

Simplicio: No doubt about it.

Teacher: Good. Now imagine that I take a cube (made of some homogeneous material) and cut it into eight smaller cubes of equal size as shown in Figure 14.1. All of the smaller cubes fall with the same

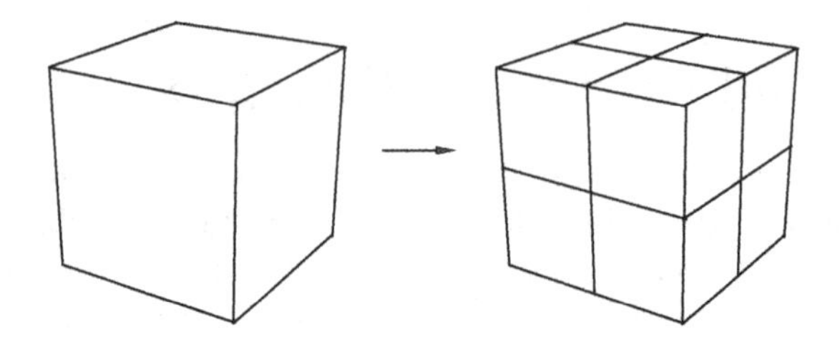

Figure 14.1: cutting up a cube.

velocity, because they have the same shape, density, and mass. Furthermore, the velocity of each of the smaller cubes is, according to your view, less than the velocity of the initial undivided cube, because the undivided cube has the same shape and density as each of the smaller cubes but an eight times larger mass. (The fact that the smaller cubes have the same shape and density as the larger cube is important, because it allows us to isolate the mass as the only distinguishing quality.)

Simplicio: I cannot argue with anything you said. Your conclusion seems entirely reasonable.

Teacher: Well, I am not so sure. What do you think happens, for example, if I loosely arrange the eight cubes to form the original larger cube? Will this ensemble of eight cubes fall with the velocity of the original larger cube or with the velocity of the smaller cubes?

Simplicio: I am beginning to see where the problem is with my theory. If the eight cubes are only loosely assembled, then the velocity of the ensemble ought to be equal to the velocity of each of its parts. After all, none of the cubes in the ensemble "knows" that there are other cubes close by. But even if we glued them back together, it is not at all clear why the resulting composite cube would want to fall at a faster rate than any of the eight smaller cubes. Where would the additional acceleration come from?

Teacher: You are absolutely right. There is no force that can provide any additional acceleration for the reassembled cube, and we are thus led to conclude that this larger cube falls at the same rate as each of the smaller ones.

Simplicio: I must admit that this conclusion appears inevitable, but what about the feather and the rock? Even if reason suggests the velocity of a falling body to be independent of its mass, there is still the undeniable experimental fact that a feather does not fall with the same velocity as a rock.

Teacher: Yes, but what causes the difference in velocity between a falling feather and a falling rock? If it's not the difference in mass, then what is it?

Sophie: I think it must be the difference in shape: due to its delicate design a feather is subject to a greater force of air resistance relative to its mass.

Teacher: You are right, that exactly is the problem. Consequently, we may expect a feather to fall with the same velocity as a rock as soon as the force of air resistance is eliminated. Indeed, experiments conducted in a vacuum confirm that in absence of air resistance all falling bodies accelerate at the same rate. Furthermore, not only is the acceleration the same for all objects, but close to the surface of the earth it is also independent of time and place (at least in very good approximation). Measurements (using meters and seconds as units) show that close to the surface of the earth the gravitational acceleration of a falling body (neglecting all other external forces such as air resistance) has the constant value

$$g \approx 9.81\, m/s^2.$$

Interestingly, the time-independence of the gravitational acceleration close to the surface of the earth allows us to derive the law of falling bodies in its original formulation.

Sophie: Why is that?

Teacher: It is not possible to explain this in a single sentence, but to get started, I would like you to recall that in general the acceleration $a(t)$ of a moving object is equal to the derivative $v'(t)$ of the velocity function (see Chapter 4, p.34). Thus, for a falling body close to the surface of the earth, we have

$$g = v'(t). \tag{14.1}$$

In trying to determine v from this equation, we are faced with the problem of *finding a function from its derivative* because we are only given the information that v has the constant derivative g. Motivated by this observation we introduce the following definition:

14.1 Definition. If $f : I \to \mathbb{R}$ is a function defined on an open interval I, then we say that a differentiable function $F : I \to \mathbb{R}$ is an *antiderivative* of f if $F' = f$, i.e.,

$$F'(x) = f(x) \text{ for all } x \in I.$$

In the language of Definition 14.1, we can say that in order to find v from equation (14.1), we need to find an antiderivative of the constant function g. Since

$$\frac{d}{dt}t = 1 \quad \text{(see Chapter 10)},$$

it follows that

$$\frac{d}{dt}gt = g. \tag{14.2}$$

Thus, it seems the problem of finding v is already solved, but we need to be a bit careful because for any constant $C \in \mathbb{R}$ we also have

$$\frac{d}{dt}(gt + C) = g.$$

This shows that there is not only one function v satisfying equation (14.1), but there are actually infinitely many such functions, corresponding to the infinitely many different values we can choose for C. Is it possible that there are even more antiderivatives of the constant function g other than those of the form $gt + C$? Fortunately, the answer is "no," as the following theorem shows:

14.2 Theorem. *If $F : I \to \mathbb{R}$ and $G : I \to \mathbb{R}$ are differentiable functions defined on an open interval I such that $F' = G'$, then there is a constant $C \in \mathbb{R}$ such that*

$$G(x) = F(x) + C \text{ for all } x \in I.$$

In other words, two functions with equal derivatives can differ only by a constant.

In essence, this theorem simply says the following:

> If F is an antiderivative of f, then any other antiderivative of f is equal to $F + C$ for some constant $C \in \mathbb{R}$.

To provide some intuitive justification, let us assume that we are given two functions F and G defined on an open interval I such that $F'(x) = G'(x)$ for all $x \in I$. Since this assumption implies that $(F - G)' = F' - G' = 0$, we may expect the function $F - G$ to be constant because its tangent slope is zero everywhere (see also the remark on p.64). Thus, there ought to be a constant $C \in \mathbb{R}$ such that $F - G = C$ or, equivalently, $F = G + C$ as desired. Alternatively, we may also argue as follows: according to the geometric interpretation of the derivative, the equation $F' = G'$ allows us to infer that the graphs of F and G have the same slope at all points $x \in I$ and therefore the same shape. In other words, the graphs must be parallel and the distance between them must be the same everywhere (see Figure 14.2). So again we arrive at the conclusion that $F - G = C$ or, equivalently, $F = G + C$ for some $C \in \mathbb{R}$. (A rigorous proof of Theorem 14.2 will be given in Chapter 40.)

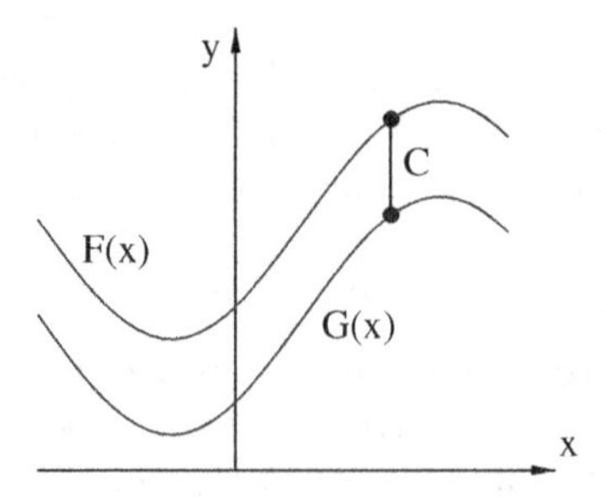

Figure 14.2: parallel graphs with equal slope.

The statement of Theorem 14.2, in conjunction with (14.1) and (14.2), allows us to infer that the velocity of a falling body must satisfy an equation of the form

$$v(t) = gt + C \text{ for some } C \in \mathbb{R}.$$

If we assume that the body is dropped at time $t = 0$, then the initial velocity is zero, i.e., $v(0) = 0$. This assumption allows us to determine the value of C, because

$$C = g \cdot 0 + C = v(0) = 0.$$

Hence

$$\boxed{v(t) = gt.} \tag{14.3}$$

Our next objective is to utilize this result for the purpose of finding a formula for the distance $s(t)$ that an object has fallen after a certain amount of time t. Since $s'(t) = v(t) = gt$ (see Chapter 4, p.34), the problem of finding $s(t)$ is equivalent to the problem of finding an antiderivative of the function gt. Using the rules for the differentiation of power functions, as discussed in Chapter 10, it is easy to verify that the derivative of $gt^2/2$ is gt. Thus, according to Theorem 14.2, there is a constant $D \in \mathbb{R}$ such that

$$s(t) = \frac{gt^2}{2} + D.$$

Since we assumed $t = 0$ to be the moment when the object is dropped, it follows that $0 = s(0) = D$, and therefore

$$\boxed{s(t) = \frac{gt^2}{2}.} \tag{14.4}$$

The appearance of the factor $g/2$ is due to the choice of meters and seconds as units of measurement. The simpler equation $s(t) = t^2$ that we derived in the Introduction results from an appropriate change in the units of measurement (see also Exercise 0.6 for a similar discussion).

To come full circle, we will now demonstrate that the law of falling bodies in Galileo's original formulation is a consequence of equation (14.4). In the first unit of time, that is, in the first second, a falling body covers the distance

$$s(1) - s(0) = \frac{g}{2}(1^2 - 0^2) = \frac{g}{2} \cdot 1.$$

In the second second the distance is

$$s(2) - s(1) = \frac{g}{2}(2^2 - 1^2) = \frac{g}{2} \cdot 3$$

and in the third second it is

$$s(3) - s(2) = \frac{g}{2}(3^2 - 2^2) = \frac{g}{2} \cdot 5.$$

Thus, it appears the distance covered in successive time intervals is indeed proportional to the odd integers 1, 3, 5.... To give a more general proof of this fact, we simply determine the distance covered by the falling body during the nth unit of time:

$$s(n) - s(n-1) = \frac{g}{2}(n^2 - (n-1)^2) = \frac{g}{2}(2n - 1).$$

Noting that $2n - 1$ is the nth odd integer, the proof is complete.

To conclude this section we wish to emphasize the following important observation: in deriving equation (14.4) from equation (14.3), we solved the problem of finding distance (or position) from velocity in a way that, on the surface, appears to be entirely disconnected from the approximation method discussed in the Introduction. To understand that these two approaches are actually closely related will be the main objective in this and the following chapters. Our discussion will culminate in the formulation of the fundamental theorem of calculus in Chapter 19. To get started with this process, we need to introduce some standard notation for sums.

Summation Notation

Summation notation consists of three elements: the Greek letter $\sum$ (capital sigma), the summation index, and the summand. It is best explained with an example:

$$\sum_{k=1}^{5}(2k+1) \text{ is read as "the sum as } k \text{ ranges from 1 to 5 of the quantity } 2k+1."$$

The sigma indicates that we are taking a sum, the index k enumerates the terms in the sum, and the value of each term is a function of k. In the example above, there are five terms and the values of these terms are obtained by replacing k successively with the numbers from 1 to 5:

$$\sum_{k=1}^{5}(2k+1) = (2 \cdot 1 + 1) + (2 \cdot 2 + 1) + (2 \cdot 3 + 1) + (2 \cdot 4 + 1) + (2 \cdot 5 + 1) = 3 + 5 + 7 + 9 + 11 = 35.$$

The summation index can actually run over arbitrary sets, but in this text it will be restricted to integers. Furthermore, the summation index is not always denoted by k as the following example illustrates:

$$\sum_{m=3}^{6}(m^2 - m) = (3^2 - 3) + (4^2 - 4) + (5^2 - 5) + (6^2 - 6) = 6 + 12 + 20 + 30 = 68.$$

14.3 Exercise. Evaluate the following sums: $\sum_{k=-2}^{3} \frac{1}{k+3}$, $\sum_{k=2}^{7} 2^k$, $\sum_{i=2}^{7} 2^k$, $\sum_{n=1}^{5} 2n$, and $\sum_{n=-1}^{4} 3$.

Finding Distance from Velocity via Approximating Sums

Let us consider an example: using the equation $v(t) = gt$ (see (14.3)) and the approximate value $g \approx 9.8\,m/s^2$, we wish to estimate the distance that an object falls within the first two seconds after it was dropped. As in the Introduction, we begin by constructing tables of lower and upper bounds for the distances that the object falls in successive intervals of time. For 10 subintervals of length 0.2 seconds we obtain:

time interval	min. velocity in m/s	max. velocity in m/s	lower bound for the distance in m	upper bound for the distance in m
$[0.0, 0.2]$	0.00	1.96	0.000	0.392
$[0.2, 0.4]$	1.96	3.92	0.392	0.784
$[0.4, 0.6]$	3.92	5.88	0.784	1.176
$[0.6, 0.8]$	5.88	7.84	1.176	1.568
$[0.8, 1.0]$	7.84	9.80	1.568	1.960
$[1.0, 1.2]$	9.80	11.76	1.960	2.352
$[1.2, 1.4]$	11.76	13.72	2.352	2.744
$[1.4, 1.6]$	13.72	15.68	2.744	3.136
$[1.6, 1.8]$	15.68	17.64	3.136	3.528
$[1.8, 2.0]$	17.64	19.60	3.528	3.920
total			17.64	21.56

To see how this table was generated, we examine the second row for the interval $[0.2, 0.4]$. The minimal velocity here is $v(0.2) = g \cdot 0.2 \approx 9.8 \cdot 0.2 = 1.96$ and the maximal velocity is $v(0.4) \approx 9.8 \cdot 0.4 = 3.92$. Consequently, the lower bound for the distance fallen is $v(0.2) \cdot (0.4 - 0.2) \approx 1.96 \cdot 0.2 = 0.392$ and the upper bound is $v(0.4) \cdot (0.4 - 0.2) \approx 0.784$.

Now we repeat the process with 40 intervals of length 0.05 seconds:

time interval	min. velocity in m/s	max. velocity in m/s	lower bound for the distance in m	upper bound for the distance in m
$[0.00, 0.05]$	0.00	1.96	0.000	0.392
$[0.05, 0.10]$	1.96	3.92	0.392	0.784
$[0.10, 0.15]$	3.92	5.88	0.784	1.176
$\vdots$	$\vdots$	$\vdots$	$\vdots$	$\vdots$
$[1.95, 2.00]$	19.11	19.60	0.9555	0.9800
total			19.11	20.09

After all this work, we still have a difference between upper and lower bound of almost one meter. To improve this result, it is helpful to generalize our approach by dividing the interval $[0, 2]$ into n subintervals of equal length $2/n$. Then the *lower bound* for the distance is

$$\left(0 + 9.8\frac{2}{n} + 9.8\frac{2 \cdot 2}{n} + 9.8\frac{2 \cdot 3}{n} + \cdots + 9.8\frac{2(n-1)}{n}\right)\frac{2}{n} = \sum_{k=0}^{n-1} 9.8\frac{2k}{n} \cdot \frac{2}{n},$$

and the *upper bound* is

$$\left(9.8\frac{2}{n} + 9.8\frac{2 \cdot 2}{n} + 9.8\frac{2 \cdot 3}{n} + \cdots + 9.8\frac{2(n-1)}{n} + 9.8\frac{2n}{n}\right)\frac{2}{n} = \sum_{k=1}^{n} 9.8\frac{2k}{n} \cdot \frac{2}{n}.$$

Using, for example, $n = 200$, we find the lower bound

$$\sum_{k=0}^{199} 9.8\frac{2k}{n} \cdot \frac{2}{n} = 19.502$$

and the corresponding upper bound is

$$\sum_{k=1}^{200} 9.8\frac{2k}{n} \cdot \frac{2}{n} = 19.698.$$

14.4 Exercise. Convince yourself that the formulae for lower and upper bounds above are correct by applying them to the case $n = 10$ and comparing your answers with the results from the first table.

In order to determine the accuracy of our estimates, we will take the difference of the lower and upper bounds, for the error in our approximation is clearly less than or equal to this difference. Comparing the two sums above, we see that they have common terms with the exception of the first term in the sum for the lower bound and the last term in the sum for the upper bound. Consequently, in subtracting the lower bound from the upper bound, all the common terms drop out and the difference is readily seen to be

$$\left(9.8\frac{2n}{n} - 0\right)\frac{2}{n} = \frac{39.2}{n}.$$

To illustrate the usefulness of this formula, let us assume that we wish to find an estimate for the distance an object falls within the first two seconds with a margin of error less than or equal to $0.01\,m$. Since the error is in general known to be less than or equal to $39.2/n$, we may solve for n the equation $39.2/n = 0.01$ to infer that an estimate with $n = 3920$ subdivisions will deviate from the actual value of the distance fallen by no more that $0.01\,m$. As we continue to choose successively larger values for n, the error in our approximation approaches zero, because

$$\lim_{n\to\infty} \frac{39.2}{n} = 0.$$

14.5 Exercise. Using approximation by summation, how many terms would be needed in the example above to ensure that our solution is within $0.002\,m$ of the actual distance?

At this point of our discussion we have established that the difference between the lower and upper estimates converges to zero as n tends to ∞, but we still have not found the exact distance an object falls during the first two seconds. With our particular problem, though, we can do better even without recourse to equation (14.4). To see this, we observe that each term of the form $9.8(2k/n)(2/n)$ in the approximating sums above can be interpreted as *the area of a rectangle* of height $9.8 \cdot 2k/n \approx g \cdot 2k/n = v(2k/n)$ and width $2/n$. As n increases, the total area of the rectangles represented by these terms approaches the area under the graph of the function $v(t) = gt \approx 9.8t$ over the interval $[0, 2]$ (see also our discussion in the Introduction and in particular Figure 5). Since the graph of v is a straight line, the area under the graph is simply the area of a right triangle with base 2 and height $g \cdot 2 \approx 9.8 \cdot 2 = 19.6$ (see Figure 14.3). Thus the area, or equivalently the total distance an object falls in the first two seconds, must be equal to $2 \cdot 19.6/2 = 19.6\,m$.

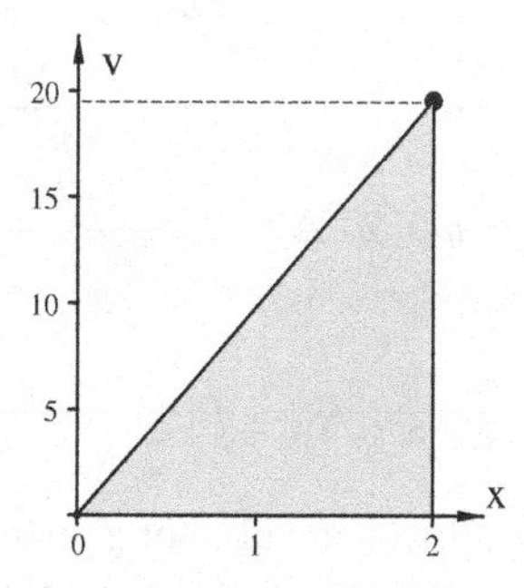

Figure 14.3: area under the graph of v.

14.6 Exercise. Find the distance that an object falls in the 4th second after it has been dropped.

As a matter of course, the method of approximation described above does not depend on the particular choice of the interval $[0, 2]$. Choosing an interval of the form $[0, t]$, the approximating sums give us estimates for the area under the graph of v over the interval $[0, t]$. Again, we end up approximating the area of a triangle, except that the base of the triangle is t instead of 2 and the height is $gt \approx 9.8t$ instead of $g \cdot 2 \approx 9.8 \cdot 2 = 19.6$. This yields

$$s(t) = \frac{gt}{2} \cdot t = \frac{gt^2}{2},$$

which is the same result as in (14.4). It is remarkable that we were thus able to derive (14.4) from (14.3) by determining (and first approximating) the area under the graph of v without any reference to the concept of antiderivatives, which featured so prominently in our initial approach.

Areas under Graphs

Motivated by our discussion in the previous section, we will now explore in general the problem of approximating areas under graphs of functions. So let us assume that $f : [a, b] \to \mathbb{R}$ is a positive continuous function (i.e., $f(x) \geq 0$ for all $x \in \mathbb{R}$). To find the area under the graph of f over the interval $[a, b]$, we use approximating rectangles as illustrated in Figure 14.4. Dividing the interval $[a, b]$

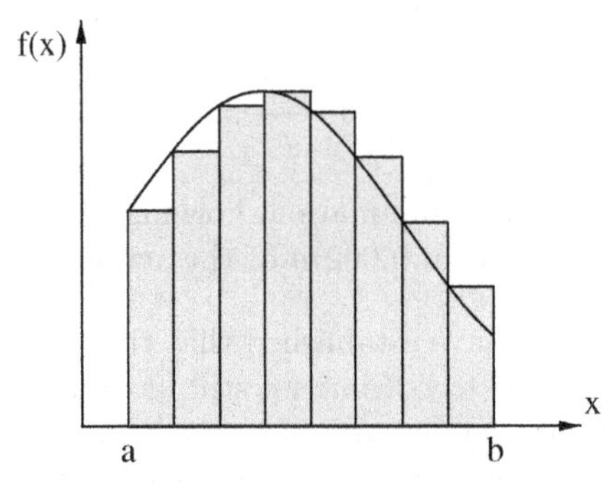

Figure 14.4: approximation of the area under the graph of f with a left sum.

into n subintervals of equal length

$$\Delta x := \frac{b-a}{n}$$

and denoting the endpoints of the subintervals by $x_0, x_1, \ldots, x_{n-1}, x_n$, it follows that

$$\begin{aligned}
x_0 &= a + 0 \cdot \Delta x = a, \\
x_1 &= a + 1 \cdot \Delta x = a + \frac{b-a}{n}, \\
x_2 &= a + 2 \cdot \Delta x = a + \frac{2(b-a)}{n}, \\
&\vdots \\
x_n &= a + n\Delta x = b.
\end{aligned}$$

Since the height of each rectangle is equal to the value of f at the left endpoint of each subinterval, we may infer that the sum of the areas of the approximating rectangles is

$$f(x_0)\Delta x + f(x_1)\Delta x + \cdots + f(x_{n-1})\Delta x = \sum_{k=0}^{n-1} f(x_k)\Delta x = \sum_{k=0}^{n-1} f\left(a + \frac{k(b-a)}{n}\right)\frac{b-a}{n}.$$

With reference to the positioning of the rectangles under the graph, this sum is referred to as the *left sum* or *left estimate*. Similarly, if the heights of the rectangles are determined by evaluating f at the right endpoint of each subinterval (see Figure 14.5), then the corresponding approximation of the area

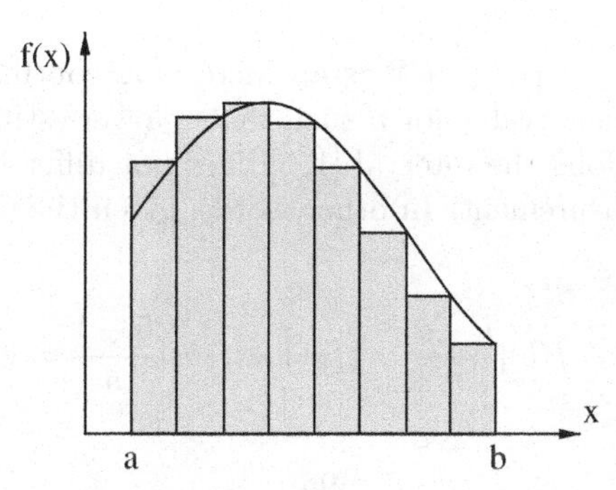

Figure 14.5: approximation of the area with a right sum.

under the graph is said to be *right sum* or *right estimate*, and its value is

$$f(x_1)\Delta x + f(x_2)\Delta x + \cdots + f(x_n)\Delta x = \sum_{k=1}^{n} f(x_k)\Delta x = \sum_{k=1}^{n} f\left(a + \frac{k(b-a)}{n}\right)\frac{b-a}{n}.$$

More generally, we can divide the interval $[a, b]$ into subintervals of differing lengths and compute the heights of the approximating rectangles by evaluating f at an arbitrary point in each of these subintervals. This approach leads us to the notion of *Riemann sums* and will be formalized in Chapter 17. If we allow f to also have negative values, then our estimates will typically also involve negative terms, and an appropriate interpretation of this case will be discussed in Chapter 17 as well.

As illustrated in Figure 14.6, the left sums of an increasing function f represent lower estimates for the area under the graph and the right sums upper estimates. Conversely, if f is decreasing, the left

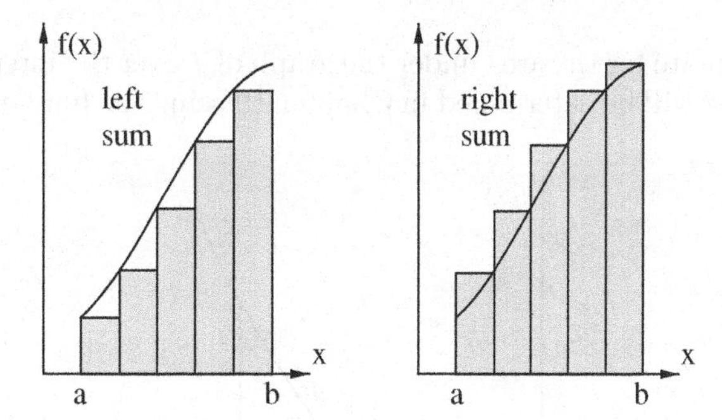

Figure 14.6: lower and upper estimates in the case that f is increasing.

sums represent upper estimates and the right sums lower estimates. In either case, the error E in our approximation can be estimated by taking the absolute value of the difference between left and right sums. So *if f is either increasing on $[a, b]$ or decreasing on $[a, b]$* then

$$E \leq \left|\sum_{k=0}^{n-1} f(x_k)\Delta x - \sum_{k=1}^{n} f(x_k)\Delta x\right| = |f(x_0)\Delta x - f(x_n)\Delta x| = |f(a) - f(b)|\Delta x,$$

or equivalently,

$$\boxed{E \leq |f(a) - f(b)|\frac{b-a}{n}.} \tag{14.5}$$

If the function f is increasing on some parts of $[a, b]$ and decreasing on others, then we cannot be sure whether a left or right sum gives us a lower or upper estimate and, in particular, the error estimate (14.5) is in this case not necessarily valid (see also Exercises 14.17 and 14.20).

14.7 Exercise. Sketch two diagrams analogous to those in Figure 14.6 for a decreasing function f, and label the diagrams as left or right sums and lower or upper estimates.

14.8 Example. Assume that $f : [1,5] \to \mathbb{R}$ is an increasing continuous function with $f(1) = 0$ and $f(5) = 24$. We wish to determine a value for n such that a lower estimate for the area under the graph of f based on n subdivisions along the interval $[1,5]$ does not differ from the actual value for the area by more than 0.04 units of measurement. In other words, given the estimate in (14.5), we require n to satisfy the inequality

$$0.04 \geq |f(a) - f(b)|\frac{b-a}{n} = |f(1) - f(5)|\frac{5-1}{n} = |0 - 24|\frac{4}{n} = \frac{96}{n},$$

or equivalently,

$$n \geq \frac{96}{0.04} = 2400.$$

Consequently, the left sum

$$\sum_{k=0}^{2399} f\left(1 + \frac{4k}{2400}\right)\frac{4}{2400}$$

gives us a lower estimate with a margin of error less than 0.04 units of measurement, as desired. To evaluate this sum requires, of course, that we are given a defining equation for f. If, for instance, we set

$$f(x) := x^2 - 1,$$

then f is increasing on $[1,5]$, $f(1) = 0$ and $f(5) = 24$. Using a calculator, we obtain

$$\sum_{k=0}^{2399} f\left(1 + \frac{4k}{2400}\right)\frac{4}{2400} = \sum_{k=0}^{2399}\left(\left(1 + \frac{4k}{2400}\right)^2 - 1\right)\frac{4}{2400} \approx 37.313.$$

This value gives us an estimate for the area under the graph of f over the interval $[1,5]$ (see Figure 14.7). The exact value of this area will be determined in Chapter 19 using the fundamental theorem of calculus.

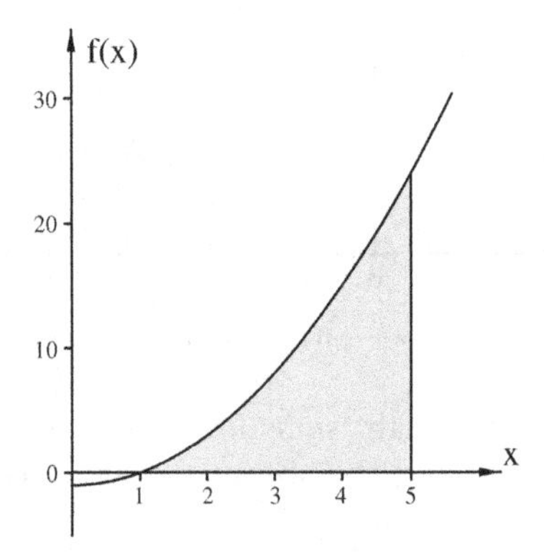

Figure 14.7: area under the graph of $f(x) = x^2 - 1$.

14.9 Exercise. Assume that $f : [1,4] \to \mathbb{R}$ is a decreasing continuous function with $f(1) = 5$ and $f(4) = 1/2$. Use the error estimate (14.5) to determine the least value for n that would guarantee a lower estimate for the area under the graph of f to differ from the actual area by no more than 0.002.

To conclude our discussion in this chapter we wish to point out that the assumption of continuity, made at the beginning of this section, is sufficient to ensure that the lower and upper estimates for the area under a graph always converge to the same limit as n tends to ∞. However, it is possible (and for practical purposes, desirable) to replace this assumption with a slightly weaker requirement, known as *piecewise continuity*, and this will be done in Chapter 17.

Additional Exercises

14.10. An object is moving along a straight line such that the velocity at time t (measured in seconds) is $v(t) = 5t^2\, m/s$. Estimate how far the object has traveled after 3 seconds using...

a) 6 subintervals (of equal length) and a left sum.

b) 12 subintervals and both a left sum and a right sum.

Furthermore,

c) give a general formula using n subintervals for both the left and the right sums,

d) find a formula for the difference between the two sums in c), and

e) use d) or the error formula in (14.5) to determine how many subintervals are needed to get an estimate that is within $1\, m$ of the actual distance.

14.11. Under the same assumptions as in Exercise 14.10, find the exact distance that the object travels during a time span from 0 to t using the method of antiderivatives.

14.12. Find an approximation for the area of the region bounded by the x-axis, the curve given by the equation $y = x^3$, and the line $x = 3$ using...

a) 6 subintervals and a left sum.

b) 12 subintervals and both a left sum and a right sum.

Furthermore,

c) give a general formula using n subintervals for both the left and right sums,

d) find a formula for the difference between the two sums in c), and

e) use the error formula in (14.5) to determine how many subintervals are needed to get an estimate that is within 1 unit of measurement of the actual area.

14.13. The *kinetic energy* of an object of mass m, moving at velocity v, is defined by the equation

$$E_{kin} := \frac{mv^2}{2}$$

and is measured in *Joule* (where $1\, J = 1\, kg\, m^2/s^2$). Assume that a rod of mass $0.6\, kg$ and length $1.2\, m$ rotates about one of its ends once every two seconds (see Figure 14.8). Find an approximation for the rod's kinetic energy by cutting it into six sections and taking the velocity at the endpoint of each section (i.e., the point farthest from the center of rotation). Note: the rod's mass is supposed to be uniformly distributed over its volume.

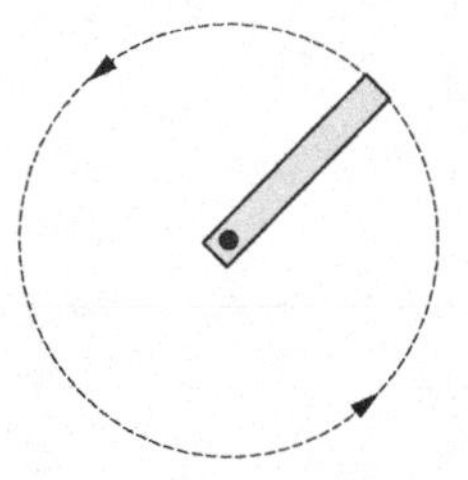

Figure 14.8: a rotating rod.

14.14. Assume that a cylindrical wheel of radius $R = 0.3\, m$ and mass $m = 2.4\, kg$ is spinning at a rate of one full revolution per second. Use the formula given in Exercise 14.13, to find an approximation for the wheel's kinetic energy by subdividing the radius into 6 subintervals and estimating the kinetic energy of each of the corresponding ring-shaped slices. Note: the wheel's mass is supposed to be uniformly distributed over its volume.

14.15. The *moment of inertia* of a mass segment of mass m that is rotating at a distance r about a given axis is defined to be mr^2. Using again six subintervals, you are to find an upper estimate for the moment of inertia of the rotating rod described in Exercise 14.13.

14.16. Use left or right sums to find a lower estimate for the area under the curve given by the equation $y = x^2$ over the interval $[1, 5]$ such that the error in the approximation is less than 0.5.

14.17. Sketch the graph of the function $f(x) := x(4 - x)$.

a) Explain why you cannot be sure that either a right sum or a left sum will give an upper estimate for the area under the graph of f over the interval $[0, 4]$.

b) Devise a method to obtain an upper estimate using a combination of left and right sums such that the error in the approximation is less than 0.1.

14.18. ★ Find an approximation for the length of the graph of the function $f(x) := 2x^2 - x^3$ over the interval $[0, 2]$ (on the x-axis). You may assume that the units on both axes are measured in inches. *Hint.* Divide the interval $[0, 2]$ into n subintervals $[x_{k-1}, x_k]$ of equal length $\Delta x = 2/n$. Then estimate the length of the curve segments corresponding to these subintervals by determining the distances between the points $(x_{k-1}, f(x_{k-1}))$ and $(x_k, f(x_k))$ on the graph. Finally, take the sum of these distances (for a large value of n) to find an approximation for the length of the graph of f.

14.19. ★ Following the same approach as in Exercise 14.18, you are to estimate the circumference of the circle of radius 1 given by the equation $x^2 + y^2 = 1$.

14.20. Give an example of a function $f : [a, b] \to \mathbb{R}$ for which the error estimate in (14.5) is not valid.

Chapter 15

Galileo Galilei and the Copernican Revolution

> The activity of God is immortality, i.e., eternal life. Therefore the movement of that which is divine must be eternal. But such is the heaven,... and for that reason to it is given the circular body whose nature it is to move always in a circle.[1] (Aristotle)

> The observed facts about earth are not only that it remains at the center, but also that it moves to the center. The place to which any fragment of earth moves must necessarily be the place to which the whole moves; and in the place to which a thing naturally moves it will naturally rest.[2] (Aristotle)

In his treatise *On the Heavens*, the Greek philosopher Aristotle (384-322 B.C.) envisioned a universe in which the earth was at rest at the center of a heavenly sphere that carried the fixed stars on their daily paths across the sky. At intermediate distances from the stars he placed the moon, the sun, and the planets on circular orbits around the earth. From a modern perspective this geocentric model of a narrowly confined universe may strike us as naive, for we have grown accustomed to the idea that our planetary home is but a tiny island in a sea of empty space. With the nearest neighbor to our sun light years away, we find ourselves in cosmic isolation, and the belief that divine providence has placed us at the physical and spiritual center of the world has long since lost its credibility. Given this modern predisposition of ours, it is easy to forget that Aristotelian cosmology actually provides an excellent framework to account for most of the phenomena that we observe in the sky. In particular, the daily and annual motion of the sun and the daily motions of the stars and planets are well explained within the geocentric paradigm. For navigational purposes the ancient model is still in use today, and it still surpasses any contemporary theories in terms of intuitive accessibility.

Simplicio: If the geocentric model was so extraordinarily successful, why did anybody ever doubt its reality?

Teacher: To answer this question, we must consider *the problem of the planets.* As we record the position of a planet in the sky, we observe a combination of two separate motions: a daily motion due to the earth's rotation about its axis, and an annual motion that is caused by the rotation of the earth and the planets about the sun. The daily motion is easily reconciled with Aristotelian cosmology if we imagine each planet to be attached to a sphere that completes one full revolution each day in synchrony with the outermost sphere of the fixed stars. By contrast, the annual motion is far more complex and far more difficult to explain within a geocentric universe, because it results from relative changes in position as the earth and the planets revolve about the sun at different orbital speeds and distances. Here we typically observe an annual drift along an approximately circular trajectory traced out at variable speed with occasional reversals in the direction of travel (see Figure 15.1).

Simplicio: Are you saying then that geocentrism began to crumble as soon as people became aware that annual planetary motions are not readily accounted for in Aristotelian cosmology?

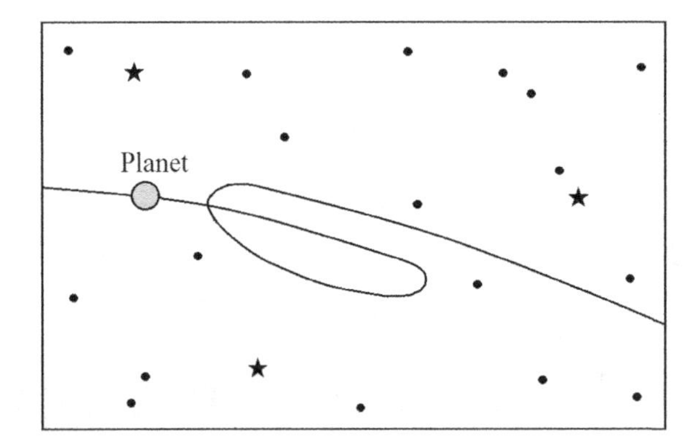

Figure 15.1: qualitative rendering of a planetary retrograde motion.

Teacher: Not at all. It took a very long time for this to happen, and for many centuries astronomers were content to invent ever more elaborate mathematical patterns to describe planetary trajectories under the traditional assumption of a central earth. Most prominently, Ptolemy of Alexandria (second century A.D.) devised in his *Almagest* a complicated system of epicycles whereby planetary trajectories were approximated by superimposing separate circular motions (see Exercise 15.1). His system achieved a relatively high degree of accuracy, and it essentially remained unchallenged as an authoritative standard for more than fourteen hundred years.

Simplicio: So what eventually brought about the change?

Teacher: Adopting a rather simplistic view, we could point to the individual effort of Nicolaus Copernicus (1473-1543) who, in his major work *On the Revolutions of the Heavenly Spheres*, proposed a heliocentric model of the universe that surpassed the Ptolemaic system in accuracy and elegance. But to suggest the issue was decided solely on scientific or even strictly empirical grounds will hardly do justice to the historical developments in their entirety. At the time of Copernicus, all of Europe appeared to be in a great transition. The new horizon of the American continent had opened up in the West, the enterprising spirit of the Renaissance had shaken off the otherworldliness of medieval scholasticism, and the authority of the Catholic Church was being undermined by the Protestant Reformation. It was a time conducive to change and revolutionary new ideas. Had a work equal in quality to that of Copernicus been written two or three hundred years earlier, it most certainly would have quickly sunk into oblivion.

Sophie: The notion that scientific theories require for their acceptance not only empirical evidence but also favorable historical circumstances is somewhat surprising. Are you suggesting that the ideal of genuine objectivity needs to be abandoned in favor of a more relativistic understanding that does not allow for truth in science to be asserted independently of its cultural context?

Teacher: I wouldn't go quite that far. For instance, cultural preferences certainly do not alter the nearly constant acceleration of a falling body, but our ability to acknowledge and properly interpret a given observation may very well depend on more than immediate sense perception. Furthermore, we also need to realize that empirical evidence usually appears unambiguous only in hindsight when decades or centuries of successful application have paved the way for a theory's general acceptance. Copernicus, for example, showed himself a true conservative in his adherence to the Aristotelian doctrine of circular (rather than elliptic) orbits, and due to this mistake, his initial model of a heliocentric universe proved less accurate than the Ptolemaic scheme. To make up for this defect, he reintroduced a profusion of epicycles that seriously diminished his theory's aesthetic appeal. The assertion, therefore, that Copernican astronomy was strongly supported by the existing observational data is simply not in keeping with the facts. Moreover, the heliocentric model seemed highly implausible in light of the traditional teaching on the mechanics of motion. For a moving earth, so it was thought, would have to be in evidence in its effects on objects on its surface. According to Aristotle, all material bodies naturally resisted motion and with a planet swiftly passing underneath would find themselves in constant danger of being left behind. Indeed, it is here, at the interface of mechanics and astronomy, that our story connects to one of the greatest pioneers of modern science: *Galileo Galilei.*

15.1 Exercise. Assume that the center of a circle of radius $r = 2$ on which a point P is marked completes one full rotation along a circle of radius $R = 4$ in four units of time (see Figure 15.2). Sketch the curve described by P under the assumption that the smaller circle, which is referred to as an *epicycle*, revolves about its own center once in every unit of time.

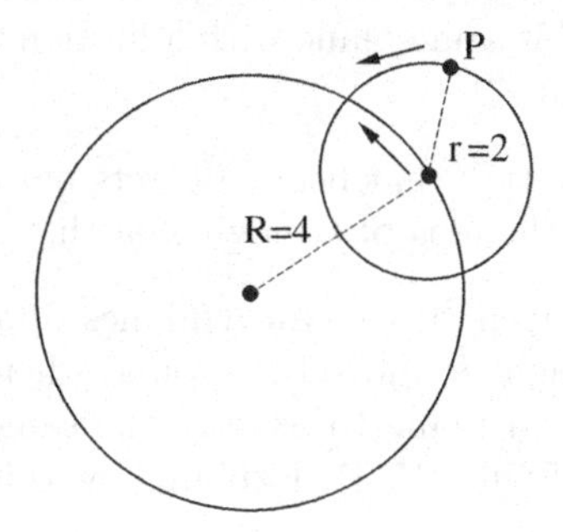

Figure 15.2: an epicycle.

Galileo was born at Pisa on the 15th of February, 1564. While little is known about his mother Giulia Ammannati, his father Vincenzio gained some notoriety for challenging the authorities with his writings on the theory of music. In the *Dialogue of Ancient and Modern Music* he wrote:

> It appears to me that they who in proof of any assertion rely simply on the weight of authority, without adducing any argument in support of it, act very absurdly. I, on the contrary, wish to be allowed freely to question and freely to answer you without any sort of adulation, as well becomes those who are in search of truth.[3]

To Vincenzio's likely regret, his son all too eagerly embraced the rebellious spirit of these words when, against his father's wishes, he abandoned the study of medicine at the University of Pisa in favor of a less prestigious career in mathematics. With a few papers on Euclidean geometry to his name, Galileo was able to secure a teaching post at the University of Pisa in 1589. True to character, he soon made himself unpopular by engaging the professors of philosophy in arguments regarding Aristotle's claim that objects fall at rates proportional to their weight.* According to legend, he climbed the Leaning Tower with cannon balls on his back to publicly conduct experiments on falling bodies. But whatever the setting of his demonstrations may have been, they did little to persuade an audience unaccustomed to the use of empirical evidence. In his dialogue on the *Two New Sciences* Galileo later commented as follows on the philosophers' stubborn denial:

> Aristotle says that a hundred-pound ball falling from a height of a hundred cubits hits the ground before a one-pound ball has fallen one cubit. I say they arrive at the same time. You find, on making the test, that the larger ball beats the smaller one by two inches. Now, behind those two inches you want to hide Aristotle's ninety-nine cubits and, speaking only of my tiny error, remain silent about his enormous mistake.[4]

In light of our earlier remarks, we may consider this episode a case in point regarding the elusiveness of objective human judgment, but it also testifies to the extraordinary clear-sightedness of Galileo's mind in reaching beyond the established patterns of thought.

In 1592, Galileo left Pisa for the chair of mathematics at the University of Padua and there began "the happiest period of his life."[5] With his intellectual powers now fully developed, his scientific work

*There are some conflicting views in the literature concerning the question whether Aristotle did indeed assert such a law. The medieval scholastic Jordanus Nemorarius, for instance, argued against this claim in his treatise *De gravi et levi.*

quickly increased in brilliance. Among his more prominent achievements were the invention of a military compass in 1597 and the formulation of the law of falling bodies in 1604. Of great significance, with respect to the aforementioned issue of refuting claims concerning the incompatibility of Copernicanism with Aristotle's teachings on motion, was his discovery of the law of inertia for horizontal movements close to the surface of the earth. He conducted the relevant experiments in 1608 and stated his conclusion in the *Two New Sciences* as follows:

> Imagine any particle projected along a horizontal plane without friction; then we know...that this particle will move along this same plane with a motion which is uniform and perpetual, provided the plane has no limits.[6]

Galileo thus rejected Aristotle's assertion that heavy objects naturally strive to be at rest and thereby removed an important objection to the idea of a moving earth.

15.2 Exercise. Assume you are given two inclined planes opposite to each other. If an object is released onto the first plane at a height h above the ground, then in absence of frictional forces it will roll down the first and up the second plane to exactly the same height h regardless of the angle of inclination of the second plane (see Figure 15.3). Explain how this observation can be used as evidence in support of Galileo's law of inertia.

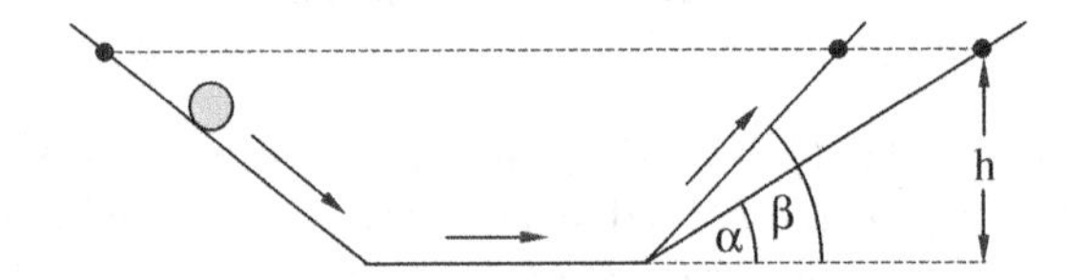

Figure 15.3: an experiment concerning the law of inertia.

To further oppose the view that the earth's motion should result in visible effects on objects on its surface, Galileo implicitly appealed to the law of inertia in a particularly lucid passage in his *Dialogue Concerning the Two Chief Systems of the World:*

> Shut yourself up with some friend in the main cabin below deck on some large ship, and have with you there some flies, butterflies and small flying animals. Have a large bowl of water with some fish in it; hang up a bottle that empties drop by drop into a wide vessel beneath it. With the ship standing still, observe carefully how the little animals fly with equal speed to all sides of the cabin. The fish swim indifferently in all directions; the drops fall into the vessel beneath; and, in throwing something to your friend, you need throw it no more strongly in one direction than another, the distances being equal; jumping with your feet together, you pass equal spaces in every direction. When you have observed all these things carefully (though there is no doubt that when the ship is standing still everything must happen in this way), have the ship proceed with any speed you like, so long as the motion is uniform and not fluctuating this way and that. You will discover not the least change in all the effects named, nor could you tell from any of them whether the ship was moving or standing still. In jumping, you will pass on the floor the same spaces as before, nor will you make larger jumps towards the stern than towards the prow even though the ship is moving quite rapidly, despite the fact that during the time that you are in the air the floor under you will be going in the direction opposite to your jump. In throwing something to your companion, you will need no more force to get it to him whether he is in the direction of the bow or the stern, with yourself situated opposite. The droplets will fall as before into the vessel beneath without dropping toward the stern, although while the drops are in the air the ship runs many spans. The fish in their water will swim towards the front of their bowl with no more effort than towards the back, and will go with equal ease to bait placed anywhere around the edge of the bowl. Finally the butterflies and flies will continue their

flights indifferently toward every side, nor will it ever happen that they are concentrated toward the stern as if tired out from keeping up with the course of the ship from which they will have been separated during long intervals by keeping themselves in the air. And if smoke is made by burning some incense, it will be seen going up in the form of a little cloud, remaining still and moving no more to one side than to the other. The cause of all these correspondences of effect is the fact that the ship's motion is common to all things contained in it and to the air also. That is why I said you should be below decks; for if this took place above in open air, which would not follow the course of the ship, more or less noticeable differences would be seen in some of the effects noted. No doubt the smoke would fall as much behind as the air itself. The flies likewise, and the butterflies, held back by the air, would be unable to follow the ship's motion if they were separated from it by a perceptible distance. But keeping themselves near it, they will follow without effort or hindrance; for the ship, being an unbroken structure, carries with it a part of the nearby air. For a similar reason we sometimes, when riding horseback, see persistent horseflies following our horses flying now to one part of their bodies and now to another. But the difference would be small as regards the falling drops, and as to the jumping and throwing it would be quite imperceptible.[7]

In 1609, Galileo was distracted from his inquiries into the laws of motion when rumors spread from the Netherlands concerning the invention of a spyglass that could bring objects at a far distance into closer view. Inspired by the news, Galileo immediately set out to build his own vastly improved version of the instrument and presented it to the Venetian Senate for its potential in military applications. Later the same year, he pointed his telescope at the night sky and there made discoveries that were to attract the attention of all of Europe. On the moon he saw a surface scarred by craters rather than the smooth heavenly perfection predicted by Aristotle. As for Jupiter, he was most amazed to find the giant planet circled by four smaller ones (the moons of Jupiter) in blatant contradiction to the traditional view that all celestial bodies must revolve about the earth. Emboldened by his success, he went on to provide strong evidence for Copernican astronomy by pointing out that the planet Venus alters its appearance through a crescent cycle in ways that cannot be accounted for under the assumption of a central earth.

When Galileo published his spectacular discoveries in the *Starry Messenger* in 1610, the issue of Copernicanism suddenly attracted greater public interest and was no longer confined to esoteric debates among astronomers. It was at this point also that opposing voices in the Catholic Church were raised with greater frequency and determination. In 1614, a young Dominican priest in Florence, by the name of Tommaso Caccini, denounced Galileo and his followers from the pulpit as "practitioners of diabolical arts" and "enemies of true religion."[8] To support their case the critics quoted from the Book of Joshua, wherein God intervenes to halt a moving sun, and also from the Psalms: "He set the earth on its foundation so that it can never be moved."[9] As a faithful Catholic, Galileo was deeply troubled by these accusations, for he considered heresy a crime "more abhorrent than death itself."[10] In a letter to his former student Benedetto Castelli he asserted the absolute veracity of Biblical revelation and "added that, though Scripture cannot err, its expounders are liable to err in many ways...when they would base themselves always on the literal meaning of the words."[11]

To understand the causes of Catholic opposition to Copernicanism is not an easy matter. In individual cases an overly zealous insistence on Biblical literalism and suspicion of secular learning certainly played a role, but such a charge cannot be leveled against the Church at large. In her long scholarly history, the Catholic Church was flexible enough to integrate the theological systems of such authentic great thinkers as Augustine and Thomas Aquinas. And in particular, the Catholic authorities did not onbject when Aquinas lend support to Aristotle's teaching concerning the earth's sphericity by employing in effect the same nonliteral mode of Scripture interpretation that Galileo endorsed as well.

If anything, the charge of anti-intellectualism can be more justly applied to certain Protestant leaders of the 16th Century who saw themselves as defenders of spiritual purity against various forms of Catholic corruption including all metaphorical or allegorical modes of Scripture interpretation. Martin Luther spoke of reason as the "devil's whore,"[12] and his "principal lieutenant,"[13] Melanchthon, is quoted as saying that

> [t]he eyes are witnesses that the heavens revolve in the space of twenty-four hours. But certain men, either from the love of novelty, or to make a display of ingenuity, have concluded that the earth moves; and they maintain that neither the eighth sphere nor the sun revolves.... Now, it is a want of honesty and decency to assert such notions publicly, and the example is pernicious. It is the part of a good mind to accept the truth as revealed by God and acquiesce in it.[14]

By contrast, the initial Catholic reaction was rather restrained. In fact, Pope Gregory XIII (1502-1585) actually helped the new ideas to gain respectability when he relied in his calendar reform on the *Prutenic Tables* that Erasmus Reinhold had constructed on the basis of the Copernican system. However, in a time when Catholic authority was already seriously challenged by the Protestant Reformation, the conservative forces within the church eventually gained a stronger voice in opposing the revolutionary spirit of the new astronomy. In the year 1600 a tribunal of the Inquisition infamously sentenced Giordano Bruno to be burned at the stake for declaring, among other more important "heresies," that the sun rather than the earth is at rest at the center of the universe.

Sixteen years later the inquisitor against Bruno, Cardinal Bellarmino, was summoned by Pope Paul V as one of several expert consultors to determine conclusively whether Copernican teaching could be reconciled with Catholic doctrine. Galileo was present in Rome in February 1616 when the cardinals of the Holy Office denounced the idea of a central sun as not only "formally heretical"[15] in direct contradiction to Holy Scripture but also as "foolish and absurd."[16] The notion that the earth is not immobile and not the center of the world was equally rejected as philosophically unsound and "erroneous in faith."[17] Having thus reestablished by decree the order of the universe, the panel communicated its conclusions to the Holy Office of the Inquisition. On February 26, Galileo was taken to the Vatican where Bellarmino announced that the heliocentric doctrine could no longer be proclaimed a fact but only a hypothesis. It was further decided that Copernicus' treatise *On the Revolutions of the Heavenly Spheres* was to be placed on the Index of Prohibited Books.

Fate appeared to take a turn for the better when in 1623 Galileo's longtime friend and admirer, Maffeo Cardinal Barberini, ascended to the See of Peter as Pope Urban VIII. Ironically though, it was under Urban's rule that Galileo was eventually tried and found guilty of heresy. In 1632, when Urban was involved in a struggle with Spain over the Thirty Years' War in Germany, Galileo attracted the pope's anger with the publication of his *Dialogue Concerning the Two Chief Systems of the World*, wherein he allowed the Aristotelian ignoramus *Simplicio* to expound Urban's philosophy of science. At the age of 68, with his health failing, Galileo was once again summoned to Rome, this time not to be admonished but to face trial at the threat of death. After four hearings before the commissary general of the Holy Office of the Inquisition, the cardinal inquisitors announced the results of their deliberations on Wednesday, June 22, 1633:

> We say, pronounce, sentence, and declare that you, Galileo, by reason of the matters which have been detailed in the trial and which you have confessed already, have rendered yourself in the judgment of this Holy Office vehemently suspected of heresy, namely of having held and believed the doctrine which is false and contrary to the Sacred and Divine Scriptures, that the Sun is the center of the world and does not move from east to west and that the Earth moves and is not the center of the world; and that one may hold and defend as probable an opinion after it has been declared and defined contrary to Holy Scripture. Consequently, you have incurred all censures and penalties enjoined and promulgated by the sacred Canons and all particular and general laws against such delinquents. We are willing to absolve you from them provided that first, with a sincere and unfeigned faith, in our presence you abjure, curse and detest the said errors and heresies, and every other error and heresy contrary to the Catholic and Apostolic Church in the manner and form we will prescribe to you....[18]

Only seven of the ten inquisitors had signed this sentence when Galileo knelt before the tribunal in the "white robe of the penitent"[19] to abjure as required.

Within days of the verdict, a clemency request by the pope's nephew Francesco Cardinal Barberini was granted and Galileo was transferred "from the dungeons of the Holy Office to the Tuscan embassy in Rome."[20] Later the same year he was first entrusted "to the custody of the archbishop of Siena"[21] and then finally given permission to return to his home in Arcetri where he stayed under house arrest until his death on January 8, 1642.

For his rejection of scholastic speculation and his insistence on experimental verification, Galileo Galilei is remembered as the first truly modern scientist and as one of the greatest pioneers in the history of ideas. From the encounter with intellectual oppression his name has emerged as a symbol for the human struggle for freedom of thought.

Chapter 16

Computing Volumes and the Hydrostatic Principle

The Hydrostatic Principle

Let us assume a solid object is placed in a body of water. If the density of the object (i.e., its mass divided by its volume) is greater than the density of water, the object will sink to the bottom. If on the other hand the density of the object is less than the density of water, the object will float on the surface. In the latter case it can be useful to know how much of the object's volume is visible above the surface and how much is hidden beneath it. For example, the captain of a ship maneuvering in the polar sea is well advised to make himself aware of the fact that 9/10 of the volume of an iceberg are under water and only 1/10 is visible above the surface. The general physical law that tells us how far a floating object is immersed in water is the so-called *hydrostatic principle*:

> Any solid lighter than a fluid will, if placed in that fluid, be so far immersed that the mass of the solid will be equal to the mass of the fluid displaced,

or equivalently,

> any solid lighter than a fluid will, if placed in that fluid, try to jump out as quickly as possible and look for a towel to get dry.

It was the Greek mathematician Archimedes of Syracuse, the father of mathematical physics, who first discovered this principle. He published this and many other *deep* results in the treatise, *On Floating Bodies*. It was undoubtedly the mathematical derivation of the hydrostatic principle that led the absentminded Archimedes to jump out of his bath and run home naked, shouting "Eureka" ("I have found it"). However, Archimedes is not best remembered for violating public standards of decency, but for being one of the three greatest mathematicians in the history of the world (the other two being Isaac Newton and Carl Friedrich Gauss).

Weighing a Ball without Using Scales

As an application of the hydrostatic principle we will devise a method for determining the mass of a ball, floating on the surface of a body of water, without using scales. We reason as follows: according to the hydrostatic principle, the mass of the ball is equal to the mass of the water that the ball displaces. Furthermore, the mass of the water displaced is equal to the density of water multiplied by the volume V of the immersed portion of the ball. Thus, in order to find the mass of the ball, we only need to determine V.

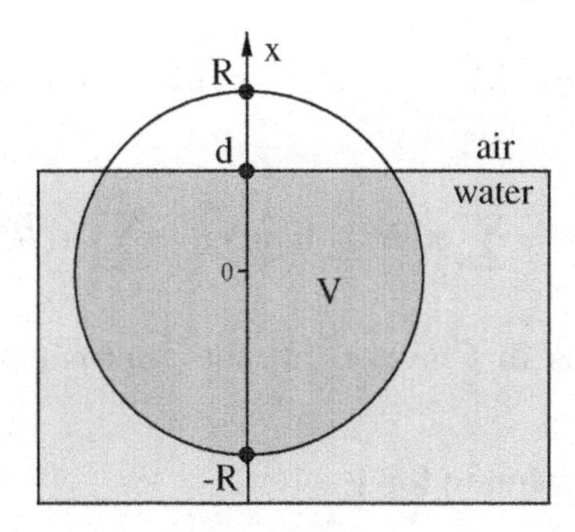

Figure 16.1: cross-sectional view of a ball immersed in water.

To get started we introduce an axis pointing upward with its origin in the center of the ball. By R we denote the radius of the ball and by d the position of the dividing line between air and water (see Figure 16.1). Motivated by our discussion of approximation methods for areas under graphs in Chapter 14, we think of V as being composed of thin horizontal slices, each of which is approximately equal in volume to a cylinder of the same height and width (see Figure 16.2 for a cross-sectional view). The sum of the volumes of these cylinders gives us an estimate for the volume V just as the sum of the areas of the approximating rectangles gave us an estimate for the area under the graph of a function.

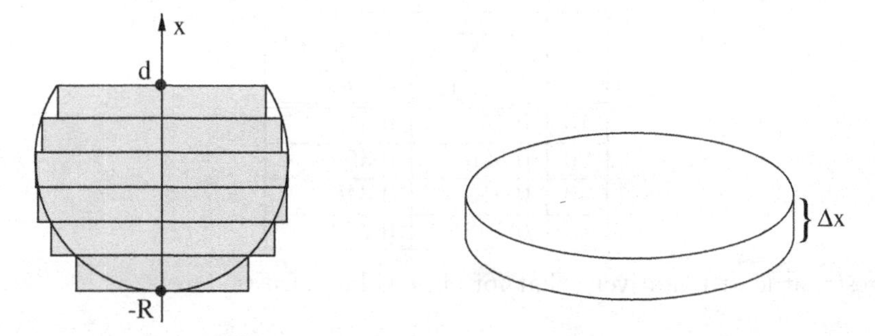

Figure 16.2: approximation of V by cylinders and approximating cylindric slice.

As indicated in Figure 16.2, we divide the interval $[-R, d]$ into n subintervals of equal length $\Delta x = (R + d)/n$. The endpoints of these subintervals are

$$x_k = -R + k\Delta x = -R + \frac{k(R + d)}{n} \text{ for } 0 \leq k \leq n. \tag{16.1}$$

Furthermore, the volume ΔV_k of the cylindric slice over the interval $[x_{k-1}, x_k]$ is equal to its height $x_k - x_{k-1} = \Delta x$ times the area of its disc-shaped base. Denoting the radius of the disc-shaped base by R_k, the theorem of Pythagoras implies that $R_k^2 = R^2 - x_k^2$ (see Figure 16.3). Thus, the area of the base

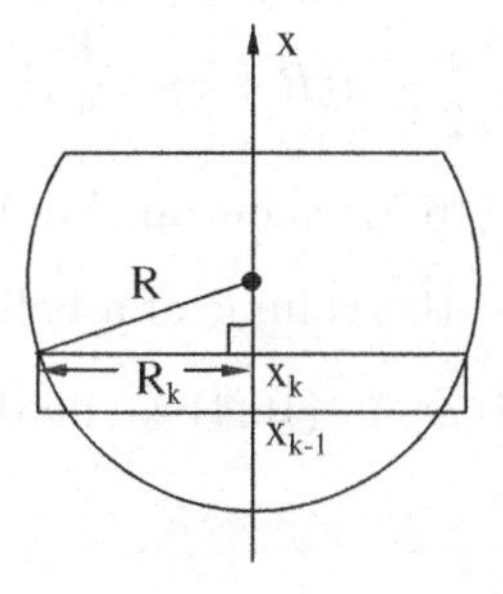

Figure 16.3: finding the radius of horizontal slices.

is $\pi R_k^2 = \pi(R^2 - x_k^2)$ and, therefore, $\Delta V_k = \pi(R^2 - x_k^2)\Delta x$. Summing up the volumes of the cylindric slices, we obtain the estimate

$$V \approx \sum_{k=1}^{n} \Delta V_k = \sum_{k=1}^{n} \pi(R^2 - x_k^2)\Delta x = \sum_{k=1}^{n} \pi(R^2 - x_k^2)\frac{R+d}{n}. \tag{16.2}$$

Note: according to our definitions in Chapter 14, the approximation above represents a right sum estimate for V.

16.1 Exercise. Find a left sum estimate for V.

Combining (16.1) with (16.2) yields

$$\begin{aligned} V &\approx \sum_{k=1}^{n} \pi\left(R^2 - \left(-R + \frac{k(R+d)}{n}\right)^2\right)\frac{R+d}{n} = \sum_{k=1}^{n} \pi\left(\frac{2kR(R+d)}{n} - \frac{k^2(R+d)^2}{n^2}\right)\frac{R+d}{n} \\ &= \frac{2\pi R(R+d)^2}{n^2}\sum_{k=1}^{n} k - \frac{\pi(R+d)^3}{n^3}\sum_{k=1}^{n} k^2. \end{aligned} \tag{16.3}$$

Since this estimate for V naturally becomes more accurate as n increases, we will now examine the convergence behavior of the terms $n^{-2}\sum_{k=1}^{n} k$ and $n^{-3}\sum_{k=1}^{n} k^2$, which appear in (16.3), as n tends to ∞. Using a calculator, we find the following values:

n	$\frac{1}{n^2}\sum_{k=1}^{n} k$	$\frac{1}{n^3}\sum_{k=1}^{n} k^2$
10	0.55	0.385
20	0.525	0.359
40	0.513	0.346
80	0.506	0.339

This table suggests, at least tentatively, that for large values of n we have

$$\frac{1}{n^2}\sum_{k=1}^{n} k \approx \frac{1}{2} \quad \text{and} \quad \frac{1}{n^3}\sum_{k=1}^{n} k^2 \approx \frac{1}{3}.$$

More precisely, we conjecture that

$$\lim_{n\to\infty} \frac{1}{n^2}\sum_{k=1}^{n} k = \frac{1}{2} \quad \text{and} \quad \lim_{n\to\infty} \frac{1}{n^3}\sum_{k=1}^{n} k^2 = \frac{1}{3}. \tag{16.4}$$

Thus, according to (16.3), the exact value of V ought to be given by the equation

$$\begin{aligned} V &= \lim_{n\to\infty}\left(\frac{2\pi R(R+d)^2}{n^2}\sum_{k=1}^{n} k - \frac{\pi(R+d)^3}{n^3}\sum_{k=1}^{n} k^2\right) \\ &= 2\pi R(R+d)^2 \cdot \frac{1}{2} - \pi(R+d)^3 \cdot \frac{1}{3} = \frac{\pi}{3}(2R-d)(R+d)^2. \end{aligned} \tag{16.5}$$

A proof of (16.4) (and, by implication, (16.5)) is outlined in the ⋆remark⋆ below.

16.2 Exercise. Use (16.5) to show that the volume of a ball of radius R is equal to $4\pi R^3/3$.

★ *Remark.* In order to verify the equations in (16.4) we need to evaluate first the sum $\sum_{k=1}^{n} k$. To do so, we observe that

$$\sum_{k=1}^{n}(k+1)^2 - \sum_{k=1}^{n} k^2 = (2^2 + 3^2 + \cdots + (n+1)^2) - (1^2 + 2^2 + \cdots + n^2) = (n+1)^2 - 1 \tag{16.6}$$

and

$$\sum_{k=1}^{n}(k+1)^2 - \sum_{k=1}^{n}k^2 = \sum_{k=1}^{n}((k+1)^2 - k^2) = \sum_{k=1}^{n}(2k+1) = 2\sum_{k=1}^{n}k + \sum_{k=1}^{n}1 = 2\sum_{k=1}^{n}k + n. \quad (16.7)$$

Combining (16.6) with (16.7), it follows that

$$2\sum_{k=1}^{n}k = (n+1)^2 - 1 - n = n^2 + n = n(n+1),$$

or equivalently,

$$\sum_{k=1}^{n}k = \frac{n(n+1)}{2}. \quad (16.8)$$

In order to find a formula for $\sum_{k=1}^{n} k^2$, we proceed in a completely analogous manner: in place of (16.6) we write the equation

$$\sum_{k=1}^{n}(k+1)^3 - \sum_{k=1}^{n}k^3 = (2^3 + 3^3 + \cdots + (n+1)^3) - (1^3 + 2^3 + \cdots + n^3) = (n+1)^3 - 1, \quad (16.9)$$

and then, using (16.8) and the easily verifiable formula $(k+1)^3 = k^3 + 3k^2 + 3k + 1$, it follows that

$$\begin{aligned}\sum_{k=1}^{n}(k+1)^3 - \sum_{k=1}^{n}k^3 &= \sum_{k=1}^{n}((k+1)^3 - k^3) = \sum_{k=1}^{n}(3k^2+3k+1) = 3\sum_{k=1}^{n}k^2 + 3\sum_{k=1}^{n}k + \sum_{k=1}^{n}1 \\ &= 3\sum_{k=1}^{n}k^2 + \frac{3n(n+1)}{2} + n.\end{aligned} \quad (16.10)$$

Combining (16.9) with (16.10) yields

$$3\sum_{k=1}^{n}k^2 = (n+1)^3 - 1 - \frac{3n(n+1)}{2} - n = \frac{n(2n+1)(n+1)}{2},$$

and therefore,

$$\sum_{k=1}^{n}k^2 = \frac{n(2n+1)(n+1)}{6}. \quad (16.11)$$

Given the results in (16.8) and (16.11), we may now employ the techniques developed in Chapter 11 (see p.91) to conclude that

$$\lim_{n\to\infty}\frac{1}{n^2}\sum_{k=1}^{n}k = \lim_{n\to\infty}\frac{n+1}{2n} = \frac{1}{2},$$

$$\lim_{n\to\infty}\frac{1}{n^3}\sum_{k=1}^{n}k^2 = \lim_{n\to\infty}\frac{(2n+1)(n+1)}{6n^2} = \frac{2}{6} = \frac{1}{3}.$$

This completes the proof of (16.4). ★

Simplicio: I was able to follow our derivation of (16.5), but how exactly are we going to use this formula to determine the mass of the ball?

Teacher: The argument is fairly simple: according to the hydrostatic principle, the mass m of the ball...

Sophie: ...is equal to the mass of the water displaced by the ball.

Teacher: Yes, and since V, as given in (16.5), is the volume of the water displaced, it follows that m is equal to the density of water multiplied by V. Consequently, in denoting the density of water by δ, we obtain

$$\boxed{m = \delta V = \frac{\pi\delta}{3}(2R - d)(R + d)^2.} \tag{16.12}$$

Simplicio: I see now where we are headed, but it would be nice to work an example.

Teacher: Okay, let's assume that a ball of radius $R = 2\,m$ (m = meters) is immersed in water so that only a spherical cap of height $1/3\,m$ is visible above the surface. In order to determine the mass m of the ball from equation (16.12), we first need to find the value of d.

Sophie: But this is easy because, according to Figure 16.1, d is simply the difference between R and the height of the cap above the surface. So we have

$$d = R - \frac{1}{3} = 2 - \frac{1}{3} = \frac{5}{3}.$$

Teacher: Correct. Now we replace R and d in equation (16.12) with the values 2 and 5/3 respectively. This yields

$$m = \frac{\pi\delta}{3}\left(2 \cdot 2 - \frac{5}{3}\right)\left(2 + \frac{5}{3}\right)^2 = \frac{847\pi\delta}{81}. \tag{16.13}$$

Simplicio: So all we have left to do is to look up the value for δ in a table and fill it in.

Teacher: That would work, but using a table is really not necessary. We should be able to infer the value of δ from common knowledge. For example, you probably know the weight of one liter of water, don't you?

Simplicio: It's about two pounds, isn't it?

Teacher: Only approximately—the exact value is $1\,kg$. Therefore, since one liter is the equivalent of 1000 cubic centimeters or 1/1000 cubic meters, it follows that

$$\delta = \frac{1\,kg}{1/1000\,m^3} = 1000\frac{kg}{m^3}.$$

Substituting for δ in equation (16.13) yields the final result:

$$m = \frac{847000\pi}{81} \approx 32851\,kg.$$

16.3 Exercise. Find the mass of a ball of radius $2\,cm$ that is floating in water with a spherical cap of height $0.4\,cm$ visible above the surface.

A General Method for Computing Volumes

The general method for finding the volume of an arbitrary solid is based on the same idea as our calculation of the volume of a ball (or part of a ball) in the previous section: we cut a given solid into thin slices, approximate the volume of each slice, and take the sum. In order to reduce the complexity of our calculations as far as possible, we will always try to slice an object in such a way that the shape of the slices is as simple as possible. Slices do not always have to be plane, they may just as well be thin hollow cylinders or spheres or assume any shape that suits the geometry of a given problem. However, to simplify our discussion, we will restrict ourselves here to the case of plane cross-sections, perpendicular to a given axis.

With reference to the set-up shown in Figure 16.4, we subdivide the interval $[a, b]$ into n subintervals of equal length $\Delta x = (b - a)/n$. Denoting by x_k the endpoints of the subintervals and by $A(x)$ the cross-sectional area at position x (perpendicular to the given axis), we obtain the following estimates

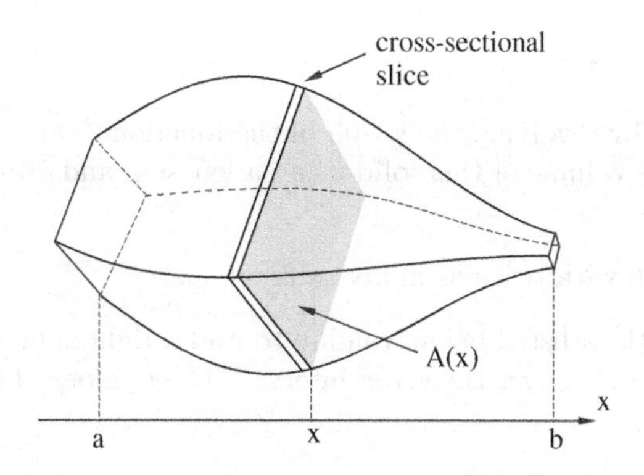

Figure 16.4: cross-sectional slice in a volume.

for the solid's volume:

$$V \approx \sum_{k=0}^{n-1} A(x_k)\Delta x \quad \text{using a left sum,}$$

$$V \approx \sum_{k=1}^{n} A(x_k)\Delta x \quad \text{using a right sum.}$$

16.4 Example. We wish to approximate the volume V of a barrel obtained by revolving the graph of the function

$$f(x) := 4 - \frac{x^2}{16}$$

about the x-axis over the interval $[-4, 4]$ (see Figure 16.5). Slicing the barrel at any point $x \in [-4, 4]$,

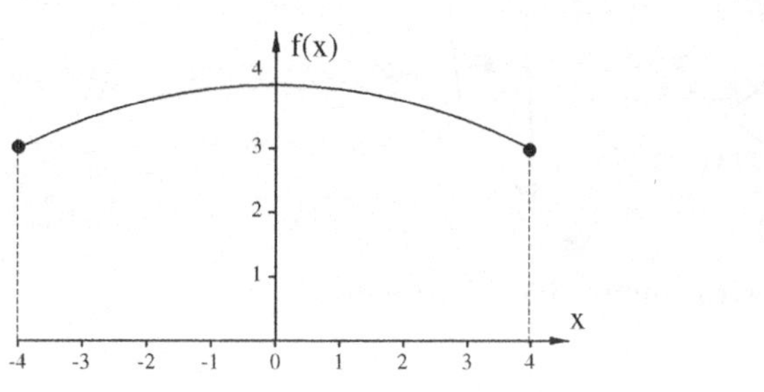

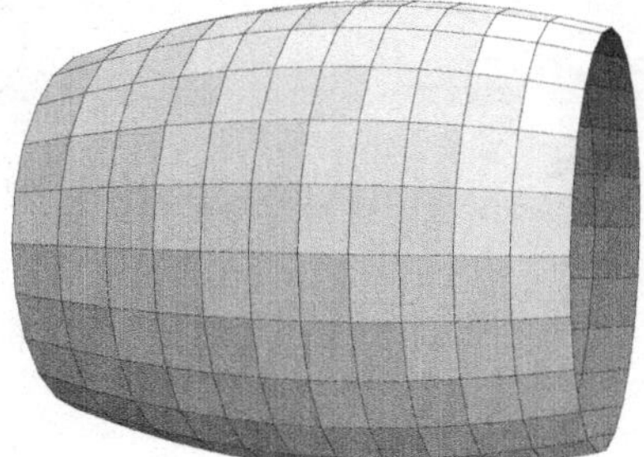

Figure 16.5: graph of $f(x) = 4 - x^2/16$ and surface of revolution generated by f.

we find a circular cross-section of radius $R = f(x) = 4 - x^2/16$. Therefore, the cross-sectional area at x is given by the formula

$$A(x) = \pi f(x)^2 = \pi \left(4 - \frac{x^2}{16}\right)^2.$$

Next, we divide the interval $[-4, 4]$ into n subintervals so that $\Delta x = 8/n$ and $x_k = -4 + k\Delta x = -4 + 8k/n$. In choosing to approximate V via a left sum, we obtain

$$V \approx \sum_{k=0}^{n-1} A(x_k)\Delta x = \sum_{k=0}^{n-1} \pi f(x_k)^2 \Delta x = \sum_{k=0}^{n-1} \pi \left(4 - \frac{1}{16}\left(-4 + \frac{8k}{n}\right)^2\right)^2 \frac{8}{n}.$$

Evaluating this sum for, say, $n = 8$ yields $V \approx 6897\pi/64 \approx 338.56$ *units*^{3}.

16.5 Exercise. Approximate the volume of the barrel by using a right sum with $n = 10$.

Additional Exercises

16.6. A solid is generated by revolving the graph of the function $f(x) = \sqrt{x}$ about the x-axis over the interval $[0,5]$. Estimate the volume of this solid using a left sum and 10 subintervals along the interval $[0,5]$.

16.7. Rework Exercise 16.6 with twice as many subintervals.

16.8. For each of the functions listed below you are to find a right sum estimate for the volume of the corresponding solid of revolution over the given interval. The number of subintervals is supposed to be 20 in each case.

a) $f(x) := 1 - x^2$ over $[-1,1]$,

b) $f(x) := (x-1)(x-2)(x-3)$ over $[1,2]$,

c) $f(x) := 1 - |x|$ over $[-1,1]$,

d) $f(x) := x^3$ over $[0,1]$.

16.9. Assume that each vertical cross-section of the solid shown in Figure 16.6 is a square.

a) At position x on the x-axis, what is the area of the corresponding cross-section?

b) Using 6 subintervals and a left sum, find an approximation for the volume of the solid.

c) Using 6 subintervals and a right sum, find an approximation for the volume of the solid.

d) Repeat b) and c) with 12 subintervals instead of 6.

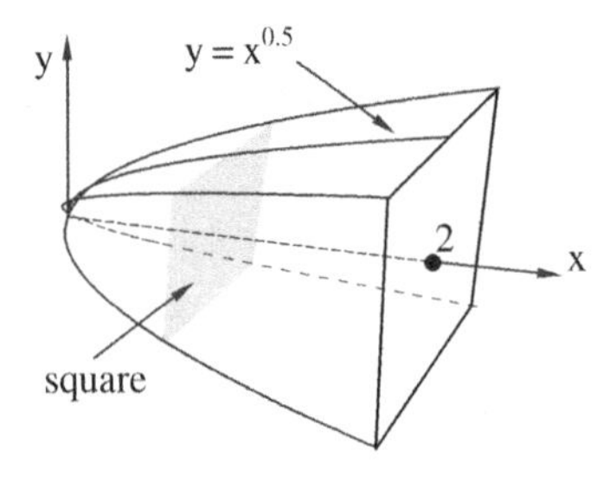

Figure 16.6: solid with square-shaped cross-sections.

16.10. A right circular cone, as shown in Figure 16.7 below, has a height of 3 feet and a radius of 3 feet at the top. Find an upper estimate for its volume by using 10 subintervals along its vertical axis of symmetry.

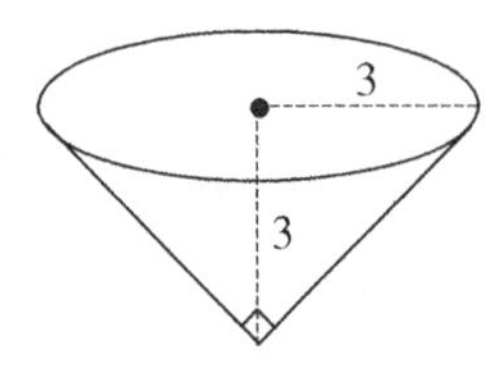

Figure 16.7: right circular cone.

16.11. Estimate the volume of the solid that satisfies the following conditions:

a) The base of the solid is a circular disc of radius 2 in the xy-plane.

b) Each cross-section perpendicular to the x-axis is a square.

Chapter 17

The Definite Integral

Riemann Sums

Expanding on our discussion of approximation via summation in Chapters 14 and 16, we are now going to explore more rigorously the second fundamental theme of calculus: accumulation (of change). We have already encountered this theme in three examples: the accumulation of changes in distance (or position) given the velocity of an object, the accumulation of areas under the graph of a function, and the accumulation of volumes in a solid given the solid's cross-sectional areas. To formalize the notion of approximating sums, which was central to all of these examples, we introduce the following definition:

17.1 Definition. Given a function $f : [a, b] \to \mathbb{R}$, an integer $n \geq 1$ and numbers $c_0, c_1, \ldots, c_n \in \mathbb{R}$ with $a = c_0 < c_1 < \cdots < c_{n-1} < c_n = b$, we set $\Delta x_k := c_k - c_{k-1}$ for $1 \leq k \leq n$. If $x_1, \ldots, x_n \in \mathbb{R}$ are values satisfying the condition $x_k \in [c_{k-1}, c_k]$ for $1 \leq k \leq n$, then

$$\sum_{k=1}^{n} f(x_k)\Delta x_k$$

is called a *Riemann sum.*

For a positive function $f : [a, b] \to \mathbb{R}$ a Riemann sum represents an approximation for the area under the graph of f over the interval $[a, b]$, because each term in the sum is equal to the area of a rectangle of height $f(x_k)$ and width Δx_k (see Figure 17.1).

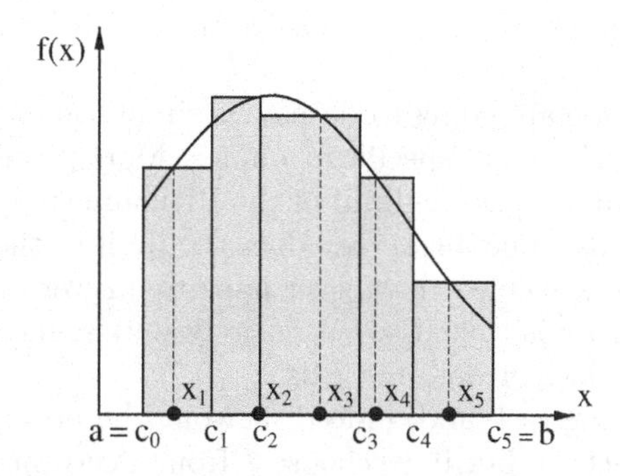

Figure 17.1: Riemann sum approximation.

As indicated in Definition 17.1, Riemann sums can be set up over subintervals of variable lengths Δx_k (see also Figure 17.1), but in most cases we will follow the practice established in Chapters 14 and 16 and work with subintervals of constant length

$$\Delta x = \frac{b - a}{n}.$$

However, regarding the choice of the evaluation points x_k, the flexibility provided by Definition 17.1, which only requires $x_k \in [c_{k-1}, c_k]$, will come in handy. For instead of setting x_k equal to the left or right endpoint of the interval $[c_{k-1}, c_k]$ (as in left or right sums) it will sometimes be preferable to choose for x_k the *midpoint* of the interval $[c_{k-1}, c_k]$. The corresponding sum is said to be a *midpoint sum*. For n subintervals, the formulae for x_k in the three cases of left, right, and midpoint sums are as follows:

$$\begin{aligned} \text{left sum:} \quad & x_k = c_{k-1} = a + (k-1)\Delta x \\ \text{right sum:} \quad & x_k = c_k = a + k\Delta x \\ \text{midpoint sum:} \quad & x_k = \frac{c_{k-1} + c_k}{2} = \frac{a + (k-1)\Delta x + a + k\Delta x}{2} = a + \frac{2k-1}{2}\Delta x. \end{aligned}$$

17.2 Example. Let us write out the midpoint sum for the function $f(x) := 3x^2 + x$ over the interval $[2, 5]$ with $n = 10$. Since $\Delta x = (b - a)/n = (5 - 2)/10 = 3/10$, the third equation above with 2 filled in for a yields

$$x_k = 2 + \frac{2k-1}{2} \cdot \frac{3}{10} = 2 + \frac{3(2k-1)}{20}.$$

Thus, the midpoint sum is

$$\sum_{k=1}^{10} f(x_k)\Delta x = \sum_{k=1}^{10} (3x_k^2 + x_k)\frac{3}{10} = \sum_{k=1}^{10} \left(3\left(2 + \frac{3(2k-1)}{20}\right)^2 + 2 + \frac{3(2k-1)}{20} \right) \frac{3}{10}.$$

17.3 Exercise. Write out the left, right, and midpoint sums for $f(x) := 5x - 5x^2$ over the interval $[1, 3]$ with $n = 4$.

The Definite Integral

Teacher: In the examples in Chapters 14 and 16, we used Riemann sums to approximate certain quantities like distances, areas, or volumes. We saw that the margin of error in our estimates became smaller as n increased and Δx decreased. In other words, the larger the number of subintervals of $[a, b]$ and the smaller the length of each subinterval, the more accurate were our estimates. Given this observation, what do we probably need to do in order to reduce the margin of error to zero?

Sophie: Since the error approaches zero as n tends to ∞, it seems natural to think that we find the exact value by taking the limit as n tends to ∞. In fact, I remember that we already discussed this strategy in the Introduction where we tried to solve the problem of finding change in position from velocity.

Teacher: You are right. What we are going to do now is simply to formalize and generalize ideas that we examined so far only in relation to specific examples. More precisely, we will define the *integral* of a function f over an interval $[a, b]$ as the limit of the Riemann sums of f as n tends to ∞. However, there is a question that needs to be answered: does the limit of the Riemann sums always exist?

Simplicio: Now that's a good one—how can we ever hope to answer a question like that? There are so many different types of functions that it seems entirely impossible to decide in general whether the limit of the Riemann sums does or does not exist.

Teacher: It does seem a daunting task and, indeed, it would be an impossible task if we allowed f to represent any arbitrary function, but if we choose f from a certain restricted class of functions then a satisfactory answer can be given. In fact, the issue of identifying permissible choices for f already came up in Chapter 14 where we mentioned that the lower and upper estimates for the area under the graph of a function converge to the same limiting value if we assume f to be continuous.

Sophie: Yes, but I also remember that we said something about the assumption of continuity being too restrictive and replacing it with a weaker assumption called "piecewise continuity."

Teacher: Sophie is right. It is possible to show that the limit of the Riemann sums of f over $[a, b]$ exists if f is *piecewise continuous* and *bounded*.

Simplicio: What do we mean by that?

Teacher: A function $f : D \subset \mathbb{R} \to \mathbb{R}$ is said to be *bounded* if there exists a constant $M \in \mathbb{R}$ such that $|f(x)| \leq M$ for all $x \in D$, and f is said to be *piecewise continuous* if it is continuous except possibly at finitely many points, that is, if there are points $z_1, \ldots, z_n \in D$ such that f is continuous on $D \setminus \{z_1, \ldots, z_n\}$.

Simplicio: If you had translated this definition into Greek, it would have made just as much sense to me as the English version.

Teacher: So what exactly did you not understand?

Simplicio: I really didn't understand anything, but for a start, where did the M come from?

Teacher: You apparently misinterpreted the logic of my statement: I didn't say where M came from or how to find it, but only that f is said to be bounded *if there exists a constant* M such that $|f(x)| \leq M$ for all $x \in D$. For instance, the function $f(x) := x$ is bounded on $D := (0,1)$, because for $M = 1$ we have $|f(x)| = |x| \leq 1 = M$ for all $x \in D = (0,1)$. By contrast, the function $g(x) := 1/x$ is *not* bounded on $(0,1)$, because $\lim_{x \to 0^+} 1/x = \infty$. So as x approaches zero from the right the values $g(x)$ increase to ∞, and it is therefore impossible to find an M such that $|g(x)| \leq M$ for all $x \in (0,1)$.

Simplicio: That actually makes some sense now. I think my problem is just that mathematical definitions always appear to be abstract and removed from intuition. It's very difficult to get used to that.

Teacher: That indeed is not easy—on the one hand we need to learn to express ourselves clearly with mathematical precision, but on the other hand we also need to develop an intuitive understanding of the underlying concepts. Since visual representations are often helpful in this process, Figure 17.2 shows you a typical graph of a bounded piecewise continuous function defined on an interval $[a, b]$.

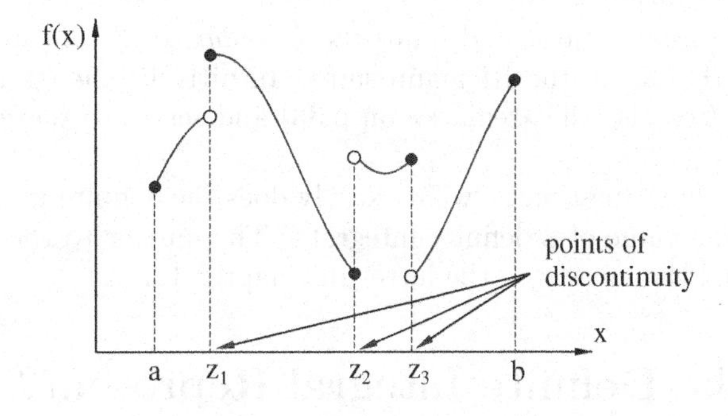

Figure 17.2: a bounded piecewise continuous function.

17.4 Theorem. *If $f : [a, b] \to \mathbb{R}$ is a bounded piecewise continuous function, then the limit of Riemann sums*

$$\lim_{\substack{n \to \infty \\ \Delta x_k \to 0}} \sum_{k=1}^{n} f(x_k) \Delta x_k$$

exists and is independent of the choice of the points $x_k \in [c_{k-1}, c_k]$.

★ *Remark.* The careful reader may have noticed that the statement of Theorem 17.4 is not entirely satisfactory, because the dependence of the Riemann sums $\sum_{k=1}^{n} f(x_k)\Delta x_k$ on n is not made sufficiently clear. To give a more rigorous formulation, we choose subdivision points $a = c_{n,0} < c_{n,1} < \cdots < c_{n,n} = b$ indexed by n such that

$$\lim_{n \to \infty} \max_{1 \leq k \leq n} (c_{n,k} - c_{n,k-1}) = 0. \tag{17.1}$$

(Note: $\max_{1 \leq k \leq n}(c_{n,k} - c_{n,k-1})$ is the maximum of the values $c_{n,k} - c_{n,k-1}$ as k ranges from 1 to n.) This condition replaces the somewhat fuzzy condition $\Delta x_k \to 0$. Then, for arbitrary $x_{n,k} \in [c_{n,k-1}, c_{n,k}]$, the theorem asserts the existence of the limit $\lim_{n \to \infty} \sum_{k=1}^{n} f(x_{n,k})(c_{n,k} - c_{n,k-1})$ and its independence of the choice of the $c_{n,k}$ and $x_{n,k}$ as long as condition (17.1) is met.

To understand why the existence of the limit in Theorem 17.4 hinges on the assumption of piecewise continuity, it is helpful to consider again the "popcorn function" as defined in the ⋆remark⋆ on p.23. In order to show that for this function the existence of the limit of the Riemann sums is crucially dependent upon the choice of the $x_{n,k}$, we first observe that every subinterval $[c_{n,k-1}, c_{n,k}]$ always contains infinitely many rational and infinitely many irrational numbers. As a consequence, we can always choose all evaluation points $x_{n,k}$ to be rational, so that $f(x_{n,k}) = 0$, or all of them to be irrational, so that $f(x_{n,k}) = 1$. In the former case the limit of the Riemann sums over an interval $[a, b]$ will be zero while in the latter it will be $b - a$. If we choose rational values for some of the $x_{n,k}$ and irrational ones for others, then the limit may either not exist at all or be equal to any number between zero and $b - a$. Think about it! ★

17.5 Definition. If $f : [a, b] \to \mathbb{R}$ is a bounded piecewise continuous function, then the *definite integral* (or *Riemann integral*) of f over $[a, b]$ is defined as

$$\int_a^b f(x)\,dx := \lim_{\substack{n\to\infty \\ \Delta x_k \to 0}} \sum_{k=1}^{n} f(x_k)\Delta x_k.$$

Remark. Using techniques that are beyond the scope of this text, it can be shown (see [Rud1]) that a function, which is continuous on a *closed* interval $[a, b]$, is also bounded on $[a, b]$. Consequently, Theorem 17.4 implies that the definite integral $\int_a^b f(x)\,dx$ exists whenever f is continuous on $[a, b]$.

The integral sign $\int$ represents an elongated S and thus indicates that integrals are limits of Sums. We read the integral notation as "integral of f from a to b with respect to x." The function f is also referred to as the *integrand*, and a and b are the *boundaries of integration*. Furthermore, the dx in the integral replaces the Δx in the Riemann sums. Intuitively, the dx is thought of as an "infinitely small" version of Δx (see also the ⋆remark⋆ on p.33) and serves to remind us that x is the *variable of integration*.

There arise two natural questions: "what exactly does the definite integral represent?" and "is there an easy way to find the value of a definite integral?" The answer to the former question will be given in the next section and the answer to the latter in Chapter 19.

What Does the Definite Integral Represent?

As illustrated in Figure 17.1, a Riemann sum of a positive (piecewise continuous and bounded) function $f : [a, b] \to \mathbb{R}$ gives us an approximation for the area under the graph of f over the interval $[a, b]$. In taking the limit as n tends to ∞ (see Definition 17.5), the error in the approximation is reduced to zero, and the Riemann sums converge to a definite integral. Thus, we may draw the following conclusion:

The definite integral $\int_a^b f(x)\,dx$ of a *positive* function f is equal to the area under the graph of f over the interval $[a, b]$.

It is important to point out that this rule is valid only for positive functions. If f is negative over parts of the interval $[a, b]$, then there will be a (partial) cancellation of negative against positive values in the Riemann sums $\sum_{k=1}^{n} f(x_k)\Delta x$, and the value of the integral will be obtained by subtracting areas below the x-axis from areas above the x-axis.

17.6 Example. Consider the function f shown in Figure 17.3. Denoting by A the area above the x-axis and by B and C the areas below the x-axis, we have

$$\int_a^b f(x)\,dx = A - B - C.$$

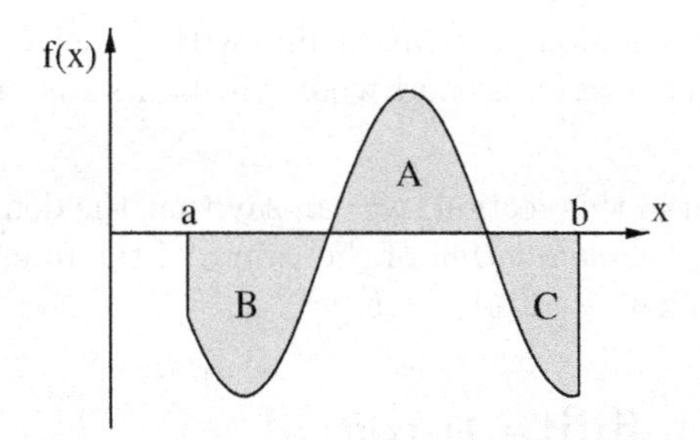

Figure 17.3: areas above and below the x-axis.

To illustrate how definite integrals of functions that are negative on parts of their domain can be useful in applications, let us examine again the problem of finding position—or change in position—from velocity. For a falling body the change in position is simply the distance fallen, because a falling body always moves in the same direction toward the surface of the earth. In general, though, the change in position may be different from the distance traveled, depending on whether a reversal in the direction of motion does or does not occur. So the total change in an object's position is to be computed by subtracting the distance traveled in the negative direction from that traveled in the positive direction. We will see in the following example that this subtraction of distances traveled in opposite directions is strictly analogous to the subtraction of areas in Example 17.6.

17.7 Example. Assume that an object travels along a straight line in such a way that its velocity at time t is given by the equation $v(t) = t - 2\ m/s$. How far does the object move in the first 5 seconds, and where does it end up relative to its starting position? To answer these questions we first notice that the velocity is negative from $t = 0$ to $t = 2$ and positive from $t = 2$ to $t = 5$ (see Figure 17.4). In other words, during the time span from $t = 0$ to $t = 2$ the object is moving backward, and from $t = 2$ to $t = 5$ it is moving forward. Recalling our discussions in Chapter 14 and in the Introduction, we know

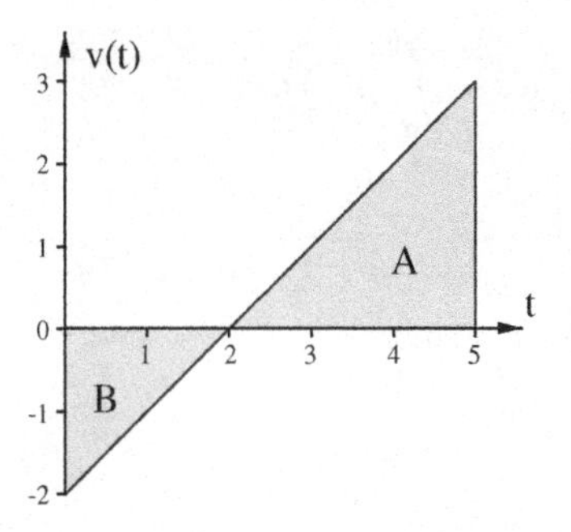

Figure 17.4: graph of $v(t) = t - 2$.

that the change in the object's position for the time interval from $t = 2$ to $t = 5$ is equal to the area A under the graph of the function v over the interval $[2, 5]$. Similarly, the negative change in position over the interval from $t = 0$ to $t = 2$ equals the negative of the area B. Thus, the *total change in positon* is

$$A - B = \int_0^5 v(t)\, dt.$$

By contrast, the *total distance traveled* is

$$B + A = \left| \int_0^2 v(t)\, dt \right| + \int_2^5 v(t)\, dt = -\int_0^2 v(t)\, dt + \int_2^5 v(t)\, dt.$$

Using elementary geometry, we find the areas of the triangular regions A and B to be 4.5 and 2 respectively. Consequently, the total change in position is 2.5 m and the total distance traveled is 6.5 m.

17.8 Exercise. An object travels along a straight line with velocity $v(t) = t - 4\ m/s$. How far does the object move within the first 5 seconds, and what will be its distance from its starting position at the end of 5 seconds?

Summarizing our discussion in this section, we can say that the definite integral of a function f over an interval $[a, b]$ gives us the *net accumulation* of the values of the function multiplied by small changes in x (symbolized by dx) between $x = a$ and $x = b$.

Properties of the Definite Integral

Before we can list the basic properties of the definite integral in full generality, we need to introduce an extension of Definition 17.5. So far we have always considered integrals over intervals $[a, b]$ with a as the lower boundary of integration and b as the upper boundary, that is, we have only considered integrals $\int_a^b f(x)\,dx$ with $a < b$. Our goal is to give a meaningful definition for an integral $\int_a^b f(x)\,dx$ with $a > b$. Considering first an approximating Riemann sum of the form $\sum_{k=1}^n f(x_k)\Delta x$, we notice that $\Delta x = (b-a)/n$ is negative whenever $a > b$. In other words, the assumption $a > b$ causes a reversal in the sign of the Riemann sum. Thus it appears reasonable to introduce the following definition:

17.9 Definition. Let $f : I \to \mathbb{R}$ be a bounded piecewise continuous function defined on an interval I. For $a, b \in I$ with $a > b$, we define

$$\int_a^b f(x)\,dx := -\int_b^a f(x)\,dx.$$

Given this definition, the properties of the definite integral, as listed in Theorem 17.10 below, are valid regardless of whether the upper boundary of integration is greater than or less than the lower boundary.

17.10 Theorem. *Let f and g be bounded piecewise continuous functions defined on an interval I. Then, for any values $a, b, c \in I$ and $\lambda \in \mathbb{R}$, we have*

a) $\displaystyle\int_a^b (f(x) + g(x))\,dx = \int_a^b f(x)\,dx + \int_a^b g(x)\,dx,$

b) $\displaystyle\int_a^b \lambda f(x)\,dx = \lambda \int_a^b f(x)\,dx,$

c) $\displaystyle\int_a^b f(x)\,dx = \int_a^c f(x)\,dx + \int_c^b f(x)\,dx,$

d) $\displaystyle\int_a^a f(x)\,dx = 0,$ *and*

e) $\displaystyle\left|\int_a^b f(x)\,dx\right| \le \left|\int_a^b |f(x)|\,dx\right|.$

Proof. We content ourselves to prove only a). (The proof of b) is very similar to that of a), the proofs of c) and d) are left as exercises, and the proof of e) can be found in [Rud1].) Using the definition of the definite integral as stated in 17.5, and the fact that the limit of a sum is equal to the sum of the limits (see Theorem 2.9), we obtain

$$\begin{aligned}\int_a^b (f(x) + g(x))\,dx &= \lim_{\substack{n\to\infty\\ \Delta x_k \to 0}} \sum_{k=1}^n (f(x_k) + g(x_k))\Delta x_k \\ &= \lim_{\substack{n\to\infty\\ \Delta x_k \to 0}} \sum_{k=1}^n f(x_k)\Delta x_k + \lim_{\substack{n\to\infty\\ \Delta x_k \to 0}} \sum_{k=1}^n g(x_k)\Delta x_k \\ &= \int_a^b f(x)\,dx + \int_a^b g(x)\,dx\end{aligned}$$

as desired. □

Remark. The proof above is not entirely rigorous, because the application of Theorem 2.9 to limits of Riemann sums is somewhat daring. For a more careful treatment of this argument the reader is referred to [Rud1].

17.11 Exercise. Prove c) and d) of Theorem 17.10.

Additional Exercises

17.12. Let $f(x) := 1/x$ for all $x \neq 0$.

a) Sketch the graph of f on the interval $[1, 2]$.

b) Divide the interval $[1, 2]$ into 5 subintervals and use left or right sums to get lower and upper estimates for the value of $\int_1^2 1/x\, dx$.

c) Write in summation notation the lower and upper esitmates for the same definite integral using n subintervals.

d) Find the difference between the upper and lower estimates in c).

e) How large a value must be chosen for n (based on the error estimate in d) to ensure that the values of the estimates in c) differ from the actual value by no more than 10^{-4}?

17.13. The graph of a function $f : [0, 2] \to \mathbb{R}$ is given below. In a group of four calculus students, each was given the same explicit formla for $f(x)$ and then asked to compute the value of $\int_0^2 f(x)\, dx$ with a margin of error less than 0.1. The results of the students' calculations are given below. Which one is correct? Explain your answer.

a) -5.8

b) 5.8

c) 58.7

d) 587.4

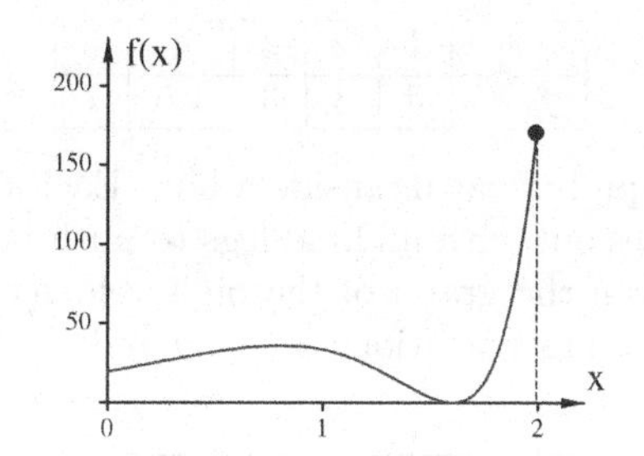

Figure 17.5: graph of f.

17.14. Given that $\int_0^1 f(x)\, dx = 4/3$, $\int_1^2 f(x)\, dx = 8/3$ and $\int_0^3 f(x)\, dx = 11/3$, find the values of the integrals $\int_0^2 f(x)\, dx$, $\int_1^3 f(x)\, dx$, and $\int_2^3 f(x)\, dx$.

17.15. Given that $\int_a^b f(x)\, dx = 18$, $\int_a^b g(x)\, dx = 5$, and $\int_a^b h(x)\, dx = -11$, evaluate the following integrals when possible. If you decide that there is not enough information given, explain your reasoning.

a) $\displaystyle\int_a^b (f(x) + g(x))\, dx$ b) $\displaystyle\int_a^b f(x)g(x)\, dx$ c) $\displaystyle\int_a^b (g(x) + h(x))\, dx$

d) $\displaystyle\int_a^b (g(x) - h(x))\, dx$ e) $\displaystyle\int_a^b \frac{g(x)}{h(x)}\, dx$ f) $\displaystyle\int_a^b (f(x) + g(x) + h(x))\, dx$

17.16. True or false? If $f : \mathbb{R} \to \mathbb{R}$ is continuous and $\int_a^b f(x)\, dx = 0$ for all $a, b \in \mathbb{R}$, then it must be true that $f(x) = 0$ for all $x \in \mathbb{R}$.

17.17. Let $f : [-5, 10] \to \mathbb{R}$ be a continuous function and let $g(x) := f(x) + 2$ for all $x \in \mathbb{R}$. If $\int_{-5}^{10} f(t)\,dt = 4$, then what is the value of $\int_{-5}^{10} g(u)\,du$?

17.18. For any real number b, find the value of $\int_0^b |2x|\,dx$. *Hint.* You need to distinguish two cases depending on whether b is greater than or less than zero.

17.19. Assume that f is a continuous function with $\int_0^1 f(x)\,dx = 2$, $\int_0^2 f(x)\,dx = 1$ and $\int_2^4 f(x)\,dx = 7$.

a) Find the values of $\int_0^4 f(x)\,dx$, $\int_1^0 f(x)\,dx$, and $\int_1^2 f(x)\,dx$.

b) Explain why there must exist an $x \in [1, 2]$ such that $f(x) < 0$.

c) Explain why there must exist a point $x \in [0, 4]$ such that $f(x) \geq 3.5$.

17.20. Let $f(t) := 2t$ and $g(t) := 2t + 1$ for all $t \in \mathbb{R}$. Complete the following tables:

a)

x	1	2	3	4	5	6
$\int_0^x f(t)\,dt$						

b)

x	1	2	3	4	5	6
$\int_0^x g(t)\,dt$						

17.21. Suppose that f and g are continuously differentiable on $\mathbb{R}$ and that for all $x \in \mathbb{R}$ we have $f(x) \leq g(x)$. Label each of the following statements as TRUE, FALSE, or POSSIBLY TRUE/POSSIBLY FALSE. Justify your answers.

a) $f'(x) \leq g'(x)$ for all $x \in \mathbb{R}$.

b) $f'(x) \geq g'(x)$ for all $x \in \mathbb{R}$.

c) $\int_0^1 f(x)\,dx \leq \int_0^1 g(x)\,dx$.

17.22. Assume that $f : \mathbb{R} \to \mathbb{R}$ is a decreasing continuous function. Given the values of f in the table below, use Riemann sums to find lower and upper estimates for $\int_1^5 f(x)\,dx$.

x	1	2	3	4	5
$f(x)$	4	3	3	1.5	1

17.23. Assume that a pig is tranquilly floating inside a box. Each time the pig bumps into one of the walls with its butt, it angrily strikes out with its hind legs to push back toward the center. Given these behavioral characteristics, and given the graph of the pig's velocity in the diagram below, you are to determine the distance between the pig's positions at $t = 0$ and $t = 5$.

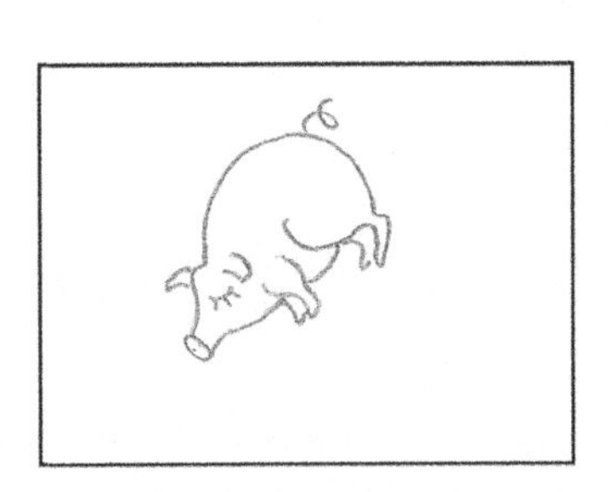
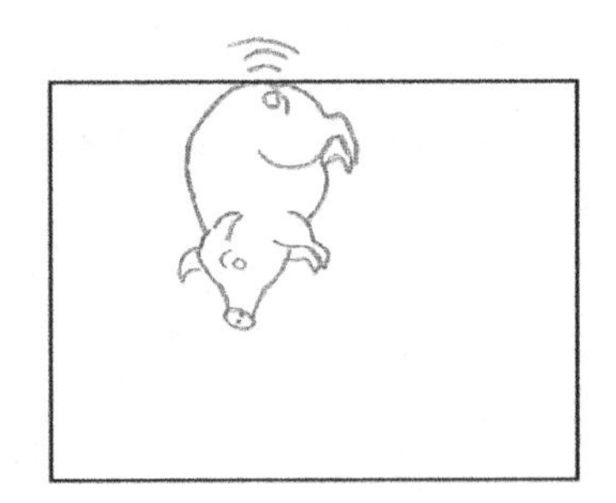
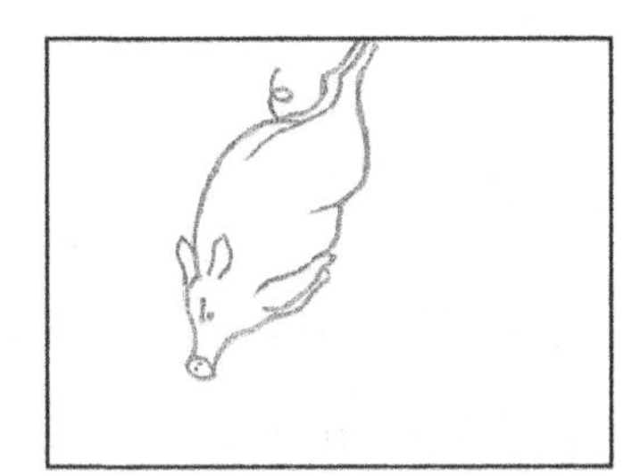

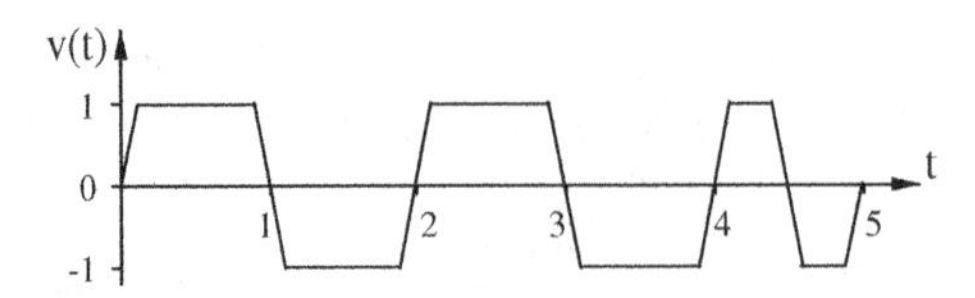

Figure 17.6: applied calculus.

Chapter 18

Bernhard Riemann and the Spirit of Pythagoras

> ...the so-called Pythagoreans, who were the first to take up mathematics, not only advanced this study, but also having been brought up in it they thought its principles were the principles of all things. Since of these principles numbers are by nature the first, and in numbers they seemed to see many resemblances to the things that exist and come into being—more than in fire and earth and water...; since, again, they saw that the modifications and the ratios of the musical scales were expressible in numbers;—since, then, all other things seemed in their whole nature to be modeled on numbers, and numbers seemed to be the first things in the whole of nature, they supposed the elements of numbers to be the elements of all things, and the whole heaven to be a musical scale and a number.[1]

These words by Aristotle describe some of the fundamental doctrines of a strange religious community that was founded by the Ionian philosopher Pythagoras in the second half of the sixth century B.C. near the city of Crotona in Southern Italy. Pythagoras was born around 570 B.C. on the island of Samos off the coast of Asia Minor (now Turkey). He is believed to have been a student of Thales in Miletus and probably also traveled to Egypt, Babylon, and possibly even India where he may have adopted some of his mystical beliefs and ascetic practices.

In Pythagorean thought, reality was ultimately grounded in abstract mathematical relations and numerical patterns. Numbers were commonly represented by geometric arrangements of pebbles or dots in the sand, and certain of these configurations assumed sacred significance. The *tetraktys*, for instance, that shows the number 10 to be the sum of 1, 2, 3 and 4 (see Figure 18.1), was given the role of a quasi-divine generating entity. The number 1, as the most basic unit, was identified with a single

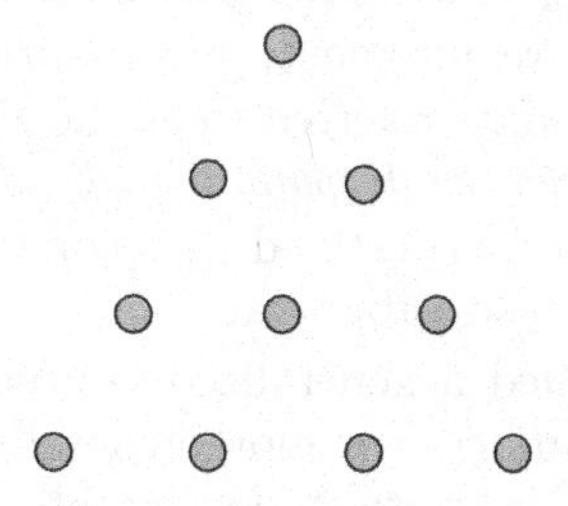

Figure 18.1: the tetraktys.

point; the number 2, represented by two points, defined a line, 3 a plane, and 4 symbolized the spacial structure of a tetrahedron. In asserting further that all material objects are composed of elementary geometric building blocks, it was therefore possible to regard numbers and numerical relations as the foundation of physical reality.

Pythagoras and his followers are famous also for their discovery of the mathematical laws underlying musical harmony. In studying tonal scales on stringed instruments, they realized, for example, that the pitch of a vibrating string is raised by one octave if its length is cut in half. This conditioning of an individual's experience of harmonious sound by numerical relations was seen to be of great significance. For it was thought to indicate that harmony was perceived when the soul had entered into a state of resonance with the abstract order of the cosmos.

Given their predisposition to search for universal truths in the ethereal realm of abstraction, it is perhaps not surprising that the Pythagoreans are generally credited with being the first to employ rigorous methods of deduction in mathematics. Most famously, the theorem of Pythagoras, which asserts the square of the hypotenuse of any rectilinear triangle to be equal to the sum of the squares of its sides, was shown to be valid by way of rational argument* (see Appendix A).

The Pythagoreans also left a rich legacy in the history of philosophy and science. Plato, for instance, derived from their ideas his conception of the soul and also his high regard for mathematics as an ideal form of knowledge. From Platonic idealism the influence then extended further to the semireligious philosophy of Plotinus, the theology of Augustine, and eventually the Renaissance mysticism of such thinkers as Giordano Bruno and Johannes Kepler. Here in particular, the Pythagorean teaching of the earth's motion around a central fire played an important role in the Copernican Revolution which brought about the transition from the geocentric to the heliocentric model of the universe (see Chapter 15). In fact, the Pythagorean cosmology was directly referred to by Copernicus in his major work *On the Revolutions of the Heavenly Spheres*, and it also motivated Kepler in his search for the fundamental laws of planetary motion (see Chapter 70). Furthermore, Kepler showed himself a genuine Neo-Pythagorean convert by passionately expressing his faith in a cosmic order that rests on divinely inspired mathematical laws. His emphasis on numerical and quantitative relations stood in stark contrast to the teleological natural philosophy of medieval scholasticism, and helped to pave the way for the rise of modern science.

Indeed, despite its obscure mysticism, the Pythagorean doctrine regarding the correspondence of material reality and mathematical abstraction has proved eerily prophetic and has been vindicated by the rapid development of science during the last four hundred years. To study a case in point, we will now move ahead in time to encounter one of the greatest mathematicians of the 19th Century: *Bernhard Riemann.*

Born the son of a Lutheran minister in Hannover, Germany in 1826, Bernhard Riemann showed his genius early when at the age of 14 he mastered the more than eight hundred pages of Legendre's treatise on number theory in little over a week. Upon entering the University of Göttingen in 1846, he at first complied with his father's wishes to pursue a course of study in theology, but soon devoted himself exclusively to his true passion—mathematics. Despite the presence of Carl Friedrich Gauss, however, the intellectual climate at Göttingen provided little inspiration for a talented young man, and Riemann therefore decided to move to Berlin "where the atmosphere was more democratic."[2] In 1849 he returned to Göttingen and two years later, after gaining his doctorate, he began the qualification process for a lecturing position. To satisfy the first requirement, he submitted a memoir on Fourier series in which he introduced the concept that is nowadays referred to as the *Riemann integral.* Concerning the second requirement, he delivered a lecture *Über die Hypothesen, die der Geometrie zu Grunde liegen (On the Hypotheses Underlying Geometry)* that was destined to become a classic of mathematical literature and even stirred the great Gauss to enthusiastic approval.

In his lecture Riemann had outlined a generalized conception of geometry that allowed for the study of spacial relations on curved surfaces (or more generally, topological spaces) of arbitrarily high dimension. In the face of such lofty abstraction, one might be tempted to consider Riemann's work a useless outgrowth of a self-serving academia but in 1916, Albert Einstein published a paper on a revolutionary new theory of gravity in which he relied on the Riemannian conception of geometry to

*The theorem itself was already known to earlier Sumerian cultures, but the Pythagoreans were the first to give a proof.

describe space and time as a four-dimensional curved continuum. Unfortunately, Riemann did not live to see these fruits of his labor, for he prematurely died of tuberculosis at the age of 39 in 1866. The mathematician Richard Dedekind described his last hours as follows:

> On the day before his death he lay beneath a fig tree, filled with joy at the glorious landscape [in Selasca, Italy], writing his last work, unfortunately left incomplete. His end came gently, without struggle or death agony; it seemed as though he followed with interest the parting of the soul from the body; his wife had to give him bread and wine, he asked her to convey his love to those at home, saying "Kiss our child." She said the Lord's prayer with him, he could no longer speak; at the words "Forgive us our trespasses" he raised his eyes devoutly, she felt his hand in hers becoming colder.[3]

The anticipation of Einstein's theory of relativity by Riemann's purely mathematical research into the foundations of geometry stands out as one of the most astonishing manifestations of the Pythagorean spirit in the history of science. In our age of technological utility and soulless quantification we are thus reminded of a distant past, and we contemplate anew the ancient dream of the divinity of number.

Chapter 19

The Fundamental Theorem of Calculus

The First Fundamental Theorem of Calculus

In Chapter 14, we observed that the problem of finding distance (or position) from velocity can be solved in two different ways—using antiderivatives or approximations via Riemann sums (i.e., integration). In this chapter we will show that these two methods are actually equivalent. The general mathematical principle underlying this equivalence is of the utmost importance and is known as the *first fundamental theorem of calculus* (FTC1). The discovery of the connection between antiderivatives and integration, or more generally between differentiation and integration, was one of the great triumphs in the development of calculus. The first to notice this relationship was Isaac Barrow, a professor of Greek, mathematics, and theology at Cambridge University. However, Barrow never fully realized the significance of his discovery and had to leave it to his student Isaac Newton to enter the history books as the discoverer of calculus. Newton shares his fame with Gottfried Wilhelm Leibniz who developed his own version of the calculus independently at almost the same time (see also Chapters 9 and 20).

To begin with, let us recall that in Chapter 14, p.124, we derived the formula $s(t) = gt^2/2$ by calculating the area under the graph of the velocity function $v(t) = gt$ over the interval $[0, t]$. Using integral notation as introduced in Chapter 17, we can express this result in the form

$$\int_0^t v(\tau)\,d\tau = \int_0^t g\tau\,d\tau = \frac{gt^2}{2} = s(t). \tag{19.1}$$

(Note: we use the letter τ here to distinguish the variable of integration from the upper boundary of integration t. This is not really necessary, but conceptually helpful.) The same result as in (19.1) was obtained on p.120 by observing that the distance function s is an antiderivative of the velocity function v. In other words, equation (19.1) implies that $\int_0^t v(\tau)\,d\tau$ is an antiderivative of v. Hence

$$\frac{d}{dt}\int_0^t v(\tau)\,d\tau = v(t). \tag{19.2}$$

The validity of this equation is not dependent upon the particular definition of v as the velocity function of a falling body. In a completely analogous fashion, equation (19.2) is true for every continuous function. More precisely, if $f : I \to \mathbb{R}$ is a continuous function defined on an open interval I, and a is an arbitrary point in I, then the integral $\int_a^x f(x)\,dx$ exists for all $x \in I$ (see the remark following Definition 17.5), and we have

$$\frac{d}{dx}\int_a^x f(t)\,dt = f(x) \text{ for all } x \in I. \tag{19.3}$$

This equation differs from (19.2) only in the lower boundary of integration (a instead of 0), the name of the function (f instead of v), and the names of the variables (x instead of t and t instead of τ). For convenience, we set $G(x) := \int_a^x f(t)\,dt$ and rewrite (19.3) in the form

$$G'(x) = f(x). \tag{19.4}$$

Remark. For a positive function f and $x \in I$ with $x > a$, the value $G(x)$ is equal to the area between the graph of f and the x-axis over the interval $[a, x]$ (see Figure 19.1). In a case where f is negative over parts of its domain, the value $G(x)$ is equal to the difference of the areas above and below the x-axis as explained in Chapter 17.

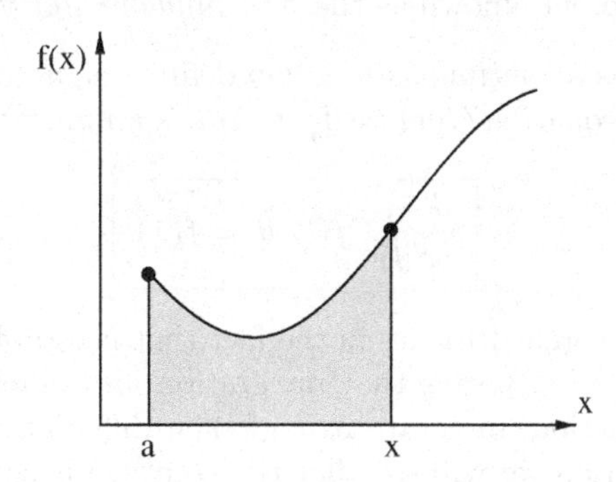

Figure 19.1: $G(x)$ interpreted as an area.

In order to prove (19.4), we will determine the derivative $G'(x)$ by evaluating the limit

$$\lim_{\Delta x \to 0} \frac{G(x + \Delta x) - G(x)}{\Delta x}.$$

Using Definition 17.9 and Theorem 17.10c, we obtain

$$G(x + \Delta x) - G(x) = \int_a^{x+\Delta x} f(t)\,dt - \int_a^x f(t)\,dt = \int_a^{x+\Delta x} f(t)\,dt + \int_x^a f(t)\,dt = \int_x^{x+\Delta x} f(t)\,dt,$$

and therefore,

$$G'(x) = \lim_{\Delta x \to 0} \frac{\int_x^{x+\Delta x} f(t)\,dt}{\Delta x}. \tag{19.5}$$

Since f is continuous (to use continuity at this point is crucial!), we may infer that

$$\lim_{\Delta x \to 0} f(x + \Delta x) = f(x) \quad \text{(see Definition 7.1)},$$

and this shows that for any sufficiently small Δx and any t between x and $x + \Delta x$, the value $f(t)$ will be very close to $f(x)$. Thus, the area under the graph of f between x and $x + \Delta x$ has approximately

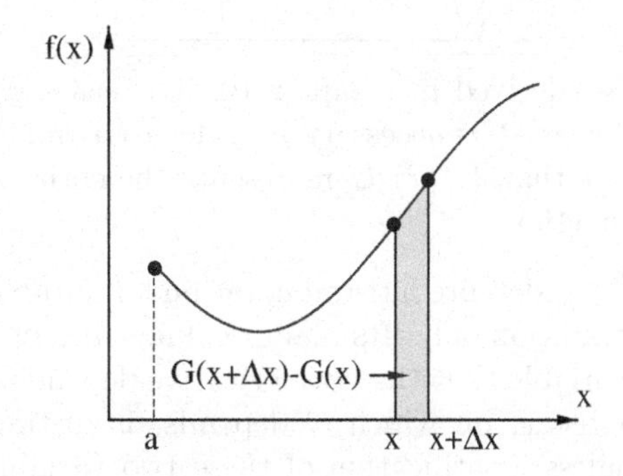

Figure 19.2: $G(x + \Delta x) - G(x)$ as an area under the graph of f.

the shape of a rectangle of height $f(x)$ and width Δx (see Figure 19.2). In other words, the integral $\int_x^{x+\Delta x} f(t)\,dt$ is approximately equal to $f(x)\Delta x$. Combining this observation with (19.5) yields

$$G'(x) \approx \lim_{\Delta x \to 0} \frac{f(x)\Delta x}{\Delta x} = f(x).$$

It can be shown (for a precise proof see [Rud1]) that this approximation is actually an exact equality, and our derivation of (19.4) (or, equivalently, (19.3)) is thus complete. To summarize our discussion, we formulate the following theorem, known as the *first fundamental theorem of calculus* (FTC1):

19.1 Theorem. *If $f : I \to \mathbb{R}$ is a continuous function defined on an open interval I, then for all $a \in I$ the function G defined by the equation $G(x) := \int_a^x f(t)\,dt$ is an antiderivative of f, i.e.,*

$$\boxed{\frac{d}{dx}\int_a^x f(t)\,dt = f(x).}$$

The great importance of Theorem 19.1 lies in the fact that it establishes a relation between integration and differentiation. It essentially says that integration and differentiation are inverse operations of each other: if we take a function, integrate it, and then differentiate the integral, we get back the function itself. In the next section we will see that the converse is also true: if we first differentiate a function and then integrate it, we get back the function itself. This mutual inverse relationship between integration and differentiation is formally reminiscent of the more familiar relationship between square roots and squares: if we take the square root of a positive number and then square it, we get back the number itself ($\sqrt{x}^{\,2} = x$ for all $x \geq 0$), and the same happens if we square the number first and then take the root ($\sqrt{x^2} = x$ for all $x \geq 0$).

Important notation. Motivated by the relation between differentiation and integration as expressed in the FTC1, we usually denote the antiderivatives of a function f by an integral without boundaries:

$$\boxed{\int f(x)\,dx.}$$

This expression is also referred to as the *indefinite integral* of f.

19.2 Example. Let f be defined by the equation $f(x) := x^2$. Since the derivative of $x^3/3$ is x^2, we may infer that any antiderivative of f is of the form $x^3/3 + C$ (see Theorem 14.2). Using the notation for antiderivatives as introduced above, we write

$$\int f(x)\,dx = \int x^2\,dx = \frac{x^3}{3} + C.$$

In general, for any integer $n \neq -1$ we have

$$\boxed{\int x^n\,dx = \frac{x^{n+1}}{n+1} + C,} \tag{19.6}$$

because using the power rule, as derived in Chapter 10, it is easy to see that x^n is the derivative of $x^{n+1}/(n+1)$. (The restriction $n \neq -1$ is necessary in order to avoid a division by zero.) In adding an arbitrary constant C, we indicate that $\int f(x)\,dx$ represents the entire set of all possible antiderivatives of f rather than just a single function.

It must be mentioned that the indefinite integral notation for antiderivatives is conceptually misleading and is used for the sake of tradition only. Its main weakness lies in its failure to keep the variable of integration separate from the variable that the antiderivative depends on: in writing $G(x) = \int_a^x f(t)\,dt$ it is made clear that the variable x, on which G depends, is distinct from the variable of integration t. By contrast, a meaningless identification of these two variables is suggested by the equation $G(x) = \int f(x)\,dx$.

The Second Fundamental Theorem of Calculus

19.3 Example. In Example 14.8 we determined the area under the graph of the function $f(x) = x^2 - 1$ over the interval $[1, 5]$ to be approximately equal to 37.313. Using the FTC1, we will now calculate the exact value of this area. To begin with, we restate the problem in a different way: for $G(x) := \int_1^x (t^2 - 1)\, dt$ we need to find the value of $G(5) = \int_1^5 (t^2 - 1)\, dt$, because the definite integral $\int_1^5 (t^2 - 1)\, dt$ is equal to the area under the graph of f over the interval $[1, 5]$. According to the FTC1, the function G is an antiderivative of f, but according to (19.6) we also know that

$$\int f(x)\, dx = \int (x^2 - 1)\, dx = \frac{x^3}{3} - x + C.$$

Therefore, Theorem 14.2 implies that there is a constant $C \in \mathbb{R}$ such that

$$\int_1^x (t^2 - 1)\, dt = G(x) = \frac{x^3}{3} - x + C.$$

To determine the value of C we use Theorem 17.10d to infer that

$$0 = \int_1^1 (t^2 - 1)\, dt = G(1) = \frac{1^3}{3} - 1 + C = -\frac{2}{3} + C.$$

Consequently, $C = 2/3$ and

$$G(x) = \frac{x^3}{3} - x + \frac{2}{3}.$$

Thus, for the area under the graph of f over the interval $[1, 5]$, we find the value $G(5) = 5^3/3 - 5 + 2/3 = 112/3 \approx 37.333$.

19.4 Exercise. For the function $f(x) := x^3 - x$, find an explicit formula for $G(x) := \int_2^x f(t)\, dt$ and determine the value of $\int_2^3 f(x)\, dx$.

To generalize the results of Example 19.3, let us assume that $f : I \to \mathbb{R}$ is a continuous function defined on an open interval I and that $F : I \to \mathbb{R}$ is an *arbitrary* antiderivative of f (in Example 19.3 we could have chosen $F(x) = x^3/3 - x$). For a given $a \in I$, the FTC1 and Theorem 14.2 imply that there is a constant $C \in \mathbb{R}$ such that for $G(x) := \int_a^x f(t)\, dt$ we have

$$G(x) = F(x) + C.$$

Since $G(a) = \int_a^a f(x)\, dx = 0$, it follows that

$$C = G(a) - F(a) = -F(a).$$

Hence,

$$\int_a^x f(t)\, dt = G(x) = F(x) - F(a). \tag{19.7}$$

This result is known as the *second fundamental theorem of calculus* (FTC2):

19.5 Theorem. *If $f : I \to \mathbb{R}$ is a continuous function defined on an open interval I and $F : I \to \mathbb{R}$ is an antiderivative of f, then*

$$\boxed{\int_a^b f(x)\, dx = F(b) - F(a) \text{ for all } a, b \in I.}$$

The difference $F(b) - F(a)$ is commonly denoted by $F(x)\Big|_a^b$.

In the previous section, we mentioned that differentiation and integration are, so to speak, inverse operations. To re-examine this relationship in light of the FTC2, we assume that $f : I \to \mathbb{R}$ is a continuously differentiable function (see Definition 7.6) defined on an open interval I. Since f is an antiderivative of f', equation (19.7) implies that for all $x, a \in I$ we have

$$\int_a^x f'(t)\,dt = f(x) - f(a).$$

Introducing the additional assumption $f(a) = 0$, we obtain $\int_a^x f'(t)\,dt = f(x)$. In other words, if we differentiate a function and then integrate the derivative, we get back the function itself (if $f(a) = 0$).

19.6 Example. To illustrate the usefulness of the FTC2, we will now revisit some of the problems discussed in Chapters 14 and 16. We begin with the problem of finding change in position from velocity for a falling body. An antiderivative of the function $v(t) = gt \approx 9.8t$ is $s(t) = 9.8t^2/2$. Therefore, the FTC2 implies that the distance a falling body covers during the first 2 seconds is

$$\int_0^2 9.8t\,dt = \left.\frac{9.8t^2}{2}\right|_0^2 = \frac{9.8 \cdot 2^2}{2} - \frac{9.8 \cdot 0^2}{2} = 19.6\,m.$$

In Chapter 16 we derived the estimate $V \approx \sum_{k=1}^n \pi(R^2 - x_k^2)\Delta x$ for the volume of the immersed portion of a ball floating on the surface of a body of water. Replacing the approximating Riemann sums in the limit $n \to \infty$ with an integral, the volume V is given by the equation

$$V = \int_{-R}^d \pi(R^2 - x^2)\,dx.$$

Using (19.6), we see that an antiderivative of $R^2 - x^2$ is $R^2x - x^3/3$, and therefore, we obtain

$$V = \pi\left.\left(R^2x - \frac{x^3}{3}\right)\right|_{-R}^d = \pi\left(dR^2 - \frac{d^3}{3} + R^3 - \frac{R^3}{3}\right) = \pi\left(dR^2 - \frac{d^3}{3} + \frac{2R^3}{3}\right).$$

This result is easily seen to be the same as (16.5), but there is a significant difference in the complexity of the two calculations: in Chapter 16 it took us two pages to prove this result, while the fundamental theorem of calculus reduces the calculation to a single line.

Given our discussion at the end of Chapter 16, it is easy to see that, in general, the volume of a solid with cross-sectional areas $A(x)$ is given by the equation

$$\boxed{V = \int_a^b A(x)\,dx.}$$

In the special case that the volume is obtained by revolving the graph of a continuous or piecewise continuous function f about the x-axis over an interval $[a, b]$, the cross-sectional areas are circles with radius $f(x)$ so that $A(x) = \pi f(x)^2$. Thus, we arrive at the following formula for the volume of a solid of revolution:

$$\boxed{V = \pi \int_a^b f(x)^2\,dx.}$$

The volume of the barrel discussed in Example 16.4 was formed by revolving the graph of $f(x) = 4 - x^2/16$ about the x-axis over the interval $[-4, 4]$. Therefore, the volume of the barrel is

$$\begin{aligned} V &= \pi \int_{-4}^4 \left(4 - \frac{x^2}{16}\right)^2 dx = \pi \int_{-4}^4 \left(16 - \frac{x^2}{2} + \frac{x^4}{256}\right) dx = \pi\left.\left(16x - \frac{x^3}{6} + \frac{x^5}{1280}\right)\right|_{-4}^4 \\ &= 2\pi\left(64 - \frac{32}{3} + \frac{4}{5}\right) \approx 340.13. \end{aligned}$$

Again, the fundamental theorem of calculus reduces the complexity of the calculation significantly.

Additional Exercises

19.7. Use the fundamental theorem of calculus to find *exact* solutions for Exercises 14.12, 16.6, 16.8, 16.9, 16.10, and 16.11.

19.8. Let F be an antiderivative of a continuous function $f : [a, b] \to \mathbb{R}$. Label each of the statements given below as TRUE, FALSE, or POSSIBLY TRUE/POSSIBLY FALSE. Justify your answers.

a) $\int_a^b f(x)\,dx = F(b) - F(a)$

b) $\int_a^b F(x)\,dx = f(b) - f(a)$

c) $\int_a^b 3 \cdot f(x)\,dx = 3(F(b) - F(a))$

d) $\int_a^b f(x)\,dx = f(t)(b - a)$ for some $t \in [a, b]$. (Here you are expected to give only an intuitive argument.)

19.9. Suppose that $f : \mathbb{R} \to \mathbb{R}$ is a continuously differentiable function. Find the value of $\int_{-2}^{2} f'(x)\,dx$ under the assumption that $f(x) = x^2 + x + 1$ for all $x \in [-3, -1]$ and $f(x) = x^3$ for all $x \in [1, 3]$.

19.10. For the function $f(x) := x^2 - x + 3$ do the following:

a) Find the value of $\int_1^3 f(x)\,dx$.

b) Find a point $t \in [1, 3]$ such that $\int_1^3 f(x)\,dx/2 = f(t)$ (see also Exercise 19.8d).

19.11. The average value of a function f over an interval $[a, b]$ is defined to be $\dfrac{1}{b-a}\displaystyle\int_a^b f(x)\,dx$.

a) Find the average values of $f_1(x) := x$, $f_2(x) := x^2$, and $f_3(x) := x^3$ over the interval $[1, 3]$.

b) For an arbitrary integer $n \neq -1$ find the average value of $f(x) := x^n$ over the interval $[1, 3]$.

19.12. Assume that $\int_c^x f(t)\,dt = 5x^3 + 40$.

a) Find a defining equation for $f(x)$.

b) Find the value of c.

19.13. Let $f(x) := \int_0^x \sqrt{16 - t^2}\,dt$.

a) Explain why $f(x) = -f(-x)$.

b) Find the value of $f(0)$.

c) What are the values of f' at $x = 2$ and $x = -2$?

19.14. Find the equation of the tangent line to the graph of the function $f(x) := \int_1^x \sqrt[3]{t^2 + 7}\,dt$ at the point $(1, f(1))$.

19.15. Find the area of the region bounded below by the x-axis and above by the parabola given by the equation $y = 2x/a^2 - x^2/a^3$ for $a > 0$. (Note: only that part of the parabola which is above the x-axis is relevant as a boundary for the region.)

19.16. For each of the functions f given below, find the value of c for which the line $x = c$ divides the area under the graph of f over the interval $[a, b]$ in such a way that the ratio of the areas to the left and to the right of the line is $2 : 1$.

a) $f(x) := (x - 2)^2$, $a = -1$, $b = 1$.

b) $f(x) := 1 - \dfrac{1}{x^2}$, $a = 1$, $b = 2$.

19.17. For the function f shown in Figure 19.3, we define $g(x) := \int_0^x f(t)\,dt$. Given this definition, you are to mark the x-coordinates of the...

- **a)** local maxima of g.
- **b)** local minima of g.
- **c)** global maximum of g.
- **d)** global minimum of g.
- **e)** points of inflection of g.

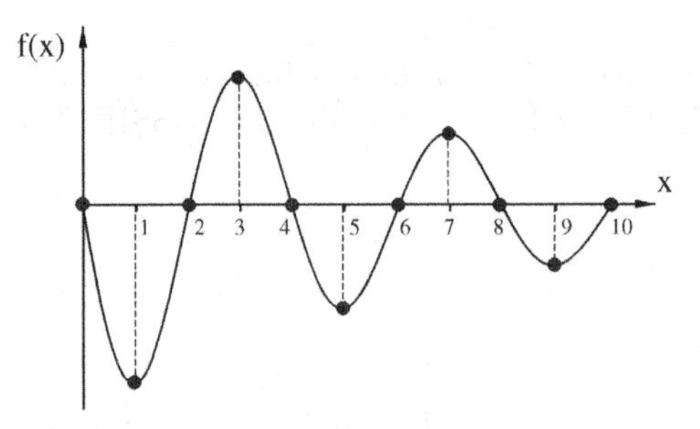

Figure 19.3: graph of $f(x)$.

19.18. For each of the limits given below, find a definite integral that is equal to the limit and then determine the value of the limit by evaluating the integral.

$$\textbf{a)}\ \lim_{n\to\infty}\frac{1}{n}\sum_{k=1}^{n}\left(\frac{k}{n}\right)^3 \qquad \textbf{b)}\ \lim_{n\to\infty}\frac{1}{n}\sum_{k=1}^{4n}\left(\frac{k}{n}\right)^3 \qquad \textbf{c)}\ \lim_{n\to\infty}\frac{1}{n}\sum_{k=n+1}^{2n}\left(\frac{k}{n}\right)^3$$

19.19. Find $f'(1)$ for $f(x) := \int_0^{x^3} t\sqrt{t+9}\,dt$.

19.20. Compute $\int_0^2 f'(t)\,dt$ under the assumption that f' is continuous on $[0, 2]$ and $f(0) = f(2)$.

19.21. Assume that f and g are continuously differentiable functions defined on $\mathbb{R}$ such that $f(x) \leq g(x)$ for all $x \in \mathbb{R}$. Show that there must exist a point $x \in \mathbb{R}$ such that $f'(x) < g'(x) + 1$.

19.22. Find the volume of the cylindrical segment shown in Figure 19.4.

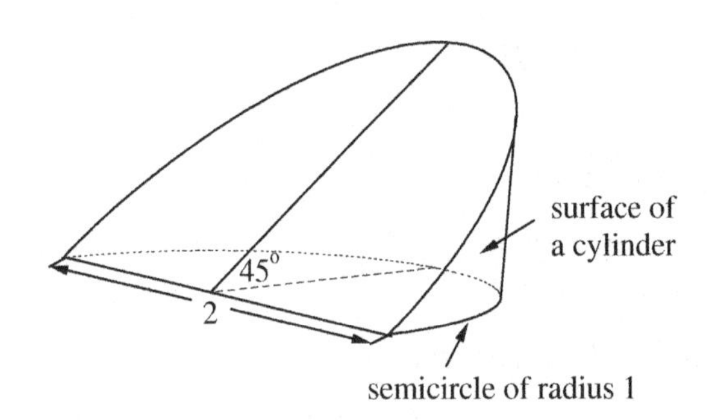

Figure 19.4: a cylindrical segment.

19.23. ★ Assume that $f : \mathbb{R} \to \mathbb{R}$ is a twice continuously differentiable function such that $f(0) = 0$ and $f''(x) \leq 0$ for all $x \in \mathbb{R}$. Use the FTC2 to prove that $f(x+y) \leq f(x) + f(y)$ for all $x, y \geq 0$.

Chapter 20

New Horizons

Sophie: Do you mind if I ask a few questions concerning the Copernican Revolution that have recently been on my mind?

Teacher: Not at all.

Sophie: At one point in our discussion, you referred, as I recall, to certain historical and cultural developments that exerted a decisive influence in the process of replacing the Aristotelian vision of a central earth with the heliocentric system of Copernicus. In this context you specifically mentioned "the enterprising spirit of the Renaissance" as a catalyst for change.

Teacher: I did indeed.

Sophie: To my knowledge, the Renaissance was a time of rediscovery, or literally, a "rebirth" of ancient skill and learning. It therefore puzzles me to think that Aristotelian cosmology came under such severe attack in an age that was much inspired by the legacy of Greek and Hellenistic culture.

Teacher: It's never easy to identify the exact causes of historical events, and whatever conclusion we finally arrive at, there is always room for doubt. In direct response to your remarks, though, I would like to urge some caution not to look upon the Renaissance chiefly as a period of restoration. I intentionally characterized the spirit of this time as "enterprising" rather than nostalgic or conservative. Classical antiquity defined a point of departure from the strictures of medievalism, not a final destination. We can see this pattern clearly in the work of Nicolaus Cusanus (1401–1464), one of the outstanding exponents of early Renaissance philosophy. He had a love for ancient manuscripts but also wished to reach beyond the confines of tradition. Being more daring even than Copernicus himself, he envisioned the universe to have no center and asserted the physical nature of the earth and the moon to be essentially identical with that of other celestial bodies. In his philosophical speculation, he frequently employed mathematical analogies that invoked the notion of infinity. For instance, the ever closer approach to the truth that he thought to be the fruit of the human search for knowledge he likened to the arbitrarily close approximation of a circle by a sequence of inscribed polygons (see the diagram on the left in Figure 20.1). Similarly, he illustrated his belief in the

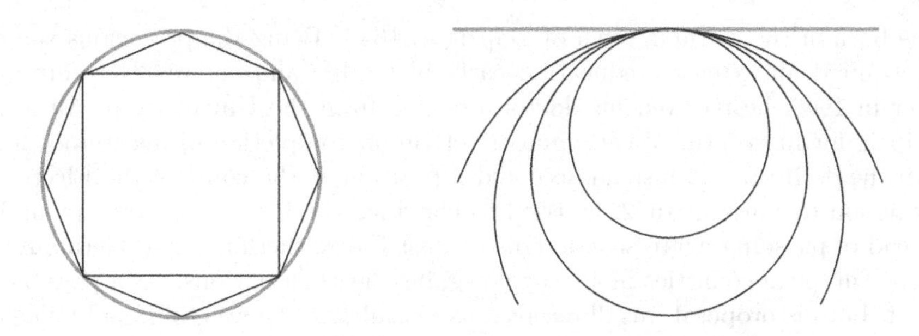

Figure 20.1: mathematical analogies employed by Nicolaus Cusanus.

reconciliation of all opposites in God with the transformation of a circle into a line in the process of increasing the circle's radius to infinity (see the diagram on the right in Figure 20.1).

Simplicio: So God is, so to speak, the point of convergence in which circle and line are restored to unity.

Teacher: That is correct and really rather striking. For not only does the mathematical character of the illustration appear prophetic in light of the development of modern science that was soon to come, but in Cusanus's appeal to the inifnite we also detect fairly overtly the visionary quality of Renaissance thought. The Hellenic spirit strove above all for moderation, for a coming to rest, so to speak, at the golden mean. But the leading minds of the Renaissance, by contrast, were driven by a desire to test the limits, to see new horizons spread out before them. Thus we notice, for example, how Sandro Botticelli places the scene of *The Annunciation* within a carefully constructed one-point perspective that opens a view on a river and a faraway mountain range. Such compositional diversions from the spiritual theme in the forefront would undoubtedly have been deemed inappropriate by a painter in medieval times. Likewise, the German artist Albrecht Dürer, one of the leading figures of the Northern Renaissance, portrayed himself at the age of 26 with a far receding landscape to his right so as to thereby suggest the scope of his ambition.

Figure 20.2: *The Annunciation* by Botticelli (1489–90) and *Self-Portrait at 26* by Albrecht Dürer (1498).

Sophie: All of these ideas and historical movements are really most absorbing. In particular, I am fascinated to see how, in the thought of Cusanus, the notion of a limit at infinity emerged and came to prominence. Afterall, limits, in a sense more narrow and more rigorously defined, are central to the calculus.

Teacher: Your point is clearly valid. For calculus indeed is distinctly modern in spirit. It is not static and finitely contained, like classical Euclidean geometry, but rather open, dynamic, and restless. Claculus is the mathematics of change, and it was in the work of *Gottfried Wilhelm Leibniz*—one of its co-inventors—that this truly modern theme of the calculus became most vividly apparent.

Leibniz was born in the German town of Leipzig in 1646. Being the precocious son of a university professor, he acquired an extensive education early in his life and entered the university at age 15. Five years later in 1666, he received his doctoral degree from the University of Altdorf, having been rejected at Leipzig for his youth. When immediately upon completion of his studies he was offered a university chair, he declined and instead accepted a position at the court of the Elector of Mainz. On a diplomatic mission to Paris in 1672, he tried to convince the French government under King Louis XIV that, instead of pursuing aggressive designs against the Netherlands and Germany, France should unite with other European countries in a struggle against heathen nations. As a possible first target he suggested Egypt, but his proposal was ill received as crusades, so he was told, had fallen out of fashion.

Still in Paris, but with his diplomatic fortunes sinking, Leibniz diverted his attention to mathematics and embarked on a course of research that eventually led to the creation of the calculus. Already in his doctoral thesis, *De Arte Combinatoria* (On the Art of Combinations), he had discussed a calculus related subject: sequences of numbers and their higher-order differences. For instance, starting from the sequence of squares

$$0, 1, 4, 9, 16, 25, 36, \ldots$$

he found the values

$$1, 3, 5, 7, 9, 11, \ldots \quad (1 = 1 - 0,\ 3 = 4 - 1 \text{ etc.})$$

for the first differences and

$$2, 2, 2, 2, 2, 2, \ldots \quad (2 = 3 - 1 = 5 - 3 \text{ etc.})$$

for the second differences. In conclusion, he stated that all third differences are zero $(0 = 2 - 2)$. Furthermore, in anticipation of the fundamental theorem of calculus, he noticed that the sum of the first differences of a finite sequence that starts from zero is equal to its last term.

20.1 Exercise. Show that the fourth differences of the sequence of cubes vanish, and justify Leibniz's assertion concerning the sum of the first differences.

To relate these observations to the study of functions, Leibniz interpreted the numbers in his sequences to be the y-values of a given function. The corresponding x-values he initially thought of as enumerating indices of the y-values, so that the change in x is 1. (Note: using the notation introduced in Chapter 4, we would write $\Delta x = 1$.) Writing l for the change in y and using the abbreviation "omn" (derived from the Latin word "omnia," which means "all") to indicate summation, he inferred that

$$\text{omn.}\, l = y$$

whenever the first y-value is zero. In present day notation this result would read

$$\sum_{k=1}^{n} \Delta y_k = \sum_{k=1}^{n} (f(x_k) - f(x_{k-1})) = f(x_n) - f(x_0) = f(x_n) - 0 = y,$$

but it also holds in the limit as Δy_k is replaced with dy:

$$\int dy = y.$$

Leibniz then proceeded to consider the function $y = f(x) = x$ and realized that for "small" l, the value of $\text{omn.}\, yl$ equals the triangular area under the graph of f. Thus, he arrived at the conclusion that

$$\text{omn.}\, yl = \frac{y^2}{2}$$

or, as we would write,

$$\int y\, dy = \frac{y^2}{2}.$$

At the next stage, he developed a simplified version of a general rule of integration that is closely related to the product rule and commonly referred to as *integration by parts* (see Chapter 35). In substituting x for l he was able to apply this rule to infer that

$$\text{omn.}\, x^2 = \frac{x^3}{3}. \tag{20.1}$$

In a manuscript dated October 29, 1675, he replaced "omn" with the familiar integral sign representing an elongated S for "sum." Thus, in place of (20.1), he now would write

$$\int x^2 = \frac{x^3}{3}.$$

In the same paper he used the letter "*d*" to indicate differences, and two weeks later introduced dy for the differential of y. Furthermore, in a rather obscure chain of reasoning, he arrived at the conclusion that "*d*" and "$\int$" are inverse operations. This insight, which captures the essence of the fundamental theorem of calculus (see Chapter 19), was probably inspired by similar ideas of Barrow, who had obtained area from antidifferentiation. However, in emphasizing the importance of summation in the process of integration, Leibniz introduced an important advance over the work of Barrow that proved crucial for future developments.

With respect to differentiation, Leibniz struggled with the problem that as long as the changes in x and y are not actually zero, the ratio dy/dx is not exactly equal to the derivative as we defined it in Chapter 4 using a limit. To overcome this difficulty, he used an argument involving similar triangles (see Figure 20.3) from which he deduced that $dy/dx = TU/SU$. However, taken as a definition for

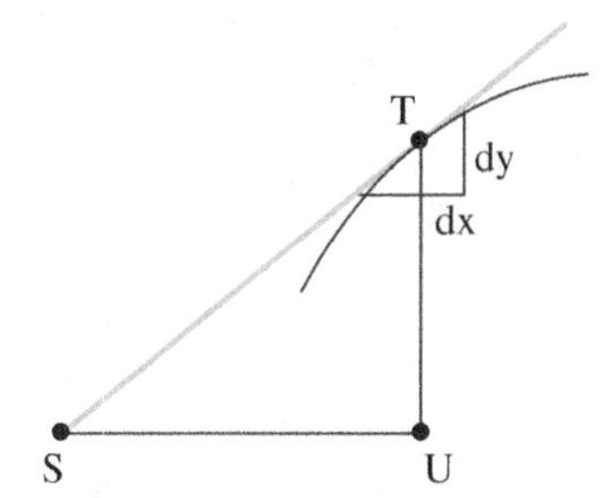

Figure 20.3: similar triangles along a tangent line.

dy/dx, this equation is not very useful, for in order to determine the so-called subtangent SU, the slope of the tangent line, or equivalently the derivative at T, must already be known.

20.2 Exercise. In the early days of calculus the subtangent was considerably more important than it is now. In Figure 20.4, SU is the subtangent of the parabola given by the equation $y = x^2$.

a) Show that the length of the subtangent is $x/2$.

b) Find the length of the subtangent of $y = x^3$.

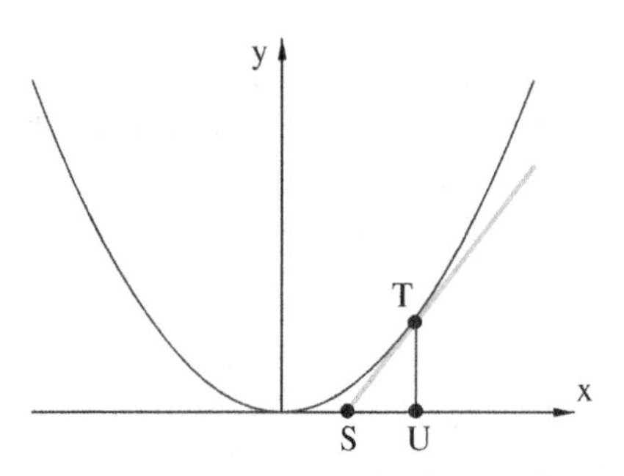

Figure 20.4: tangent line at the graph of $y = x^2$.

By July 11, 1677, in spite of his difficulties in defining dy/dx, Leibniz was able to correctly state the general rules for the differentiation of sums, differences, products, and quotients of functions as well as for powers and roots of functions, the latter being special cases of what we now refer to as the chain rule.

Leibniz first published his results on the calculus in the *Acta Eruditorum* in 1684. When Newton followed suit with his *Principia* in 1687, an unfortunate controversy over priority in discovery arose that was unprofitable to all sides. Since Newton had communicated results concerning the calculus to friends since 1665, Leibniz was suspected of plagiarism. However, from the historical record there can be no doubt that Newton and Leibniz worked independently. Both need to be given credit for developing the partial results of such thinkers as Fermat and Barrow into a universal tool for the study of functions.

Interestingly, neither Newton nor Leibniz are best remembered for their seminal contributions to the history of mathematics. Newton did his greatest work in physics, while Leibniz is most famous for his philosophical ideas. Already in the *De Arte Combinatoria* of 1666, he had attempted to develop a universal method of reasoning. His goal was to establish a rigorous logical system by means of which disputes in all realms of human knowledge could be settled by performing logical calculations in accordance with certain standard rules. In fact, had Leibniz found the time to organize and publish his results, he most likely would have entered the history books as the inventor of symbolic logic, a discipline that later came to prominence in positivistic philosophy and early 20th-century research into the foundations of mathematics.

In addition, Leibniz also did important work in theology, history, philology, geology, and library science. He was a diplomat and lawyer by profession, but also served as an engineer in the mines of the Harz Mountains and even busied himself with technological inventions involving clocks, windmills, and hydraulic presses. That is to say, his versatility was utterly astounding.

In the final years of his life, Leibniz fell out of favor with the house of Hannover and died lonely and neglected in 1716. The man, who possibly for the last time in European history had acquired universal knowledge, was buried in almost total anonymity. Only the French Academy of Sciences took note with an obituary in honor of his achievements.

PART III

SPECIAL FUNCTIONS

Chapter 21

The Natural Logarithm and Newton's Law of Cooling (Part 1)

Newton's Law of Cooling (Part 1)

Teacher: Today we will discuss a classical application of the calculus—the problem of heat propagation. As an example, consider a potato at room temperature ($25^\circ\, C$) that is placed in a heated oven ($205^\circ\, C$) to bake. Given the additional information that after 12 minutes the temperature of the potato has reached $175^\circ\, C$, we wish to find an equation that describes the potato's temperature at any point in time t.

Sophie: So we need to determine the temperature of the potato as a function of t?

Teacher: Yes, and to express this functional dependence in our notation, we write $T(t)$ for the temperature of the potato at time t. Before we get into the details of the calculation, though, I would like you to sketch an approximate graph of T based on your intuitive understanding of the heating process.

Simplicio: Wouldn't we need some more information before we can sketch a graph?

Teacher: If we wanted to sketch a precise graph, we would indeed have to find first a defining equation for $T(t)$, but at this point I only want you to discuss the shape of the graph at a qualitative level.

Simplicio: Well, we can certainly say that T is increasing, because the temperature of a cold potato placed in a hot oven increases over time.

Teacher: That is a good observation, but I think we can say a bit more. For instance, how will the rate of change of $T(t)$ vary over time?

Sophie: I think that the rate of change will be decreasing, because at the beginning the difference in temperature between the potato and the oven is very large so that the temperature will increase quickly; but as the temperature of the potato approaches the temperature of the oven, the rate of change in $T(t)$ will get smaller.

Teacher: We actually have a name for functions with a decreasing rate of change. We call them...

Sophie: ...concave down.

Teacher: Correct. There is another aspect of your answer, Sophie, that is also interesting: you said that "the temperature of the potato *approaches* the temperature of the oven." In geometric terms, this simply means that the graph of T has a horizontal asymptote at $205^\circ\, C$—the temperature of the oven. A qualitative sketch of the graph of T, consistent with these observations, is shown in Figure 21.1.

Simplicio: The graph looks reasonable, but how are we going to get from here to an exact equation for $T(t)$?

Teacher: That is our next problem. However, we cannot expect to solve this problem using only deductive mathematical reasoning because we are dealing with a physical phenomenon. Instead, our solution needs to be based on an empirically verifiable physical law, which in our case is *Newton's*

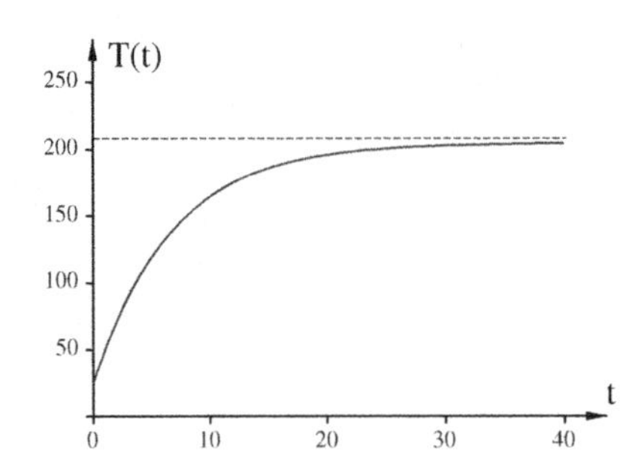

Figure 21.1: graph of T.

law of cooling:

> The temperature of an object changes at a rate proportional to the difference between the object's temperature and the surrounding temperature.

Sophie: That makes sense. For in the example of the potato in the oven, we said that the rate of change of $T(t)$ decreases as $T(t)$ approaches the surrounding temperature of the oven (or equivalently, as the difference between $T(t)$ and the surrounding temperature of the oven approaches zero). But I don't quite understand why the rate of change in $T(t)$ is simply proportional to the difference between $T(t)$ and the surrounding temperature. Couldn't we just as well have said that perhaps the rate of change of $T(t)$ is proportional to the cube of the difference between $T(t)$ and the surrounding temperature?

Teacher: You need to remember that Newton's *law* of cooling is not a mathematical proposition but a law of physics. The available experimental evidence suggests that the rate of change of $T(t)$ is indeed proportional to the difference between $T(t)$ and the surrounding temperature and not, as you proposed, to the cube of this difference. We simply have to accept this as a fact.

Sophie: Thank you, that clarifies it.

Teacher: In order for us to be able to apply Newton's law of cooling, we need to express this law in the form of an equation, and for this purpose we first need to find a mathematical expression for the rate of change in the object's temperature, that is, for the rate of change in $T(t)$.

Simplicio: Isn't the rate of change simply the change in the temperature $T(t)$ over the change in t?

Teacher: That would be the *average* rate of change, but that is not really what we want. Think about it: if we measure the difference between the object's temperature and the surrounding temperature at one particular moment in time t, then we also need to determine the rate of change in the object's temperature at that same moment in time. In other words, we need to determine the *instantaneous* rate of change in $T(t)$...

Sophie: ...which is the derivative $T'(t)$.

Teacher: Correct. So if we denote by H the surrounding temperature (which in the case of the potato in the oven is equal to $205° C$), then, according to Newton's law of cooling, there is a constant of proportionality k such that

$$\boxed{T'(t) = k(H - T(t)),}$$

or equivalently,

$$\frac{T'(t)}{H - T(t)} = k.$$

Integrating both sides of this equation yields

$$\int \frac{T'(t)}{H - T(t)}\, dt = \int k\, dt = kt + C. \tag{21.1}$$

The integral on the left-hand side of this equation can be solved by using the antiderivative of the function $f(x) := 1/x$, because if $F'(x) = 1/x$, then the chain rule implies that

$$-\frac{d}{dt}F(H-T(t)) = -F'(H-T(t))\frac{d}{dt}(H-T(t)) = f(H-T(t))T'(t) = \frac{T'(t)}{H-T(t)}.$$

In other words,

$$\boxed{-F(H-T(t)) \text{ is an antiderivative of } \frac{T'(t)}{H-T(t)} \text{ if } F \text{ is an antiderivative of } f(x) = \frac{1}{x}.} \tag{21.2}$$

Unfortunately, here we encounter a little problem: the antiderivative of $1/x = x^{-1}$ cannot be determined from (19.6), because for $n = -1$ we run into a division by zero in the term $x^{n+1}/(n+1)$. For this reason, we will introduce the antiderivative of $1/x$ in the next section directly via the FTC1, and then derive its most important properties from our knowledge of the definite integral.

The Natural Logarithm

If f is a continuous function defined on an open interval I then, according to the FTC1, the function $F(x) := \int_a^x f(t)\,dt$ is an antiderivative of f for every $a \in I$. In particular, $\int_1^x 1/t\,dt$ is an antiderivative of the function $f(x) := 1/x$ defined on $(0,\infty)$, because $1/x$ is continuous on $(0,\infty)$ (see Figure 21.2). The antiderivative $\int_1^x 1/t\,dt$ is commonly denoted by $\ln(x)$ and referred to as the *natural logarithm* of

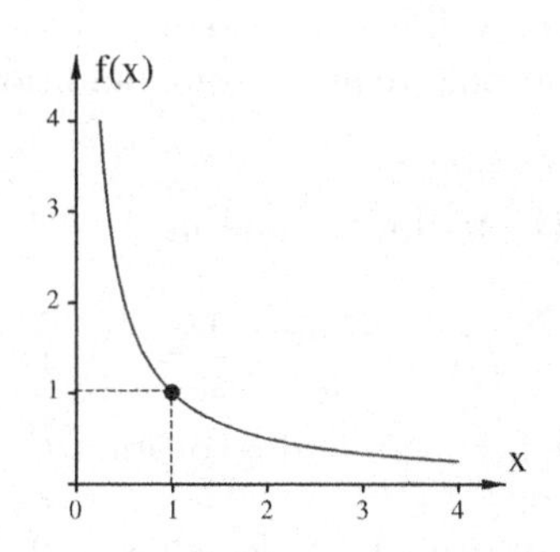

Figure 21.2: graph of $f(x) = 1/x$.

x. So we define

$$\boxed{\ln(x) := \int_1^x \frac{1}{t}\,dt \text{ for all } x > 0.}$$

The restriction $x > 0$ is necessary because, as x approaches zero from the right, the values of $1/x$ tend to positive infinity (see Figure 21.2) so that for $x \leq 0$ the integral $\int_1^x 1/t\,dt$ is undefined (a general discussion of this type of situation will be given in Chapter 38).

From the definition above and the graph of $f(x) = 1/x$ as shown in Figure 21.2, we can immediately derive the following properties of $\ln(x)$:

$$\frac{d}{dx}\ln(x) = \frac{1}{x}, \tag{21.3}$$

$$\ln(1) = 0, \tag{21.4}$$

$$\ln(x) > 0 \text{ whenever } x > 1, \tag{21.5}$$

$$\ln(x) < 0 \text{ whenever } 0 < x < 1. \tag{21.6}$$

Proof. As we already mentioned, (21.3) is a direct consequence of the FTC1. Property (21.4) follows from Theorem 17.10d by observing that $\ln(1) = \int_1^1 1/t\,dt = 0$. Furthermore, if $x > 1$, then $\ln(x) =$

$\int_1^x 1/t\,dt$ is greater than zero because $1/t$ is positive on $[1, x]$, and this proves (21.5). Finally, to prove (21.6), we notice that for $0 < x < 1$ we have $\ln(x) = \int_1^x 1/t\,dt = -\int_x^1 1/t\,dt < 0$ (see Definition 17.9) because $1/t$ is positive on $[x, 1]$. □

Other important properties of the natural logarithm function are listed in the following theorem (readers who are not familiar with the basic definitions and properties of exponentials are referred to Appendix B):

21.1 Theorem. *If a and b are positive real numbers, then*

a) $\ln(ab) = \ln(a) + \ln(b)$,

b) $\ln(a/b) = \ln(a) - \ln(b)$,

c) $\ln(1/a) = -\ln(a)$, *and*

d) $\ln(a^r) = r\ln(a)$ *for all rational numbers r.*

Furthermore,

e) $\ln(x)$ *is a strictly increasing function,*

f) $\ln(x)$ *is concave down, and*

g) $\lim_{x\to\infty}\ln(x) = \infty$ *and* $\lim_{x\to 0^+}\ln(x) = -\infty$.

Remark. A number is said to be *rational* if it can be written as a fraction of two integers (see also the ⋆remark⋆ on p.23). Moreover, the validity of statement c) above can be extended from the rational numbers to all real numbers, but this requires some additional work and will have to wait until Chapters 23 and 24.

Proof. **a)** For $x \in (0, \infty)$ we set $f(x) := \ln(ax)$ and $g(x) := \ln(x)$. Then, using the chain rule and (21.3), we may infer that

$$f'(x) = \frac{a}{ax} = \frac{1}{x} = g'(x).$$

Therefore, according to Theorem 14.2, we can find a constant $C \in \mathbb{R}$ such that

$$f(x) = g(x) + C \text{ for all } x \in (0, \infty).$$

To determine C, we set $x = 1$. This yields

$$\ln(a) = f(1) = g(1) + C = \ln(1) + C = C \quad \text{(by (21.4))}.$$

Consequently, $\ln(ax) = \ln(x) + \ln(a)$ for all $x \in (0, \infty)$, and a) now follows by setting $x = b$.
b) Using a), we obtain

$$\ln\left(\frac{a}{b}\right) = \ln\left(\frac{a}{b}\right) + \ln(b) - \ln(b) = \ln\left(\left(\frac{a}{b}\right)b\right) - \ln(b) = \ln(a) - \ln(b).$$

c) This is a trivial consequence of b) and (21.4).
d) See Exercise 21.2.
e) and **f)** Since

$$\frac{d}{dx}\ln(x) = \frac{1}{x} > 0$$

for all $x \in (0, \infty)$, Theorem 4.11 implies that $\ln(x)$ is strictly increasing. The fact that $\ln(x)$ is concave down follows from Theorem 11.10 by observing that

$$\frac{d^2}{dx^2}\ln(x) = -\frac{1}{x^2} < 0.$$

g) Using d), we may infer that $\ln(2^n) = n\ln(2)$ for all integers $n > 0$. Since $\ln(2) > 0$ by (21.5), it is obvious that as n increases to ∞, $n\ln(2)$ increases to ∞ as well. Since we already know that $\ln(x)$ is strictly increasing, it follows that

$$\lim_{x\to\infty} \ln(x) = \lim_{n\to\infty} \ln(2^n) = \infty.$$

Furthermore, d) implies that $\ln(1/x) = \ln(x^{-1}) = -\ln(x)$, and therefore

$$\lim_{x\to 0+} \ln(x) = \lim_{x\to\infty} \ln\left(\frac{1}{x}\right) = -\lim_{x\to\infty} \ln(x) = -\infty.$$

This completes the proof of Theorem 21.1. □

21.2 Exercise. Prove Theorem 21.1d. *Hint.* Use Theorem 21.1a to show first that $\ln(x^n) = n\ln(x)$ for all integers n and $\ln(x^{1/m}) = (1/m)\ln(x)$ for all positive integers m. Then prove c) for $r = n/m$.

Consistent with the properties listed above, the graph of $\ln(x)$ is shown in Figure 21.3.

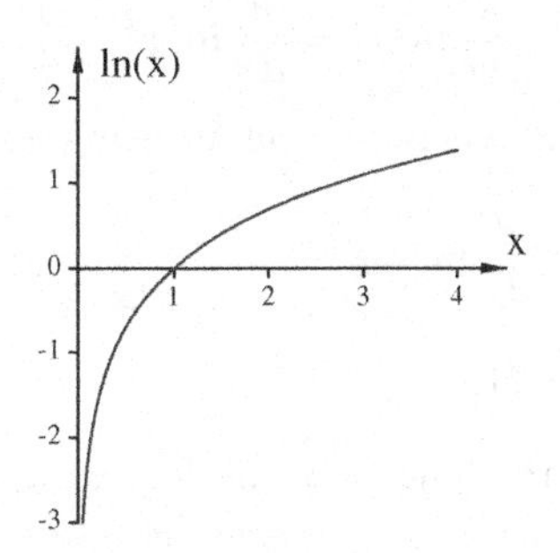

Figure 21.3: graph of $\ln(x)$.

21.3 Example. Theorem 21.1 is frequently useful in algebraic simplifications as the following examples are intended to illustrate:

a) Using Theorem 21.1a, c, d, the expression $2\ln(1/4) + \ln(8)$ simplifies as follows:

$$2\ln(1/4) + \ln(8) = \ln((1/4)^2) + \ln(8) = \ln(1/16) + \ln(8) = \ln(8/16) = \ln(1/2) = -\ln(2).$$

b) To simplify the term $\ln(x^2 - 4) - \ln(x+2)$, we apply Theorem 21.1b. This yields

$$\ln(x^2 - 4) - \ln(x+2) = \ln\left(\frac{x^2 - 4}{x+2}\right) = \ln\left(\frac{(x+2)(x-2)}{x+2}\right) = \ln(x+2).$$

The natural logarithm function is defined to be the antiderivative of $1/x$ only on the interval $(0, \infty)$, whereas the maximal domain of the function $1/x$ is $\mathbb{R} \setminus \{0\}$ (the set of all real numbers different from zero). It is therefore desirable to also find an antiderivative of $1/x$ on $(-\infty, 0)$. To this end, we examine the graph of the function $\ln|x|$ as shown in Figure 21.4. The symmetry of the graph of $\ln|x|$ with respect to the vertical axis implies that the slope at any point $(x, \ln|x|)$ with $x < 0$ is the negative of the slope at the point $(-x, \ln|-x|)$. (Note: if $x < 0$, then $-x > 0$.) This suggests that $\ln|x|$ is an antiderivative of $1/x$ on $\mathbb{R} \setminus \{0\}$, because for $f(x) = 1/x$ we have $f(x) = -f(-x)$.

21.4 Theorem. *The function $f : \mathbb{R} \setminus \{0\} \to \mathbb{R}$, defined by the equation $f(x) := \ln|x|$, is an antiderivative of $1/x$ on $\mathbb{R} \setminus \{0\}$, i.e.,*

$$f'(x) = \frac{d}{dx} \ln|x| = \frac{1}{x} \text{ for all } x \in \mathbb{R} \setminus \{0\},$$

or equivalently,

$$\int \frac{1}{x}\, dx = \ln|x| + C.$$

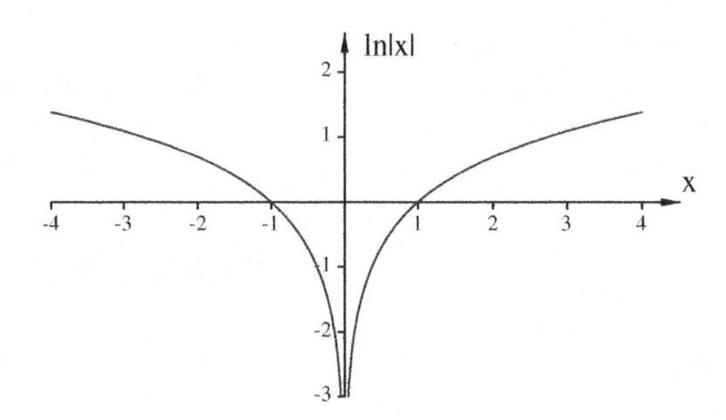

Figure 21.4: graph of $\ln|x|$.

Proof. We need to show that the derivative of $f(x) = \ln|x|$ is $1/x$ for all $x \neq 0$. For $x > 0$, (21.3) implies that

$$\frac{d}{dx}\ln|x| = \frac{d}{dx}\ln(x) = \frac{1}{x},$$

as desired. If $x < 0$, then $|x| = -x$ and therefore, again using (21.3) and the chain rule, we obtain

$$\frac{d}{dx}\ln|x| = \frac{d}{dx}\ln(-x) = \frac{1}{-x}\cdot\frac{d}{dx}(-x) = \frac{-1}{-x} = \frac{1}{x}.$$

This completes the proof of Theorem 21.4. □

21.5 Exercise. Use the chain rule to show that the following equation is valid for all $a, b \in \mathbb{R}$ with $a \neq 0$:

$$\boxed{\int \frac{1}{ax+b}\,dx = \frac{1}{a}\ln|ax+b| + C.}$$

21.6 Exercise. Use Theorem 21.1 to explain why the result of Exercise 21.5 can also be written in the form

$$\int \frac{1}{ax+b}\,dx = \frac{1}{a}\ln\left|x+\frac{b}{a}\right| + C.$$

Application of $\ln(x)$ to the Problem of Heat Propagation

According to Theorem 21.4, the antiderivative F in (21.2) is equal to $\ln|x|$. Therefore, by combining (21.1) with (21.2), we obtain

$$\boxed{-\ln|H - T(t)| = kt + C} \tag{21.7}$$

With the techniques that are to be discussed in Chapter 22, it will be possible to solve this equation for $T(t)$, but even in absence of an explicit solution we can already use equation (21.7) to determine the values of the constants H, k, and C. The surrounding temperature H is the temperature of the oven and by assumption equal to $205°\,C$. Since the initial temperature of the potato at time $t = 0$ is $25°\,C$, it follows that

$$-\ln(180) = -\ln|205 - 25| = -\ln|H - T(0)| = C.$$

To determine k, we use the fact that at time $t = 12\,min$ the temperature is assumed to be $T(12) = 175°\,C$. This yields

$$-\ln(30) = -\ln|205 - 175| = -\ln|H - T(12)| = k\cdot 12 + C = 12k - \ln(180),$$

or equivalently,

$$k = \frac{\ln(180) - \ln(30)}{12} = \frac{\ln(180/30)}{12} = \frac{\ln(6)}{12} \quad \text{(by Theorem 21.1b).}$$

In summary:

$$H = 205, \ C = -\ln(180), \ \text{and} \ k = \frac{\ln(6)}{12}. \tag{21.8}$$

Additional Exercises

21.7. Find the maximal domain of each of the following functions:

a) $\ln(x^2 - 5x + 6)$ **b)** $\sqrt{\ln(x+4)}$ **c)** $\ln|x - 4|$ **d)** $\ln(x^3 - 8)$

21.8. Differentiate each of the following functions:

a) $\ln(x^2 - 5x + 6)$ **b)** $x^2 \ln(x+6)$ **c)** $\ln(x)/x$ **d)** $(\ln(x+6))^2$
e) $\ln(\sqrt{x})$ **f)** $\ln(5x+6)$ **g)** $\ln(3x-5)$ **h)** $\ln(ax+b)$

21.9. Determine the following antiderivatives:

a) $\displaystyle\int \frac{3}{x}\, dx$ **b)** $\displaystyle\int \frac{2x}{1+x^2}\, dx$ **c)** $\displaystyle\int \frac{4}{1-5x}\, dx$
d) $\displaystyle\int \frac{x+1}{x-3}\, dx$ **e)** $\displaystyle\int \frac{x}{2x+3}\, dx$ **f)** $\displaystyle\int \frac{3x^2}{x^3-3}\, dx$

Hint. The results in b) and f) will be of the form $\ln|g(x)| + C$. To determine $g(x)$, you need to set the given function in the integral equal to the derivative of $\ln|g(x)|$, which is to be computed using the chain rule. Furthermore, the integral in d) can be solved by adding and subtracting 3 in the numerator $x+1$ and then splitting up the fraction into a sum of the form $1 + a/(x-3)$. A similar trick can also be applied in e).

21.10. Express the following statements as mathematical equations:

a) The rate of change in temperature is proportional to the square of pressure.
b) The force applied is proportional to the rate of change in velocity.
c) The slope of the function is proportional to the cube of the dependent variable.

21.11. Find the equation of the line tangent to the graph of $f(x) := \ln(x)$ at $x_0 = 1$.

21.12. Show that the area under the graph of the function $f(x) := 1/x$ over the interval $[1/2, 2]$ is equal to $\ln(4)$.

21.13. Find the equation of the normal line to the graph of $f(x) := x\ln(x)$ at $x_0 = 2$, that is, the line perpendicular to the tangent line to the graph at $(x_0, f(x_0))$.

21.14. Simplify the following expressions as far as possible:

a) $\ln(6) - \ln(3)$ **b)** $\ln(x^2 - 1) - \ln(x - 1)$ **c)** $\ln(\sqrt{x}) + \ln(x)/2$ **d)** $\ln(9) + 2\ln(1/3)$

21.15. In the statement "$\ln(x^2) = 2\ln(x)$ for all $x \in \mathbb{R} \setminus \{0\}$," there is a mistake. Identify the mistake and then correct it by filling in the blank in the statement "$\ln(x^2) = \dots$ for all $x \in \mathbb{R} \setminus \{0\}$."

21.16. Given a differentiable function f, find the derivatives of the following functions:

a) $\ln(1 + f(x))$ **b)** $f(1 + \ln(x))$ **c)** $\ln(1 + \ln(f(x)))$ **d)** $f(\ln(\ln(x) + f(x)))$

Chapter 22

Inverse Functions, the Exponential Function, and the Law of Cooling (Part 2)

Inverse Functions

Teacher: Before we return to the potato in the oven that we studied in the previous lesson, we need to discuss some basic facts concerning *inverse functions*. To begin with, we will play a little game. Simplicio, I want you to think of a number.

Simplicio: Just any number?

Teacher: Yes, and don't tell us which one you picked. Are you ready?

Simplicio: Yes.

Teacher: Now take your number, multiply it two times by itself, and tell us the result.

Simplicio: 27.

Teacher: Can we tell from the result 27 which number Simplicio picked initially?

Sophie: Of course, he picked 3, because $3^3 = 27$.

Teacher: Now let me play the game with Sophie. Please pick a number.

Sophie: Got one.

Teacher: Multiply it two times by itself and tell me the result.

Sophie: -8.

Teacher: Simplicio, what was the number that Sophie picked initially?

Simplicio: -2, because $(-2)^3 = -8$.

Teacher: Excellent. Now let me change the rules of the game a bit: Simplicio, I want you to pick a number, square it, and tell us the result.

Simplicio: I got 4.

Teacher: What was the number that Simplicio picked this time?

Sophie: We cannot be sure, because the square of both -2 and 2 is equal to 4. So all we can say is that Simplicio must have picked 2 *or* -2.

Teacher: That is exactly right. To analyze the difference between the two games from a theoretical point of view, we first notice that in both cases, the rules of the game establish a functional relationship between input and output values. For example, according to the rules of the first game, we take a number x and produce the output value $f(x) := x^3$. Then, given only this output value, we are able to determine the value of x because x is simply the cube root of $f(x)$. In other words, the function $f : \mathbb{R} \to \mathbb{R}$ with $f(x) := x^3$ has the property that to each output value y in the range of f there corresponds exactly one input value x in the domain of f with $f(x) = y$. By contrast, the output value $g(x) := x^2$ in the second game does not allow us to deduce the value of x from $g(x)$ because x is equal to either $\sqrt{g(x)}$ or $-\sqrt{g(x)}$. However, this uncertainty regarding the value of x can be removed if we are given some additional information. If, for example, Simplicio had said that he had picked a *positive* number the square of which is 4, then we could have been sure that the number he had picked was 2. In mathematical terminology, the restriction of x to positive values is equivalent to a restriction of the domain of the function g to $[0, \infty)$. So just as the function f, the function $g : [0, \infty) \to \mathbb{R}$ with $g(x) := x^2$ has the property that to each output value y in the range of g there corresponds exactly one input value x in the domain of g such that $g(x) = y$.

22.1 Definition. A function $f : A \to B$ is said to be *invertible* if to every y in the range of f there corresponds exactly one value $x \in A$ with $f(x) = y$. Equivalently, we can say that f is invertible if for every y in the range of f the equation $f(x) = y$ has exactly one solution $x \in A$.

It is important to point out that when deciding whether a function is invertible, it is not enough to know only the functional relationship between input and output values—we also need to know the domain of the function. Given the discussion above, we see, for example, that the equation $f(x) = x^2$ defines an invertible function on $[0, \infty)$, but not on $\mathbb{R}$.

For real-valued functions defined on subsets of $\mathbb{R}$ (i.e., functions of the form $f : D \subset \mathbb{R} \to \mathbb{R}$), there is a nice geometric interpretation of invertibility known as the *horizontal line test*:

For all $y \in \mathbb{R}$ the equation $f(x) = y$ has at most one solution (that is, f is invertible), if and only if every line parallel to the x-axis intersects the graph of f at most once.

The horizontal line test for the function $f(x) := x^3$ defined on $\mathbb{R}$ is illustrated in Figure 22.1.

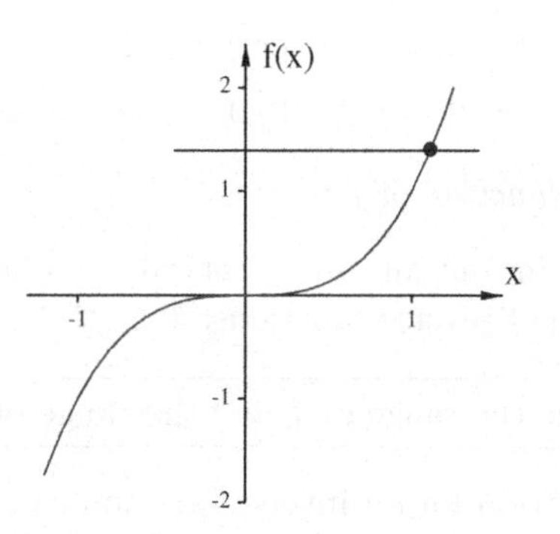

Figure 22.1: horizontal line test for $f(x) = x^3$.

If a function is not invertible, then there are some values y for which we can find more than one solution of the equation $f(x) = y$. In other words, there are lines parallel to the x-axis that intersect the graph of f more than once, and f, therefore, does not pass the horizontal line test. This, for instance, is the case for the function $f(x) := x^2$ defined on $\mathbb{R}$ (see Figure 22.2).

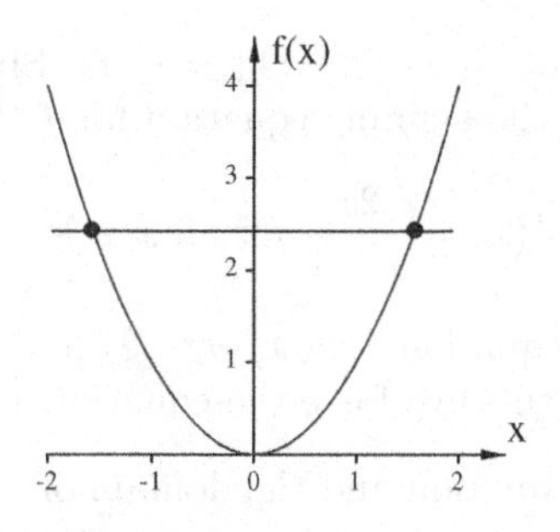

Figure 22.2: horizontal line test for $f(x) = x^2$.

22.2 Exercise. Decide which of the following functions are invertible:

a) $f : [0, \infty) \to \mathbb{R}$; $f(x) := \sqrt{x}$

b) $f : \mathbb{R} \to \mathbb{R}$; $f(x) := x^2(x-1)$

c) $f : [-2, -1] \to \mathbb{R}$; $f(x) := x^2$

d) $f : [1, \infty) \to \mathbb{R}$; $f(x) := x^2(x-1)$

22.3 Exercise. Explain why a function of the form $f : \{1, 2, 3\} \to \{1, 2\}$ cannot be invertible.

If a function $f : A \to B$ is invertible, then for every y in the range of f there is exactly one element x in the domain corresponding to it in the sense that $f(x) = y$. This means that there is a reverse functional dependence of the input values x in the domain on the output values y in the range—each y in the range $R(f)$ is assigned exactly one x in the domain. As an example, let us consider the function $f : [0, \infty) \to \mathbb{R}$ defined by the equation $f(x) := x^2$. Given a value $y = f(x)$ in the range $R(f) = [0, \infty)$, we see that x is equal to $\sqrt{y}$. So the function $g : [0, \infty) \to [0, \infty)$ with $g(y) := \sqrt{y}$ satisfies the equation $f(g(y)) = y$ for all $y \in [0, \infty)$ (simply because $\sqrt{y}^2 = y$). Furthermore, for every x in the domain of f (i.e., $x \in [0, \infty)$) we have $g(f(x)) = x$ because if $x \geq 0$ then $\sqrt{x^2} = |x| = x$. Thus, we say that g is the *inverse function* of f. The inverse function is usually denoted by f^{-1} (be careful *not* to misinterpret this notation as $1/f$).

22.4 Theorem. *If A and B are sets and $f : A \to B$ is an invertible function, then there is a function $f^{-1} : R(f) \to A$ such that*

$$(f^{-1} \circ f)(x) = f^{-1}(f(x)) = x \text{ for all } x \in A \text{ and}$$
$$(f \circ f^{-1})(y) = f(f^{-1}(y)) = y \text{ for all } y \in R(f).$$

f^{-1} is referred to as the inverse function of f.

For clarification we wish to point out an important rule, implicit to the statement of Theorem 22.4, concerning the domain and range of inverse functions:

The domain of f^{-1} is the range of f and the range of f^{-1} is the domain of f.

22.5 Example. An explicit equation for an inverse function can sometimes be found by solving for x the equation $f(x) = y$. To illustrate this point, let us define $f : \mathbb{R} \setminus \{2\} \to \mathbb{R}$ by the equation

$$f(x) := \frac{x}{x-2}.$$

Solving the equation $y = x/(x-2)$ for x yields

$$x = \frac{2y}{y-1}.$$

Consequently, $f^{-1}(y) = 2y/(y-1)$ for all $y \in \mathbb{R} \setminus \{1\} = R(f)$. Since we usually denote the independent variable by x instead of y, we write the defining equation for f^{-1} in the form

$$f^{-1}(x) = \frac{2x}{x-1} \text{ for all } x \in \mathbb{R} \setminus \{1\}.$$

Note: instead of solving for x the equation $y = x/(x-2)$ and then replacing y with x, we can also switch x and y at the outset and then solve for y the equation $x = y/(y-2)$.

22.6 Exercise. Find the defining equation and the domain of f^{-1} for the function $f : \mathbb{R} \setminus \{-2\} \to \mathbb{R}$ defined by the equation

$$f(x) := \frac{3x}{x+2}.$$

The graph of an inverse function f^{-1} is obtained by reflecting the graph of the original function f across the diagonal line $y = x$ (see Figure 22.3). This can be understood as follows: let (x, y) be a point on the graph of f, i.e., $f(x) = y$. Then (y, x) is a point on the graph of f^{-1}, because $x = f^{-1}(f(x)) = f^{-1}(y)$. It is not difficult to see that for any $x, y \in \mathbb{R}$ the positions of the points (x, y) and (y, x) are symmetric relative to the diagonal line given by the equation $y = x$. Consequently, the graph of f^{-1} is indeed a reflection of the graph of f across this diagonal line.

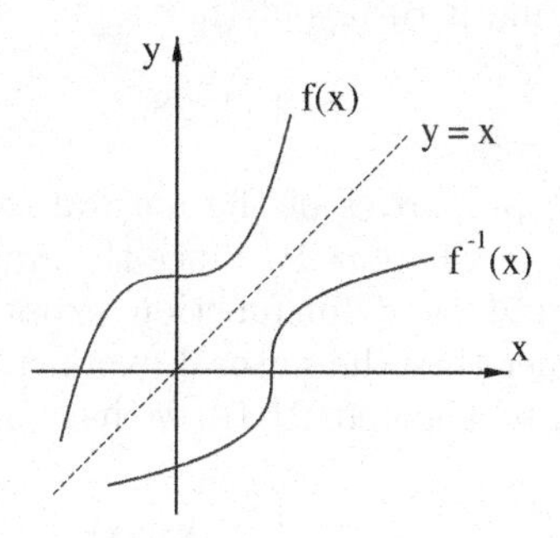

Figure 22.3: graph of an inverse function.

The next theorem tells us how to find the derivative of an inverse function f^{-1} from the derivative of f.

22.7 Theorem. *If $f : I \to \mathbb{R}$ is an invertible and differentiable function defined on an open interval I such that $f'(x) \neq 0$ for all $x \in I$, then f^{-1} is differentiable as well and*

$$(f^{-1})'(x) = \frac{1}{f'(f^{-1}(x))} \quad \text{for all } x \in R(f).$$

Proof. To prove the differentiability of f^{-1} is beyond the scope of this text (see [Rud1]), but in accepting by faith the existence of the derivative of f^{-1}, the values of $(f^{-1})'$ are readily determined: using the chain rule and the equation $f(f^{-1}(x)) = x$, we obtain

$$1 = \frac{d}{dx}x = \frac{d}{dx}f(f^{-1}(x)) = f'(f^{-1}(x))(f^{-1})'(x),$$

and therefore,

$$(f^{-1})'(x) = \frac{1}{f'(f^{-1}(x))}.$$

□

22.8 Exercise. Use Theorem 22.7 with $f(x) := x^2$ to verify that $\frac{d}{dx}\sqrt{x} = \frac{1}{2\sqrt{x}}$ (see Example 4.6).

The Natural Exponential Function

Since the graph of the natural logarithm function as shown in Figure 21.3 passes the horizontal line test, we may infer that $\ln(x)$ is invertible on its domain $(0, \infty)$. Given this observation, we may introduce the following definition:

22.9 Definition. The *natural exponential function* is defined to be the inverse function of the natural logarithm function and is denoted by $\exp(x)$. The domain of the natural exponential function is $\mathbb{R}$ and its range is $(0, \infty)$. In particular, we have the defining equations

$$\exp(\ln(x)) = x \text{ for all } x \in (0, \infty),$$
$$\ln(\exp(x)) = x \text{ for all } x \in \mathbb{R}.$$

Furthermore, the value $\exp(1)$ is known as *Euler's number* and denoted by the letter e.

The value of Euler's number can be determined via numeric approximations from the definition of the natural logarithm function. Since $e = \exp(1)$, it follows that $\ln(e) = \ln(\exp(1)) = 1$. Therefore, e satisfies the equation

$$\int_1^e \frac{1}{x}\, dx = 1.$$

Given this equation, the value of e can be determined to an arbitrarily high degree of accuracy via approximations by Riemann sums, and it turns out that

$$e \approx 2.71828.$$

Our next objective is to derive the properties of the natural exponential function from those of the natural logarithm function as stated in Chapter 21. Since the graph of the natural exponential function is a reflection of the graph of the natural logarithm function across the diagonal line given by the equation $y = x$ (see Figure 22.4), we may infer that the natural exponential function is strictly increasing and concave up. Furthermore, according to Theorem 21.1d, we have $\ln(x^r) = r\ln(x)$ for all rational numbers

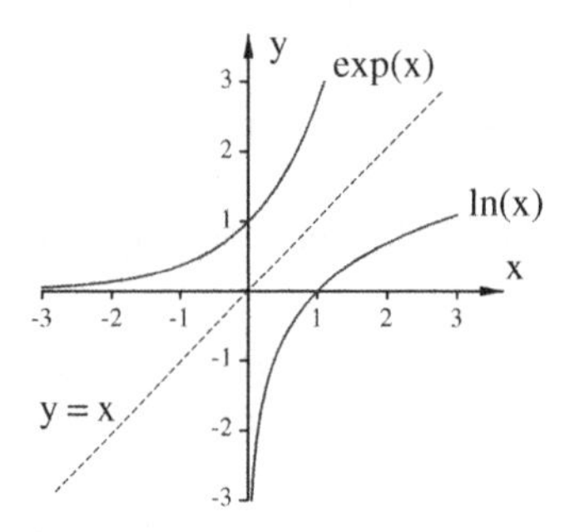

Figure 22.4: graph of $\exp(x)$.

r and all $x \in (0, \infty)$. Hence

$$e^r = \exp(\ln(e^r)) = \exp(r\ln(e)) = \exp(r \cdot 1) = \exp(r). \tag{22.1}$$

In words: for all rational numbers r, the value of the natural exponential function at r is e raised to the power r. This observation justifies the name natural *exponential* function for $\exp(x)$, and it also explains why $\ln(x)$ is referred to as the natural *logarithm* function. For $\ln(x)$ does exactly what a logarithm is supposed to do: from a given value $y = e^r$, it produces the exponent r because $\ln(y) = \ln(e^r) = r\ln(e) = r$. Moreover, equation (22.1) lends plausibility to the following *definition* of exponential expressions of the form e^x for real numbers x that are not necessarily rational:

$$\boxed{e^x := \exp(x).}$$

Given this definition, it follows that

$$\begin{aligned} e^{\ln(x)} &= x \text{ for all } x \in (0, \infty) \text{ and} \\ \ln(e^x) &= x \text{ for all } x \in \mathbb{R} \text{ (see Definition 22.9).} \end{aligned} \tag{22.2}$$

Equivalently, we may infer that for all $x \in \mathbb{R}$ and all $y \in (0, \infty)$ the following statement is valid:

$$\boxed{e^x = y \text{ if and only if } x = \ln(y).} \tag{22.3}$$

The next theorem lists some simple properties of the natural exponential function that the reader is probably well familiar with from elementary lessons on the subject.

22.10 Theorem. *Let $a, b \in \mathbb{R}$. Then*

a) $e^a e^b = e^{a+b}$,

b) $\dfrac{e^a}{e^b} = e^{a-b}$, *and*

c) $e^0 = 1$.

Proof. **a)** This is a consequence of Theorem 21.1a, (22.2), and (22.3): setting $A := e^a$ and $B := e^b$, it follows that $\ln(A) = a$ and $\ln(B) = b$, and therefore,

$$e^a e^b = AB = e^{\ln(AB)} = e^{\ln(A)+\ln(B)} = e^{a+b}.$$

b) See Exercise 22.11.
c) This is an immediate consequence of (21.4) and (22.3). □

22.11 Exercise. Prove Theorem 22.10b.

To determine the derivative of the natural exponential function, we recall that $1/x$ is the derivative of $\ln(x)$ and apply Theorem 22.7 to $f(x) = \ln(x)$ and $f^{-1}(x) = e^x$:

$$\frac{d}{dx}e^x = \frac{d}{dx}f^{-1}(x) = \frac{1}{f'(f^{-1}(x))} = \frac{1}{1/f^{-1}(x)} = f^{-1}(x) = e^x.$$

In a somewhat informal manner the same result can be deduced from the fact that the equation $y = e^x$ is equivalent to the equation $x = \ln(y)$:

$$1 = \frac{d}{dx}x = \frac{d}{dx}\ln(y) = \frac{1}{y}\cdot\frac{dy}{dx} = \frac{y'}{y}.$$

Multiplying by y yields $y = y'$. This establishes the very remarkable fact that the derivative of the natural exponential function is the natural exponential function itself:

$$\boxed{\frac{d}{dx}e^x = e^x.}$$

22.12 Exercise. Find the following derivatives: $\dfrac{d}{dx}e^{(x^2)}$, $\dfrac{d}{dx}x^2e^{2x-1}$, and $\dfrac{d}{dx}\ln(e^{\ln|x+2|})$.

The Law of Cooling (Part 2)

Given the equation

$$-\ln|H - T(t)| = kt + C,$$

which we arrived at in (21.7), we are now able to determine $T(t)$ by first multiplying this equation by -1 and then applying the natural exponential function to both sides. This yields

$$|H - T(t)| = e^{-kt-C} = e^{-kt}e^{-C},$$

and therefore

$$H - T(t) = \pm e^{-C}e^{-kt}$$

or, equivalently,

$$T(t) = H \pm e^{-C}e^{-kt}.$$

Substituting the values for H, C, and k as listed in (21.8), we see that

$$T(t) = 205 \pm e^{\ln(180)}e^{-\ln(6)t/12} = 205 \pm 180e^{-\ln(6)t/12}.$$

In order to decide whether we have a positive or negative sign in front of the term $180e^{-\ln(6)t/12}$, we need to recall that $25 = T(0) = 205 \pm 180e^0 = 205 \pm 180$. Since $25 = 205 - 180$, our final result is

$$T(t) = 205 - 180e^{-\ln(6)t/12} \approx 205 - 180e^{-0.1493t}.$$

Additional Exercises

22.13. Find the maximal domain of each of the functions given below. Then, given these domains, decide which of the functions are invertible and explicitly state the domain of the inverse function in case it exists.

a) $f(x) = x^3 - 1$ **b)** $f(x) = 3x - 1$ **c)** $f(x) = x^2 + 2$
d) $f(x) = 1 + \frac{1}{x}$ **e)** $f(x) = \sqrt{3x - 5}$ **f)** $f(x) = \frac{x+1}{x-2}$

22.14. For each of the following functions find $(f^{-1})'(x)$.

a) $f(x) = x^5$ **b)** $f(x) = \frac{1}{x-1}$ **c)** $f(x) = \frac{1}{x^3}$ **d)** $f(x) = x^3 - 1$

22.15. Find the derivative of each of the following functions:

a) e^{2x} **b)** e^{3x-2} **c)** $e^{3x} - 2$ **d)** $e^{\ln(x)}$ **e)** $\ln(e^{-x})$
f) e^{5x^2} **g)** xe^x **h)** x^2e^{4x} **i)** $\frac{e^x + e^{-x}}{e^x - e^{-x}}$ **j)** $\frac{e^{2x}}{e^{2x} + 1}$

22.16. Determine the following antiderivatives:

a) $\int e^{4x}\,dx$ **b)** $\int e^{5x+1}\,dx$ **c)** $\int e^{3-2x}\,dx$ **d)** $\int e^{ax+b}\,dx$
e) $\int \frac{e^{4x}}{e^{4x}-1}\,dx$ **f)** $\int xe^{x^2}\,dx$ **g)** $\int e^x e^{e^x}\,dx$ **h)** $\int e^{\ln(x)}\,dx$

Hint. The antiderivatives e), f), and g) will be of the form $a\ln|f(x)| + C$, $be^{g(x)} + C$, and $e^{h(x)} + C$ respectively. Apply the chain rule to each of these expressions and set the result equal to the corresponding integrand. Furthermore, the integral in h) looks complicated but is actually trivial. Think about it!

22.17. Find a function $P(t)$ such that $P'(t) = 6(2000 - P(t))/5$ and $P(0) = 100$. *Hint.* You need to emulate the calculation that allowed us to determine $T(t)$ from the equation $T'(t) = k(H - T(t))$.

22.18. Find a function $Y(t)$ such that $Y'(t) = 4(Y(t) - 50)/5$ and $Y(0) = 100$.

22.19. Given the graph of $f(x) := e^x$ as shown in Figure 22.4, sketch the graphs of the following functions:

a) e^{2x} **b)** $2e^x$ **c)** $e^x + 2$ **d)** e^{x+2}
e) $2e^{2x}$ **f)** e^{-x} **g)** $-e^{-x}$ **h)** $(e^x + e^{-x})/2$

22.20. Sketch the graph of $f(x) := x^2e^{-x}$, identifying all local extrema and points of inflection.

22.21. The concentration of a drug t minutes after it has been injected into the bloodstream is given by the equation $C(t) = K(e^{-bt} - e^{-at})/(a - b)$, where $K > 0$ and $a > b > 0$. At what time does the concentration reach its maximum?

22.22. Suppose that the concentration of a chemical in parts per million is given by the equation $C(t) = 100(50 + te^{-t/10})$ for $0 \le t \le 50$.

a) Find the largest and smallest concentration during the given interval.

b) At what time does the growth rate of the concentration assume its minimal value?

22.23. Find all local maxima and minima and all points of inflection of the function $f(x) := x^2\ln(x)$ for $x > 0$. Then sketch the graph of f.

22.24. ★ Assume that a 240 gallon aquarium tank is equipped with an overflow system that allows for an inflow of replacement water at a rate of 0.8 gallons per hour.

a) Find a formula for the concentration of fresh water in the tank under the assumption that the overflow system is activated at time $t = 0$.

b) How long will it take until 90% of the water has been replaced?

22.25. Show that $\ln\left(\frac{x+y}{2}\right) \geq \frac{\ln(x)+\ln(y)}{2}$ for all $x, y \in (0, \infty)$.

Chapter 23

The General Exponential Function and Fruit Flies

The Size of a Fruit Fly Population

We wish to develop a model for the growth pattern of a fruit fly population. For this purpose, we assume that at time $t = 0$ we have 8 fruit flies (4 male, 4 female) and that it takes the time T (maybe seven days) for each pair of fruit flies to produce a new generation of, say, 2 fruit flies that are then ready for reproduction themselves. Given these assumptions, we may infer that at time T we have the original 8 flies plus 2 flies for each of the four pairs. Thus the number of flies is $8 + 4 \cdot 2 = 16$, corresponding to 8 pairs of male and female flies (for simplicity we assume that the numbers of males and females are always exactly equal and that all pairs reproduce simultaneously at perfectly regular intervals). As the process continues, the population increases to a total of $16 + 8 \cdot 2 = 32$ flies or 16 pairs at time $2T$ and then to $32 + 16 \cdot 2 = 64$ flies at time $3T$. Denoting by $P(t)$ the number of flies at time t, we summarize these results as follows:

$$\begin{aligned} P(0) &= 8 = 8 \cdot 2^0, \\ P(T) &= 16 = 8 \cdot 2^1, \\ P(2T) &= 32 = 8 \cdot 2^2, \\ P(3T) &= 64 = 8 \cdot 2^3, \\ &\vdots \end{aligned} \tag{23.1}$$

Eventually, of course, the growth will level off for lack of food or other limiting factors, but for a small period of time (such as a few days), we do indeed observe a *geometric growth* in the sense that the quotient of the number of fruit flies for two successive generations is constant:

$$2 = \frac{P(T)}{P(0)} = \frac{P(2T)}{P(T)} = \frac{P(3T)}{P(2T)} = \dots$$

Examining the values in (23.1), it is easy to see that

$$P(nT) = 8 \cdot 2^n \tag{23.2}$$

for any nonnegative integer n. In order to apply calculus to the analysis of our fruit fly population, it is necessary to define P for all points in time t—not only for $t = nT$. Given the result in (23.2), it is reasonable to attempt the definition

$$P(t) := 8 \cdot 2^{t/T} \text{ for all } t \in \mathbb{R}, \tag{23.3}$$

because then it would follow that $P(nT) = 8 \cdot 2^n$ for all integers n. In other words, the extension of P in (23.3) would generate the same values at the times $t = nT$ as the formula (23.2). The trouble is that so far we do not have a definition for an exponential expression of the form $2^{t/T}$ or in general a^x for $a > 0$, because in Chapter 22 we only defined exponentials to the base e.

The General Exponential Function

Teacher: Let us discuss the problem of assigning a value to an exponential expression of the form a^x for $a > 0$ and $x \in \mathbb{R}$. What, for example, is the value of $2^{3.5}$?

Simplicio: We know that for any positive integer n, the exponential 2^n is defined as the n-fold product of 2 with itself. By analogy, $2^{3.5}$ ought to be equal to 2 multiplied with itself 3.5 times.

Sophie: But Simplicio, how can you possibly multiply a number 3.5 times with itself? We can multiply a number 3 times with itself or perhaps 4 times, but not 3.5 times. It just doesn't make sense.

Teacher: Sophie is right. The standard definition

$$a^n := \underbrace{a \cdot \cdots \cdot a}_{n \text{ times}}$$

can only be applied if n is a positive integer. As explained in Appendix B, we can easily obtain a generalization to the case of rational exponents $r = n/m$ using the equation $a^r := \sqrt[m]{a^n}$, but in a calculus context this definition of exponentials in terms of n-fold products and mth roots is cumbersome and incomplete because it does not extend to irrational exponents $x \in \mathbb{R}$. However, the fact that we have a definition for a^r for rational r allows us to take an indirect approach to the problem of defining a^x for arbitrary $x \in \mathbb{R}$. According to Theorem 21.1d, the equation $\ln(a^r) = r\ln(a)$ holds for all rational r. Consequently, it seems reasonable to require it to be valid also for irrational $x \in \mathbb{R}$. In other words, a^x ought to be defined in such a way that

$$\boxed{\ln(a^x) = x\ln(a) \text{ for all } x \in \mathbb{R}.} \tag{23.4}$$

Simplicio: Isn't it a bit weird to identify which properties a^x should satisfy *before* a definition for a^x has actually been given?

Teacher: Think about it this way: whatever definition we come up with has to be compatible with the results that we have already established. In particular, any meaningful definition for a^x will have to be consistent with the fact that $\ln(a^r) = r\ln(a)$ for all rational numbers r.

Simplicio: Okay, but I still don't see where exactly we are headed. When we said that we wanted to define a^x, I expected that we would end up with a defining equation $a^x := \ldots$, but all we have now is the property in (23.4).

Teacher: You are right, but fortunately such an explicit definition for a^x is easily obtained by applying the natural exponential function to both sides of the equation $\ln(a^x) = x\ln(a)$. This yields $\exp(\ln(a^x)) = \exp(x\ln(a))$, and since the natural exponential function is the inverse of the natural logarithm function, we may infer that $a^x = \exp(x\ln(a)) = e^{x\ln(a)}$.

23.1 Definition. For $a > 0$ and $x \in \mathbb{R}$, we define the *general exponential function* via the equation

$$a^x := e^{x\ln(a)}.$$

The most important properties of the general exponential function are listed in the following theorem:

23.2 Theorem. *Let* $a, b > 0$ *and* $x, y \in \mathbb{R}$. *Then*

a) $a^x a^y = a^{x+y}$,

b) $a^x / a^y = a^{x-y}$,

c) $(a^x)^y = a^{xy}$,

d) $a^0 = 1$,

e) $(ab)^x = a^x b^x$,

f) $(a/b)^x = a^x/b^x$, *and*

g) *for* $a \neq 1$ *the range of the exponential function* a^x *is* $(0, \infty)$ *and for* $a = 1$ *the range is* $\{1\}$.

Proof. **a)** $a^x a^y = e^{x \ln(a)} e^{y \ln(a)} = e^{(x+y)\ln(a)} = a^{x+y}$.
c) Since $\ln(a^x) = \ln(e^{x \ln(a)}) = x \ln(a)$, it follows that $(a^x)^y = e^{y \ln(a^x)} = e^{xy \ln(a)} = a^{xy}$.
e) $(ab)^x = e^{x \ln(ab)} = e^{x \ln(a) + x \ln(b)} = e^{x \ln(a)} e^{x \ln(b)} = a^x b^x$.
g) For $a = 1$ we have $\ln(a) = \ln(1) = 0$, and therefore $a^x = e^{x \ln(1)} = e^0 = 1$. Consequently, the range of a^x is in this case equal to $\{1\}$. For $a \neq 1$, we need to show that for all $y \in (0, \infty)$ there exists an $x \in \mathbb{R}$ such that $y = a^x$. So let us choose an arbitrary $y \in (0, \infty)$. Since a is different from 1, $\ln(a)$ is different from zero and $x := \ln(y)/\ln(a)$ is well defined. Using now the definition of a^x, we may infer that

$$a^x = e^{x \ln(a)} = e^{\ln(y) \ln(a)/\ln(a)} = e^{\ln(y)} = y$$

as desired. □

23.3 Exercise. Prove Theorem 23.2b, d, f.

Remark. Theorem 23.2c shows in particular that for all $a, b \in \mathbb{R}$ we have

$$(e^a)^b = e^{ab}.$$

This property of the natural exponential function was not included in Theorem 20.9, because at that stage the expression $(e^a)^b$ was still undefined.

Figure 23.1 shows the graph of the exponential function $f(x) := a^x$ for three different values of a.

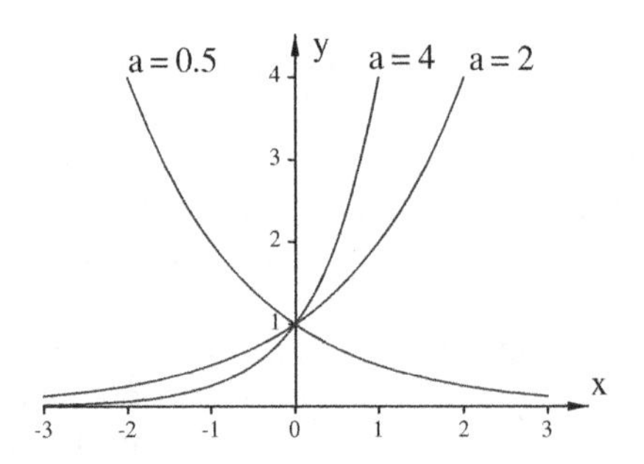

Figure 23.1: graphs of exponential functions $f(x) = a^x$.

Having defined the general exponential function, we are now able to make sense of the attempted definition of $P(t)$ in (23.3):

$$P(t) = 8 \cdot 2^{t/T} = 8 \cdot e^{(t/T)\ln(2)}.$$

This function is a *model* for the increase of the fruit fly population during the initial period of geometric growth. Of course, the number of fruit flies at any point in time is always an integer, which is not the case for $P(t)$. So in general, $P(t)$ is only an approximation for the actual size of the fruit fly population, but for practical purposes this is usually sufficient. The graph of P is shown in Figure 23.2.

To generalize the results of the preceding discussion, we assume that we are given a population of size $P(t)$ with a reproductive rate per pair r and an initial size P_0. Then, in denoting again by T the

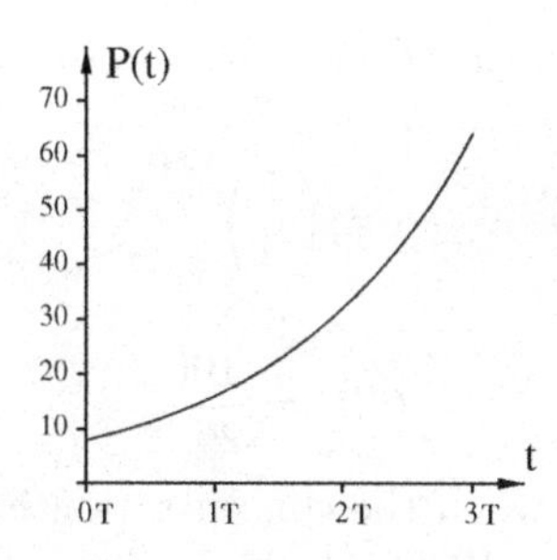

Figure 23.2: graph of $P(t) = 8 \cdot 2^{t/T}$.

time it takes to produce a new generation, we obtain

$$P(0) = P_0 = P_0\left(1 + \frac{r}{2}\right)^0,$$
$$P(T) = P_0 + \frac{1}{2}P_0 r = P_0\left(1 + \frac{r}{2}\right)^1,$$
$$P(2T) = P_0\left(1 + \frac{r}{2}\right) + \frac{1}{2}P_0\left(1 + \frac{r}{2}\right)r = P_0\left(1 + \frac{r}{2}\right)^2,$$
$$P(3T) = P_0\left(1 + \frac{r}{2}\right)^2 + \frac{1}{2}P_0\left(1 + \frac{r}{2}\right)^2 r = P_0\left(1 + \frac{r}{2}\right)^3,$$
$$\vdots$$
$$P(nT) = P_0\left(1 + \frac{r}{2}\right)^n.$$

Therefore, with t in place of nT, it follows that

$$\boxed{P(t) = P_0\left(1 + \frac{r}{2}\right)^{t/T}.} \tag{23.5}$$

23.4 Exercise. Find $P(t)$ given an initial population of 6, a reproductive time of 8 days, and a reproductive rate per pair of 7.

23.5 Example. In order to discuss a problem involving decay rather than growth, we consider the following question concerning the concentration of a prescription drug in the blood stream: how long does it take until there is only 1% of the original concentration remaining? The results of several measurements of the concentration at different times are shown in the table below:

time in hours	% of original dose remaining
0	100.00
4	50.00
8	25.00
12	12.50
16	6.25
20	3.13
24	1.56
28	0.78

We see that every 4 hours the concentration is reduced by 50%. This observation suggests that we are dealing with an exponential decay, and it therefore appears plausible to think that the concentration $C(t)$ might be a function of the form

$$C(t) = C_0 a^t \text{ with } C_0 = C(0) = 100.$$

To determine the base a, we use the fact that $C(4) = 50$ and solve for a the equation $50 = C(4) = 100a^4$. This yields $a^4 = 1/2$ and

$$a = \left(\frac{1}{2}\right)^{1/4}.$$

Hence

$$C(t) = \frac{100}{2^{t/4}}.$$

In order to find the moment in time t when the concentration is reduced to 1% of its original amount, we need to solve for t the equation $1 = 100/2^{t/4}$. Multiplying this equation by $2^{t/4}$ and applying the natural logarithm to both sides, we obtain $(t/4)\ln(2) = \ln(2^{t/4}) = \ln(100)$, and therefore

$$t = \frac{4\ln(100)}{\ln(2)} \approx 26.575\,h.$$

The Rate of Growth of a Fruit Fly Population

In the language of calculus, the rate of change of $P(t)$ at time t is the derivative $P'(t)$. Consequently, in order to determine the growth rate of a fruit fly population, we need to be able to calculate the derivative of a general exponential function. Using the chain rule and the fact that the derivative of e^x is e^x itself, we observe that

$$\frac{d}{dx}a^x = \frac{d}{dx}e^{x\ln(a)} = e^{x\ln(a)}\frac{d}{dx}(x\ln(a)) = a^x\ln(a).$$

Thus, for all $x \in \mathbb{R}$ and $a > 0$ we have

$$\boxed{\frac{d}{dx}a^x = a^x\ln(a).} \tag{23.6}$$

Using this formula and the chain rule, we can now determine the derivative of the function P in (23.5):

$$P'(t) = P_0\left(1+\frac{r}{2}\right)^{t/T}\ln\left(1+\frac{r}{2}\right)\frac{d}{dt}\frac{t}{T} = \frac{P_0}{T}\left(1+\frac{r}{2}\right)^{t/T}\ln\left(1+\frac{r}{2}\right).$$

This result shows that the rate of growth is large if T is small and r is large. Intuitively, this is not surprising, because if it takes only a short amount of time to produce a new generation (T small) and each pair of males and females produces a large number of offspring (r large), then the population will grow very rapidly.

23.6 Exercise. Use (23.6) to find the derivative of $P(t) = 8 \cdot 2^{t/T}$.

23.7 Exercise. Find the following derivatives: $\frac{d}{dx}3^{x-1}$, $\frac{d}{dx}\left(\frac{1}{2}\right)^{3x}$, and $\frac{d}{dx}2^{(2^x)}$.

Additional Exercises

23.8. A colony of bacteria doubles in size every 10 minutes. At noon on a certain day, the colony completely filled a petri dish. At what time (to the nearest hour, minute, and second) was the dish covered to...

a) 50%? **b)** 25%? **c)** 10%? **d)** 2%?

23.9. Sketch the graph of the function $f(x) := 2^x$, and find the equation of the function whose graph results from reflecting the graph of f...

a) about the x-axis.

b) about the y-axis.

c) about the line $y = 5$.

d) about the line $x = 2$.

e) first about the x-axis, then about the y-axis.

f) first about the y-axis, then about the x-axis.

23.10. Assume that a radioactive substance decays with a half-life of 15 hours. So every 15 hours 50% of the substance decays into nonradioactive isotopes. How long will it take until only 10% of the original amount is left?

23.11. At a ticket price of \$25 a theatre company attracts on average 200 spectators to each performance. Furthermore, each increase in the ticket price by \$5 results in a reduction of the audience by 10%.

a) Find a formula for the number of spectators as a function of the ticket price.

b) Find a formula for revenue generated by each performance as a function of the ticket price.

c) How much should the theatre company charge for each ticket in order to maximize revenue?

Hint. The number of spectators $N(t)$ is an exponential function of the form $N_0 a^t$. For additional inspiration, take a look at Example 23.5.

23.12. Differentiate each of the following functions:

a) 2^{2x}	b) 3^{2-3x}	c) $5^{3x^2} - 2$	d) $4^{\ln(x)}$	e) $\ln(1.5^{-x})$
f) $(1/2)^{2x^3}$	g) xe^{4x}	h) $x^2 6^{4x-1}$	i) $x^2 + 2^x$	j) $(3^{3x} + 1)^{-1}$

23.13. Suppose that the concentration of a chemical in parts per million is given by the equation $C(t) = 500(10 + t2^{-t/4})$ for $0 \leq t \leq 20$.

a) Find the largest and smallest concentration during the given interval.

b) At what time does the growth rate of the concentration assume its minimal value?

23.14. Determine the following antiderivatives:

$$\text{a)} \int 2^{4x}\,dx \quad \text{b)} \int \left(\frac{1}{2}\right)^{5x+1} dx \quad \text{c)} \int \frac{3^{4x}}{(3^{4x} - 1)^2}\,dx \quad \text{d)} \int x2^{x^2}\,dx$$

Hint. To solve c) you should examine the result of Exercise 23.12j, and for d) you need to find a constant a and a function $g(x)$ such that the derivative of $a2^{g(x)}$ is equal to the integrand $x2^{x^2}$.

23.15. A rich young ruler decides to give away 50% of his fortune on the last day of each month. Assuming that he has no income, and that he starts with \$2 billion, how long will it take until he has given away more than \$1.9 billion? *Hint.* Take a look at Example 23.5.

23.16. Assume that $g : \mathbb{R} \to \mathbb{R}$ is an increasing differentiable function such that $g(x) > 1/\ln(2)$ for all $x \in \mathbb{R}$. Prove that the function $2^{g(x)}/g(x)$ is increasing as well.

23.17. Assume that f is a general exponential function of the form $f(x) = P_0 a^x$.

a) Determine P_0 and a from the assumption that $f(2) = 40$ and $f(4) = 30$.

b) Use the results of a) to find the maximal value of the function $g(x) := xf(x)$.

23.18. Assume that you currently earn \$40,000 per year. Your income increases at a rate of 3% each year, and 10% of your annual income is invested in a retirement fund that generates an annual interest of 6%. How much money will you have in your retirement fund if you continue working for 40 years, and how much will this money be worth in terms of present-day purchasing power if the annual rate of inflation is 2.5%?

Chapter 24

The General Logarithm and Power Functions

The General Logarithm Function

In Chapter 22, we defined the natural exponential function as the inverse of the natural logarithm function, which in turn had been defined in Chapter 21 via the FTC1 as the antiderivative of $1/x$. In the present chapter, we will reverse this approach and define the general logarithm function as the inverse of the general exponential function. To do so, we observe that the general exponential function $f(x) := a^x$ with $0 < a \neq 1$ is indeed invertible, because the graph of f passes the horizontal line test (see Figure 23.1). (Note: for $a = 1$ we have $f(x) = 1^x = 1$ so that f in this case is clearly not invertible.)

Remark. The fact that the graph of f passes the horizontal line test also follows from the observation that f is strictly increasing for $a > 1$ and strictly decreasing for $0 < a < 1$. This is so because for $a > 1$ we have $f'(x) = a^x \ln(a) > 0$ for all $x \in \mathbb{R}$, and for $0 < a < 1$ we have $f'(x) = a^x \ln(a) < 0$ (see (21.5), (21.6), Theorem 23.2g, and Theorem 4.11).

24.1 Definition. For $0 < a \neq 1$, we define the *general logarithm function to the base* a as the inverse function of the general exponential function $f(x) := a^x$. Furthermore, for every $x \in (0, \infty)$, the value of the natural logarithm function at x is denoted by $\log_a(x)$. Note: the domain of the general logarithm function is $(0, \infty)$ and its range is $\mathbb{R}$ because $\mathbb{R}$ and $(0, \infty)$ respectively are the domain and range of the general exponential function a^x for $0 < a \neq 1$ (see also Chapter 22, p.178).

Alternatively, the definition of $\log_a(x)$ can be stated as follows: for all $x \in (0, \infty)$ and all $y \in \mathbb{R}$ we have

$$\boxed{\log_a(x) = y \text{ if and only if } a^y = x.}$$

Given this alternative formulation, we can easily express the general logarithm function in terms of the natural logarithm function, for in applying the natural logarithm to the equation $e^{y \ln(a)} = a^y = x$, we obtain $y \ln(a) = \ln(x)$, and solving for y yields $y = \ln(x)/\ln(a)$. Hence

$$\boxed{\log_a(x) = \frac{\ln(x)}{\ln(a)}.} \tag{24.1}$$

Since $\ln(e) = 1$, this result also shows that

$$\boxed{\ln(x) = \log_e(x).}$$

The most important properties of the general logarithm function are stated in the following theorem:

24.2 Theorem. *Let $a, b, x, y \in (0, \infty)$ and assume that $a, b \neq 1$. Then*

a) $\log_a(xy) = \log_a(x) + \log_a(y)$,

b) $\log_a(x/y) = \log_a(x) - \log_a(y)$,

c) $\log_a(x^y) = y \log_a(x)$,

d) $\log_a(1) = 0$, *and*

e) $\log_b(x) = \dfrac{\log_a(x)}{\log_a(b)}$.

Proof. **a)** $\log_a(xy) = \dfrac{\ln(xy)}{\ln(a)} = \dfrac{\ln(x) + \ln(y)}{\ln(a)} = \log_a(x) + \log_a(y)$.

c) $\log_a(x^y) = \dfrac{\ln(e^{y \ln(x)})}{\ln(a)} = \dfrac{y \ln(x)}{\ln(a)} = y \log_a(x)$.

e) $\dfrac{\log_a(x)}{\log_a(b)} = \dfrac{\ln(x)/\ln(a)}{\ln(b)/\ln(a)} = \dfrac{\ln(x)}{\ln(b)} = \log_b(x)$. □

24.3 Exercise. Prove Theorem 24.2b, d.

24.4 Exercise. Use Theorem 24.2e to show that $\log_a(b) = \dfrac{1}{\log_b(a)}$ for all $a, b \in (0, \infty) \setminus \{1\}$.

24.5 Exercise. Sketch the graph of $\log_a(x)$ for various values of a, and formulate a general rule concerning the dependence of the shape of the graph on the value of a.

Given equation (24.1), we can easily find the derivative of the general logarithm function by using the fact that the derivative of $\ln(x)$ is $1/x$: for $a, x \in (0, \infty)$ with $a \neq 1$, we have

$$\boxed{\frac{d}{dx} \log_a(x) = \frac{1}{\ln(a)x}.}$$

24.6 Exercise. Find the following derivatives: $\dfrac{d}{dx} \log_4(x^2 - 3)$, $\dfrac{d}{dx} \log_{1/2}(3x)$, and $\dfrac{d}{dx} \log_2(\log_5(2^x))$.

Power Functions

In Chapter 10, we established the following rule for the differentiation of power functions with integer exponents:

$$\frac{d}{dx} x^n = n x^{n-1}. \tag{24.2}$$

Now we will consider the case where the exponent is not necessarily an integer, and show that a formula analogous to (24.2) is still valid. For $a \in \mathbb{R}$, we define a *power function* $f : (0, \infty) \to \mathbb{R}$ via the equation

$$\boxed{f(x) := x^a.}$$

Notice that a power function differs from an exponential function in that the variable is in the base rather than in the exponent. Several graphs of power functions for different values of a are shown in Figure 24.1.

Simplicio: I don't quite understand why in our definition of the power function $f(x) = x^a$, we restricted the domain of f to the positive real numbers. After all, an exponential with a negative base like, for example, $(-2)^3$ is easily evaluated—we simply form the product $(-2) \cdot (-2) \cdot (-2)$ to obtain -8. So it seems plausible to allow also negative values for x in the exponential x^a.

Teacher: That's a good point. However, with respect to the example you gave, I would like to direct your attention to the fact that $(-2)^3$ is well defined only because the exponent 3 is an *integer*. If

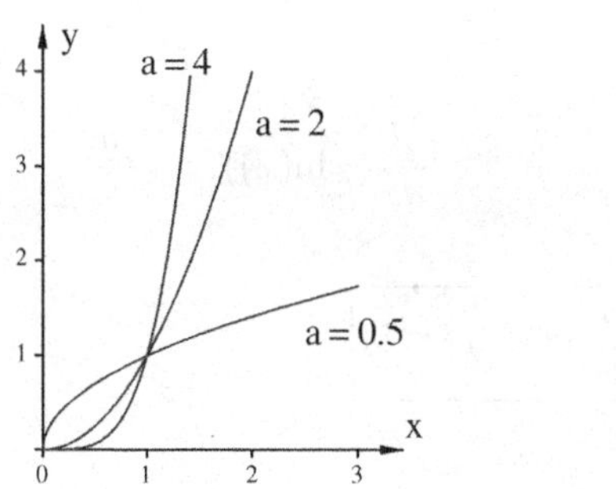

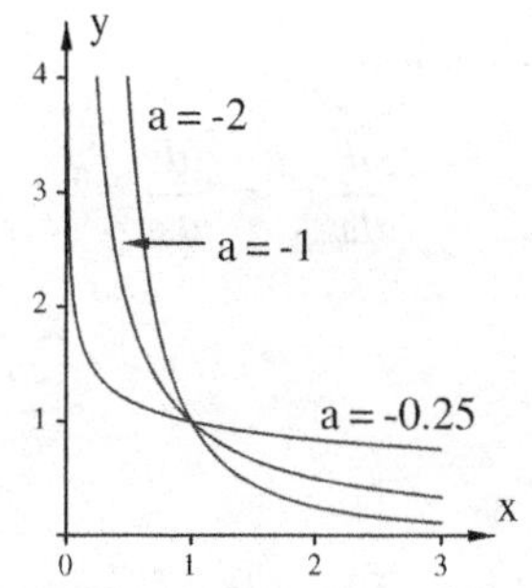

Figure 24.1: graphs of power functions $f(x) = x^a$.

the exponent is not an integer, then we have to appeal to Definition 23.1, which, when applied to our case, says that $x^a = e^{a\ln(x)}$. Consequently, x^a is defined only for positive values of x because the domain of the natural logarithm $\ln(x)$ is $(0, \infty)$.

Simplicio: Perhaps our general definition of exponentials needs some revision.

Teacher: No, our definition is fine, but let me give you another example to illustrate the problem: consider the case $a = 1/2$. Since $x^{1/2} = \sqrt{x}$...

Simplicio: May I interrupt for a moment?

Teacher: Sure.

Simplicio: I know this is an elementary fact, but actually, why do we say that $x^{1/2} = \sqrt{x}$?

Teacher: You can answer this question yourself. Just think of how you find the square root of x.

Simplicio: Well, I would have to find a number that, multiplied by itself, equals x.

Teacher: So then, is it true that $x^{1/2}$ multiplied by itself equals x?

Simplicio: Ah, I see: if we multiply $x^{1/2}$ by itself, then the exponents are added so that $x^{1/2}x^{1/2} = x^1 = x$, and therefore, $x^{1/2} = \sqrt{x}$. That makes sense.

Teacher: Back to our discussion: since $x^{1/2} = \sqrt{x}$, it is clear that $x^{1/2}$ can only be defined for positive real numbers, because square roots of negative numbers do not exist. So the restriction of x to positive real numbers is not the result of an inadequate definition of x^a via the natural logarithm and exponential functions, but is also readily seen to be necessary in simple cases where a is not an integer.

Sophie: You said that negative values for x are permissible only in the special case where a is an integer, but that doesn't seem to be true because $x^{1/3}$, which is the third root of x, is certainly well defined for positive *and* negative values of x. For instance, the third root of -1 is -1, because $(-1)^3 = -1$.

Teacher: Your point is well taken, but there is a little problem: if, as you say, $x^{1/3}$ is the third root of x, then would you agree that $-1 = (-1)^{1/3}$?

Sophie: Certainly, I already gave that example myself.

Teacher: Would you also agree that, according to the rules stated in Theorem 23.2, we have $(-1)^{1/3} = (-1)^{2/6} = \left((-1)^2\right)^{1/6}$?

Sophie: It seems entirely plausible.

Teacher: I am not so sure, because clearly $(-1)^2 = 1$, and therefore we are forced to conclude that $-1 = (-1)^{1/3} = 1^{1/6} = 1$.

Sophie: I must admit, the result looks unconvincing. So what went wrong?

Teacher: In setting $(-1)^{2/6}$ equal to $\left((-1)^2\right)^{1/6}$, we assumed that a negative number raised to a non-integer power is a well-defined expression to which the ordinary rules for exponentials apply in the same way as for exponentials with a positive base. Unfortunately, though, this is not the case. For $x < 0$, the use of the exponential notation $x^{1/3}$ is a mere convention. We are not dealing here with a genuine exponential to which the standard rules, as stated in Theorem 23.2, can be applied. So we really need to be very careful whenever we are using the exponential notation x^a for negative values of x.

Our next objective is to determine the derivative of a power function. As mentioned above, x^a is equal to $e^{a\ln(x)}$, and therefore,

$$\frac{d}{dx}x^a = \frac{d}{dx}e^{a\ln(x)} = e^{a\ln(x)}\frac{d}{dx}(a\ln(x)) = x^a\frac{a}{x}.$$

Hence

$$\boxed{\frac{d}{dx}x^a = ax^{a-1}.}$$

Functions of the Form $f(x)^{g(x)}$

As a generalization of both power and exponential functions, we will now consider the case where the exponent *and* the base of an exponential expression depend on x. So let us assume that $f : I \to (0,\infty)$ and $g : I \to \mathbb{R}$ are differentiable functions defined on an open interval I, and let $h : I \to \mathbb{R}$ be defined by the equation

$$h(x) := f(x)^{g(x)} \text{ for all } x \in I.$$

To determine the derivative of h we use the chain rule, the product rule, and the definition of the general exponential function:

$$\begin{aligned}\frac{d}{dx}f(x)^{g(x)} &= \frac{d}{dx}e^{g(x)\ln(f(x))} = e^{g(x)\ln(f(x))}\frac{d}{dx}(g(x)\ln(f(x)))\\ &= f(x)^{g(x)}\left(g'(x)\ln(f(x)) + \frac{g(x)f'(x)}{f(x)}\right).\end{aligned}$$

Hence

$$\boxed{\frac{d}{dx}f(x)^{g(x)} = f(x)^{g(x)}\left(g'(x)\ln(f(x)) + \frac{g(x)f'(x)}{f(x)}\right).}$$

24.7 Exercise. Find the following derivatives: $\frac{d}{dx}x^{\ln(x)}$, $\frac{d}{dx}(x^2-3)^{\sqrt{2}-x}$, and $\frac{d}{dx}x^e$.

Additional Exercises

24.8. Evaluate the following expressions without using a calculator:

a) $\log_5(25)$ **b)** $\log_{\sqrt{2}}(4)$ **c)** $\log_6(20) + \log_6(1.8)$
d) $\log_4(8)$ **e)** $2^{\log_2(7)}$ **f)** $4^{\log_2(3)}$

24.9. Find the derivative of each of the following functions:

a) $\log_2(3x-1)$ **b)** $2^{\log_5(x)}$ **c)** $\left(\log_5(3x^2)\right)^{x^2}$ **d)** $e^{\log_4(x)}$ **e)** $\log_3(9^{-x})$
f) $x^2\log_{10}(x^2)$ **g)** x^{4x} **h)** $x^2(6x)^{4x-1}$ **i)** $x^e + e^x$ **j)** $x^\pi/\log_{10}(\pi x)$

24.10. Determine the following antiderivatives:

a) $\int x^{4e}\,dx$ **b)** $\int \log_2(5^x)\,dx$ **c)** $\int \frac{1}{x^\pi}\,dx$

24.11. For a given $a > 0$, we set $f_a(x) := \log_2(ax)$. Sketch the graph of f_a for various values of a, and formulate a statement regarding the dependence of the shape of the graph on the value of the parameter a.

24.12. Let $f(x) := \dfrac{\ln(x)}{x} - \dfrac{1}{e}$ for $x > 0$.

a) Find the intervals on which f is increasing and those on which it is decreasing.

b) Find all positive numbers x for which $x^e < e^x$.

24.13. Use Riemann sums to estimate the value of $\int_1^a \log_a(x)\,dx$ for $a = 2,\ 3,\ 4,\ 5,\ e$.

24.14. Find the equation of the tangent line to the graph of $f(x) := \log_a(x)$ at $(1, f(1))$ for an arbitrary value $a \in (0, \infty) \setminus \{1\}$.

24.15. Find the derivative of the function $f(x) := x^{1/\ln(x)}$ and use your result to show that $x^{1/\ln(x)} = e$ for all $x \in (0, \infty) \setminus \{1\}$.

24.16. Show that $x^{1/\ln(x)} = e$ for all $x \in (0, \infty) \setminus \{1\}$ by using Exercise 24.4. Do not use differentiation as in Exercise 24.15.

Chapter 25

Honeycombs (Part 1)

The intricate interdependencies and high levels of organization in insect colonies present us with some of the most wonderful examples for the miraculous creativity of nature. Among social insects, the honeybee stands out not only for the sophistication of its methods of communication, such as the differential dance as a means of locating flower sources, but also for its skills as a craftsman. Bees construct honeycombs as double layers of congruent hexagonal tubes that are open on one side and closed with three equilateral parallelogram faces on the other (see Figure 25.1). Careful measurements have shown with a high level of consistency that the parallelogram faces are joined together in such a way that the angle θ between the faces and the tube center line is approximately 54°.

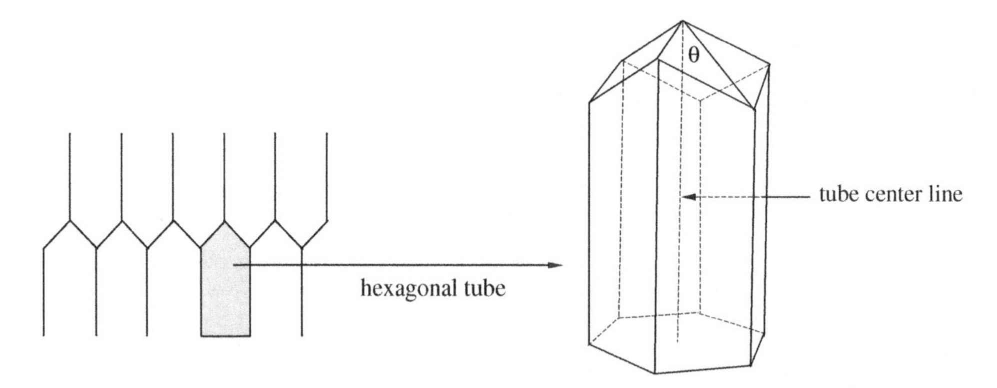

Figure 25.1: honeycomb and hexagonal tube.

25.1 Exercise. Examine a real honeycomb for the characteristics described above.

Teacher: In order to see whether the honeybee really deserves the title of "master craftsman," we will employ our calculus skills to determine for which angle θ the amount of material needed for the construction of the honeycomb is minimal, and then compare our result with the observed angle of 54°.

Simplicio: I am not sure I understand the problem.

Teacher: Take a look at Figure 25.2. Here we see how the shape of a hexagonal tube covered with three parallelogram faces at one end depends on the angle θ. Since the amount of material needed for the construction is clearly proportional to the surface area of the tube, it follows that we need to find the angle θ for which the surface area is minimal.

Sophie: Am I right in assuming that the ideal value for θ will be equal or at least very close to the value that we actually observe in nature?

Simplicio: How did you figure that out?

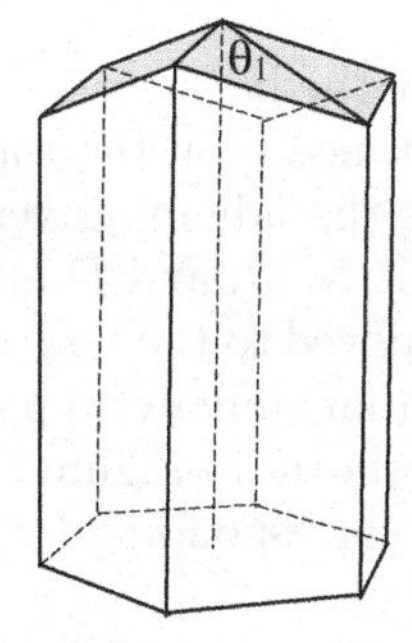

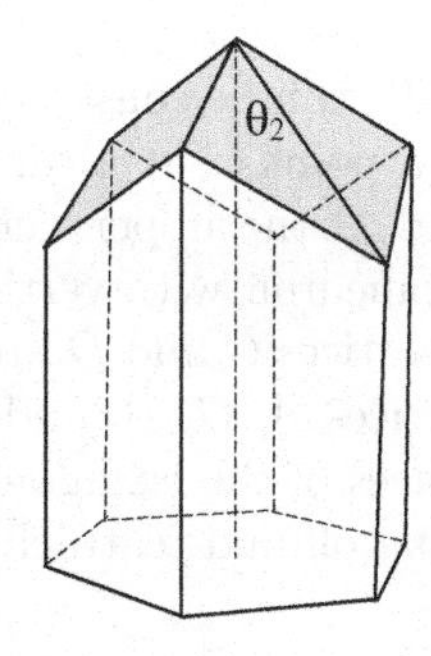

Figure 25.2: surface area in dependence on θ.

Sophie: I didn't. I just thought the teacher probably wouldn't suggest this problem for discussion if it didn't come out exactly right.

Teacher: Indeed, we will see that the theoretically best choice for θ is very close to $54°$. In other words, bees are really pretty good at making the most of the available building material.

Simplicio: But this is truly remarkable. How do we account for such a phenomenon?

Teacher: I am not sure, but in simplified form the standard explanation according to the theory of evolution can be stated as follows: of two hives competing for survival in the same habitat, the one with a population more highly skilled at using the available resources will have a competitive advantage and pass on the genetic secret of its success to its more numerous offspring. Since the building material for the honeycombs is one of the relevant resources, it is conceivable that the less skillful workers in one of the hives will gradually be replaced by the more efficient craftsmen emerging from the other. Eventually, as this process of genetically-based adaptation and selection continues over long periods of time, a breed of maximally efficient honeybees evolves that even stands the test of calculus.

Sophie: This explanation makes some sense, but I have a hard time imagining that bees constructing honeycombs with $\theta = 54°$ would have a measurable competitive advantage over breeds working with, say, $\theta = 50°$. Isn't there possibly an explanatory gap in our argument?

Teacher: I think your point deserves serious consideration, but we need to keep in mind that the time spans involved are presumably vast, and even minor selective advantages may therefore have a significant effect.*

A Partial Solution

Teacher: In order to determine the surface area of the hexagonal tube in Figure 25.1, we first need to

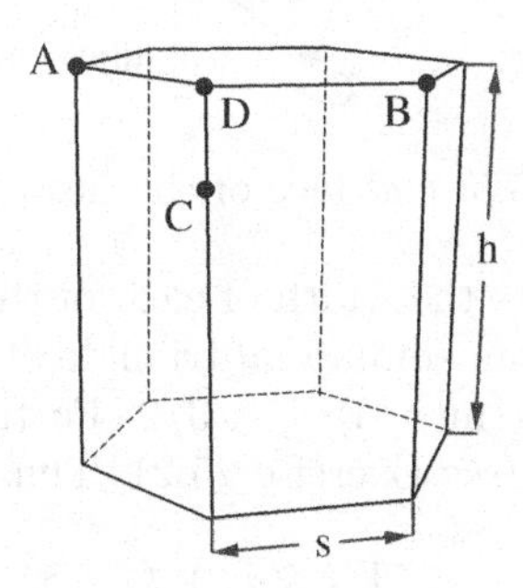

Figure 25.3: a hexagonal prism.

*The question of whether evolutionary change occurs gradually over vast spans of time, as Darwin conjectured in his original formulation of the theory of evolution, or in brief spurts of activity followed by long periods of stasis is subject to a continuing debate among evolutionary biologists. For an exposition of the latter hypothesis the reader is referred to [Gou].

understand how such a tube can be constructed from the flat hexagonal prism shown in Figure 25.3. To begin with, we mark a point C at a certain distance from the top on one of the vertical edges and denote by A, B, and D the upper vertices of the adjoining vertical faces (see Figure 25.3). Then we slice off the tetrahedron with vertices at A, B, C, and D and let it rotate by 180 degrees about the line $\overline{AB}$. The vertices C and D are thus moved to the positions E and F respectively (see Figure 25.4), and the vertices A, C, B, and E form an equilateral parallelogram (i.e., a rhombus). The remaining two end faces of the hexagonal tube shown in Figure 25.1 are constructed in exactly the same manner by slicing off and reattaching the corresponding tetrahedra.

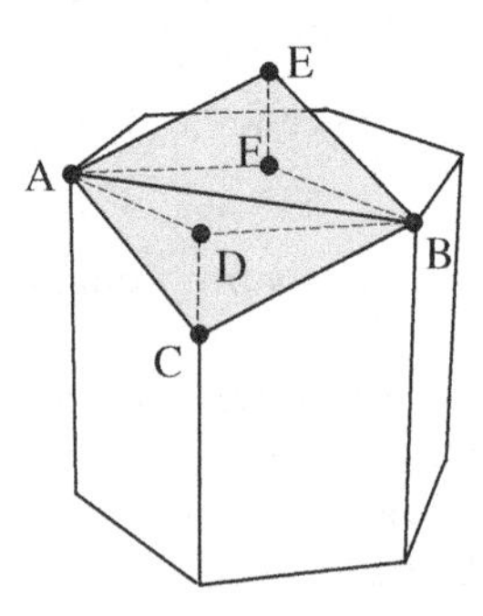

Figure 25.4: constructing an end face.

Sophie: Do I understand correctly that our cutting-and-pasting procedure leaves the total volume unchanged?

Teacher: Indeed, this observation is of some significance, because it shows that the surface area is in itself already a sufficient measure of efficiency.

Sophie: In other words, in determining the best possible construction, we do not have to weigh decreases in area against conceivable simultaneous decreases in volume.

Teacher: That is correct; but let us proceed now to compute the surface area of the parallelogram $ACBE$. In Figure 25.5, this parallelogram is shown with two diagonals $\overline{AB}$ and $\overline{CE}$ that bisect each other at P. Thus, the parallelogram is divided into four congruent right triangles with side

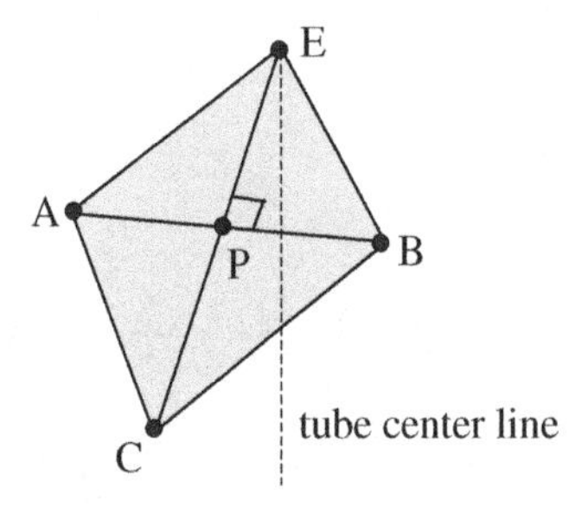

Figure 25.5: end face of the hexagonal tube.

lengths $\overline{AP}$ and $\overline{PE}$. Denoting by s the width of each of the side faces (see Figure 25.3), we notice that the points A, D, and F form an equilateral triangle of side length s. Consequently, the height $\overline{AP}$ of this triangle is $s\sin(60^\circ) = s\sin(\pi/3) = s\sqrt{3}/2$. Furthermore, $\sin(\theta) = \overline{PF}/\overline{PE} = (s/2)/\overline{PE}$, or equivalently, $\overline{PE} = s/(2\sin(\theta))$ (see Exercise 25.2). Thus, the area A_f of each end face is

$$A_f = 4 \cdot \frac{1}{2}\left(\frac{\sqrt{3}}{2}s\right)\left(\frac{s}{2\sin(\theta)}\right).$$

Therefore, the surface area of the entire rhombic cover is

$$3A_f = \frac{3\sqrt{3}\,s^2}{2\sin(\theta)}.$$

The remaining surface area of the hexagonal tube consists of the six congruent trapezoidal faces at the sides. Denoting by h the height of the original hexagonal tube (see Figure 25.3), it follows that the area A_t of each trapezoidal face is

$$A_t = \frac{s(h - \overline{CD} + h)}{2} \quad \text{(see Exercise 25.3).}$$

Since $\overline{CD} = s\cot(\theta)/2$ (see Exercise 25.2), we may infer that

$$A_t = \frac{s}{2}\left(2h - \frac{s}{2}\cot(\theta)\right).$$

Therefore, the total area A is given by the equation

$$\boxed{A = 6A_t + 3A_f = 6hs + \frac{3s^2(\sqrt{3} - \cos(\theta))}{2\sin(\theta)}.} \tag{25.1}$$

Now we are left with the problem of minimizing A as a function of θ by solving for θ the equation $dA/d\theta = 0$ (see Chapters 11 and 14). In order to be able to determine the derivative of A we will discuss in the next chapter the differentiation of trigonometric functions and then apply our results to complete the solution of the honeycomb problem in Chapter 27.

★ Can We Do Better?

When faced with the amazing complexity and efficiency of organic structures, we are frequently forced to concede that human reason is no match for the ingenuity of nature and that we have little to offer by way of improvement. For a long time it was therefore believed that the most economical polygonal cover of a hexagonal prism that is sufficiently symmetric to allow for the construction of a double layer of cells (see Figure 25.1) is the arrangement of three rhombic faces found in a honeycomb. Colin MacLaurin, for example, asserted in 1743 that "[t]he cells, by being hexagonal, are the most capacious, in proportion to their surface, of any regular figures that leave no interstices between them, and at the same time

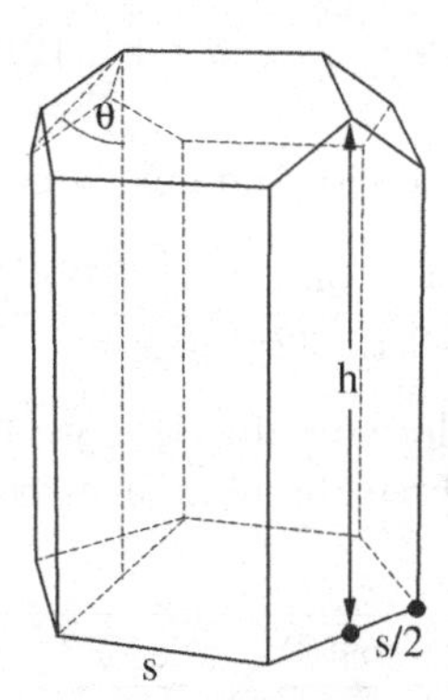

Figure 25.6: a more economical cover.

admit of the most perfect bases"[1] (emphasis ours). However, in 1964 the Hungarian mathematician L. Fejes Toth discovered that a slight reduction in surface area can be achieved for covers composed of two parallelograms and two hexagons (see Figure 25.6). The outlines of these covers on a layer of adjacent cells are shown in a cross-sectional projective view in the diagram on the right in Figure 25.7. For comparison, the diagram on the left shows the standard structure in an actual honeycomb.

A fairly tedious but essentially straightforward computation (see Exercise 25.3) shows that the total surface area of the hexagonal tube in Figure 25.6 in dependence on the parameters h, s, and θ is given

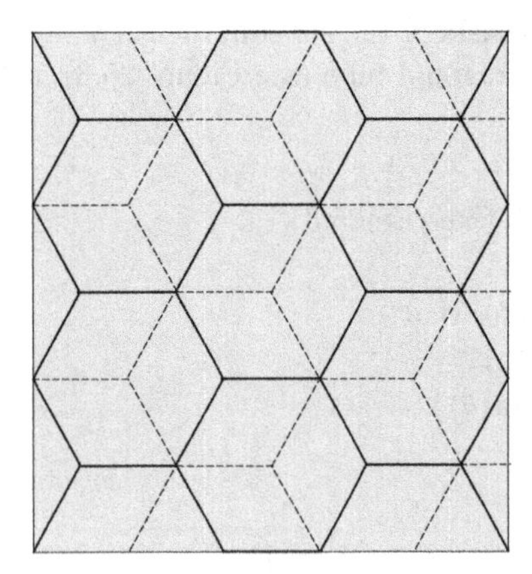

Figure 25.7: cross-sectional projective views of layers of adjacent cells.

by the formula

$$\boxed{A = 6hs + \frac{s^2(5\sqrt{3 - 2\cos^2(\theta)} + \sqrt{3} - 4\cos(\theta))}{4\sin(\theta)}.} \tag{25.2}$$

As in the case of equation (25.1), we are now left with the problem of minimizing A as a function of θ. It turns out that the minimum is assumed when $\cos(\theta)$ approximately equals 0.745004 (see Exercise 26.14). Substituting this value for $\cos(\theta)$ in (25.2) and observing that $\sin(\theta) = \sqrt{1 - \cos^2(\theta)} \approx 0.66706$, it follows that

$$A \approx 6hs + 2.10842s^2.$$

In order to compare this result to the one we found in (25.1) for an actual cell in a honeycomb, we must ask the reader to accept for the moment that the expression for A in (25.1) is minimized when $\cos(\theta)$ equals $1/\sqrt{3}$ (a proof will be given in Chapter 27). Substituting this value for $\cos(\theta)$ in (25.1), the minimal area of a cell in a honeycomb is readily seen to be

$$6hs + \frac{3s^2}{\sqrt{2}} \approx 6hs + 2.12132s^2.$$

Thus, the relative reduction in surface area for the hexagonal tube in Figure 25.6 is

$$\frac{(6hs + 2.12132s^2) - (6hs + 2.10842s^2)}{6hs + 2.12132s^2} = \frac{0.0129s}{6h + 2.12132s}.$$

To obtain an upper estimate for this relative reduction, we need to determine the least possible value of h for a given value of s. Since the three rhombic faces can be placed atop the hexagonal prism in Figure 25.3 only if $h \geq \overline{CD}$, and since

$$\overline{CD} = \frac{s\cot(\theta)}{2} = \frac{s\cos(\theta)}{2\sin(\theta)} = \frac{s/\sqrt{3}}{2\sqrt{2}/\sqrt{3}} = \frac{s}{2\sqrt{2}} \approx 0.35355s,$$

we may infer that

$$\frac{0.0129s}{6h + 2.12132s} \leq \frac{0.0129s}{6 \cdot 0.35355s + 2.12132s} \approx 0.00304.$$

In other words, if ever a more sophisticated race of honeybees were to evolve that constructs polygonal cell covers by adjoining two hexagons to two parallelograms at just the right angles, the amount of building material saved would be no more than about 0.3% of the amount used by present-day varieties. If we had taken into account that in a regular honeycomb h is considerably larger than the minimal value of $0.35355s$, the relative savings would have been found to be even less significant.

Remark. Just as the rhombic cover on a regular honeycomb cell, the polygonal cover in Figure 25.6 can be produced from the original flat hexagonal prism in Figure 25.3 by slicing off and then reattaching certain tetrahedra and polyhedra. Consequently, the two competing constructions in Figures 25.1 and 25.6 represent cells of equal volume, and a direct comparison of surface areas is therefore a sufficient measure of efficiency.

Additional Exercises

25.2. Use elementary geometry and trigonometry (see Chapter 6) to carefully explain why $\overline{CD} = s\cot(\theta)/2$.

25.3. Explain why the area of a trapezoid, as shown in Figure 25.8, is equal to $h(a+b)/2$.

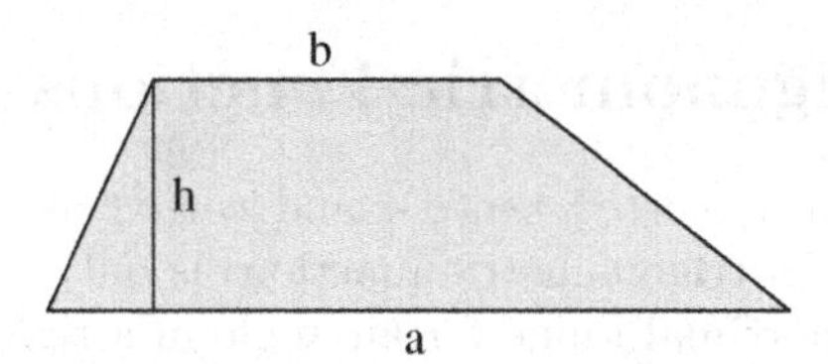

Figure 25.8: a trapezoid.

25.4. ★ Verify the formula given in (25.2).

Chapter 26

Trigonometric Functions

Definition of the Trigonometric Functions

In Chapter 6 we introduced $\sin(\alpha), \cos(\alpha), \tan(\alpha)$, and $\cot(\alpha)$ as quotients of side lengths of right triangles. This way of defining the trigonometric functions is only meaningful for values of α between 0 and $\pi/2$, because that is the maximal range for an angle in a right triangle. In order to extend the domain of the trigonometric functions in a natural way to the entire set of real numbers, we consider a unit circle (i.e., a circle of radius 1) with its center at the origin of an xy-coordinate system. To every angle t, measured in radians, there corresponds a point $P(t) = (x(t), y(t))$ on the unit circle such that t is the angle between the positive x-axis and the line segment from the origin $(0,0)$ to the point $P(t)$ (see Figure 26.1). (Note: for angles $t > 2\pi$, we have to complete more than one full loop around the unit circle in order to get to $P(t)$, and for negative angles we reverse the direction of travel and move clockwise rather than counterclockwise.)

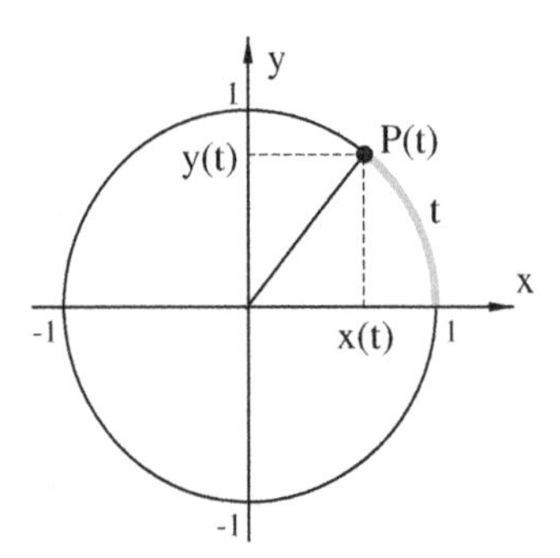

Figure 26.1: the unit circle.

For all $t \in \mathbb{R}$ we define

$$\boxed{\begin{aligned} \sin(t) &:= y(t), \\ \cos(t) &:= x(t). \end{aligned}} \tag{26.1}$$

In order to convince ourselves that these extended definitions of sine and cosine are consistent with the right-triangle definitions given in Chapter 6, we observe that for $0 \leq t \leq \pi/2$ the points $(0,0)$, $(x(t), 0)$, and $P(t) = (x(t), y(t))$ form a right triangle with hypotenuse 1 and side lengths $x(t)$ and $y(t)$. Thus, according to the right-triangle definitions, we may infer that

$$\sin(t) = \frac{y(t)}{1} = y(t) \text{ and } \cos(t) = \frac{x(t)}{1} = x(t),$$

as desired. Furthermore, the theorem of Pythagoras (see Appendix A) applied to the right triangle with vertices at $(0,0)$, $(x(t), 0)$, and $P(t) = (x(t), y(t))$ implies that $1 = x(t)^2 + y(t)^2$. Using (26.1) to replace

$x(t)$ and $y(t)$ with $\cos(t)$ and $\sin(t)$ respectively, we obtain the *trigonometric theorem of Pythagoras:*

$$\boxed{\cos^2(t) + \sin^2(t) = 1.}$$

Given the graphs of sine and cosine in Figure 26.2, we also see that $\sin(t) = 0$ if and only if t is an

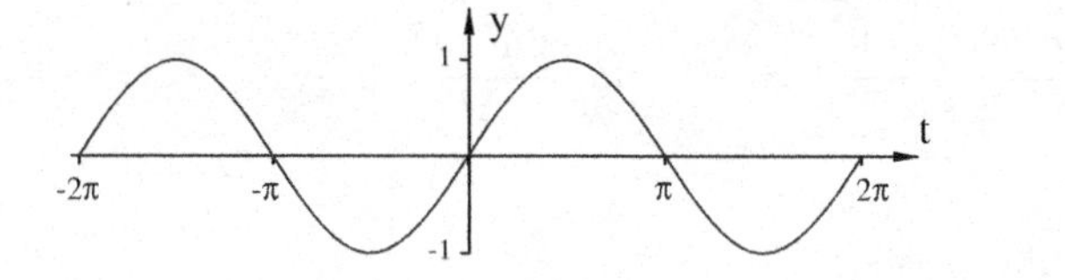

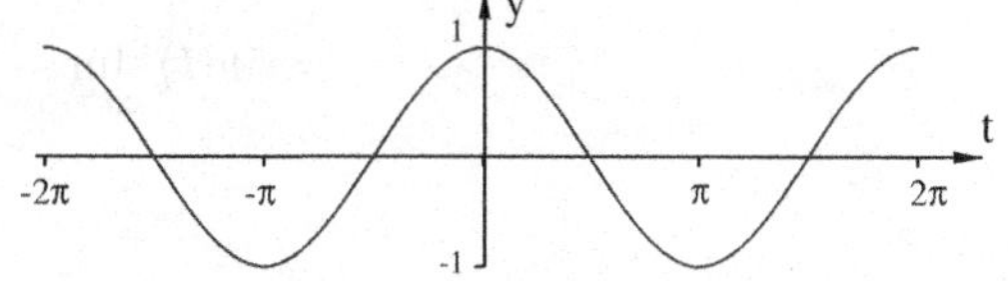

Figure 26.2: graphs of $\sin(t)$ and $\cos(t)$.

integer multiple of π, and $\cos(t) = 0$ if and only if t is an odd integer multiple of $\pi/2$, i.e.,

$$\sin(n\pi) = 0 \text{ and } \cos((2n+1)\pi/2) = 0 \text{ for all integers } n.$$

Taking these observations into account, we define the tangent and cotangent functions as follows:

$$\tan(t) := \frac{\sin(t)}{\cos(t)}$$

for all $t \in \mathbb{R}$ that are not odd integer multiples of $\pi/2$ (i.e., $t \neq (2\pi + 1)\pi/2$ for all integers n), and

$$\cot(t) := \frac{\cos(t)}{\sin(t)}$$

for all $t \in \mathbb{R}$ that are not integer multiples of π (i.e., $t \neq n\pi$ for all integers n). The graphs of the tangent and cotangent functions are shown in Figure 26.3.

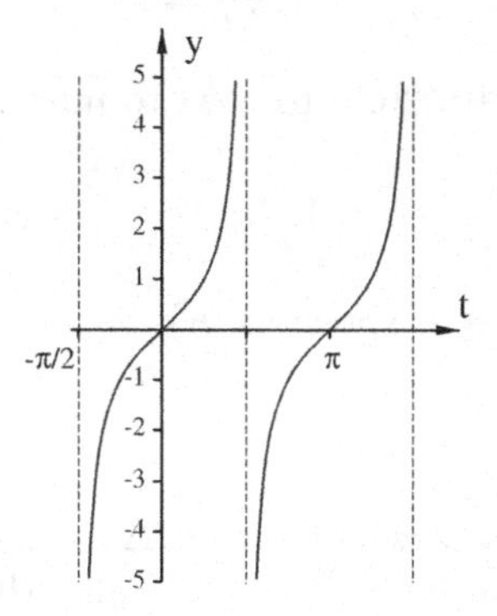

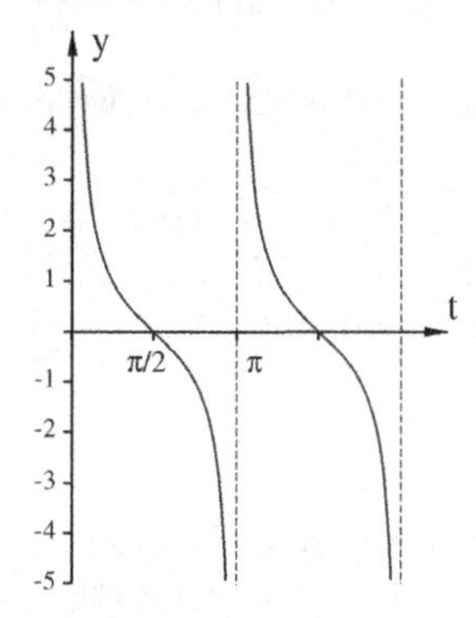

Figure 26.3: graphs of $\tan(t)$ and $\cot(t)$.

26.1 Exercise. Use the trigonometric theorem of Pythagoras to show that

$$1 + \tan^2(t) = \frac{1}{\cos^2(t)} \quad \text{and} \quad 1 + \cot^2(t) = \frac{1}{\sin^2(t)}.$$

Differentiation of Trigonometric Functions

According to the addition laws for sine and cosine (see Appendix A), the following identities are valid for all $t, \Delta t \in \mathbb{R}$:

$$\sin(t + \Delta t) = \sin(t)\cos(\Delta t) + \sin(\Delta t)\cos(t),$$
$$\cos(t + \Delta t) = \cos(t)\cos(\Delta t) - \sin(t)\sin(\Delta t).$$

Using these equations, we may compute the derivatives of sine and cosine as follows:

$$\begin{aligned}\frac{d}{dt}\sin(t) &= \lim_{\Delta t\to 0}\frac{\sin(t+\Delta t)-\sin(t)}{\Delta t}\\ &= \lim_{\Delta t\to 0}\frac{\sin(t)\cos(\Delta t)+\sin(\Delta t)\cos(t)-\sin(t)}{\Delta t}\\ &= \sin(t)\lim_{\Delta t\to 0}\frac{\cos(\Delta t)-1}{\Delta t}+\cos(t)\lim_{\Delta t\to 0}\frac{\sin(\Delta t)}{\Delta t}\end{aligned} \tag{26.2}$$

and

$$\begin{aligned}\frac{d}{dt}\cos(t) &= \lim_{\Delta t\to 0}\frac{\cos(t+\Delta t)-\cos(t)}{\Delta t}\\ &= \lim_{\Delta t\to 0}\frac{\cos(t)\cos(\Delta t)-\sin(t)\sin(\Delta t)-\cos(t)}{\Delta t}\\ &= \cos(t)\lim_{\Delta t\to 0}\frac{\cos(\Delta t)-1}{\Delta t}-\sin(t)\lim_{\Delta t\to 0}\frac{\sin(\Delta t)}{\Delta t}.\end{aligned} \tag{26.3}$$

In order to determine $\lim_{\Delta t\to 0}\sin(\Delta t)/\Delta t$, we use the estimate

$$\sin(\Delta t)\le \Delta t\le \tan(\Delta t) \text{ for all } \Delta t\in[0,\pi/2], \tag{26.4}$$

the validity of which is strongly suggested by the geometric relations shown in Figure 26.4 (for a rigorous proof of (26.4) the reader is referred to Exercise 26.15). Multiplying the inequality $\Delta t\le\tan(\Delta t)$ by $\cos(\Delta t)$ yields

$$\Delta t\cos(\Delta t)\le\sin(\Delta t),$$

and combining this estimate with the first inequality in (26.4), we obtain

$$\cos(\Delta t)=\frac{\Delta t\cos(\Delta t)}{\Delta t}\le\frac{\sin(\Delta t)}{\Delta t}\le\frac{\Delta t}{\Delta t}=1 \tag{26.5}$$

for all $\Delta t\in(0,\pi/2]$. For $-\pi/2\le\Delta t<0$, we may apply (26.5) to $-\Delta t$ to infer that

$$\cos(-\Delta t)\le\frac{\sin(-\Delta t)}{-\Delta t}\le 1.$$

Since $\cos(-\Delta t)=\cos(\Delta t)$ and $\sin(-\Delta t)=-\sin(\Delta t)$ (see Appendix A), it again follows that

$$\cos(\Delta t)\le\frac{\sin(\Delta t)}{\Delta t}\le 1.$$

In other words, the estimate (26.5) is valid regardless of whether $0<\Delta t\le\pi/2$ or $-\pi/2\le\Delta t<0$. Since cosine is obviously a continuous function, and since $\cos(0)=1$, we may thus conclude that

$$1=\lim_{\Delta t\to 0}\cos(\Delta t)\le\lim_{\Delta t\to 0}\frac{\sin(\Delta t)}{\Delta t}\le 1.$$

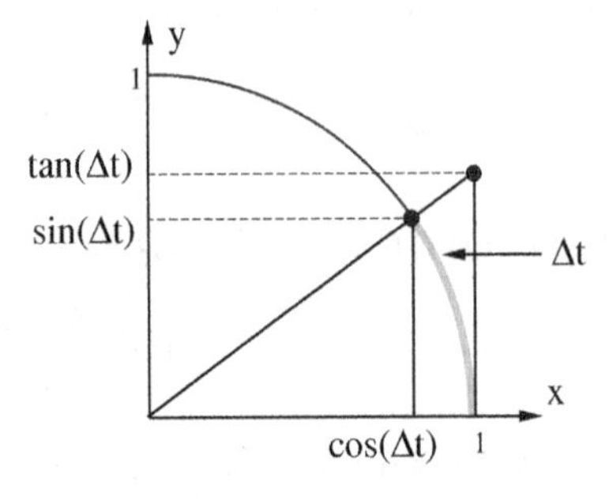

Figure 26.4: estimating Δt.

Consequently,

$$\lim_{\Delta t\to 0}\frac{\sin(\Delta t)}{\Delta t}=1. \tag{26.6}$$

In order to determine $\lim_{\Delta t\to 0}(\cos(\Delta t)-1)/\Delta t$ (see (26.2) and (26.3)), we apply (26.6) in conjunction with the trigonometric theorem of Pythagoras:

$$\begin{aligned}\lim_{\Delta t\to 0}\frac{\cos(\Delta t)-1}{\Delta t} &= \lim_{\Delta t\to 0}\frac{(\cos(\Delta t)-1)(\cos(\Delta t)+1)}{\Delta t(\cos(\Delta t)+1)}\\ &= \left(\lim_{\Delta t\to 0}\frac{\cos^2(\Delta t)-1}{\Delta t}\right)\lim_{\Delta t\to 0}\frac{1}{\cos(\Delta t)+1}\\ &= \left(\lim_{\Delta t\to 0}\frac{-\sin^2(\Delta t)}{\Delta t}\right)\cdot\frac{1}{2}\\ &= -\frac{1}{2}\left(\lim_{\Delta t\to 0}\sin(\Delta t)\right)\left(\lim_{\Delta t\to 0}\frac{\sin(\Delta t)}{\Delta t}\right)\\ &= -\frac{0\cdot 1}{2}=0.\end{aligned} \tag{26.7}$$

Combining (26.2) and (26.3) with (26.6) and (26.7) yields

$$\boxed{\begin{aligned}\frac{d}{dt}\sin(t)&=\cos(t),\\ \frac{d}{dt}\cos(t)&=-\sin(t).\end{aligned}} \tag{26.8}$$

Alternatively, once the derivative of $\sin(t)$ has been determined, the derivative of $\cos(t)$ can also be computed from the chain rule using the trigonometric formulae $\sin(\pi/2-t)=\cos(t)$ and $\cos(\pi/2-t)=\sin(t)$ (see Appendix A):

$$\frac{d}{dt}\cos(t)=\frac{d}{dt}\sin\left(\frac{\pi}{2}-t\right)=-\cos\left(\frac{\pi}{2}-t\right)=-\sin(t).$$

26.2 Exercise. Apply the quotient rule to show that

$$\boxed{\begin{aligned}\frac{d}{dt}\tan(t)&=\frac{1}{\cos^2(t)}=1+\tan^2(t),\\ \frac{d}{dt}\cot(t)&=-\frac{1}{\sin^2(t)}=-1-\cot^2(t).\end{aligned}}$$

26.3 Exercise. Determine the following derivatives: $\frac{d}{dt}\sin(3t^2)$, $\frac{d}{dt}e^{\cos(t)}$, and $\frac{d}{dt}\cos^3(4t-1)$.

26.4 Exercise. Find the antiderivatives $\int\sin(5t)\,dt$ and $\int(1+\tan^2(t))\,dt$.

Additional Exercises

26.5. Differentiate each of the following functions:

a) $x\sin(x)$ **b)** $\sin^2(x)$ **c)** $\tan(e^{3x})$ **d)** $e^{4x}\sin(2x)$ **e)** $x^{\cos(x)}$
f) $x^2\cos(3x)$ **g)** $\sin(x^2)$ **h)** $2^{-x}\cos(5x)$ **i)** $2x\sin(e^{4x})$ **j)** $\ln|\sin(4x)+\cos(3x)|$

26.6. Determine the following antiderivatives:

a) $\int 2\sin(3x)\,dx$ **b)** $\int x\sin(3x^2)\,dx$ **c)** $\int (2+\tan^2(2x))\,dx$ **d)** $\int \cos(\pi x)\,dx$

e) $\int \cos(2-3x)\,dx$ **f)** $\int \sin(x)\cos(x)\,dx$ **g)** $\int (\sin^2(4x)+\cos^2(4x))\,dx$ **h)** $\int x\cot^2(x^2)\,dx$

Hint. For b), f), and h) you should try to think of a function that equals the given integrand when differentiated using the chain rule.

26.7. Sketch the graph of each of the following functions on the interval $[-2\pi, 2\pi]$:

a) $\sin(x)\cos(x)$ **b)** $x/2+\cos(x)$ **c)** $\sin^3(x)$
d) $\sin(x^2)$ **e)** $\sin(x)+\cos(x)$ **f)** $x\sin(x)/2$

26.8. Find the equation of the line tangent to the graph of $f(x) := \dfrac{x+\sin(x)}{\cos(x)}$ at $(0, f(0))$.

26.9. Given that $\sin(x)\cos(x) = e^y$, determine dy/dx in terms of x.

26.10. Assume that a projectile's initial velocity is v meters per second. If the projectile is shot into the air at an angle θ radians relative to the horizontal, then its range is given by the equation $R = v^2\sin(2\theta)/g$, where g is the gravitational acceleration close to the surface of the earth. For which angle θ is the range largest?

26.11. According to Pouiseuilles Law, if blood flows into a straight blood vessel of radius r branching off of another straight blood vessel with radius R at an angle θ radians, the total resistance T of the blood in the branching vessel is given by the equation

$$T = C\left(\frac{a - b^4\cot(\theta)}{R^4} + \frac{b^4}{r^4\sin(\theta)}\right),$$

where a, b, and C are constants and $r < R$. Show that the total resistance is minimal when $\cos(\theta) = (r/R)^4$.

26.12. Find the derivative of each of the following functions (with respect to x):

a) $\displaystyle\int_1^x \frac{\sin(t)}{t}\,dt$ **b)** $\displaystyle\int_1^{\cos(x)} \frac{1}{t}\,dt$ **c)** $\displaystyle\int_1^{x^2} e^{\tan(t)}\,dt$ **d)** $\displaystyle\int_1^3 \sin(t^3)\,dt$

Hint. Apply the chain rule in b) and c).

26.13. In order to find a formula for the derivative of the sine function, we determined earlier in this chapter the limit $\lim_{x\to 0}\sin(x)/x$. What is the value of this limit and what is the derivative of $\sin(x)$ if x is measured in degrees rather than radians?

26.14. ★ Show that the function A in (25.2) assumes its minimal value for $\cos(\theta) \approx 0.745004$. *Hint.* Use a computer (or standard approximation techniques such as Newton's method) to solve for $\cos(\theta)$ the equation $dA/d\theta = 0$.

26.15. Prove the statement in (26.4). *Hint.* In Figure 26.4 determine the area of the triangle with vertices at $(0,0)$, $(1,0)$, and $(\cos(\Delta t), \sin(\Delta t))$, the area of the circular sector over the angle Δt, and the area of the triangle with vertices at $(0,0)$, $(1,0)$, and $(1, \tan(\Delta t))$.

Chapter 27

Trigonometric Inverse Functions and Honeycombs (Part 2)

Definition of the Trigonometric Inverse Functions

Since none of the graphs in Figures 26.2 and 26.3 passes the horizontal line test, we may infer that none of the trigonometric functions defined in Chapter 22 is invertible. We need to remember, though, that a function's invertibility or noninvertibility is crucially dependent upon the chosen domain. While the trigonometric functions are not invertible on their maximal domains, it is easy to see from the graphs in Chapter 26 that the sine and tangent functions are invertible on the interval from $-\pi/2$ to $\pi/2$ and the cosine and cotangent functions are invertible on the interval from 0 to π (these intervals are not the only possible choices, but they are the standard choices, and every scientific calculator is programmed to use them).

27.1 Definition. **a)** The *sine inverse function*, which is also referred to as *arcsine*, is defined to be the inverse function of

$$f_1 : [-\pi/2, \pi/2] \to \mathbb{R}$$
$$t \mapsto \sin(t).$$

The values of the sine inverse function are denoted by $\sin^{-1}(x)$ or alternatively by $\arcsin(x)$. The domain of the sine inverse function is $[-1, 1]$ (which is the range of f_1) and its range is $[-\pi/2, \pi/2]$ (which is the domain of f_1).

b) The *cosine inverse function* or *arccosine* is the inverse function of

$$f_2 : [0, \pi] \to \mathbb{R}$$
$$t \mapsto \cos(t).$$

The values of the cosine inverse function are denoted by $\cos^{-1}(x)$ or alternatively by $\arccos(x)$. The domain of the cosine inverse function is $[-1, 1]$ and its range is $[0, \pi]$.

c) The *tangent inverse function* or *arctangent* is the inverse function of

$$g_1 : (-\pi/2, \pi/2) \to \mathbb{R}$$
$$t \mapsto \tan(t).$$

The values of the tangent inverse function are denoted by $\tan^{-1}(x)$ or alternatively by $\arctan(x)$. The domain of the tangent inverse function is $\mathbb{R}$ (which is the range of g_1) and the range is $(-\pi/2, \pi/2)$ (which is the domain of g_1).

d) The *cotangent inverse function* or *arccotangent* is the inverse function of

$$g_2 : (0, \pi) \to \mathbb{R}$$
$$t \mapsto \cot(t).$$

The values of the cotangent inverse function are denoted by $\cot^{-1}(x)$ or alternatively by $\text{arccot}(x)$. The domain of the cotangent inverse function is $\mathbb{R}$ and its range is $(0, \pi)$.

The graphs of the trigonometric inverse functions, as shown in Figures 27.1 and 27.2, are obtained by reflecting the graphs of the corresponding trigonometric functions across the diagonal line described by the equation $y = x$ (see Chapter 22).

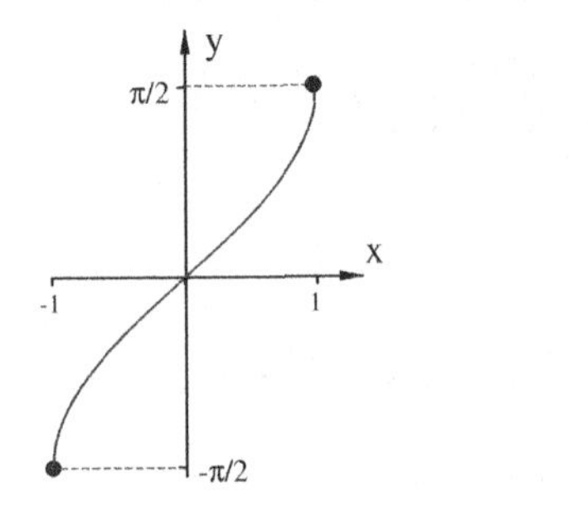

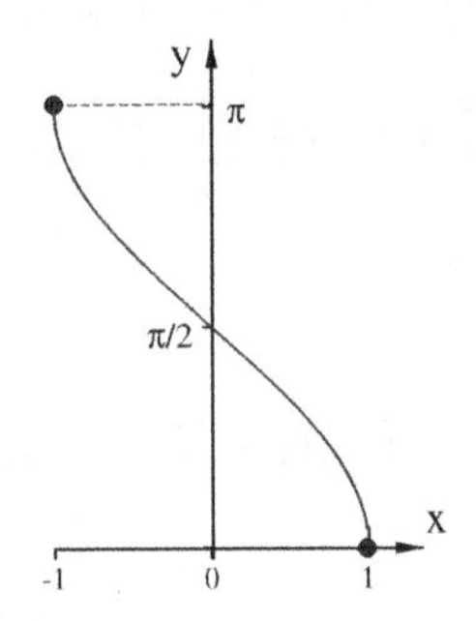

Figure 27.1: graphs of $\arcsin(x)$ and $\arccos(x)$.

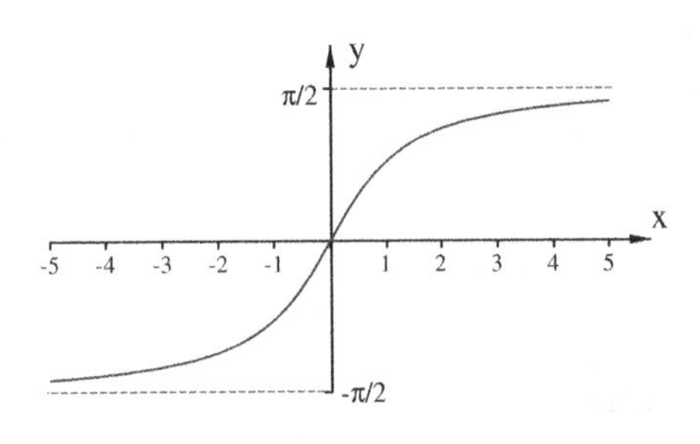

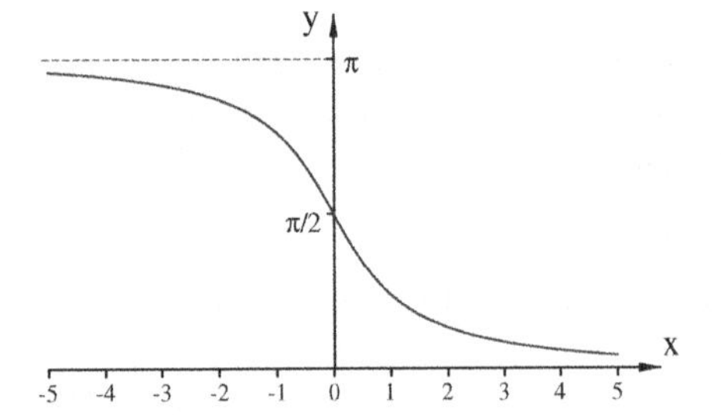

Figure 27.2: graphs of $\arctan(x)$ and $\text{arccot}(x)$.

Teacher: To test your comprehension of the definitions in 27.1, let me ask you the following question: what is the value of $\arcsin(\sin(5\pi/6))$?

Simplicio: According to Theorem 22.4, the fundamental relationship between a function and its inverse is expressed in the equation $f^{-1}(f(x)) = x$. In our case, f is the sine function and f^{-1} is arcsine. So I conclude that for $x = 5\pi/6$ we have $\arcsin(\sin(5\pi/6)) = f^{-1}(f(5\pi/6)) = 5\pi/6$.

Teacher: You made exactly the mistake I was hoping for.

Simplicio: Why is that?

Teacher: In your answer you asserted that arcsine is the inverse of the sine function.

Simplicio: Isn't that how we defined arcsine?

Teacher: No, it couldn't be, because the sine function defined on $\mathbb{R}$ is not invertible—only the restricted sine function f_1 is invertible.

Simplicio: I am getting confused.

Teacher: How about you, Sophie?

Sophie: I must admit I would have given the same wrong answer as Simplicio, but now that you have pointed out the distinction between the sine function and its restriction f_1, I think our problem is that $5\pi/6$ is not contained in the domain of f_1. Simplicio used the general equation $f^{-1}(f(x)) = x$ to derive his answer, but as stated in Theorem 22.4, this equation makes sense only if x is in the domain of f. In our case f corresponds to f_1, and therefore the equation $\arcsin(\sin(x)) = f_1^{-1}(f_1(x)) = x$ is valid only for $x \in [-\pi/2, \pi/2]$, that is, for values x that are in the domain of f_1. Furthermore, since the range of f_1^{-1} is the domain of f_1, it cannot be true that $\arcsin(\sin(5\pi/6)) = f_1^{-1}(\sin(5\pi/6)) = 5\pi/6$, because $5\pi/6$ is not contained in the range $R(f_1^{-1}) = [-\pi/2, \pi/2]$.

Teacher: Your argument is entirely correct, and it illustrates well how important it is to think of a function not only in terms of an equation that assigns output values $f(x)$ to given input values x, but also in terms of domain and range. In reminding us that the sine function is invertible only on its restricted domain $[-\pi/2, \pi/2]$, Sophie was able to explain why $\arcsin(\sin(5\pi/6))$ is not equal to $5\pi/6$.

Simplicio: I understand now where I made a mistake, but the question we posed initially still has not been settled—we still don't know the correct value of $\arcsin(\sin(5\pi/6))$.

Teacher: Let me give you a hint: the value of sine at $5\pi/6$ is equal to $1/2$ and so is the value of sine at $\pi/6$ (the reader should check this by using the results of Exercise 6.2 and the definition of the sine function as given in Chapter 26).

Simplicio: What you apparently are suggesting is that

$$\arcsin\left(\sin\left(\frac{5\pi}{6}\right)\right) = \arcsin\left(\frac{1}{2}\right) = \arcsin\left(\sin\left(\frac{\pi}{6}\right)\right),$$

but how do we proceed?

Sophie: It's easy: since $\pi/6$ is contained in the interval $[-\pi/2, \pi/2]$, it follows that $\arcsin(\sin(\pi/6)) = f_1^{-1}(f_1(\pi/6)) = \pi/6$. In other words, your equation, Simplicio, implies that

$$\arcsin\left(\sin\left(\frac{5\pi}{6}\right)\right) = \frac{\pi}{6}.$$

Teacher: Indeed, that is the correct answer. Central to our discussion was the observation that the simplification $\arcsin(\sin(x)) = x$ is valid only if x is contained in the domain of f_1, that is, in the interval $[-\pi/2, \pi/2]$. Similarly, in reversing the order of sine and arcsine, the equation $\sin(\arcsin(x)) = x$ is valid only if x is in the domain of f_1^{-1}. In summary:

$$\begin{aligned}\arcsin(\sin(x)) &= x \text{ for all } x \in [-\pi/2, \pi/2],\\ \sin(\arcsin(x)) &= x \text{ for all } x \in [-1, 1].\end{aligned}$$

27.2 Exercise. State analogues to the two statements above for cosine, tangent, and cotangent.

27.3 Exercise. Evaluate the following expressions without using a calculator: $\sin(\arcsin(-0.1))$, $\arcsin(\sin(7\pi/6))$, $\arccos(\cos(-\pi/6))$, and $\arctan(\tan(\pi))$.

Honeycombs (Part 2)

Given the derivatives of the trigonometric functions in Chapter 26 and the definitions of the trigonometric inverse functions in 27.1, we are now able to complete the solution of the honeycomb problem. In order to minimize the surface area A of the honeycomb which, according to (25.1), satisfies the equation

$$A = 6hs + \frac{3s^2(\sqrt{3} - \cos(\theta))}{2\sin(\theta)},$$

we determine the derivative of A via the quotient rule:

$$\frac{dA}{d\theta} = \frac{3s^2(\sin^2(\theta) - (\sqrt{3} - \cos(\theta))\cos(\theta))}{2\sin^2(\theta)} = \frac{3s^2(1 - \sqrt{3}\cos(\theta))}{2\sin^2(\theta)}. \tag{27.1}$$

In setting this derivative equal to zero, we see that the critical points of A are solutions of the equation

$$\cos(\theta) = \frac{1}{\sqrt{3}}.$$

Applying the cosine inverse function to both sides of this equation yields

$$\theta = \arccos\left(\frac{1}{\sqrt{3}}\right) \approx 0.9553\, rad \approx 54.74^\circ.$$

This result shows that the theoretically ideal value for θ is indeed remarkably close to the observed value of approximately 54°.

27.4 Exercise. Verify equation (27.1).

27.5 Exercise. Use the second derivative test to show that A assumes a minimum at $\theta = \arccos(1/\sqrt{3})$.

Remark. In the process of minimizing A we never specified the volume of the hexagonal tube, but only assumed s and h to be given constants. Consequently, the angle 54.74° minimizes the surface area of any hexagonal tube capped at one end with three equilateral parallelograms.

Differentiation of the Trigonometric Inverse Functions

Using the derivative of the sine function and the rule for the differentiation of inverse functions (see Theorem 22.7) we obtain

$$\frac{d}{dx}\arcsin(x) = \frac{1}{\cos(\arcsin(x))}. \tag{27.2}$$

27.6 Exercise. Show that

$$\frac{d}{dx}\arccos(x) = \frac{1}{-\sin(\arccos(x))}.$$

To simplify the formulae in (27.2) and Exercise 27.6, we apply the trigonometric theorem of Pythagoras: since $\arcsin(x) \in [-\pi/2, \pi/2]$, it follows that $\cos(\arcsin(x)) \geq 0$ and therefore,

$$\cos(\arcsin(x)) = \sqrt{1 - \sin^2(\arcsin(x))} = \sqrt{1 - x^2} \text{ for all } x \in [-1, 1].$$

Similarly, it can be shown that

$$\sin(\arccos(x)) = \sqrt{1 - x^2} \text{ for all } x \in [-1, 1].$$

Combining these observations with (27.2) and Exercise 27.6 yields

$$\boxed{\begin{aligned} \frac{d}{dx}\arcsin(x) &= \frac{1}{\sqrt{1-x^2}}, \\ \frac{d}{dx}\arccos(x) &= \frac{-1}{\sqrt{1-x^2}}. \end{aligned}} \tag{27.3}$$

27.7 Exercise. Proceeding in a completely analogous fashion, use the rule for the differentiation of inverse functions and the formulae for the derivatives of tangent and cotangent (see Exercise 26.2) to show that

$$\boxed{\begin{aligned} \frac{d}{dx}\arctan(x) &= \frac{1}{1+x^2}, \\ \frac{d}{dx}\operatorname{arccot}(x) &= \frac{-1}{1+x^2}. \end{aligned}}$$

27.8 Exercise. According to the formulae for the derivatives of the sine inverse and cosine inverse functions, we have

$$\frac{d}{dx}\arcsin(x) = -\frac{d}{dx}\arccos(x).$$

Use this observation to prove that

$$\arcsin(x) + \arccos(x) = \frac{\pi}{2} \text{ for all } x \in [-1, 1].$$

Additional Exercises

27.9. Differentiate each of the following functions:

a) $x \arcsin(x)$ **b)** $\arcsin^2(x)$ **c)** $\arctan(e^{3x})$
d) $e^{4x} \arcsin(2x)$ **e)** $x^{\arccos(x)}$ **f)** $x^2 \arccos(3x)$
g) $2^{-x} \arccos(5x^2)$ **h)** $2x \arcsin(e^{4x})$ **i)** $\ln|\arcsin(x) + \arccos(x)|$

27.10. Given a differentiable function f, determine the derivatives of the following functions:

a) $\arcsin(\tan(f(x)))$ **b)** $\sin(\arccos(f(x)))$ **c)** $\sin(\arcsin(f(x)))$ **d)** $\tan(\cot(f(x)))$

27.11. Determine the following antiderivatives:

a) $\displaystyle\int \frac{2}{1+x^2}\,dx$ **b)** $\displaystyle\int \frac{2}{\sqrt{1-x^2}}\,dx$ **c)** $\displaystyle\int \frac{5x}{1+x^2}\,dx$
d) $\displaystyle\int \frac{3}{1+4x^2}\,dx$ **e)** $\displaystyle\int \frac{2}{\sqrt{1-4x^2}}\,dx$ **f)** $\displaystyle\int \frac{-2x}{\sqrt{1-x^2}}\,dx$

Hint. Use the chain rule for c) and f).

27.12. Without using a calculator, evaluate each of the following expressions:

a) $\arcsin(0.5)$ **b)** $\cos(\arcsin(0.5))$ **c)** $\sin(\arctan(3))$
d) $\arccos(-\sqrt{3}/2)$ **e)** $\arcsin(\sin(-2))$ **f)** $\tan(\arcsin(3/4))$

27.13. Determine which of the following expressions are well defined:

a) $\sin(\arcsin(1.5))$ **b)** $\sin(\arcsin(-0.5))$ **c)** $\cos(\arccos(-1.5))$ **d)** $\tan(\arctan(1000))$

27.14. Find the area under the graph of the function $f(x) := \dfrac{1}{1+x^2}$ over the interval $[-1, 1]$.

27.15. Use differentiation to show that

$$\arctan(x) + \arctan\left(\frac{1}{x}\right) = \begin{cases} \pi/2 & \text{if } x > 0, \\ -\pi/2 & \text{if } x < 0. \end{cases}$$

27.16. A boat is being pulled toward a dock by means of a winch. The winch is located at the end of the dock and is 10 feet above the level at which the tow rope is attached to the bow of the boat. The rope is being pulled at a rate of 1 ft/s. Determine the rate at which the angle of elevation from the bow of the boat to the end of the dock is changing when the length of the rope is 30 feet.

27.17. Show that the function $f : [-\pi/4, \pi/4] \to \mathbb{R}$ with $f(x) := \sin(x) + \cos(x)$ is stricly increasing and therefore invertible. Then find a formula for $f^{-1}(x)$ and determine the derivative of f^{-1}. *Hint.* To find a formula for $f^{-1}(x)$, you may want to square both sides of the equation $y = \sin(x) + \cos(x)$ and then apply the double angle formula $\sin(2x) = 2\sin(x)\cos(x)$.

Chapter 28

Explaining the Rainbow

> And God said, "This is the sign of the covenant I am making between me and you and every living creature with you, a covenant for all generations to come: I have set my rainbow in the clouds, and it will be the sign of the covenant between me and the earth. Whenever I bring clouds over the earth and the rainbow appears in the clouds, I will remember my covenant between me and you and all living creatures of every kind."[1]

From the dawn of civilization to the 21st century, the rainbow has never failed to inspire the admiration of human observers with its brilliant display of colors and perfect circular shape. In our modern scientific world we have lost to a certain extent the sense of awe and reverence that people in biblical times may have felt at this marvelous sight, but this is not so much an indication for the demystifying effect of science as it is evidence for our lack of imagination. Every scientific "explanation" only shifts the mystery to a deeper level, and our ability to formulate mathematical models for certain natural phenomena such as rainbows is in itself profoundly mysterious. In this spirit then we may set out to "explain" the rainbow without fear of diminishing our appreciation for this wonder of nature.

The Position of a Rainbow in the Sky

Rainbows are best observed shortly after a rainfall when sunlight breaks through a cloud cover and strikes raindrops still in the air against the dark background of a receding storm front. Conditions are most favorable when the sun is low in the sky just after sunrise or right before sunset, for at these times the light rays from the sun are nearly parallel to the ground and therefore best positioned to reach the

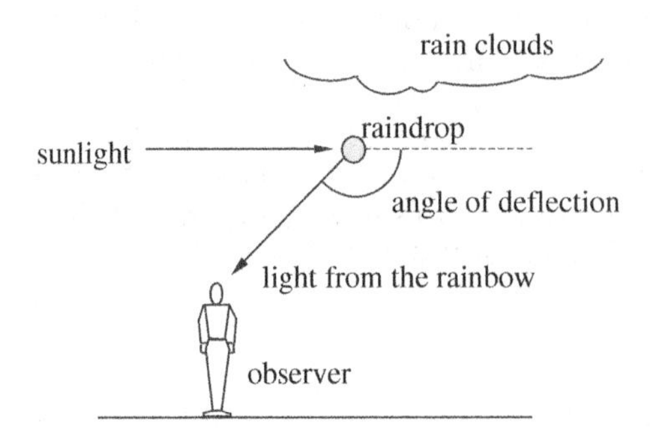

Figure 28.1: observing a rainbow.

raindrops under a cloud cover (see Figure 28.1). The position of the sun also affects the angle at which a rainbow is observed. With the sun low in the sky, light from a rainbow reaches us via a clockwise deflection of about 137.5° from the horizontal (see the next section for an explanation) and is observed at the corresponding *rainbow angle* of $180^\circ - 137.5^\circ = 42.5^\circ$ (see the diagram on the left in Figure 28.2). If the sun is higher up so that its rays are inclined at an angle δ, the same deflection by about 137.5° will cause a rainbow to appear at an angle of observation of $42.5^\circ - \delta$ (see the diagram on the right in Figure 28.2).

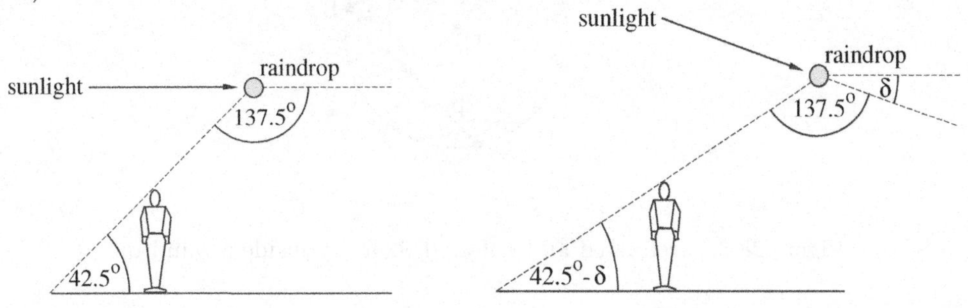

Figure 28.2: the angle of observation in dependence on the position of the sun.

Remark. It is understood that the diagrams in Figure 28.2 show only a cross-sectional view. In both cases, we must imagine the observer at the vertex of a cone on the surface of which the rainbow forms a circular arc. The axis of symmetry of this cone is parallel to the direction of travel of light from the sun.

28.1 Exercise. What is the permissible range of values for the angle of inclination δ if a rainbow is to be visible from the ground?

The Rainbow Angle

Sophie: I was trying to analyze what sort of interaction between light and water could cause the deflection that we discussed in regard to Figure 28.2 and, I must admit, I was a bit puzzled. At first I thought it was simply a matter of light being reflected off a raindrop's surface, but this left me wondering how a single reflection could account for the characteristic display of colors in narrow circular arcs. After all, given a raindrop's approximately spherical shape, one would naturally expect reflections to occur in all possible directions, thus creating a broad rather than a narrow band of increased brightness in the sky. I then considered a light ray that enters a raindrop on one side and re-emerges on the other. In this case the law of refraction (as stated in (12.4)) may allow us to explain the appearance of colors, because the refractive index of water (see p.102) is different for light of different wavelengths (see the discussion on colors in the next section), but the total deflection would certainly be less than the asserted 137.5 degrees.

Teacher: Your analysis sounds very promising, but you stopped just short of drawing the obvious conclusion: if the appearance of colors can be explained from the law of refraction, and the large deflection angle from the law of reflection, then it seems that refractions and reflections must occur in unison. In Figure 28.3 we are shown a light ray that is refracted upon entering a raindrop, then reflected at the opposite side and finally refracted a second time before passing back into air. At each of the points A, B, and C, the light ray loses some of its intensity because one part of it is always reflected and the other refracted (the parts that do not reach the observer are not shown in Figure 28.3). This loss of intensity also explains why secondary rainbows are usually hard to detect, because the light that produces them must undergo one additional reflection (see the discussion at the end of this chapter).

Simplicio: Secondary bows, I am sure, are a fascinating topic, but I am still struggling with the basics, for I still don't understand exactly how a deflection angle of 137.5 degrees can be computed for the path shown in Figure 28.3.

Teacher: Before we can address this issue, Simplicio, we need to understand how the laws of reflection and refraction determine the shape of this path. In Chapter 6, we saw that for curved surfaces the

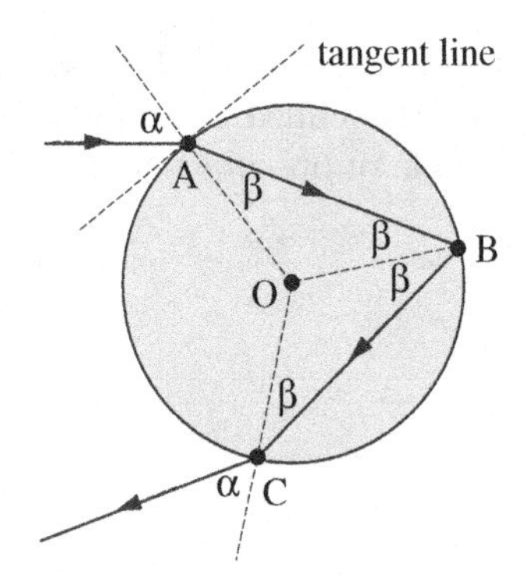

Figure 28.3: a refracted and reflected light ray inside a raindrop.

law of reflection is to be formulated with reference to tangent lines at certain cross-sections of the surface. Such a tangent line is drawn through A in Figure 28.3, and both α and β are measured relative to the line perpendicular to this tangent line. Since the raindrop is assumed to be spherical in shape, the line perpendicular to the tangent line at each of the points A, B, and C is simply the line connecting the respective point with the center O. Furthermore, the equality of the distances $\overline{AO}$ and $\overline{BO}$ (both are equal to the radius of the sphere) implies that the angles $\measuredangle OAB$ and $\measuredangle ABO$ are equal as well. (Note: for three arbitrary points P, Q, and R, we denote by $\measuredangle PQR$ the angle at Q enclosed by the line segments $\overline{QP}$ and $\overline{QR}$.) Similarly, we may infer that $\measuredangle OBC = \measuredangle BCO$, and the equality of $\measuredangle OBC$ and $\measuredangle ABO$ follows from the law of reflection (see Chapter 12, p.101). These observations explain why in Figure 28.3 the angles $\measuredangle OAB$, $\measuredangle ABO$, $\measuredangle OBC$, and $\measuredangle BCO$ are all denoted by β, and why the angle of incidence α at A has the same value as the angle of refraction at C (see also the remark on p.103). Furthermore, at point A the (clockwise) change in the direction of the light ray is $\alpha - \beta$, at B the deflection is $\pi - 2\beta$, and at C it is again $\alpha - \beta$. Consequently, the total deflection angle is

$$D(\alpha) = \alpha - \beta + \pi - 2\beta + \alpha - \beta = \pi + 2\alpha - 4\beta. \tag{28.1}$$

Sophie: This formula appears to suggest that D is dependent on α *and* β, but in writing "$D(\alpha)$," you indicated a functional dependence only on α. Why is that?

Teacher: The dependence of D on β can be eliminated if we use the law of refraction to express β in terms of α. According to (12.4), we have

$$\sin(\beta) = \frac{\sin(\alpha)}{k},$$

or equivalently,

$$\beta = \arcsin\left(\frac{\sin(\alpha)}{k}\right).$$

Substituting this expression for β in equation (28.1) yields

$$D(\alpha) = \pi + 2\alpha - 4\arcsin\left(\frac{\sin(\alpha)}{k}\right). \tag{28.2}$$

For $k \approx 1.33$ (see p.102), the graph of D is shown in Figure 28.4.

Sophie: I am still a bit confused regarding the functional dependence of D on α. For the fact that we observe a rainbow at a very definite angle in the sky, rather than spread out over a wide region, seems to imply that D is constant or at least very nearly so. How then is it possible that the graph in Figure 28.4 shows D covering a fairly wide range of values (on the vertical axis)?

Teacher: The answer to this, and to Simplicio's earlier question concerning the computation of the deflection angle of 137.5°, is hidden in the fact that D assumes a minimum at the angle α_0 shown in Figure 28.4.

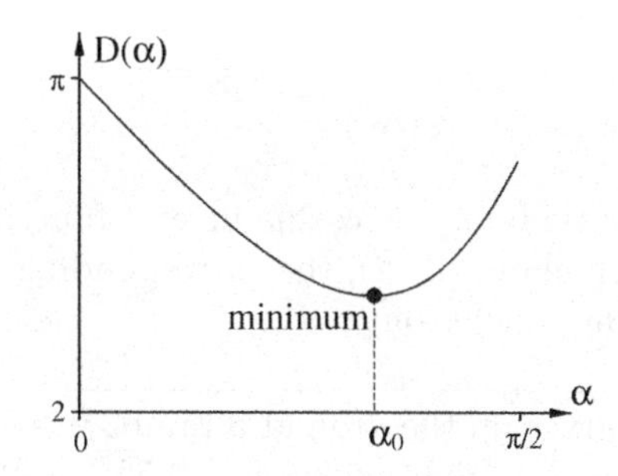

Figure 28.4: graph of D.

Sophie: I must admit, I cannot see any connection at all between the position of a rainbow in the sky and the existence of a minimum at α_0.

Simplicio: I am also totally lost, but that's hardly surprising.

Teacher (showing his true self): Given your performance in this course so far, Simplicio, I'm afraid you are right.

Simplicio (the model student): Thank you, Teacher—it's good to be honest. Thank you!

Teacher (back to normal): Don't be too harsh on yourself, Simplicio. You really are quite a talented fellow, and your lack of understanding in this case is due only to the extraordinary difficulty of the problem at hand. (Truth be told, a student with a mind as keen as our dear Simplicio would give most teachers reason to rejoyce, but the tragic ending in the Epilogue requires that Simplicio be effectively demoralized. In other words, the bashing he receives is a dramaturgic need.)

Teacher: To see how the existence of the minimum at α_0 affects the appearance of a rainbow in the sky, let us consider a small interval J of values close to $D(\alpha_0)$ as shown in Figure 28.5. The corresponding range for α is the interval I. Since the slope of D close to α_0 is approximately zero, the interval I

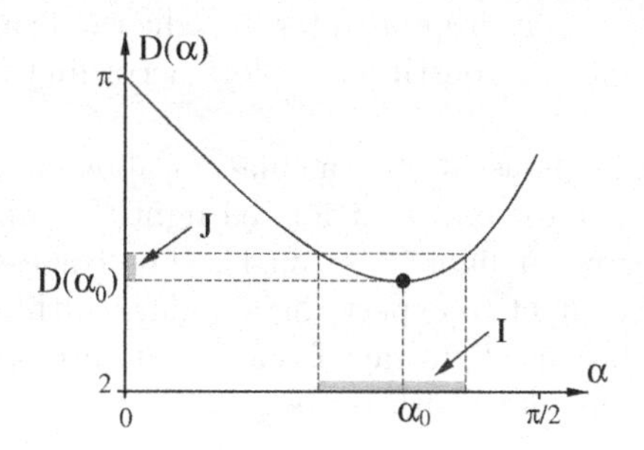

Figure 28.5: graph of D.

is significantly larger than J (see Figure 28.5). In other words, a relatively large band of light rays with angles of incidence close to zero is deflected into a small range of values near $D(\alpha_0)$. Thus, we observe a *focusing effect* which significantly increases the intensity of light at deflection angles approximately equal to $D(\alpha_0)$, and only in this direction is the intensity sufficient to be perceptible to the human eye. In order to determine α_0, we need to compute the derivative of D and set it equal to zero (see our discussion in Chapters 11 and 12). Using the results in (26.8), (27.3), and (28.2), we may apply the chain rule to infer that

$$0 = D'(\alpha_0) = 2 - \frac{4\cos(\alpha_0)}{k\sqrt{1-\sin^2(\alpha_0)/k^2}}.$$

Since $\sin^2(\alpha_0) = 1 - \cos^2(\alpha_0)$ (by the trigonometric theorem of Pythagoras), it follows that

$$4\cos^2(\alpha_0) = k^2\left(1 - \frac{\sin^2(\alpha_0)}{k^2}\right) = k^2 - 1 + \cos^2(\alpha_0),$$

and therefore,

$$\cos(\alpha_0) = \sqrt{\frac{k^2 - 1}{3}}. \tag{28.3}$$

Using the value 1.33 for k and applying the cosine inverse function, we find α_0 to be approximately equal to 59.56°. Given the formula in (28.2), the corresponding deflection angle is $D(\alpha_0) \approx 137.5°$. Thus, we arrive at the following conclusion:

For an observer on the ground a rainbow is visible at the *rainbow angle* of approximately $42.5° = 180° - 137.5°$.

The Colors of the Rainbow

In a treatise published in 1873, the British physicist James Clark Maxwell proposed that all electromagnetic phenomena can be described by four fundamental equations, which nowadays are known as Maxwell's equations. Since certain solutions of these equations modeled a wavelike propagation through space at the speed of light, Maxwell concluded that light is, in fact, an electromagnetic wave (see our discussion in Chapter 80) and as such characterized by a well-defined wavelength. The visible light from the sun, which we perceive as white, is composed of electromagnetic waves with wavelengths ranging from 4,000 to 7,000 angstroms. Light with wavelengths between 6,470 and 7,000 angstroms is perceived as red, and at the other end of the spectrum between 4,000 and 4,240 angstroms it is perceived as violet.

For colors to occur in a rainbow it is crucial that the speed of light in a given medium is different for different wavelengths, because this implies that the ratio $k = c_a/c_w$, which is central to the formulation of the law of refraction, is dependent on the wavelength as well. The value of k for red light is approximately 1.3318 and for violet light it is 1.3435. Equation (12.4) therefore allows us to infer that for a given angle of incidence, the angle of refraction is smaller for red light than for violet light. Consequently, upon passing through a drop of water, a light ray from the sun (such as the one shown in Figure 28.3) is decomposed into its constituent colors according to the different wavelengths in its spectrum.

Following exactly the same steps as in our calculation above and using the different values for k for different wavelengths, it is easy to show that for red light the minimal angle of deflection $D(\alpha_0)$ is approximately 137.7°, while for violet light it is 139.4°. The corresponding rainbow angles are 42.3° and 40.6° respectively. Thus, we ought to expect the angular width of a rainbow to be about 1.7°, but this is actually not quite right, because light rays from the sun are not perfectly parallel. Adjusting for this error, it turns out that the angular width is about 2.2°.

The Secondary Bow

Under ideal conditions it is sometimes possible to observe two rainbows at the same time. The light rays that account for the formation of the so-called *secondary bow* go through a total of two refractions and two reflections, as shown in Figure 28.6, and are generally very low in intensity. The total (counterclockwise) deflection angle for the ray shown in Figure 28.6 is readily seen to be

$$(\alpha - \beta) + (\pi - 2\beta) + (\pi - 2\beta) + (\alpha - \beta) = 2\pi + 2\alpha - 6\beta.$$

In complete analogy to (28.3) it is not difficult to show that the critical point of the corresponding deflection function satisfies the equation

$$\cos(\alpha_0) = \sqrt{\frac{k^2 - 1}{8}}.$$

28.2 Exercise. Verify this formula.

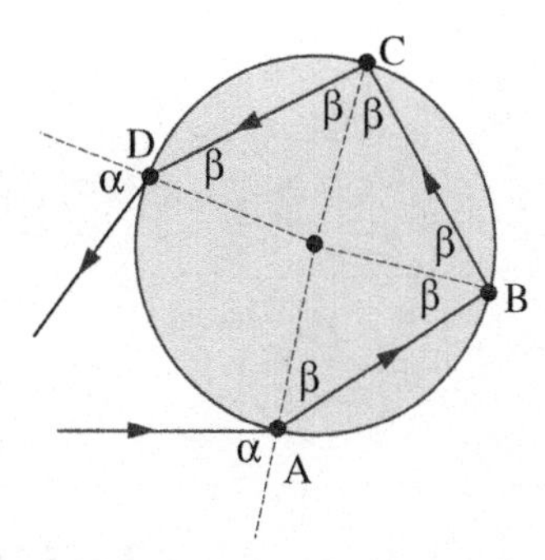

Figure 28.6: a light ray creating a secondary bow.

Using the value $k \approx 1.33$, we find that in this case α_0 is approximately equal to 71.94° and the corresponding deflection angle is 230.1°. Thus, the secondary bow is visible at an angle of about $230.1° - 180° = 50.1°$.

Additional Exercises

28.3. As we observe a pencil partially immersed in water, we notice that it does not appear to be straight, but rather bent or broken at the surface (see Figure 28.7). Explain this phenomenon using the law of refraction.

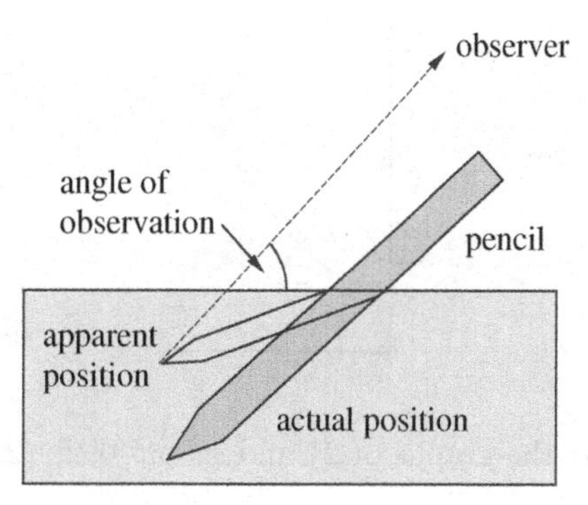

Figure 28.7: a refraction experiment.

28.4. ★ Considering again the experiment in Exercise 28.3, and taking into account the shape of the graph of the sine function over the interval $[0, \pi/2]$ (in reference, of course, to the law of refraction), explain why the "bending of the pencil" is significantly more pronounced for smaller angles of observation (see Figure 28.7) than for larger ones.

28.5. How far above the ground would a plane have to fly if a rainbow located under rain clouds half a mile ahead were to appear as a full circle? You may assume the sun to be very low in the sky.

Chapter 29

Relative Growth and Decay

Comparison of Growth Rates

Having introduced various types of functions in the preceding chapters, there arises a natural question: what are the relative magnitudes of all these functions? Of particular interest in this context is a comparison of growth (or decay) rates as x tends to ∞. As an example, let us consider the exponential function $f(x) := 2^x$ and the power function $g(x) := x^2$. Figure 29.1 shows the graphs of f and g on three different intervals. We see that on the interval $(0, 2)$ the values of the exponential function f are

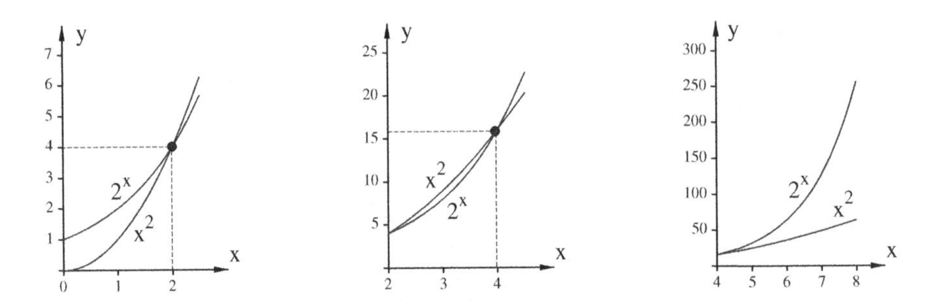

Figure 29.1: the graphs of 2^x and x^2 on $[0, 2]$, $[2, 4]$, and $[4, 8]$.

greater than the values of the power function g, and that at $x = 2$ the two functions are equal. Then on the interval $(2, 4)$ the roles are reversed and the power function is greater than the exponential function with equality at $x = 4$. Finally, on the interval $(4, 8)$ the exponential function f is again greater than g and is apparently now growing much more rapidly.

29.1 Exercise. With the help of a graphing calculator or a computer, analyze the relative magnitude of the functions $f(x) := 4^x$ and $g(x) := x^4$.

The graphical representation of the functions f and g in Figure 29.1 gives us an impression of their relative growth for increasing values of x, but pictorial evidence alone is certainly not very satisfactory. A more rigorous analytic approach is therefore called for. In this context, the first step is to clearly formulate the assertion that we are trying to establish. Considering the diagram on the right in Figure 29.1, it appears reasonable to expect that the values $f(x)$ are very large compared to $g(x)$ if the values of x are large compared to 4, i.e., $f(x) \gg g(x)$ if $x \gg 4$. In other words, we expect that

$$\frac{f(x)}{g(x)} \gg 1 \text{ if } x \gg 4.$$

From this point of view it makes sense to regard the problem of comparing the relative growth rates of

f and g as equivalent to the problem of finding the limit

$$L = \lim_{x \to \infty} \frac{f(x)}{g(x)} \quad \text{(if it exists).}$$

If $L = 0$, we say that g *dominates* f, and similarly if $L = \infty$, we say that f *dominates* g. Given the rule for evaluating the limit of the quotient of two functions as stated in Theorem 2.9, we might be tempted to think that L can be determined as the quotient of $\lim_{x\to\infty} f(x)$ and $\lim_{x\to\infty} g(x)$, but the problem is of course that for $f(x) = 2^x$ and $g(x) = x^2$ we have $\lim_{x\to\infty} f(x) = \lim_{x\to\infty} g(x) = \infty$, and the quotient ∞/∞ is not a well defined expression—it is a so-called *indeterminate form*. In the next section, we will develop a method that allows us to evaluate limits of this type.

Indeterminate Forms and L'Hôpital's Rule

In several examples throughout the text we have seen that if a function f is not continuous at a point x_0, or is undefined at x_0, the limit of f at x_0 (if it exists) cannot be determined by simply evaluating f at x_0. In such cases, we have typically relied on algebraic simplifications or estimates based on geometric arguments as we did, for instance, in computing

$$\lim_{\Delta x \to 0} \frac{\sin(\Delta x)}{\Delta x} \quad \text{(see Chapter 26).} \tag{29.1}$$

29.2 Exercise. Evaluate the following limits: $\lim_{x\to 1} \frac{x^2-1}{x-1}$, $\lim_{x\to 1} \frac{x^3-1}{x-1}$, $\lim_{x\to 1} \frac{x-1}{x^2-1}$, and $\lim_{x\to 1} \frac{(x-1)^2}{x^2-1}$.

A direct evaluation of the term $\sin(\Delta x)/\Delta x$ at $\Delta x = 0$ yields the indeterminate form $0/0$. Other indeterminate forms are ∞/∞, $\infty - \infty$, $0 \cdot \infty$, 0^0, 1^∞, and ∞^0. Notice that these are not arithmetic expressions, but rather symbolic representations of certain types of limits. To understand why indeterminate forms do not represent specific numerical values, let us consider the limits $\lim_{x\to 0^+} 0^x$ and $\lim_{x\to 0^+} x^0$. Both of these limits are of the form 0^0, but the value of the former is 0, while the latter equals 1. In fact, *the "value" of an indeterminate form can be any arbitrary number.* Let us choose, for instance, a value $\lambda \in \mathbb{R}$ and consider the limit

$$\lim_{x\to 0} \frac{\lambda x}{x}.$$

This limit is of the form $0/0$ and its value is the *arbitrary number* λ (simply cancel out the x).

Our strategy for evaluating limits representing indeterminate forms will be to transform any such limit into a limit of the form $0/0$ or ∞/∞ and then to apply a general theorem known as *L'Hôpital's Rule* (named after the French aristocrat, Marquis de L'Hôpital, who was a sponsoring benefactor of the actual discoverer of the rule). The simplest case of this rule is stated in the following theorem:

29.3 Theorem. *Let $f : I \to \mathbb{R}$ and $g : I \to \mathbb{R}$ be two differentiable functions defined on an open interval I such that for some $x_0 \in I$ we have $f(x_0) = g(x_0) = 0$. If $g'(x_0) \neq 0$ and $g(x) \neq 0$ for all $x \in I \setminus \{x_0\}$, then*

$$\lim_{x\to x_0} \frac{f(x)}{g(x)} = \frac{f'(x_0)}{g'(x_0)}.$$

Proof. Since f and g are both differentiable, it follows that

$$\lim_{x\to x_0} \frac{f(x) - f(x_0)}{x - x_0} = f'(x_0),$$
$$\lim_{x\to x_0} \frac{g(x) - g(x_0)}{x - x_0} = g'(x_0).$$

Using the assumption $f(x_0) = g(x_0) = 0$ and Theorem 2.9, we obtain

$$\lim_{x\to x_0} \frac{f(x)}{g(x)} = \lim_{x\to x_0} \frac{f(x)-f(x_0)}{g(x)-g(x_0)} = \lim_{x\to x_0} \frac{\dfrac{f(x)-f(x_0)}{x-x_0}}{\dfrac{g(x)-g(x_0)}{x-x_0}} = \frac{\lim\limits_{x\to x_0} \dfrac{f(x)-f(x_0)}{x-x_0}}{\lim\limits_{x\to x_0} \dfrac{g(x)-g(x_0)}{x-x_0}} = \frac{f'(x_0)}{g'(x_0)}.$$

This completes the proof of the theorem. □

29.4 Example. Let us consider the limit

$$\lim_{x\to 0} \frac{\sin(x)}{x}.$$

(This is the same limit as in (29.1) except that Δx has been replaced with x to simplify our notation.) We already showed in Chapter 26 that this limit is equal to 1, and we will now verify this result by using L'Hôpital's Rule. Setting $f(x) := \sin(x)$ and $g(x) := x$, it follows that $f(0) = g(0) = 0$. Moreover, $g(x) \neq 0$ for all $x \in \mathbb{R} \setminus \{0\}$ and $g'(0) = g'(x) = 1 \neq 0$. Therefore, the conditions in Theorem 29.3 are satisfied, and since $f'(x) = \cos(x)$, we obtain

$$\lim_{x\to 0} \frac{\sin(x)}{x} = \frac{f'(0)}{g'(0)} = \frac{\cos(0)}{1} = 1. \tag{29.2}$$

In order to understand why the statement of Theorem 29.3 is actually not very satisfactory, it is helpful to consider the limit

$$\lim_{x\to 0} \frac{1-\cos(x)}{x^2}. \tag{29.3}$$

Setting $f(x) := 1 - \cos(x)$ and $g(x) := x^2$, it is easy to see that $f(0) = g(0) = 0 = f'(0) = g'(0)$. Consequently, Theorem 29.3 cannot be applied, because $f'(0)/g'(0)$ is here an indeterminate expression of the form $0/0$. In a situation like this, we typically rely on a stronger version of L'Hôpital's Rule, which is stated in Theorem 29.5. The proof of this theorem is rather intricate and will be omitted (see [Rud1]).

29.5 Theorem. a) *Assume that $f : I \to \mathbb{R}$ and $g : I \to \mathbb{R}$ are functions defined on an open interval I such that f and g are differentiable on $I \setminus \{x_0\}$ for some $x_0 \in I$ and $g'(x) \neq 0$ for all $x \in I \setminus \{x_0\}$. If $\lim_{x\to x_0} f'(x)/g'(x)$ exists and $\lim_{x\to x_0} f(x) = \lim_{x\to x_0} g(x) = 0$ or $\lim_{x\to x_0} f(x) = \lim_{x\to x_0} g(x) = \infty$, then $\lim_{x\to x_0} f(x)/g(x)$ exists as well and*

$$\lim_{x\to x_0} \frac{f(x)}{g(x)} = \lim_{x\to x_0} \frac{f'(x)}{g'(x)}.$$

b) *For a given $c \in \mathbb{R}$, we assume that $f : (c,\infty) \to \mathbb{R}$ and $g : (c,\infty) \to \mathbb{R}$ are differentiable functions such that $g'(x) \neq 0$ for all $x \in (c,\infty)$. If $\lim_{x\to\infty} f'(x)/g'(x)$ exists and $\lim_{x\to\infty} f(x) = \lim_{x\to\infty} g(x) = 0$ or $\lim_{x\to\infty} f(x) = \lim_{x\to\infty} g(x) = \infty$, then $\lim_{x\to\infty} f(x)/g(x)$ exists as well and*

$$\lim_{x\to \infty} \frac{f(x)}{g(x)} = \lim_{x\to \infty} \frac{f'(x)}{g'(x)}.$$

29.6 Example. To illustrate the usefulness of this theorem, let us consider again the limit in (29.3) with $f(x) = 1 - \cos(x)$ and $g(x) = x^2$. According to Theorem 29.5a, we may infer that

$$\lim_{x\to 0} \frac{1-\cos(x)}{x^2} = \lim_{x\to 0} \frac{f'(x)}{g'(x)} = \lim_{x\to 0} \frac{\sin(x)}{2x} = \frac{1}{2} \lim_{x\to 0} \frac{\sin(x)}{x} = \frac{1}{2} \quad \text{(by (29.2))}.$$

It was possible to deduce this result from Theorem 29.5 but not from Theorem 29.3 because, in contrast to Theorem 29.3, the statement of Theorem 29.5 does not require $g'(x_0)$ to be different from zero but rather requires only the limit of f'/g' at x_0 to exist. In general, if the higher-order derivatives of

f and g exist up to, say, degree n, then according to Theorem 29.5, the existence of the limit of $f^{(n)}(x)/g^{(n)}(x)$ implies the existence of the limit of $f^{(n-1)}(x)/g^{(n-1)}(x)$ (assuming, of course, that this limit is of the form ∞/∞ or $0/0$). This in turn implies the existence of the limit of $f^{(n-2)}(x)/g^{(n-2)}(x)$, and continuing in this manner we arrive at the conclusion that the limit of $f(x)/g(x)$ exists as well. Furthermore, Theorem 29.5 also asserts that all of these limits are equal, that is

$$\lim_{x\to x_0 \text{ or } \infty} \frac{f^{(n)}(x)}{g^{(n)}(x)} = \lim_{x\to x_0 \text{ or } \infty} \frac{f^{(n)}(x)}{g^{(n)}(x)} = \cdots = \lim_{x\to x_0 \text{ or } \infty} \frac{f'(x)}{g'(x)} = \lim_{x\to x_0 \text{ or } \infty} \frac{f(x)}{g(x)}.$$

Examples of Applications of L'Hôpital's Rule

To illustrate how the techniques developed in the previous section are applied in practice, we will now consider examples of limits representing the different types of indeterminate forms.

Type 1: $\frac{0}{0}$

a)

$$\lim_{x\to 0} \frac{e^x - 1}{\sin(2x)} = \lim_{x\to 0} \frac{d/dx\,(e^x - 1)}{d/dx\,\sin(2x)} = \lim_{x\to 0} \frac{e^x}{2\cos(2x)} = \frac{1}{2}.$$

b)

$$\lim_{x\to\infty} \frac{e^{-x}}{x^{-2}} = \lim_{x\to\infty} \frac{x^2}{e^x} = \lim_{x\to\infty} \frac{2x}{e^x} = \lim_{x\to\infty} \frac{2}{e^x} = 0.$$

In b) we first transformed the original limit $\lim_{x\to\infty} e^{-x}/x^{-2}$, which is of the form $0/0$, into the limit $\lim_{x\to\infty} x^2/e^x$, which is of the form ∞/∞, and then we applied Theorem 29.5b twice. Notice that our calculation implies that $e^x \gg x^2$ as x tends to ∞. In general, repeated application of L'Hôpital's Rule shows that for all $a, b \in \mathbb{R}$ with $b > 1$ we have $b^x \gg x^a$ as x tends to ∞ (see Theorem 29.8 below).

29.7 Exercise. Evaluate $\lim_{x\to 0} \frac{\sin(x)}{x^2 + x}$ and $\lim_{x\to 2} \frac{x - 2\cos(\pi x)}{x^2 - 4}$.

Type 2: $\frac{\infty}{\infty}$

a)

$$\lim_{x\to\pi/2^-} \frac{\tan(x)}{-\ln(\cos(x))} = \lim_{x\to\pi/2^-} \frac{1/\cos^2(x)}{\sin(x)/\cos(x)} = \lim_{x\to\pi/2^-} \frac{1}{\cos(x)\sin(x)} = 1\cdot \lim_{x\to\pi/2^-} \frac{1}{\cos(x)} = \infty.$$

b) If $a, c > 0$ and $c \neq 1$, then

$$\lim_{x\to\infty} \frac{\log_c(x)}{x^a} = \lim_{x\to\infty} \frac{1/(x\ln(c))}{ax^{a-1}} = \lim_{x\to\infty} \frac{1}{a\ln(c)x^a} = 0.$$

Motivated by the results in Type 1b and Type 2b we formulate the following theorem:

29.8 Theorem. *Let a, b, and c be real numbers such that $a, c > 0$, $c \neq 1$, and $b > 1$. Then*

$$\lim_{x\to\infty} \frac{x^a}{b^x} = 0 = \lim_{x\to\infty} \frac{\log_c(x)}{x^a}.$$

In other words, for large x we have $b^x \gg x^a \gg |\log_c(x)|$ (note: we insert here the absolute-value signs, because $\log_c(x) < 0$ if $x > 1 > c$).

29.9 Exercise. Prove the first equality in Theorem 29.8: $\lim_{x\to\infty} \frac{x^a}{b^x} = 0$.

Type 3: $0 \cdot \infty$

a) We wish to evaluate

$$\lim_{x \to 0+} -x \ln(x).$$

This limit is of the type $0 \cdot \infty$, because $\lim_{x\to 0+} x = 0$ and $\lim_{x\to 0+} -\ln(x) = \infty$. To directly apply Theorem 29.5 to a limit of this form is not possible. However, there is no serious problem because

$$\lim_{x\to 0+} -x\ln(x) = \lim_{x\to 0+} \frac{\ln(x)}{1/x},$$

and the limit on the right-hand side is now of the form ∞/∞. Consequently, Theorem 29.5b allows us to infer that

$$\lim_{x\to 0+} -x\ln(x) = \lim_{x\to 0+} \frac{-\ln(x)}{1/x} = \lim_{x\to 0+} \frac{-1/x}{-1/x^2} = \lim_{x\to 0+} x = 0.$$

b) In the next example, we transform a limit of the type $0 \cdot \infty$ into a limit of the type $0/0$.

$$\lim_{x\to\infty} x\ln\left(\frac{x-1}{x+1}\right) = \lim_{x\to\infty} \frac{\ln((x-1)/(x+1))}{1/x} = \lim_{x\to\infty} \frac{2/(x^2-1)}{-1/x^2} = \lim_{x\to\infty} \frac{-2x^2}{x^2-1} = \lim_{x\to\infty} \frac{-4x}{2x} = -2.$$

29.10 Exercise. Evaluate $\lim_{x\to 0+} \sin(x)\ln(\sin(x))$ and $\lim_{x\to\infty} e^{-x}\ln(x)$.

Type 4: $\infty - \infty$

To compute

$$\lim_{x\to 0} \left(\frac{1}{x} - \frac{1}{\sin(x)}\right)$$

we add up the fractions, and observe that the resulting

$$\lim_{x\to 0} \frac{\sin(x) - x}{x\sin(x)}$$

is of the form $0/0$. Therefore, we may apply Theorem 29.5a to conclude that

$$\lim_{x\to 0} \frac{\sin(x)-x}{x\sin(x)} = \lim_{x\to 0} \frac{\cos(x)-1}{\sin(x)+x\cos(x)} = \lim_{x\to 0} \frac{-\sin(x)}{2\cos(x) - x\sin(x)} = 0.$$

29.11 Exercise. Evaluate $\lim_{x\to\infty} \left(\sqrt{x^2+3x} - x\right)$ and $\lim_{x\to 2+} \left(\frac{1}{x-2} - \frac{1}{\ln(x-1)}\right)$.

Type 5: ∞^0

Let us determine the limit

$$L = \lim_{x\to\infty} \ln(x)^{1/x}.$$

To transform indeterminate forms of exponential type we always use the natural logarithm: since the natural logarithm function is continuous, it follows that

$$\ln(L) = \lim_{x\to\infty} \ln\left(\ln(x)^{1/x}\right) = \lim_{x\to\infty} \frac{\ln(\ln(x))}{x}.$$

The limit on the right is of the form ∞/∞. Hence

$$\lim_{x\to\infty} \frac{\ln(\ln(x))}{x} = \lim_{x\to\infty} \frac{1/(x\ln(x))}{1} = 0.$$

This shows that $\ln(L) = 0$, and therefore $L = e^{\ln(L)} = e^0 = 1$.

29.12 Exercise. Evaluate $\lim_{x\to 0^+}\left(\frac{1}{x}\right)^x$.

Type 6: 0^0

We wish to determine the limit

$$L = \lim_{x\to 0^+} x^{1/\ln(-\ln(x))}.$$

Using the natural logarithm, as in the previous example, yields

$$\ln(L) = \lim_{x\to 0^+} \ln\left(x^{1/\ln(-\ln(x))}\right) = \lim_{x\to 0^+} \frac{\ln(x)}{\ln(-\ln(x))} = \lim_{x\to 0^+} \frac{1/x}{1/(x\ln(x))} = \lim_{x\to 0^+} \ln(x) = -\infty.$$

Hence $L = e^{-\infty} = 0$.

29.13 Exercise. Evaluate $\lim_{x\to 0^+} x^{\sin(x)}$.

Type 7: 1^∞

To discuss this type of limit, we will consider the problem of calculating interest on a bank account. If we invest \$1 at 5% interest for one year, we will have \$1.05 at the end of the year. This is called simple interest. Perhaps more interesting (and certainly more profitable) is the case of *compound interest.* In compounding, the interest is partially applied at regular periods throughout the year rather than just once at the end of the year. This means that any interest paid at a certain time of the year earns itself interest for the rest of the year. Let us consider some examples:

Compound twice a year (semiannually):

Period 1: We start with \$1 and add half of the yearly interest, which is 2.5%. So after 6 months we have $\$(1 + 0.025 \cdot 1) = \$1.025$.

Period 2: We now start with \$1.025 and add again 2.5% so that the total amount after one year is $\$(1.025 + 0.025 \cdot 1.025) = \$1.025^2 = \$1.050625$.

Compound four times a year (quarterly):

Period 1: We start with \$1 and add one fourth of 5%(=1.25%) at the end of three months. This yields $\$(1 + 0.0125 \cdot 1) = \$1.0125$.

Period 2: We start with \$1.0125 and add 1.25% to obtain $\$(1.0125 + 0.0125 \cdot 1.0125) = \$1.0125^2 \approx \$1.025156$ after six months.

Period 3: We start with $\$1.0125^2$ and add 1.25%, resulting in $\$(1.0125^2 + 0.0125 \cdot 1.0125^2) = \$1.0125^3 \approx \$1.037971$ after nine months.

Period 4: We start with $\$1.0125^3$ and add 1.25% to obtain $\$(1.0125^3 + 0.0125 \cdot 1.0125^3) = \$1.0125^4 \approx \$1.050945$ after one year.

In summary, for an initial investment of \$1 and a yearly interest rate of 5%, we have the following returns after one year:

compounding periods	amount after one year
1	$\$1.05^1 = \$1.05$
2	$\$1.025^2 = \$1.050625$
4	$\$1.0125^4 \approx \$1.050945$

In general, if the the number of compounding periods is denoted by n, then we generate $\$(1+0.05/n)^n$ after one year. As n increases, we approach what is known as *continuous compounding*. In this case the return after one year is the value of the limit

$$\lim_{n\to\infty}\left(1+\frac{0.05}{n}\right)^n,$$

which is of the form 1^∞. Denoting this limit by L and writing the more familiar variable x in place of n, we obtain

$$\begin{aligned}\ln(L) &= \lim_{x\to\infty}\ln\left(\left(1+\frac{0.05}{x}\right)^x\right) = \lim_{x\to\infty} x\ln\left(1+\frac{0.05}{x}\right) = \lim_{x\to\infty}\frac{\ln(1+0.05/x)}{1/x}\\ &= \lim_{x\to\infty}\frac{-(0.05/x^2)/(1+0.05/x)}{-1/x^2} = \lim_{x\to\infty}\frac{0.05}{1+0.05/x} = 0.05.\end{aligned}$$

Thus after one year of continuous compounding, our return is $\$L = \$e^{0.05} \approx \$1.051271$.

If we perform a completely analogous calculation for an arbitrary rate y in place of 0.05, then we arrive at the following remarkable formula:

$$\boxed{\lim_{x\to\infty}\left(1+\frac{y}{x}\right)^x = e^y \text{ for all } y\in\mathbb{R}.} \tag{29.4}$$

29.14 Exercise. Verify this formula.

Given the result in (29.4), it is not difficult to see that in general the return on an initial investment $\$P_0$ after continuously compounding an annual interest rate y for a span of t years is

$$\boxed{\$P_0e^{yt}.}$$

29.15 Exercise. Evaluate $\lim_{x\to 0}(\cos(x))^{1/x^2}$.

Additional Exercises

29.16. Evaluate the following limits:

a) $\lim_{x\to 0}|x|^x$ **b)** $\lim_{x\to 2}\frac{x^2-4}{x-2}$ **c)** $\lim_{x\to 1}\frac{\sin^2(x-1)}{x-1}$ **d)** $\lim_{x\to\infty} x\sin\left(\frac{1}{x}\right)$

e) $\lim_{x\to 0^+} x\log_{10}(x)$ **f)** $\lim_{x\to 0}(1+x)^{1/x}$ **g)** $\lim_{x\to 0}\frac{2^x-1}{x}$ **h)** $\lim_{x\to 0^+}\sin^x(x)$

i) $\lim_{x\to 0}\frac{\sin(\pi x)}{x}$ **j)** $\lim_{x\to 0}\frac{500x^2}{\sin(x)+500x^2}$ **k)** $\lim_{x\to\infty}\frac{e^x-e^{2x}}{e^x+e^{-x}}$ **l)** $\lim_{x\to 0^+}\left(\frac{1}{x}+\ln(x)\right)$

m) $\lim_{x\to\pi/2^-}(\tan(x))^{\cos(x)}$ **n)** $\lim_{x\to 0}\frac{2e^x-x^2-2x-2}{x^4+x^3}$ **o)** $\lim_{x\to\infty}\frac{x+\cos(x)}{x}$ **p)** $\lim_{x\to\infty}(1+x^2)^{1/x}$

29.17. Find an integer $k>0$ for which $\lim_{x\to 0}\frac{1-\cos(\sin(x))}{x^k}$ exists, and then compute the value of the limit.

29.18. What is wrong with the following calculation?

$$\lim_{x\to 0}\frac{x+\sin(x)}{x+\cos(x)} = \lim_{x\to 0}\frac{d/dx\,(x+\sin(x))}{d/dx\,(x+\cos(x))} = \lim_{x\to 0}\frac{1+\cos(x)}{1-\sin(x)} = \frac{1+1}{1-0} = 2.$$

29.19. What will be your return on an initial investment of \$200 after ten years of continuously compounding an annual interest rate of 6%? What will be your return if interest is compounded quarterly?

29.20. Order the functions given below according to the speed at which they increase to infinity in the limit $x \to \infty$. The function that increases the slowest should be listed first and the one that increases the fastest, last.

a) $f_1(x) = x^x$ **b)** $f_2(x) = x^2$ **c)** $f_3(x) = e^x$
d) $f_4(x) = (\ln(x))^{\ln(x)}$ **e)** $f_5(x) = \ln(x)$ **f)** $f_6(x) = \sqrt{x}$

29.21. Assume that f and g are differentiable functions defined on $\mathbb{R}$ that satisfy the following conditions: $\lim_{x\to\infty} f(x) = \lim_{x\to\infty} g(x) = \infty$ and $\lim_{x\to\infty} f'(x)e^{f(x)}/(g'(x)e^{g(x)}) = 2.346$. Use L'Hôpital's rule to show that $\lim_{x\to\infty} f'(x)/g'(x) = 1$.

29.22. Assume that f and g are differentiable functions defined on $\mathbb{R}$ that satisfy the following conditions: $\lim_{x\to\infty} f(x) = \lim_{x\to\infty} g(x) = \infty$ and $\lim_{x\to\infty} f'(x)/g'(x) = 5.981$. Use L'Hôpital's rule to show that $\lim_{x\to\infty} \ln(f(x))/\ln(g(x)) = 1$.

29.23. Assume that $f : \mathbb{R} \to \mathbb{R}$ and $g : \mathbb{R} \to \mathbb{R}$ are differentiable functions such that $\lim_{x\to\infty} f(x) = \lim_{x\to\infty} g(x) = \infty$ and

$$\lim_{x\to\infty} \frac{f'(x)2^{f(x)}}{g'(x)3^{g(x)}} = 1.$$

Find the value of $\lim_{x\to\infty} 2^{f(x)}/3^{g(x)}$.

29.24. What is wrong with the following conclusion?

$$\lim_{x\to 0} \frac{x^2\sin(1/x)}{x} = \lim_{x\to 0} \frac{2x\sin(1/x) - \cos(1/x)}{1},$$

and since $\lim_{x\to 0} \cos(1/x)$ does not exist, it follows that $\lim_{x\to 0} x^2\sin(1/x)/x$ does not exist either.

PART IV

METHODS OF INTEGRATION

Chapter 30

Integration by Substitution and Rocket Motion (Part 2)

Integration by Substitution

In order to solve the equations of rocket motion as derived in Chapter 8, we need to develop a general method of integration, known as *integration by substitution.* To get started, we assume that for open intervals $I, J \subset \mathbb{R}$ we are given a continuous function $f : J \to \mathbb{R}$ and a continuously differentiable function $g : I \to J$ (see Definition 7.6). If $F : J \to \mathbb{R}$ is an antiderivative of f, then the chain rule implies that

$$(F \circ g)'(x) = \frac{d}{dx}F(g(x)) = f(g(x))g'(x) \text{ for all } x \in I.$$

In other words, $F \circ g$ is an antiderivative of $(f \circ g)g'$. According to the FTC2, we therefore obtain

$$\int_a^b f(g(x))g'(x)\,dx = F(g(x))\Big|_a^b = F(g(b)) - F(g(a)) = F(u)\Big|_{g(a)}^{g(b)} \quad \text{for all } a, b \in I.$$

Using again the FTC2, we may express the last term of this equation as the integral of f in the boundaries from $g(a)$ to $g(b)$. In order to emphasize that the boundaries have changed (from a and b to $g(a)$ and $g(b)$) we use a different letter for the variable of integration and write u instead of x. This yields

$$\int_a^b f(g(x))g'(x)\,dx = \int_{g(a)}^{g(b)} f(u)\,du.$$

To introduce the variable u is not necessary, but intuitively helpful because it suggests that we are essentially *substituting* u for $g(x)$. We will explain this idea more clearly in some of the examples below.

The equation above is frequently useful in reducing the complexity of an integrand: instead of the integral of the typically complicated function $f(g(x))g'(x)$, we only need to find the integral of the typically simpler function $f(u)$. The following theorem summarizes the preceding discussion and establishes the technique of *integration by substitution* for definite and indefinite integrals:

30.1 Theorem. *If for open intervals $I, J \subset \mathbb{R}$ we are given functions $f : J \to \mathbb{R}$ and $g : I \to J$ that are respectively continuous and continuously differentiable, then for all $a, b \in I$ we have*

$$\int_a^b f(g(x))g'(x)\,dx = \int_{g(a)}^{g(b)} f(u)\,du.$$

Furthermore, if $F : J \to \mathbb{R}$ is an antiderivative of f, then

$$\int f(g(x))g'(x)\,dx = F(g(x)) + C.$$

Remark. According to Theorems 7.5 and 10.10, the assumptions of continuity on f and continuous differentiability on g imply that $(f \circ g)g'$ is continuous. Consequently, the integral $\int_a^b f(g(x))g'(x)\,dx$ exists (see also the remark following Definition 17.5).

30.2 Example. We wish to find the integral of the function $2xe^{(x^2)}$ in the boundaries from 0 to 2. Setting $f(x) := e^x$ and $g(x) := x^2$, it follows that

$$f(g(x))g'(x) = e^{g(x)}g'(x) = e^{(x^2)}2x.$$

Having thus written the integrand $2xe^{(x^2)}$ in the form $f(g(x))g'(x)$, we may apply Theorem 30.1 to conclude that

$$\int_0^2 2xe^{(x^2)}\,dx = \int_{g(0)}^{g(2)} f(u)\,du = \int_0^4 e^u\,du = e^u\Big|_0^4 = e^4 - 1.$$

Furthermore, since $F(x) := e^x$ is an antiderivative of f (see Chapter 22, p.181), the second equation in Theorem 30.1 implies that

$$\int 2xe^{(x^2)}\,dx = F(g(x)) + C = e^{(x^2)} + C.$$

Given this example, the reader may wonder what should be done if we cannot find functions f and g that allow us to write the integrand in the form $f(g(x))g'(x)$. The answer is "nothing" or "try something else." Integration by substitution is our best bet when we don't know how to solve an integral, but there is no guarantee that it always works, and it is not always easy to use.

In practical applications of integration by substitution it is usually convenient to begin by *substituting* a new variable u for $g(x)$. In the case of Example 30.2, we proceed as follows: first we set $u := x^2$ and determine the derivative

$$\frac{du}{dx} = 2x.$$

Then, in a symbolic calculation with infinitesimals (see the ⋆remark⋆ on p.33), we solve this equation for du and obtain

$$du = 2x\,dx.$$

With the substitutions u for x^2 and du for $2x\,dx$, we are now able to perform the integration:

$$\int_0^2 2xe^{(x^2)}\,dx = \int_0^2 \underbrace{e^{(x^2)}}_{e^u}\underbrace{2x\,dx}_{du} = \int_{0^2}^{2^2} e^u\,du = \int_0^4 e^u\,du = e^4 - 1.$$

Regarding the problem of appropriately defining u, the following *golden rule of integration by substitution* is very helpful:

Define u in such a way that the derivative of u appears as a factor in the integrand.	(30.1)

Note: in the example above the derivative $du/dx = 2x$ was a factor in the integrand $\mathbf{2xe^{(x^2)}}$.

30.3 Exercise. Find the values of the integrals $\int_0^2 xe^{(2x^2)}\,dx$ and $\int_0^2 3x^2e^{(x^3)}\,dx$. Note: use the substitution $u = 2x^2$ in the first integral and $u = x^3$ in the second.

30.4 Example. We wish to find the value of the definite integral

$$\int_0^{\pi/2} \cos(x)\sin(x)\,dx.$$

With reference to the golden rule in (30.1), we set $u := \sin(x)$, because then the derivative

$$\frac{du}{dx} = \cos(x)$$

appears as a factor in the integrand $\cos(x)\sin(x)$. Replacing now $\sin(x)$ with u and $\cos(x)\,dx$ with du, we may infer that

$$\int_0^{\pi/2} \cos(x)\sin(x)\,dx = \int_{\sin(0)}^{\sin(\pi/2)} u\,du = \frac{u^2}{2}\bigg|_0^1 = \frac{1}{2}.$$

30.5 Exercise. Find the value of the definite integral $\displaystyle\int_0^{\pi/4} \frac{\tan(x)}{\cos^2(x)}\,dx$.

30.6 Example. We wish to find the antiderivative of the function $h : (0,\infty) \setminus \{1\} \to \mathbb{R}$ defined by the equation

$$h(x) := \frac{1}{x\ln(x)}.$$

In order to make the right choice for u we observe that the derivative of $\ln(x)$ is $1/x$. In light of the golden rule in (30.1) it is therefore reasonable to set $u := \ln(x)$. This yields

$$\frac{du}{dx} = \frac{1}{x}, \quad du = \frac{1}{x}\,dx$$

and

$$\int h(x)\,dx = \int \frac{1}{x\ln(x)}\,dx = \int \frac{1}{u}\,du = \ln|u| + C = \ln|\ln(x)| + C.$$

Note: in the last step we substituted $\ln(x)$ back in for u, because we needed to find the antiderivative of h as a function of x.

Alternatively, we may use the second statement of Theorem 30.1 with $g(x) := \ln(x)$, $f(x) := 1/x$, $F(x) := \ln|x|$, and $f(g(x))g'(x) = 1/(x\ln(x)) = h(x)$ to infer that

$$\int h(x)\,dx = F(g(x)) + C = \ln|\ln(x)| + C.$$

30.7 Exercise. Find the antiderivative of $h(x) := \dfrac{\ln^2(x)}{x}$.

Rocket Motion (Part 2)

In Chapter 8 we derived the following equations describing the accelerated motion of a rocket:

$$m(t)v'(t) = -m'(t)u \quad \text{(no external forces)}, \tag{30.2}$$

$$m(t)v'(t) = F(t) - m'(t)u \quad \text{(external force } F(t)). \tag{30.3}$$

Here we denoted by $m(t)$ the mass of the rocket, by $v(t)$ its velocity, and by u the (constant) speed at which the burnt propellant gas is ejected through the boosters. To continue this discussion, we first assume that there are no external forces. In this case, the rocket moves at a constant velocity v_0 until at some point in time t_0, the engines are turned on and the rocket begins to accelerate. The increase in velocity during the burning cycle of the boosters is described by (30.2). Rewriting equation (30.2) in the form

$$v'(t) = -\frac{m'(t)u}{m(t)},$$

we see that v is an antiderivative of $-m'(t)u/m(t)$. Therefore, we may apply the FTC2 to conclude that

$$v(t) - v_0 = v(t) - v(t_0) = \int_{t_0}^{t} v'(\tau)\,d\tau = -\int_{t_0}^{t} \frac{m'(\tau)u}{m(\tau)}\,d\tau. \tag{30.4}$$

(Note: the variable of integration is denoted by τ in order to keep it distinct from the upper boundary of integration t.) To solve the integral on the right-hand side in (30.4), we use integration by substitution.

Since we have already used the letter u to denote the speed of the propellant gas, we choose w to denote the substitution variable. Setting $w := m(\tau)$, we obtain

$$\frac{dw}{d\tau} = m'(\tau) \quad \text{and} \quad m'(\tau)\,d\tau = dw.$$

Hence

$$\begin{aligned}
-\int_{t_0}^{t} \frac{m'(\tau)u}{m(\tau)}\,d\tau &= -\int_{m(t_0)}^{m(t)} \frac{u}{w}\,dw = -u\ln|w|\Big|_{m(t_0)}^{m(t)} = -u(\ln|m(t)| - \ln|m(t_0)|) \\
&= -u(\ln(m(t)) - \ln(m(t_0))) \;\; (\text{because } m(t) \geq 0) \qquad (30.5) \\
&= u\ln\left(\frac{m(t_0)}{m(t)}\right).
\end{aligned}$$

Combining (30.5) with (30.4) yields

$$\boxed{v(t) = v_0 + u\ln\left(\frac{m(t_0)}{m(t)}\right).} \qquad (30.6)$$

30.8 Example. Let us assume that at time $t = 0$ a rocket's mass is $m(0) = 100000\,kg$ and its velocity is $v_0 = 20000\,mi/h$. We wish to determine the velocity v after $1000\,kg$ of fuel have been burned and ejected through the boosters at speed $u = 10000\,mi/h$. Since the remaining mass is $m(t) = 100000 - 1000 = 99000\,kg$, we may use equation (30.6) to conclude that

$$v = 20000 + 10000\ln\left(\frac{100000}{99000}\right) \approx 20100.5\,\frac{mi}{h}.$$

This calculation shows that the rocket's final velocity does not depend on the time it takes to burn the fuel but only on the initial velocity, the initial mass, the remaining mass, and the value of u. In other words, it doesn't matter for how long the boosters are burning—only the total mass of fuel burned and the speed of emission are of interest.

30.9 Exercise. A rocket in interstellar space has a total mass m, half of which is fuel. By how much will the rocket's velocity increase if all fuel is burned and ejected at speed $u = 20000\,mi/h$?

30.10 Example. In absence of external forces, a rocket assumes its maximal velocity v_{max} at the moment when all available fuel has been burned. In order to determine v_{max}, we denote by M_R the rocket's mass without fuel and by M_F the mass of fuel at time t_0 (as before, t_0 is the point in time when the boosters are turned on). Denoting by T the time it takes to burn all available fuel, we see that $m(t_0) = M_R + M_F$ and $m(T) = M_R$. Thus, the maximal velocity is

$$v_{max} = v_0 + u\ln\left(\frac{M_R + M_F}{M_R}\right). \qquad (30.7)$$

30.11 Example. Let us consider the case of a rocket that is accelerating away from the surface of the earth under the influence of gravity. If the rocket does not move too far out into space, the gravitational acceleration g remains approximately constant (see also our discussion of the law of falling bodies in the Introduction and in Chapter 14), and the gravitational force on the rocket is given by the equation

$$F(t) = -m(t)g.$$

The negative sign indicates that the positive direction of the rocket's motion is opposite to the direction of the gravitational force, that is, opposite to the direction toward the center of the earth. Given equation (30.3) and the formula for $F(t)$ above, we may infer that

$$v'(t) = -g - \frac{m'(t)u}{m(t)}.$$

Integrating both sides of this equation in the boundaries from t_0 to t yields

$$v(t) - v_0 = \int_{t_0}^{t} v(\tau)\,d\tau = -\int_{t_0}^{t} g\,d\tau - \int_{t_0}^{t} \frac{m'(\tau)u}{m(\tau)}\,d\tau = -g\cdot(t-t_0) + u\ln\left(\frac{m(t_0)}{m(t)}\right).$$

If the rocket is on the ground at time $t_0 = 0$ (i.e., $v_0 = 0$) with mass $M_R + M_F$, then

$$v(t) = -gt + u\ln\left(\frac{M_R + M_F}{m(t)}\right). \tag{30.8}$$

30.12 Exercise. Is it possible to use equation (30.8) to derive a formula for the maximal velocity of a rocket under the influence of the gravitational field of the earth such that v_{max} is independent of t as in equation (30.7)? Explain your answer.

Additional Exercises

30.13. Use integration by substitution to solve the definite and indefinite integrals given below.

a) $\displaystyle\int_0^1 \frac{\arctan(x)}{1+x^2}\,dx$ **b)** $\displaystyle\int_0^1 \frac{x}{2x^2-3}\,dx$ **c)** $\displaystyle\int \cos(x)\sin^3(x)\,dx$

d) $\displaystyle\int \frac{\sin(x)}{\cos^3(x)}\,dx$ **e)** $\displaystyle\int \frac{1}{x\ln(x)\ln^2(\ln(x))}\,dx$ **f)** $\displaystyle\int \frac{\ln(1/x^2)}{x}\,dx$ (for $x > 0$)

g) $\displaystyle\int_0^1 e^x e^{(e^x)}\,dx$ **h)** $\displaystyle\int \frac{\arcsin(x-1)}{\sqrt{2x-x^2}}\,dx$ **i)** $\displaystyle\int_0^{\pi/2} \cos^3(x)\,dx$

j) $\displaystyle\int_0^{\pi/2} \cos^5(x)\,dx$ **k)** $\displaystyle\int \cot(x)\,dx$ **l)** $\displaystyle\int \cot^3(x)\,dx$

m) $\displaystyle\int x\sqrt{x+2}\,dx$ **n)** $\displaystyle\int_0^1 \frac{x}{\sqrt{1+x}}\,dx$ **o)** $\displaystyle\int \frac{1}{1+\sqrt{x}}\,dx$

Hint. Consider using trigonometric identities for those integrals that involve trigonometric functions (not necessarily in all cases, but in some).

30.14. Below you are given two equations relating a function f to its derivative f'. Use integration by substitution to determine f (up to a constant).

a) $f'(x) = 1 + f(x)^2$

b) $3f'(x) = xf(x)$

Hint. Divide both sides of the equation in a) by $1 + f(x)^2$ and then integrate. A similar approach also works for b).

30.15. Evaluate $\int_1^2 f(3x-1)\,dx$ from the assumption that $\int_2^5 f(u)\,du = 10$.

30.16. Evaluate $\int_0^{36} f(v)\,dv$ from the assumption that $\int_0^3 xf(4x^2)\,dx = 10$.

30.17. Evaluate $\int_0^2 f(u)\,du$ from the assumption that $\int_0^{\pi/2} \cos(x)f(2\sin(x))\,dx = 4$.

30.18. Find the value of $\int_{-1}^1 xf(x^2)\,dx$ for an arbitrary continuous function $f : [-1,1] \to \mathbb{R}$.

30.19. Assume that a rocket is on the ground at time $t = 0$ with $m(0) = 1$. Use the equation of rocket motion for the nearly constant gravitational field close to the surface of the earth to find the minimal value for $|m'(0)|$ that will allow the rocket to lift off the ground under the assumption $u = 50$.

30.20. ★ Use integration by substitution to find the antiderivative of $1/\sin(x)$. Then apply your result to determine the antiderivative of $1/\cos(x)$. *Hint.* To find the antiderivative of $1/\sin(x)$, it is helpful to apply the double angle formula for sine (see Appendix A) to the equation $\sin(x) = \sin(2 \cdot x/2)$. To determine the antiderivative of $1/\cos(x)$, you should use the formula $\cos(x) = \sin(\pi/2 - x)$.

30.21. Determine the following antiderivatives:

$$\text{a)} \int \frac{1}{1+x^2}\,dx \quad \text{b)} \int \frac{x}{1+x^2}\,dx \quad \text{c)} \int \frac{x^2}{1+x^2}\,dx$$

Hint. To solve c), you should think of ways to algebraically manipulate the integrand $x^2/(1+x^2)$.

30.22. Determine the indefinite integrals given below. (Note: your answer will in each case depend on $f(x)$.)

$$\text{a)} \int \frac{f'(x)}{1+f(x)^2}\,dx \quad \text{b)} \int f'(x)(1-f(x))\,dx \quad \text{c)} \int \frac{f(x)f'(x)}{1+f(x)}\,dx$$

30.23. Determine the indefinite integrals listed below under the assumption that $F' = f$. (Note: in each case your answer will involve the function F.)

$$\text{a)} \int x^2 f(x^3)\,dx \quad \text{b)} \int \frac{f(\ln(x))}{x}\,dx \quad \text{c)} \int x^2 \cos(x^3) f(\sin(x^3))\,dx$$

30.24. Which of the following values or quantities affect the maximal velocity that a rocket in interstellar space can reach?

a) The initial velocity.

b) The mass of fuel on board.

c) The speed of fuel emission.

d) The time that it takes to burn the fuel.

Chapter 31

Inverse Integration by Substitution and Designing a Radar Antenna (Part 2)

A Nonstandard Substitution

In Chapter 6 we derived the following defining equation for a function f, the graph of which describes the cross-sectional shape of a radar antenna:

$$f'(x) = \frac{f(x)}{x} \pm \sqrt{\frac{f(x)^2}{x^2} + 1}. \tag{31.1}$$

We showed that $f(x) = x^2 - 1/4$ is one particular solution of this equation (see p.54), and our task in the present chapter will be to find *all* possible solutions (see also Exercise 6.4). In order to simplify equation (31.1) we introduce the function

$$h(x) := \frac{f(x)}{x}.$$

Using the quotient rule, we obtain

$$h'(x) = \frac{d}{dx}\frac{f(x)}{x} = \frac{xf'(x) - f(x)}{x^2} = \frac{1}{x}\left(f'(x) - \frac{f(x)}{x}\right),$$

and therefore,

$$xh'(x) = f'(x) - \frac{f(x)}{x}.$$

Given this result, equation (31.1) can be rewritten in the form

$$xh'(x) = \pm\sqrt{h(x)^2 + 1},$$

or equivalently,

$$\frac{h'(x)}{\sqrt{h(x)^2 + 1}} = \pm\frac{1}{x}. \tag{31.2}$$

Remark. The reader should regard the derivation of (31.2) from (31.1) as a case study in mathematical creativity. Although generalizations are possible (see Chapter 43), the idea of substituting $h(x)$ for $f(x)/x$ is itself not derived from any general principle and requires initially a certain amount of ingenuity. After all, we cannot expect to have rules for every situation! Frequently, we need to rely on our own original insight in finding the key to a solution. This indeed is the true challenge (and joy) of mathematics.

The first step in solving equation (31.2) for $h(x)$ is to integrate both sides. This yields

$$\int \frac{h'(x)}{\sqrt{h(x)^2+1}}\,dx = \pm\int \frac{1}{x}\,dx = \pm \ln|x| + C, \tag{31.3}$$

and with the substitutions $u := h(x)$ and $du = h'(x)\,dx$ we obtain

$$\int \frac{h'(x)}{\sqrt{h(x)^2+1}}\,dx = \int \frac{1}{\sqrt{u^2+1}}\,du. \tag{31.4}$$

Thus, we are left with the problem of finding the antiderivative of the function $1/\sqrt{u^2+1}$. To efficiently handle this problem, we will introduce hyperbolic functions in the next section and then discuss a modification of integration by substitution that we will refer to as *inverse integration by substitution.*

Hyperbolic Functions

Before we get to the actual subject matter in this section, we wish to point out that the main difficulty in dealing with *hyperbolic functions* is emotional and psychological in nature. For many students the word "hyperbolic" has an air of desperation and failed exams attached to it that is as persistent as it is unjustified. Hyperbolic functions are without question less difficult to handle than, for example, trigonometric functions, and the reader is therefore advised to view the current section as an easy intermission before we continue with more difficult topics.

31.1 Definition. For $x \in \mathbb{R}$ we define the *hyperbolic sine* and *hyperbolic cosine* functions via the equations

$$\sinh(x) := \frac{e^x - e^{-x}}{2},$$
$$\cosh(x) := \frac{e^x + e^{-x}}{2}.$$

To study the properties of these functions we begin by considering values $x \geq 0$. For $x = 0$ it is easy to check that $\sinh(x) = \sinh(0) = 0$ and $\cosh(x) = \cosh(0) = 1$. As x increases from 0 to large positive values, the term e^{-x} approaches zero. Consequently, we have

$$\sinh(x) \lessapprox \frac{e^x}{2} \lessapprox \cosh(x) \text{ for } x \gg 0.$$

In other words, as x increases, the graph of the hyperbolic sine approaches the graph of $e^x/2$ from below and the graph of the hyperbolic cosine approaches the graph of $e^x/2$ from above. Further information concerning the hyperbolic functions can be derived from the following identities (see Exercise 31.2):

$$\sinh(-x) = -\sinh(x), \tag{31.5}$$
$$\cosh(-x) = \cosh(x). \tag{31.6}$$

The first of these equations implies that the graph of the hyperbolic sine remains unchanged when rotated by 180° about the origin of an xy-coordinate system, and the second says that the graph of the hyperbolic cosine is symmetric relative to the y-axis (see Figure 31.1).

31.2 Exercise. Use Definition 31.1 to verify equations (31.5) and (31.6).

The most important identity for hyperbolic functions is the *hyperbolic theorem of Pythagoras*:

$$\boxed{\cosh^2(x) - \sinh^2(x) = 1.} \tag{31.7}$$

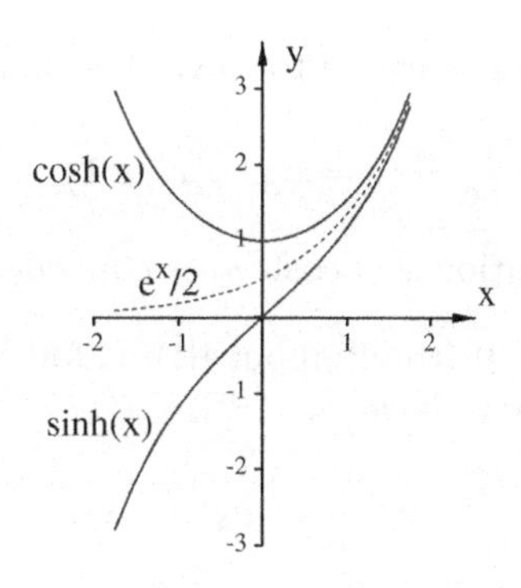

Figure 31.1: graphs of $\sinh(x)$ and $\cosh(x)$.

31.3 Exercise. Prove the hyperbolic theorem of Pythagoras. *Hint.* Replace $\sinh(x)$ and $\cosh(x)$ with the terms given in Definition 31.1, and then use elementary algebra to simplify the resulting expression.

★ *Remark.* For readers familiar with analytic geometry (see also Appendix C), we wish to provide a rationale for labeling the functions sinh and cosh as "hyperbolic." Since $\cosh^2(t) - \sinh^2(t) = 1$ (by the hyperbolic theorem of Pythagoras), the point $(\cosh(t), \sinh(t))$ is located on the horizontal hyperbola described by the equation $x^2 - y^2 = 1$. As t ranges from $-\infty$ to ∞, the point $(\cosh(t), \sinh(t))$ traces out one complete branch of this hyperbola—the one on the positive side of the x-axis (see also Figure 31.3 on p.239). ★

The derivatives of the hyperbolic sine and cosine obey the following simple rules:

$$\boxed{\frac{d}{dx}\sinh(x) = \cosh(x),} \tag{31.8}$$

and

$$\boxed{\frac{d}{dx}\cosh(x) = \sinh(x).} \tag{31.9}$$

31.4 Exercise. Use Definition 31.1 to verify the equations in (31.8) and (31.9).

We now turn our attention to the subject of *hyperbolic inverse functions.* Given the graphs in Figure 31.1, we notice that the hyperbolic sine is an invertible function on $\mathbb{R}$ because its graph passes the horizontal line test. Since the domain and range of the hyperbolic sine are both equal to $\mathbb{R}$, it follows that the domain and range of the corresponding inverse function are equal to $\mathbb{R}$ as well (see Chapter 22). Unfortunately, matters are slightly more complicated for the hyperbolic cosine, because, according to Figure 31.1, the graph of cosh does not pass the horizontal line test. In other words, cosh, defined on $\mathbb{R}$, is not an invertible function. To deal with this difficulty we will *restrict the domain of* cosh *to* $[0, \infty)$. So whenever we speak of the *inverse function of the hyperbolic cosine*, we actually refer to the inverse function of

$$\begin{aligned} f : [0, \infty) &\to \mathbb{R} \\ x &\mapsto \cosh(x). \end{aligned} \tag{31.10}$$

Since the domain and range of f are $[0, \infty)$ and $[1, \infty)$ respectively, it follows that the domain of f^{-1} is $[1, \infty)$ and its range is $[0, \infty)$.

In order to find explicit formulae for the hyperbolic inverse functions, we will now solve for y the equations $x = \sinh(y)$ and $x = \cosh(y)$ (see also the note at the end of Example 22.5). Beginning with the hyperbolic sine, we write the equation $x = \sinh(y) = (e^y - e^{-y})/2$ in the form $2x = e^y - e^{-y}$. Subtracting $2x$ and multiplying by e^y yields

$$0 = (e^y)^2 - 2xe^y - 1.$$

Using the quadratic formula (see Appendix D) to solve this equation for e^y, we obtain

$$e^y = x \pm \sqrt{x^2+1}. \tag{31.11}$$

31.5 Exercise. Show that the equation $x = \cosh(y)$ is equivalent to $e^y = x \pm \sqrt{x^2-1}$.

Since $e^y > 0$ and $x - \sqrt{x^2+1} < 0$ (think about it!), it follows that the sign in front of the square root in (31.11) must be positive. So we have

$$e^y = x + \sqrt{x^2+1}.$$

Applying the natural logarithm to both sides yields

$$y = \ln\left(x + \sqrt{x^2+1}\right).$$

Thus, we have found the defining equation for the inverse function of the hyperbolic sine. To obtain an analogous formula for the inverse function of the hyperbolic cosine (i.e., the inverse function of f in (31.10)), we use the result of Exercise 31.5 to infer that the equation $x = \cosh(y)$ is equivalent to

$$y = \ln\left(x \pm \sqrt{x^2-1}\right).$$

Due to the $\pm$ sign, we now have two solutions: $y_1 = \ln(x + \sqrt{x^2-1})$ and $y_2 = \ln(x - \sqrt{x^2-1})$ (see Figure 31.2). Since the range of the inverse function of f in (31.10) is $[0, \infty)$, the positive solution y_1 is

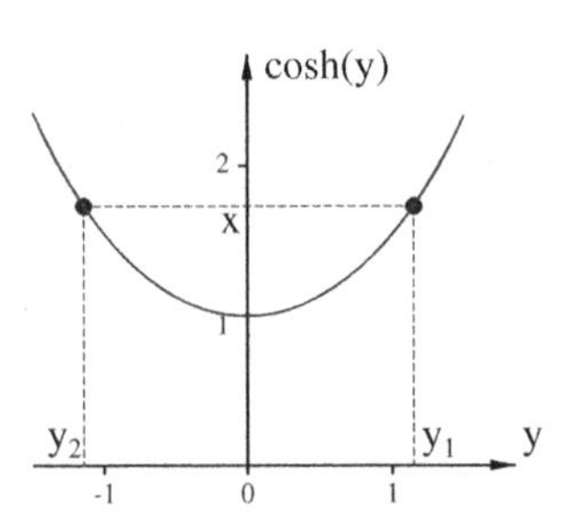

Figure 31.2: solutions of $\cosh(y) = x$.

the one we need to pick. Consequently, the defining equation for the inverse function of the hyperbolic cosine is

$$y = \ln\left(x + \sqrt{x^2-1}\right).$$

We summarize our results in the following theorem:

31.6 Theorem. **a)** *The hyperbolic sine function, defined on* $\mathbb{R}$, *is invertible, and the values of its inverse function are denoted by* $\sinh^{-1}(x)$ *or* $\operatorname{arsinh}(x)$ *(hyperbolic area sine). The values of* $\operatorname{arsinh}(x)$ *are given by the equation*

$$\operatorname{arsinh}(x) = \ln\left(x + \sqrt{x^2+1}\right) \quad \textit{for all } x \in \mathbb{R},$$

and the domain and range of the hyperbolic area sine are both equal to $\mathbb{R}$.

b) *The function*

$$f : [0, \infty) \to \mathbb{R}$$
$$x \mapsto \cosh(x)$$

is invertible, and the values of its inverse function are denoted by $\cosh^{-1}(x)$ *or* $\operatorname{arcosh}(x)$ *(hyperbolic area cosine). The values of* $\operatorname{arcosh}(x)$ *are given by the equation*

$$\operatorname{arcosh}(x) = \ln\left(x + \sqrt{x^2 - 1}\right) \quad \textit{for all } x \in [1, \infty),$$

and the domain and range of the hyperbolic area cosine are $[1, \infty)$ *and* $[0, \infty)$ *respectively.*

★ *Remark.* The use of the prefix "area" in denoting the hyperbolic inverse functions is motivated by the following geometric fact: let (x, y) be a point in the first quadrant (i.e., $x, y \geq 0$) on the hyperbola given by the equation $x^2 - y^2 = 1$ (see Figure 31.3), and let $t \in \mathbb{R}$ such that $(x, y) = (\cosh(t), \sinh(t))$ or equivalently $t = \operatorname{arcosh}(x) = \operatorname{arsinh}(y)$ (see the ⋆remark⋆ on p.237). Then the difference of the *areas* A and B shown in Figure 31.3 on the left, is equal to t (see Exercise 31.7). Furthermore, the area of the sector C shown on the right is equal to $t/2$ because $C = A - (A + B)/2 = (A - B)/2$.

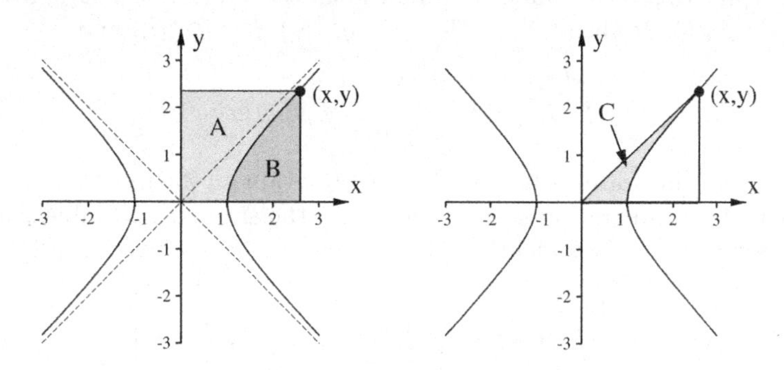

Figure 31.3: area property of the hyperbolic inverse functions.

31.7 Exercise. Let $t \in [0, \infty)$ and $(x, y) := (\cosh(t), \sinh(t))$. Prove that the areas A and B, shown in Figure 31.3, satisfy the equation $A - B = t$. *Hint.* Use integration to express the difference $A - B$ as a function of x (or t), and prove that $A - B - t$ is constant by demonstrating that the derivative of $A - B - t$ with respect to x (or t) is zero. Then set x equal to 1 (or t equal to 0) to show that $A - B - t = 0$. ★

Inverse Integration by Substitution

31.8 Example. We wish to find the value of the definite integral

$$\int_0^1 \frac{1}{\sqrt{1 + x^2}}\, dx.$$

With $x := \sinh(t)$ and $dx = \cosh(t)\, dt$ (see (31.8)) the rule of integration by substitution, as stated in Theorem 30.1, implies that

$$\int_{\operatorname{arsinh}(0)}^{\operatorname{arsinh}(1)} \frac{\cosh(t)}{\sqrt{1 + \sinh^2(t)}}\, dt = \int_0^1 \frac{1}{\sqrt{1 + x^2}}\, dx.$$

(Note: in order to find the boundaries for the integral on the right, we used the fact that $\sinh(\operatorname{arsinh}(x)) = x$ for all $x \in \mathbb{R}$.) The trick is now to eliminate the root in the denominator of the integrand on the left-hand side by using the hyperbolic theorem of Pythagoras in (31.7). This yields

$$\int_0^1 \frac{1}{\sqrt{1 + x^2}}\, dx = \int_{\operatorname{arsinh}(0)}^{\operatorname{arsinh}(1)} \frac{\cosh(t)}{\sqrt{\cosh^2(t)}}\, dt = \int_{\operatorname{arsinh}(0)}^{\operatorname{arsinh}(1)} \frac{\cosh(t)}{|\cosh(t)|}\, dt.$$

Since $\cosh(t) > 0$ for all $t \in \mathbb{R}$, it follows that $|\cosh(t)| = \cosh(t)$. Furthermore, the defining equation for $\operatorname{arsinh}(x)$ in Theorem 31.6 shows that $\operatorname{arsinh}(0) = 0$ and $\operatorname{arsinh}(1) = \ln(1+\sqrt{2})$. Hence

$$\int_0^1 \frac{1}{\sqrt{1+x^2}}\,dx = \int_0^{\ln(1+\sqrt{2})} 1\,dt = \ln(1+\sqrt{2}).$$

This example illustrates that the basic idea of *inverse integration by substitution* is actually very simple—we use ordinary integration by substitution in the "reverse direction." To make this more precise, let us recall the following important result from Chapter 30: if for open intervals $I, J \subset \mathbb{R}$ we are given a continuous function $f : J \to \mathbb{R}$ and a continuously differentiable function $g : I \to J$, then

$$\int_a^b f(g(x))g'(x)\,dx = \int_{g(a)}^{g(b)} f(u)\,du \text{ for all } a, b \in I.$$

We will now reinterpret this result under the additional assumption that g is invertible. To do so, we set $c := g(a)$ and $d := g(b)$. It then follows that $a = g^{-1}(c)$, $b = g^{-1}(d)$ and

$$\int_c^d f(u)\,du = \int_{g^{-1}(c)}^{g^{-1}(d)} f(g(x))g'(x)\,dx.$$

In order to indicate that this formula is typically used to evaluate the integral of f rather than $(f \circ g)g'$, we replace the variable of integration u on the left with the standard variable x and write t instead of x in the integral on the right. This yields

$$\boxed{\int_c^d f(x)\,dx = \int_{g^{-1}(c)}^{g^{-1}(d)} f(g(t))g'(t)\,dt.} \tag{31.12}$$

To derive an analogous formula for indefinite integrals, we assume that H is an antiderivative of $(f \circ g)g'$. Then, using the chain rule and the rule for the differentiation of inverse functions (see Theorem 22.7), we obtain

$$\frac{d}{dx}H(g^{-1}(x)) = f(g(g^{-1}(x)))g'(g^{-1}(x))\frac{d}{dx}g^{-1}(x) = f(x)\frac{g'(g^{-1}(x))}{g'(g^{-1}(x))} = f(x).$$

This proves that $H \circ g^{-1}$ is an antiderivative of f. Hence

$$\boxed{\int f(x)\,dx = H(g^{-1}(x)) + C.} \tag{31.13}$$

31.9 Example. Expanding on our discussion in Example 31.8, we wish to determine the antiderivative of the function $f : \mathbb{R} \to \mathbb{R}$ defined by the equation

$$f(x) := \frac{1}{\sqrt{x^2+1}}.$$

Setting $g(t) := \sinh(t)$ yields

$$f(g(t))g'(t) = \frac{g'(t)}{\sqrt{g(t)^2+1}} = \frac{\cosh(t)}{\sqrt{\cosh^2(t)}} = \frac{\cosh(t)}{|\cosh(t)|} = 1 \quad (\text{because } \cosh(t) \geq 0).$$

Since $H(t) := t$ is an antiderivative of the constant function 1, we may use (31.13) to infer that

$$\int \frac{1}{\sqrt{x^2+1}}\,dx = H(g^{-1}(x)) + C = g^{-1}(x) + C = \operatorname{arsinh}(x) + C.$$

The same result can also be obtained by directly substituting $\sinh(t)$ for x and $\cosh(t)\,dt$ for dx:

$$\int \frac{1}{\sqrt{x^2+1}}\,dx = \int \frac{1}{\sqrt{\sinh^2(t)+1}}\cosh(t)\,dt = \int \frac{\cosh(t)}{\sqrt{\cosh^2(t)}}\,dt = \int 1\,dt = t + C = \operatorname{arsinh}(x) + C.$$

31.10 Example. Let us evaluate the definite integral

$$\int_0^1 \sqrt{x^2+1}\,dx.$$

As above, we define $g(t) := \sinh(t)$ and then use (31.12) to conclude that

$$\int_0^1 \sqrt{x^2+1}\,dx = \int_{g^{-1}(0)}^{g^{-1}(1)} \sqrt{g(t)^2+1}\,g'(t)\,dt = \int_{\operatorname{arsinh}(0)}^{\operatorname{arsinh}(1)} \cosh^2(t)\,dt.$$

Alternatively, we can substitute $\sinh(t)$ for x and $\cosh(t)\,dt$ for dx:

$$\int_0^1 \sqrt{x^2+1}\,dx = \int_{\operatorname{arsinh}(0)}^{\operatorname{arsinh}(1)} \sqrt{\sinh^2(t)+1}\,\cosh(t)\,dt = \int_{\operatorname{arsinh}(0)}^{\operatorname{arsinh}(1)} \cosh^2(t)\,dt.$$

Given that $\operatorname{arsinh}(0) = 0$ and $\operatorname{arsinh}(1) = \ln(1+\sqrt{2})$ (see Theorem 31.6), we thus obtain

$$\begin{aligned}
\int_0^1 \sqrt{x^2+1}\,dx &= \int_0^{\ln(1+\sqrt{2})} \cosh^2(t)\,dt = \int_0^{\ln(1+\sqrt{2})} \frac{1}{4}(e^t+e^{-t})^2\,dt \\
&= \frac{1}{4}\int_0^{\ln(1+\sqrt{2})} (e^{2t}+2+e^{-2t})\,dt = \frac{1}{4}\left(\frac{e^{2t}}{2}+2t-\frac{e^{-2t}}{2}\right)\Bigg|_0^{\ln(1+\sqrt{2})} \\
&= \frac{1}{8}\left((1+\sqrt{2})^2+4\ln(1+\sqrt{2})-\frac{1}{(1+\sqrt{2})^2}\right) = \frac{\sqrt{2}+\ln(1+\sqrt{2})}{2}.
\end{aligned}$$

31.11 Exercise. Verify the last equality in this calculation.

31.12 Exercise. Find the antiderivative of the function $f(x) := \sqrt{x^2+1}^{\,3}$ and evaluate the integral $\int_0^1 f(x)\,dx$.

31.13 Example. Another type of substitution that appears rather frequently involves trigonometric functions. To give an example, we consider the integral

$$\int_0^{1/2} \frac{1}{\sqrt{1-x^2}^{\,3}}\,dx.$$

Using the trigonometric theorem of Pythagoras (see Chapter 26) and the substitution $x := \sin(t)$ with $dx = \cos(t)\,dt$, it follows that

$$\begin{aligned}
\int_0^{1/2} \frac{1}{\sqrt{1-x^2}^{\,3}}\,dx &= \int_{\arcsin(0)}^{\arcsin(1/2)} \frac{1}{\sqrt{1-\sin^2(t)}^{\,3}}\cos(t)\,dt = \int_0^{\pi/6} \frac{\cos(t)}{\sqrt{\cos^2(t)}^{\,3}}\,dt \\
&= \int_0^{\pi/6} \frac{\cos(t)}{|\cos(t)|^3}\,dt = \int_0^{\pi/6} \frac{1}{\cos^2(t)}\,dt \quad (\text{because } \cos(t) \geq 0 \text{ for all } t \in [0,\pi/6]) \\
&= \tan(t)\Big|_0^{\pi/6} = \frac{1}{\sqrt{3}}.
\end{aligned}$$

As in the case of substitutions involving hyperbolic functions, the crucial step in the preceding calculation was the elimination of the square root in the denominator.

31.14 Exercise. Find the value of the integral $\int_0^{1/2} \sqrt{1-x^2}\,dx$. *Hint.* The double angle formula $\cos(2t) = 2\cos^2(t) - 1$ will be helpful.

A list of frequently occurring *standard substitutions*, including all substitutions in the preceding examples, is given in the following table:

term in the integrand	hyperbolic substitution	trigonometric substitution
$\sqrt{a^2+x^2}$	$x = a\sinh(t)$	$x = a\tan(t)$
$\sqrt{a^2-x^2}$	$x = a\dfrac{\sinh(t)}{\cosh(t)}$	$x = a\sin(t)$
$\sqrt{x^2-a^2}$	$x = a\cosh(t)$	$x = \dfrac{a}{\cos(t)}$

According to this table, an integral involving a term like $\sqrt{4-x^2} = \sqrt{2^2-x^2}$ might be solvable with the substitutions $x = 2\sinh(t)/\cosh(t)$ or $x = 2\sin(t)$. Which of these substitutions is more convenient depends on the specifics of the problem and, of course, there is no guarantee that either one will work.

31.15 Example. ★ In order to illustrate that occasionally nonstandard substitutions (rather than hyperbolic or trigonometric substitutions) can be helpful as well, let us consider the problem of finding the antiderivative of the function $f : (0,\infty) \to \mathbb{R}$ defined by the equation

$$f(x) := \frac{1}{x\sqrt{x^2+1}}.$$

Introducing the substitution $x := 1/t$ with $dx = -1/t^2\,dt$, we obtain

$$\begin{aligned}\int \frac{1}{x\sqrt{x^2+1}}\,dx &= -\int \frac{1}{(1/t)\sqrt{(1/t^2)+1}}\frac{1}{t^2}\,dt = -\int \frac{1}{t\sqrt{(1+t^2)/t^2}}\,dt = -\int \frac{|t|}{t\sqrt{1+t^2}}\,dt \\ &= -\int \frac{t}{t\sqrt{1+t^2}}\,dt \quad (\text{because } x \in (0,\infty) \text{ implies } t > 0) \\ &= -\int \frac{1}{\sqrt{1+t^2}}\,dt = -\operatorname{arsinh}(t) + C \quad (\text{see Example 31.9}) \\ &= -\operatorname{arsinh}\left(\frac{1}{x}\right) + C. \text{ ★}\end{aligned}$$

Designing a Radar Antenna (Part 2)

To continue our discussion from the first section of this chapter, we combine the result of Example 31.9 with (31.3) and (31.4). This yields

$$\pm\ln|x| + C = \int \frac{h'(x)}{\sqrt{h(x)^2+1}}\,dx = \int \frac{1}{\sqrt{u^2+1}}\,du = \operatorname{arsinh}(u) + D.$$

(We denote the constant of integration on the right by D to keep it distinct from C on the left.) Recalling now that $u = h(x) = f(x)/x$, we obtain

$$\operatorname{arsinh}\left(\frac{f(x)}{x}\right) = \pm\ln|x| + C - D.$$

Since the cross-sectional shape of a radar antenna, as described by the graph of f, is symmetric with respect to the y-axis (see also Figure 6.2), it is sufficient to determine $f(x)$ only for positive values of x. So let us assume that $x > 0$. Then $|x| = x$, and in setting $q := C - D$, it follows that

$$\operatorname{arsinh}\left(\frac{f(x)}{x}\right) = \pm\ln(x) + q$$

or equivalently,

$$\frac{f(x)}{x} = \sinh(\pm\ln(x) + q) = \frac{1}{2}\left(x^{\pm 1}e^q - \frac{1}{x^{\pm 1}e^q}\right).$$

Multiplying by x and defining $p := e^q$, we may infer that

$$f(x) = \frac{1}{2}\left(px^{1\pm1} - \frac{x}{px^{\pm1}}\right).$$

Thus, there are two possible formulae for $f(x)$:

$$f(x) = \frac{1}{2}\left(px^2 - \frac{1}{p}\right) \text{ or } f(x) = \frac{1}{2}\left(p - \frac{x^2}{p}\right).$$

Both of these equations correctly describe the cross-sectional shape of a radar antenna, but only the one on the left is consistent with our initial set-up in Figure 6.2 where we chose to orient the radar antenna toward the positive direction of the y-axis. By contrast, the graph corresponding to the second equation is a downward parabola that models a radar antenna oriented toward the negative direction of the y-axis. Consequently, the cross-sectional shape of a radar antenna, as shown in Figure 6.2, is in general given by the graph of a function f of the form

$$\boxed{f(x) = \frac{1}{2}\left(px^2 - \frac{1}{p}\right) \text{ for some } p > 0.}$$

(Note: this equation is valid also for negative values of x because for all $p > 0$, the parabolic graph of f is symmetric with respect to the y-axis.) Figure 31.4 shows several graphs of f for different values of p.

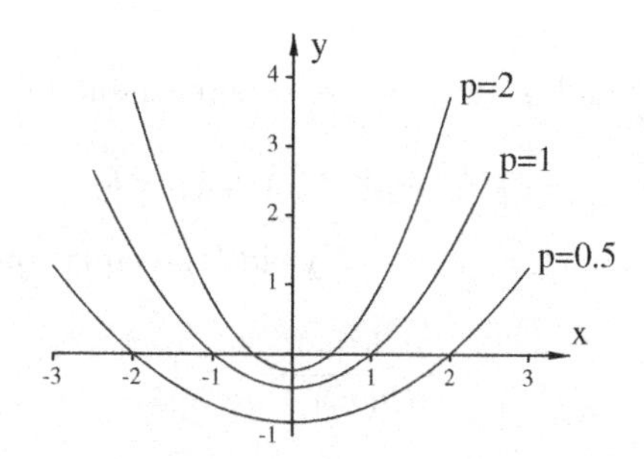

Figure 31.4: cross-sectional shapes of radar antennas.

Remark. The focusing property that motivates the design of a radar antenna is important also in other technical applications. For instance, if we replace the receiver of a radar antenna with a light bulb and cover the inside of the antenna with a reflecting material, then the resulting *parabolic mirror* reflects all light rays coming from the light bulb in the direction parallel to the mirror's axis of symmetry. This inverse focusing effect is commonly used in the construction of electric light sources such as headlights in cars.

Additional Exercises

31.16. Assume that for a differentiable function $f : \mathbb{R} \to \mathbb{R}$ we have

$$f'(x) = \sin^2\left(\frac{f(x)}{x}\right) + \frac{f(x)}{x} \text{ for all } x \neq 0.$$

Show that the substitution $h(x) := f(x)/x$ yields the equation

$$\frac{h'(x)}{\sin^2(h(x))} = \frac{1}{x}.$$

Then use this result to determine $h(x)$, and thereby $f(x)$, via integration by substitution (up to a constant).

31.17. Assume that $f : \mathbb{R} \to \mathbb{R}$ and $g : \mathbb{R} \to \mathbb{R}$ are differentiable functions such that

$$f'(x) = g\left(\frac{f(x)}{x}\right).$$

Show that the substitution $h(x) := f(x)/x$ yields the equation

$$\frac{h'(x)}{g(h(x)) - h(x)} = \frac{1}{x}.$$

31.18. Assume that $f : \mathbb{R} \to \mathbb{R}$ is a differentiable function such that

$$f'(x) = (f(x) + x + 1)\ln(f(x) + x + 1) - 1.$$

Show that the substitution $h(x) := f(x) + x + 1$ yields the equation

$$\frac{h'(x)}{h(x)\ln(h(x))} = 1.$$

Then use this result to determine $h(x)$, and thereby $f(x)$, via integration by substitution (up to a constant).

31.19. Assume that $f : \mathbb{R} \to \mathbb{R}$ and $g : \mathbb{R} \to \mathbb{R}$ are differentiable functions such that for some $a, b \in \mathbb{R}$ we have

$$f'(x) = g(f(x) + ax + b).$$

Show that the substitution $h(x) := f(x) + ax + b$ yields the equation

$$\frac{h'(x)}{g(h(x)) + a} = 1.$$

31.20. Use inverse integration by substitution to determine the following definite and indefinite integrals:

$$\textbf{a)} \int_0^4 \sqrt{16 + x^2}^{\,3}\, dx \quad \textbf{b)} \int \sqrt{x^2 - 1}^{\,3}\, dx \text{ for } x > 1 \quad \textbf{c)} \int_0^2 \frac{1}{\sqrt{4 + x^2}^{\,3}}\, dx$$

$$\textbf{d)} \int \sqrt{1 - x^2}\, dx \qquad \textbf{e)} \int \frac{1}{\sqrt{4 - x^2}^{\,5}}\, dx \qquad \textbf{f)} \int_0^{\pi/4} \frac{1}{\cos^3(x)}\, dx$$

Hint. For f), you may want to write $1/\cos^3(x)$ as $\sqrt{1 + \tan^2(x)}/\cos^2(x)$ and then use a combination of ordinary integration by substitution and inverse integration by substitution to evaluate the integral.

31.21. Evaluate the following definite integrals:

$$\textbf{a)} \int_0^1 x\sqrt{1 - x^2}\, dx \quad \textbf{b)} \int_0^1 x^2\sqrt{1 - x^2}\, dx \quad \textbf{c)} \int_0^1 \frac{x}{\sqrt{1 - x^2}}\, dx \quad \textbf{d)} \int_0^1 \frac{x^2}{\sqrt{1 - x^2}}\, dx$$

Hint. For some of these integrals inverse integration by substitution may not be needed.

31.22. Work Exercise 31.7 in a different way: set up definite integrals that are equal to the areas A and B respectively and use inverse integration by substitution to evaluate these integrals.

31.23. For the definite and indefinite integrals given below, identify f, g, H, c, and d as they appear in equations (31.12) and (31.13) (whichever applies). Then use (31.12) and (31.13) to solve the integrals.

a) $\displaystyle\int_0^2 \frac{1}{\sqrt{4+x^2}}\,dx$ b) $\displaystyle\int \frac{1}{\sqrt{4+x^2}}\,dx$

31.24. ★ Find the antiderivative of the function $f(x) = \dfrac{x^2}{\sqrt{x^2+2x+2}}$. *Hint.* Add and subtract.

31.25. Prove the following identities:

a) $\sinh(x+y) = \sinh(x)\cosh(y) + \cosh(x)\sinh(y)$,

b) $\cosh(x+y) = \cosh(x)\cosh(y) + \sinh(x)\sinh(y)$,

c) $\sinh(2x) = 2\sinh(x)\cosh(x)$,

d) $\cosh(2x) = \cosh^2(x) + \sinh^2(x) = 2\cosh^2(x) - 1 = 1 + 2\sinh^2(x)$.

31.26. Use the formulae in Exercise 31.25 to determine the antiderivative of $\sqrt{x^2+1}$.

31.27. Evaluate $\int_0^{\pi/6} f(2\cos(t))\cos(t)\,dt$ from the assumption that $\int_0^1 f(\sqrt{4-x^2})\,dx = 10$.

31.28. Evaluate $\int_0^6 f(\sqrt{9+x^2})\,dx$ from the assumption that $\int_0^{\ln(2+\sqrt{5})} f(3\cosh(t))\cosh(t)\,dt = 10$.

Chapter 32

★ Ellipses and Kidney Stones

The Elliptic Mirror

In Chapter 31 we showed that the cross-sectional shape of a radar antenna in an xy-coordinate system is given by an equation of the form

$$y = f(x) = \frac{1}{2}\left(px^2 - \frac{1}{p}\right).$$

For any value $p \neq 0$, the curve described by this equation is a *parabola*. In analytic geometry parabolas are identified as members of a larger class of curves—the so-called *conic sections* (see Appendix C)—that also includes ellipses and hyperbolas. In analogy to our discussion of the reflection property of radar antennas, our goal in this chapter is to study a reflection property of ellipses that has found interesting applications in medicine.

The standard equation for an *ellipse* in horizontal position with major radius a and minor radius b is

$$\boxed{\frac{x^2}{a^2} + \frac{y^2}{b^2} = 1.} \tag{32.1}$$

The two *foci* of the ellipse are $F_1 = (c, 0)$ and $F_2 = (-c, 0)$, where the value of c is related to a and b via the equation

$$\boxed{a^2 = b^2 + c^2.} \tag{32.2}$$

A typical graph of an ellipse described by equation (32.1) is shown in Figure 32.1.

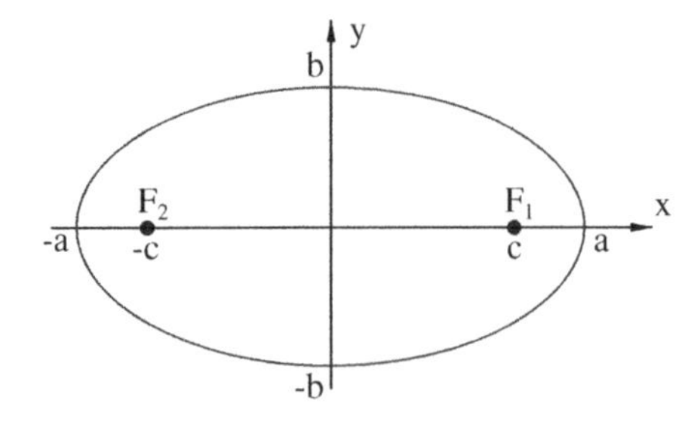

Figure 32.1: an ellipse.

To construct an *elliptic mirror*, we revolve the ellipse in Figure 32.1 about the x-axis and cover the inside of the resulting *ellipsoid* with a reflecting material. Our objective is now to prove that a light ray emitted at one of the foci is reflected off the mirror's surface in such a way that it passes through the other focus. To begin with, we consider Figure 32.2, which shows a light ray inside an elliptic mirror emitted from the focus at F_2. According to the general law of reflection (see Chapter 6), the angles of the incoming light ray and the reflected light ray relative to the tangent line at the point of reflection (x_0, y_0) are equal. In order to find the slope of the tangent line at (x_0, y_0), we need to determine the

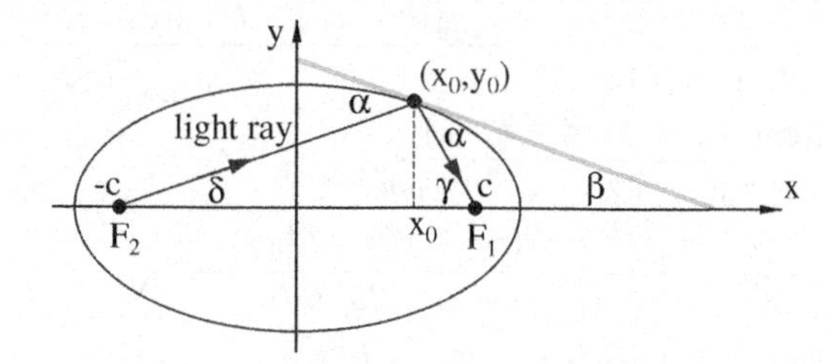

Figure 32.2: light ray reflected inside an ellipse.

derivative of the function f whose graph is the upper half of the ellipse (the part above the x-axis). To obtain a defining equation for f we solve equation (32.1) for y. This yields

$$f(x) = y = \frac{b}{a}\sqrt{a^2 - x^2} \text{ for all } x \in [-a, a].$$

Since the slope of the tangent line at $(x_0, y_0) = (x_0, f(x_0))$ is $\tan(\pi - \beta) = -\tan(\beta)$, it follows that

$$\tan(\beta) = -f'(x_0) = \frac{bx_0}{a\sqrt{a^2 - x_0^2}} = \frac{b^2 x_0}{a^2 y_0}. \tag{32.3}$$

Given Figure 32.2, elementary geometry shows that $\beta + \delta + \pi - \alpha = \pi = \pi - \gamma + \alpha + \beta$ and, therefore,

$$\gamma = 2\beta + \delta. \tag{32.4}$$

Observing further that the slope of the line representing the reflected light ray is $\tan(\pi - \gamma) = -\tan(\gamma)$, we may conclude that the equation of this line is

$$y = -\tan(\gamma)(x - x_0) + y_0.$$

In order to prove the reflection property of the elliptic mirror, we need to demonstrate that the reflected light ray passes through the focus at $F_1 = (c, 0)$. In other words, we need to show that

$$0 = -\tan(\gamma)(c - x_0) + y_0,$$

or equivalently,

$$\frac{y_0}{c - x_0} = \tan(\gamma). \tag{32.5}$$

To verify this equation we first express $\tan(\gamma)$ in terms of $\tan(\beta)$ and $\tan(\delta)$ by means of the addition law for tangent, the double angle formula for tangent (see Appendix A), and equation (32.4):

$$\tan(\gamma) = \tan(2\beta + \delta) = \frac{\tan(2\beta) + \tan(\delta)}{1 - \tan(2\beta)\tan(\delta)} = \frac{\dfrac{2\tan(\beta)}{1 - \tan^2(\beta)} + \tan(\delta)}{1 - \dfrac{2\tan(\beta)\tan(\delta)}{1 - \tan^2(\beta)}}$$
$$= \frac{2\tan(\beta) + \tan(\delta)(1 - \tan^2(\beta))}{1 - \tan^2(\beta) - 2\tan(\beta)\tan(\delta)}. \tag{32.6}$$

Furthermore, according to Figure 32.2, we have

$$\tan(\delta) = \frac{y_0}{c + x_0}.$$

Combining this equation with (32.3) and (32.6), we obtain

$$
\begin{aligned}
\tan(\gamma) &= \frac{\dfrac{2b^2x_0}{a^2y_0} + \dfrac{y_0}{c+x_0}\left(1 - \dfrac{b^4x_0^2}{a^4y_0^2}\right)}{1 - \dfrac{b^4x_0^2}{a^4y_0^2} - \dfrac{2b^2x_0}{a^2(c+x_0)}} = \frac{\dfrac{2a^2b^2x_0(c+x_0) + a^4y_0^2 - b^4x_0^2}{a^4y_0(c+x_0)}}{\dfrac{a^4y_0^2(c+x_0) - b^4x_0^2(c+x_0) - 2a^2b^2x_0y_0^2}{a^4y_0^2(c+x_0)}} \\
&= \frac{y_0(2a^2b^2x_0(c+x_0) + a^4y_0^2 - b^4x_0^2)}{a^4y_0^2(c+x_0) - b^4x_0^2(c+x_0) - 2a^2b^2x_0y_0^2} \\
&= \frac{y_0(2a^2b^2x_0(c+x_0) + a^2b^2(a^2 - x_0^2) - b^4x_0^2)}{a^2b^2(a^2 - x_0^2)(c+x_0) - b^4x_0^2(c+x_0) - 2b^4x_0(a^2 - x_0^2)} \\
&= \frac{y_0(2a^2x_0(c+x_0) + a^2(a^2 - x_0^2) - b^2x_0^2)}{a^2(a^2 - x_0^2)(c+x_0) - b^2x_0^2(c+x_0) - 2b^2x_0(a^2 - x_0^2)} \\
&= \frac{y_0(2a^2cx_0 + a^2x_0^2 + a^4 - b^2x_0^2)}{a^4c - a^2cx_0^2 + a^4x_0 - a^2x_0^3 - b^2cx_0^2 - 2a^2b^2x_0 + b^2x_0^3} \\
&= \frac{y_0(2a^2cx_0 + c^2x_0^2 + a^4)}{a^4c - a^2cx_0^2 + a^4x_0 - a^2x_0^3 - (a^2 - c^2)cx_0^2 - 2a^2(a^2 - c^2)x_0 + (a^2 - c^2)x_0^3} \qquad \text{(by (32.2))} \\
&= \frac{y_0(a^2 + cx_0)^2}{a^4c - 2a^2cx_0^2 - a^4x_0 + c^3x_0^2 + 2a^2c^2x_0 - c^2x_0^3} \\
&= \frac{y_0(a^2 + cx_0)^2}{a^4(c - x_0) + 2a^2cx_0(c - x_0) + c^2x_0^2(c - x_0)} \\
&= \frac{y_0(a^2 + cx_0)^2}{(c - x_0)(a^2 + cx_0)^2} \\
&= \frac{y_0}{c - x_0}.
\end{aligned}
$$

This calculation proves (32.5) and thus establishes the reflection property of elliptic mirrors, as desired.

Kidney Stones

In modern medical technology, the reflection property of elliptic mirrors has been successfully applied to the treatment of kidney stones. In focusing electrically generated waves inside an elliptic bathtub, it is possible to remove kidney stones without surgery. To accomplish this feat, the patient must be positioned inside the tub in such a way that the kidney stone is located exactly at one of the foci. Then a shock wave generator is placed in the other focus so that the waves it produces will meet at the position of the kidney stone after reflecting off the bathtub's surface (see Figure 32.3). Under the

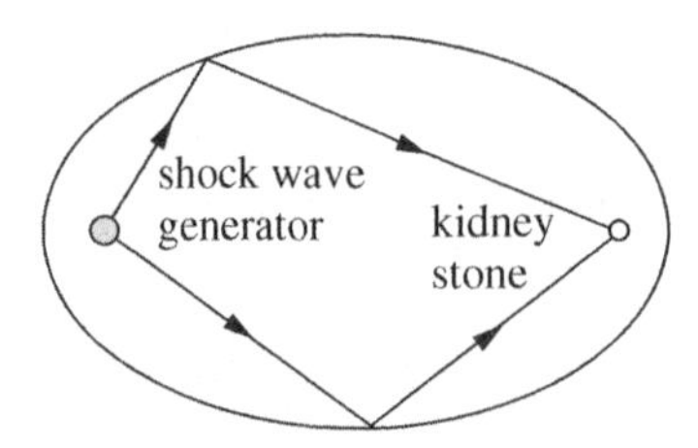

Figure 32.3: elliptic mirror with shock wave generator.

impact of the focused shock waves the stone crumbles, thereby releasing the patient from a painful ailment.

It is interesting to notice that shock waves produced at a certain point in time will reach the kidney stone simultaneously because the sum of the distances from the two foci is the same for all points on the surface of the elliptic bathtub (see Exercise 32.2).

Additional Exercises

32.1. Find the equation of an ellipse with minor radius $b = 2$ and foci at $(\pm 3, 0)$.

32.2. Show that for any point (x, y) on the ellipse shown in Figure 32.1, the sum of the distances to the foci F_1 and F_2 is equal to $2a$.

32.3. Project: A hyperbola in standard position (see Appendix C and Figure 32.4) is given by the equation

$$\frac{x^2}{a^2} - \frac{y^2}{b^2} = 1.$$

The foci F_1 and F_2 are at the points $(c, 0)$ and $(-c, 0)$ respectively, where

$$c^2 = a^2 + b^2.$$

Prove the following reflection property of *hyperbolic mirrors:* a light ray emitted at the focus F_2 is reflected off the outside of the hyperbolic mirror such that the extension of the reflected ray passes through the second focus F_1 (see Figure 32.4).

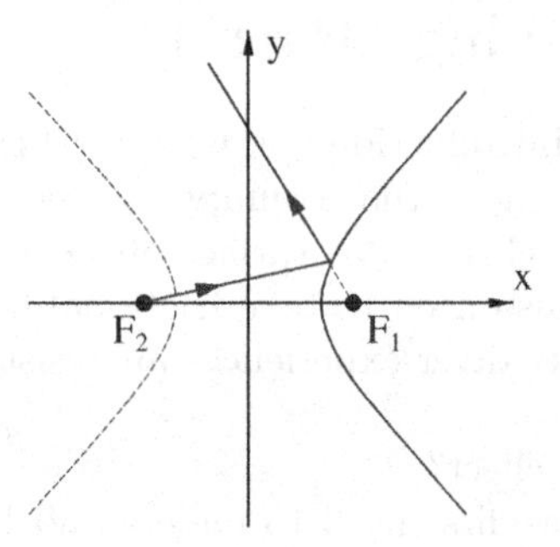

Figure 32.4: reflection at a hyperbolic mirror.

Chapter 33

Integration of Rational Functions and the Physics of Sky Diving

The Physics of Sky Diving (Part 1)

Teacher: In earlier lessons (see the Introduction and Chapter 14), we derived equations for the position and velocity of a falling body under the assumption that the only acting force is the (almost) constant gravitational attraction close to the surface of the earth. In reality, though, things are a bit more complicated, because we also need to take frictional forces into account. For instance, upon jumping out of an airplane, a sky diver experiences air resistance as a retarding force that opposes the gravitational pull of the earth.

Simplicio: How do we quantify this effect?

Teacher: To answer this question, we first need to understand how the force of air resistance changes as a sky diver's velocity increases.

Sophie: I would think that it increases as well—the greater the velocity, the greater the force of air resistance.

Teacher: Indeed, experimental evidence shows that the force of air resistance increases with the square of the velocity. To make this more precise, we assume that the downward motion of a sky diver (or in general, a falling body) is measured on a coordinate axis with the positive direction pointing *downward*. Then, in denoting by $v(t)$ and m the sky diver's velocity and mass respectively, the (constant) gravitational force is equal to mg, and the force of air resistance at time t is $-cv(t)^2$, where c is a constant of proportionality. Thus, the total force at time t is given by the equation*

$$F(t) = mg - cv(t)^2. \tag{33.1}$$

The value of c depends on the position that the sky diver assumes during the descent. For example, in head-down position the air resistance and the corresponding value of c are smaller than in spread eagle position (see also the ⋆remark⋆ below). To simplify our discussion, we will always assume that the sky diver maintains the same position in the air so that the value of c remains constant.

33.1 Exercise. Do you think that the value of c for a child will be different from that for an adult? Explain your answer.

Simplicio: Why is there a negative sign in equation (33.1) in front of the term $cv(t)^2$ but not in front of mg?

Teacher: The signs are different because the corresponding forces have different orientations—the gravitational force points toward the surface of the earth, which in our setup is the positive direction, while the force of air resistance points upward in the negative direction.

*Equation (33.1) is valid only if the falling body is not too small and its velocity is neither close to the speed of sound nor vanishingly small.

Sophie: I am trying to understand why the force of air resistance increases with the square of the velocity. It somehow seems more natural to think that the increase in force would be proportional to $v(t)$ rather than $v(t)^2$.

Teacher: To address this question we need to observe that at a microscopic level the force of air resistance is caused by collisions of a falling body with molecules in the air and...

Sophie: That's exactly what I thought, but as I followed this line of reasoning, I was forced to conclude that the number of collisions increases in proportion to $v(t)$ rather than $v(t)^2$. After all, an object moving twice as fast as another object covers twice as much distance in the same amount of time and presumably collides with roughly twice as many molecules in the air.

Teacher: I agree that the number of collisions increases in proportion to $v(t)$, but you overlooked the fact that the impact of the average collision increases with $v(t)$ as well.

Sophie: That makes sense, but then I have a different question: how exactly do we quantify this increase in the average impact to arrive at the conclusion that the force of air resistance is proportional to $v(t)^2$ rather than, for example, $v(t)^3$ or $v(t)^{3/2}$?

Teacher: We would have to use conservation laws for energy and momentum (see Exercise 33.13), but this is leading us too far. Instead of pursuing this subject further, I would rather direct your attention to another interesting fact: according to equation (33.1), the force of air resistance cannot increase without limit, because if $cv(t)^2 = mg$, then the total force $F(t)$ is zero, and there is no more acceleration. Consequently, a sky diver will approach a *terminal velocity* v_{max} that is given by the equation $cv_{max}^2 = mg$ or, equivalently,

$$v_{max} = \sqrt{\frac{mg}{c}}. \tag{33.2}$$

In spread eagle position, the terminal velocity is about $120\, mi/h$.

★ *Remark.* Given equation (33.2), it is tempting to think that the terminal velocity is proportional to $\sqrt{m}$, but this is actually not true because c indirectly depends on m as well. Intuitively this is not surprising, for a person with a greater mass will typically also have a greater surface area that offers more resistance to the air. Since the complex shape of a sky diver is mathematically difficult to handle, we will simplify matters by analyzing instead the motion of a ball of radius R that is dropped from a certain height above the ground. We assume that the ball is filled with a material of uniform density δ (you may want to think of a marble rather than a ball if you wish) and denote by V the ball's volume. Then its mass is given by the equation

$$m = \delta V. \tag{33.3}$$

Air resistance is caused by air molecules colliding with the ball, and the number of collisions increases in proportion to the ball's surface area A. Therefore, the force of air resistance $-cv(t)^2$ increases in proportion to A as well, and in denoting the corresponding constant of proportionality by λ, we may write

$$c = \lambda A. \tag{33.4}$$

Furthermore, the ball's volume and surface area are given by the following well known formulae:

$$V = \frac{4\pi R^3}{3} \quad \text{and} \quad A = 4\pi R^2.$$

Combining these equations with (33.3) and (33.4), we obtain

$$m = \frac{4\pi\delta R^3}{3} \quad \text{and} \quad c = 4\pi\lambda R^2.$$

Solving the equation on the left for R and substituting the resulting term for R in the equation on the right yields

$$c = 4\pi\lambda \left(\frac{3m}{4\pi\delta}\right)^{2/3}.$$

For simplification we set $\mu := 4\pi\lambda(3/(4\pi\delta))^{2/3}$ so that

$$c = \mu m^{2/3}.$$

Replacing now c in (33.2) with $\mu m^{2/3}$ yields

$$v_{max} = \sqrt{\frac{mg}{\mu m^{2/3}}} = \left(\sqrt{\frac{g}{\mu}}\right)\sqrt[6]{m}.$$

Thus, our calculation suggests that the terminal velocity is proportional to the *sixth root* of m rather than the square root of m. ★

Having found a formula for the terminal velocity, we will now turn our attention to the more difficult task of determining the velocity $v(t)$ at an arbitrary moment in time t. According to Newton's fundamental law of mechanics (see Chapter 8, p.64), we have

$$F(t) = ma(t) = mv'(t).$$

(Note: we do not need to use the more general form of Newton's law, $F(t) = p'(t)$, because unlike the mass of a rocket that is burning fuel (see Chapter 8), the mass of a sky diver remains constant.) Combining the equation above with (33.1), we obtain

$$mv'(t) = mg - cv(t)^2.$$

Hence

$$\frac{v'(t)}{(mg/c) - v(t)^2} = \frac{c}{m}.$$

Using (32.2), this equation can be rewritten in the form

$$\frac{v'(t)}{v_{max}^2 - v(t)^2} = \frac{g}{v_{max}^2}.$$

Integrating both sides in the boundaries from 0 to t yields

$$\int_0^t \frac{v'(\tau)}{v_{max}^2 - v(\tau)^2}\,d\tau = \int_0^t \frac{g}{v_{max}^2}\,d\tau = \frac{g}{v_{max}^2}\,t.$$

The integral on the left-hand side can be simplified with the substitutions $u := v(t)$ and $du = v'(t)\,dt$:

$$\int_0^t \frac{v'(\tau)}{v_{max}^2 - v(\tau)^2}\,d\tau = \int_{v(0)}^{v(t)} \frac{1}{v_{max}^2 - u^2}\,du.$$

Under the assumption that the (vertical) velocity of the sky diver at time $t = 0$ is zero (i.e., $v(0) = 0$), we finally arrive at the equation

$$\int_0^{v(t)} \frac{1}{v_{max}^2 - u^2}\,du = \frac{g}{v_{max}^2}\,t. \qquad (33.5)$$

The function $f(u) = 1/(v_{max}^2 - u^2)$ in the integrand is a so-called rational function. In the next section, we will learn how to find antiderivatives of functions of this type.

Integration of Rational Functions

A *rational function* is by definition a quotient of two polynomials. More precisely, f is said to be rational if there are nonnegative integers n and m and coefficients $a_i, b_j \in \mathbb{R}$ such that

$$f(x) = \frac{a_n x^n + \cdots + a_1 x + a_0}{b_m x^m + \cdots + b_1 x + b_0}$$

for all $x \in \mathbb{R}$ with $b_m x^m + \cdots + b_1 x + b_0 \neq 0$.

33.2 Example.

$$f(x) := \frac{x^4 + 3x - 1}{x^3 - 2x}$$

is a rational function with $n = 4$ and $m = 3$.

It is possible to develop a general strategy for integrating arbitrary rational functions, but in this text we will be content to restrict ourselves to functions of the form

$$f(x) = \frac{ax + b}{x^2 + px + q}.$$

In order to integrate functions of this type, we need to distinguish three cases depending on the number of roots of the denominator $x^2 + px + q$. We will discuss one example for each case.

Case 1: $x^2 + px + q$ has exactly one root.

Let us determine the antiderivative of the function f defined by the equation

$$f(x) := \frac{2x + 1}{x^2 - 2x + 1}.$$

The only root of the denominator $x^2 - 2x + 1 = (x - 1)^2$ is $x = 1$. Using the substitution $u := x - 1$ (or equivalently, $x = u + 1$) with $du = dx$, we obtain

$$\begin{aligned} \int f(x)\,dx &= \int \frac{2x + 1}{(x - 1)^2}\,dx = \int \frac{2(u + 1) + 1}{u^2}\,du \\ &= \int \frac{2}{u}\,du + \int \frac{3}{u^2}\,du = 2\ln|u| - \frac{3}{u} + C \\ &= 2\ln|x - 1| - \frac{3}{x - 1} + C. \end{aligned}$$

33.3 Exercise. Find the value of the definite integral $\displaystyle\int_{-1}^{1} \frac{x}{x^2 + 4x + 4}\,dx$.

33.4 Exercise. Solve the indefinite integral $\displaystyle\int \frac{1}{2x^2 - 8x + 8}\,dx$.

Case 2: $x^2 + px + q$ has two distinct roots.

Let f be defined by the equation

$$f(x) := \frac{2x + 1}{x^2 - x - 2}.$$

The two distinct roots of the denominator $x^2 - x - 2 = (x - 2)(x + 1)$ are 2 and -1. In order to integrate f, we need to use a general algebraic technique known as *partial fractions decomposition* (see Chapter 49 for a complete discussion of this subject). More precisely, we need to determine constants A and B such that

$$f(x) = \frac{2x + 1}{(x - 2)(x + 1)} = \frac{A}{x - 2} + \frac{B}{x + 1}. \tag{33.6}$$

It can be shown that such constants always exist, but instead of deriving a general formula for A and B for arbitrary factorizations in the denominator (which is not difficult at all), we will be content to demonstrate how A and B can be determined in the current example. Multiplying both sides of equation (33.6) by $(x - 2)(x + 1)$, we obtain

$$2x + 1 = A(x + 1) + B(x - 2).$$

Since this equation should be valid for all values of x, we may, in particular, substitute the values of the roots $x = 2$ and $x = -1$. This yields the following equations:

$$\begin{aligned} 5 = 2 \cdot 2 + 1 = A(2+1) + B(2-2) = 3A \quad (\text{for } x = 2), \\ -1 = 2 \cdot (-1) + 1 = A(-1+1) + B(-1-2) = -3B \quad (\text{for } x = -1). \end{aligned}$$

In solving for A and B respectively, we find the values $A = 5/3$ and $B = 1/3$.

Alternatively, we can find A and B by comparing coefficients in the equation

$$2x + 1 = A(x+1) + B(x-2) = (A+B)x + A - 2B.$$

Since x is arbitrary, the coefficients on both sides must be equal. Hence

$$\begin{aligned} 2 &= A + B, \\ 1 &= A - 2B. \end{aligned}$$

Solving again for A and B, we obtain the same result as above: $A = 5/3$ and $B = 1/3$. Substituting these values for A and B in (33.6) yields

$$f(x) = \frac{5/3}{x-2} + \frac{1/3}{x+1}.$$

Given this equation, it is now easy to find the antiderivative of f:

$$\int f(x)\,dx = \int \frac{5/3}{x-2}\,dx + \int \frac{1/3}{x+1}\,dx = \frac{5}{3}\ln|x-2| + \frac{1}{3}\ln|x+1| + C.$$

33.5 Exercise. Evaluate the definite integral $\displaystyle\int_0^1 \frac{1}{x^2-2}\,dx$.

Case 3: $x^2 + px + q$ *has no roots.*

Let f be defined by the equation

$$f(x) := \frac{3x+1}{x^2+2x+5}.$$

Whenever the denominator (in our case $x^2 + 2x + 5$) does not have any (real) roots, the first step is always to complete the square (see Appendix D):

$$x^2 + 2x + 5 = (x+1)^2 + 4.$$

Substituting first u for $x + 1$ (or equivalently, $u - 1$ for x) with $du = dx$, and then v for $u/2$ with $du = 2\,dv$, we see that

$$\begin{aligned} \int f(x)\,dx &= \int \frac{3x+1}{(x+1)^2+4}\,dx = \int \frac{3(u-1)+1}{u^2+4}\,du = \int \frac{3u}{u^2+4}\,du - \int \frac{2}{u^2+4}\,du \\ &= \frac{1}{2}\int \frac{3(u/2)}{(u/2)^2+1}\,du - \frac{1}{2}\int \frac{1}{(u/2)^2+1}\,du = \int \frac{3v}{v^2+1}\,dv - \int \frac{1}{v^2+1}\,dv. \end{aligned} \tag{33.7}$$

According to Exercise 26.2, we have

$$\frac{d}{dv}\arctan(v) = \frac{1}{v^2+1},$$

and therefore,

$$\int \frac{1}{v^2+1}\,dv = \arctan(v) + C_1. \tag{33.8}$$

Furthermore, the substitution $w := v^2 + 1$ with $dw = 2v\,dv$ yields

$$\int \frac{3v}{v^2+1}\,dv = \int \frac{3}{2w}\,dw = \frac{3}{2}\ln|w| + C_2 = \frac{3}{2}\ln|v^2+1| + C_2. \tag{33.9}$$

Setting $C := C_2 - C_1 - 3\ln(4)/2$ and combining (33.7) with (33.8) and (33.9), we may infer that

$$\begin{aligned}\int f(x)\,dx &= \frac{3}{2}\ln|v^2+1| - \arctan(v) + C_2 - C_1 = \frac{3}{2}\ln\left|\left(\frac{u}{2}\right)^2 + 1\right| - \arctan\left(\frac{u}{2}\right) + C_2 - C_1 \\ &= \frac{3}{2}\ln|u^2+4| - \arctan\left(\frac{u}{2}\right) + C_2 - C_1 - \frac{3}{2}\ln(4) \\ &= \frac{3}{2}\ln(x^2+2x+5) - \arctan\left(\frac{x+1}{2}\right) + C.\end{aligned}$$

Note: we were allowed to take away the absolute value signs around $x^2 + 2x + 5$ because this term has no roots and is always positive.

33.6 Exercise. Find the value of the definite integral $\displaystyle\int_{-1}^{1} \frac{2x+1}{x^2+4x+8}\,dx$.

The Physics of Sky Diving (Part 2)

With the techniques developed in the previous section, we are now able to solve the integral in equation (33.5). The quadratic term $v_{max}^2 - u^2 = -(u^2 - v_{max}) = -(u - v_{max})(u + v_{max})$ has the two distinct roots v_{max} and $-v_{max}$. In order to obtain the corresponding partial fractions decomposition, we need to determine constants A and B such that

$$\frac{1}{u^2 - v_{max}^2} = \frac{A}{u - v_{max}} + \frac{B}{u + v_{max}}$$

or equivalently,

$$1 = A(u + v_{max}) + B(u - v_{max}).$$

Setting first $u = v_{max}$ and then $u = -v_{max}$, the constants A and B are easily seen to have the following values:

$$A = \frac{1}{2v_{max}} \quad \text{and} \quad B = -\frac{1}{2v_{max}}.$$

Hence

$$\begin{aligned}\int_0^{v(t)} \frac{1}{v_{max}^2 - u^2}\,du &= -\int_0^{v(t)} \frac{1}{u^2 - v_{max}^2}\,du = -\left(\int_0^{v(t)} \frac{1/(2v_{max})}{u - v_{max}}\,du - \int_0^{v(t)} \frac{1/(2v_{max})}{u + v_{max}}\,du\right) \\ &= -\frac{1}{2v_{max}}\left(\ln|u - v_{max}| - \ln|u + v_{max}|\right)\Big|_0^{v(t)} = -\frac{1}{2v_{max}}\ln\left|\frac{v_{max} - u}{v_{max} + u}\right|\Bigg|_0^{v(t)} \\ &= -\frac{1}{2v_{max}}\ln\left(\frac{v_{max} - v(t)}{v_{max} + v(t)}\right).\end{aligned}$$

To eliminate the absolute value signs in the last step is permissible because, as the upper limit of a sky diver's velocity, v_{max} must satisfy the inequality $v_{max} \pm u = v_{max} \pm v(t) \geq 0$. Combining the result of our integration above with equation (33.5) yields

$$-\frac{1}{2v_{max}}\ln\left(\frac{v_{max} - v(t)}{v_{max} + v(t)}\right) = \frac{g}{v_{max}^2}\,t.$$

33.7 Exercise. Show that this equation is equivalent to the equation

$$\frac{v_{max} - v(t)}{v_{max} + v(t)} = e^{-2gt/v_{max}}.$$

In solving the equation in Exercise 33.7 for $v(t)$, we arrive at the following conclusion:

$$\boxed{v(t) = v_{max}\frac{1 - e^{-2gt/v_{max}}}{1 + e^{-2gt/v_{max}}} = v_{max}\frac{\sinh(gt/v_{max})}{\cosh(gt/v_{max})}.} \tag{33.10}$$

33.8 Exercise. Verify this equation.

33.9 Example. We wish to find the downward velocity that a sky diver reaches 10 seconds after jumping off an airplane under the assumption that the terminal velocity is $v_{max} = 120\, mi/h \approx 54\, m/s$. Substituting the values $t = 10\, s$, $v_{max} = 54\, m/s$, and $g = 9.8 m/s^2$ in equation (33.10) yields

$$v(10) \approx 51.21\, m/s.$$

This shows that after only ten seconds, the sky diver is already falling at a rate of 94.8% of the terminal velocity. We further notice that

$$v(t) < \lim_{t\to\infty} v(t) = v_{max}\frac{1 - \lim_{t\to\infty} e^{-2gt/v_{max}}}{1 + \lim_{t\to\infty} e^{-2gt/v_{max}}} = v_{max}.$$

Thus, at least in theory, the sky diver never quite reaches the terminal velocity but only approaches it arbitrarily closely as t tends to infinity.

Additional Exercises

33.10. Determine the following definite and indefinite integrals:

a) $\displaystyle\int_0^1 \frac{x-1}{x^2+x-20}\,dx$ **b)** $\displaystyle\int \frac{x}{x^2-6x+10}\,dx$ **c)** $\displaystyle\int_0^2 \frac{x-1}{2x^2+4x+2}\,dx$

d) $\displaystyle\int \frac{2x+1}{4x^2-8x+20}\,dx$ **e)** $\displaystyle\int \frac{x^2+3x+1}{x^2-2x+2}\,dx$ **f)** $\displaystyle\int \frac{x}{2x^2-12x+18}\,dx$

g) $\displaystyle\int \frac{x^2}{x^2-2x-3}\,dx$ **h)** $\displaystyle\int \frac{2x+1}{x^2-8x+16}\,dx$ **i)** $\displaystyle\int \frac{x^2+1}{x^2-8x+16}\,dx$

33.11. Determine the following antiderivatives:

a) $\displaystyle\int \frac{1}{\sqrt{x}(x-\sqrt{x})}\,dx$ **b)** $\displaystyle\int \frac{1}{\sqrt{x}^3-2x+2\sqrt{x}}\,dx$ **c)** $\displaystyle\int_0^2 \frac{\sqrt{x}}{x+2\sqrt{x}+1}\,dx$

33.12. Suppose a moving object is subject to a force $F(t) = mg - cv(t)$. Use Newton's fundamental law of mechanics and integration by substitution to determine $v(t)$ under the additional assumption that $v(t_0) = v_0$.

33.13. ★ The purpose of the sequence of problems stated below is to make plausible the physical law of the proportionality of the force of air resistance to the square of the velocity of a falling body. (A note of caution: The "derivation" below makes no claims to rigor. For given the turbulent nature of air flow around a falling body, it is very difficult to develop a truly satisfactory theoretical model of the force of air resistance.)

a) Assume that in absence of any external forces two objects with masses m and M are moving along a straight line with respective velocities u_0 and v_0 until they meet in a perfectly elastic collision (i.e., no kinetic energy is converted into thermal energy). Denoting the velocities after the collision by u and v, the law of the preservation of momentum implies that

$$mu_0 + Mv_0 = mu + Mv.$$

Furthermore, according to the law of the preservation of energy, which in this case involves only kinetic energies, we have

$$\frac{mu_0^2}{2} + \frac{Mv_0^2}{2} = \frac{mu^2}{2} + \frac{Mv^2}{2}.$$

Use the equations above to find formulae that respectively express v and u only in terms of m, M, u_0, and v_0.

b) Assume that a body of mass M and velocity v_0 collides with a single air molecule of mass m that was at rest prior to the collision (i.e., $u_0 = 0$). Use a) to determine the change in momentum of the body of mass M due to the collision with the air molecule. (Note: The assumption that the air molecule was at rest prior to the collision is in a sense not very realistic, for even if there is no wind at all, there will always be movements of air molecules at a microscopic level. However, in the statistical average, the effects of these movements in still air will cancel each other out because the velocities of the individual molecules can be assumed to be uniformly distributed with regard to the possible directions of motion in three-dimensional space.)

c) Explain why the number of collisions with air molecules during a small time interval $[t, t + \Delta t]$ is (approximately) proportional to $v(t)\Delta t$. You may assume the density of air to be constant.

d) Use the results of b) and c) to explain why the force of air resistance is proportional to $-v(t)^2$. *Hint.* The momentum $p(t) = Mv(t)$ of the falling body satisfies the approximate equation $F(t) \approx (p(t + \Delta t) - p(t))/\Delta t$.

33.14. The problems listed below are intended as a review of the ordinary law of falling bodies that does *not* take into account any restraining frictional forces (see in particular our discussion in Chapter 14).

a) Find the velocity $v(t)$ of a falling body of mass m under the influence of gravity but in absence of air resistance. You may assume that the initial velocity is $v(0) = v_0$. (In your answer, the velocity will be a function of t and it will also be dependent upon v_0.)

b) Use your result in a) to find the position $x(t)$ of a falling body (in absence of air resistance) under the assumption that $x(0) = x_0$ (so x_0 is the height above the ground from which the body begins its descent).

c) What is the terminal velocity of a falling body in absence of air resistance? (This is a bit of a trick question.)

d) What is the maximal height that an object will reach when it is shot vertically into the air from the ground with an initial speed of $10\,m/s$.

e) If an object is dropped from a height of 10 m, how long does it take for it to reach the ground?

33.15. Determine the velocity $v(t)$ of a given object from the assumption that...

a) $v'(t) = v(t)^2 + 4v(t) + 8$ and $v(0) = 2\,m/s$. *Hint.* Divide both sides of the given equation by $v(t)^2 + 4v(t) + 8$.

b) $v'(t) = t(v(t)^2 + 4v(t) + 3)$ and $v(0) = 1\,m/s$.

Chapter 34

Tossing Coins (Part 1)

Counting Heads and Tails

Teacher: Suppose that in tossing a coin five times you get four heads and only one tail. An observer witnessing the scene draws the conclusion that coin-tossing experiments are in general more likely to produce heads than tails. How would you respond to such a claim contradicting our common sense assessment that heads and tails occur with equal probability?

Simplicio: I would just try a little longer. After a sufficiently large number of trials we are sure to get approximately equal numbers of heads and tails.

Teacher: Would you feel confident to enter a wager that after, say, 100 tosses you will have 50 heads and 50 tails?

Simplicio: I said "*approximately* equal numbers of heads and tails," not *exactly* equal numbers. After 100 tosses we might just as well have 48 heads and 52 tails instead of exactly 50 heads and 50 tails.

Sophie: I agree with Simplicio. The ratio of the number of heads over the total number of trials will be close to 1/2, but not necessarily equal to 1/2.

Teacher: And what would you expect to happen to this ratio as the number of trials increases further from 100 to, say, 1,000?

Sophie: The ratio would most likely be even closer to 1/2. But only in the *limit*, as the number of trials tends to ∞, would I expect the value to be exactly equal to 1/2.

Teacher: This is a good way of looking at the problem, because the notion of a limit is indeed very helpful in formalizing our intuitive expectation that heads and tails are equally likely to occur. To see this more clearly, we respectively denote by $N_n(H)$ and $N_n(T)$ the numbers of Heads and Tails after n coin tosses. Then, our long term expectation for the outcomes of a coin-tossing experiment is expressed by the equation

$$\lim_{n\to\infty} \frac{N_n(H)}{n} = \lim_{n\to\infty} \frac{N_n(T)}{n} = \frac{1}{2}. \tag{34.1}$$

That is, as the number of trials increases to ∞, the ratio of the number of heads (or tails) over the total number of trials approaches 1/2.

Sophie: I am a bit confused regarding the nature of this statement. Does it represent a conjecture, a mathematical theorem, or an empirical fact?

Teacher: That is an excellent question. Of the three alternatives that you mentioned, I think we can safely eliminate the last one because to establish the existence of the limits in (34.1) empirically would require an infinite number of trials. To consider (34.1) as a conjecture or mathematical theorem is possible if we can create a mathematical model within which the validity of (34.1) admits of verification.

Sophie: I don't quite understand how we can ever hope to come to a definite conclusion regarding the truth content of a statement that, by its very nature, involves the uncertainties of a random process.

Teacher: We need to be careful here to distinguish between the problem of establishing the validity of a mathematical proposition within a specific mathematical model and the far more general question of linking the model with the reality that is being modeled. Within the mathematical model that we are going to discuss in this chapter and the next, it will indeed be possible to prove that the convergence of the ratios $N_n(H)/n$ and $N_n(T)/n$ to $1/2$ occurs with probability 1 (i.e., is certain), but to endow this conclusion with some confidence regarding its relevance to observations made in actual coin-tossing experiments is essentially a matter of faith. Not blind faith, to be sure, because experiments involving large numbers of trials do certainly tend to support the claim in (34.1), but faith all the same.

Permutations and Probabilities

Teacher: Given the preceding discussion, it appears we are in agreement that in coin-tossing experiments we are more likely to get approximately equal numbers of heads and tails than, for example, only heads.

Simplicio: That sounds reasonable.

Teacher: Does it follow then that after, say, five trials the outcome $HHTTH$ would be more likely to occur than the outcome $HHHHH$.

Simplicio: No doubt about it.

Teacher: How do you justify your answer?

Simplicio: We just said that getting equal numbers of heads and tails is more probable than getting only heads. Just look at the sequence $HHTTH$: here the first two trials resulted in an H, and it therefore is certainly more likely for the third trial to yield a T rather than again an H as in the outcome $HHHHH$.

Teacher: So you are basically saying that after two equal outcomes the coin is getting bored and decides for a change, to show us a different side. I may believe in free will, but not necessarily in regard to coins.

Simplicio: That's because you are suffering from a serious lack of imagination. I recommend that you let some fresh air into your brain and stop talking big about calculus for a while.

Brief intermission. In an effort to reestablish his authority, we see the teacher involved in a fistfight, frantically pulling Simplicio's hair and fending off blows to his own lower jaw. With his honor restored but badly bruised, the teacher then returns to his notes and continues his deliberations...

Teacher: Reviewing the arguments that have just been exchanged, we are forced to draw the following conclusion: the outcomes $HHTTH$ and $HHHHH$ are equally likely.

Sophie: How do we reconcile this assertion with the statement that approximately equal numbers of heads and tails are more probable than heads only?

Teacher: The crucial question is whether we take the order of the individual outcomes into account or not. The table in (34.2) shows that an outcome with three letters H and two letters T (such as $HHTTH$) can occur in exactly 10 different arrangements. By contrast, there is only one arrangement with five letters H—the outcome $HHHHH$. Therefore, getting three heads and two tails (disregarding the order) is ten times more likely than getting only heads, whereas the individual outcome $HHTTH$ (in this particular order) occurs with the same probability as $HHHHH$.

H	H	H	H	H	H	T	T	T	T
H	H	H	T	T	T	H	H	H	T
H	T	T	H	H	T	H	H	T	H
T	H	T	H	T	H	H	T	H	H
T	T	H	T	H	H	T	H	H	H

(34.2)

Simplicio: That makes sense, but aren't there any easier ways to compare probabilities? I hope we will not always have to list all possible arrangements as in (34.2).

Teacher: In order to derive convenient formulae for calculating probabilities, it is helpful to determine first the total number of outcomes after n coin tosses. For $n = 1$ there are only two outcomes: H and T. For $n = 2$ there are already 4 possibilities, because we can have first an H and then again an H, or first an H and then a T, or we can have first a T and then an H or a T (see the table below).

H	H	T	T
H	T	H	T

For three coin tosses ($n = 3$), each of the possible outcomes after two tosses can again be followed by an H or a T so that there is a total of eight different outcomes:

H	H	H	H	T	T	T	T
H	H	T	T	H	H	T	T
H	T	H	T	H	T	H	T

(34.3)

It is clear how the pattern continues: each of the eight possible outcomes after three tosses can be followed by an H or a T, and we thus have 16 possible outcomes after four coin tosses. Based on these observations, we may formulate the following general rule:

> After tossing a coin n times there are 2^n possible outcomes, each of which occurs with probability $1/2^n$.

Given this result, what do you think is the *probability* of getting three heads and two tails after five coin tosses?

Sophie: Since the total number of outcomes after five trials is $2^5 = 32$, and since the number of arrangements of three letters H and two letters T is 10 (see the table in (34.2)), I would say that the probability is $10/32 = 5/16$.

Simplicio: So the chance of getting three heads and two tails in five trials is $5/16 \cdot 100\% = 31.25\%$.

Teacher: Correct, but probabilities are always expressed as numbers between 0 and 1. Therefore, 5/16 is the better answer.

34.1 Exercise. Use the table in (34.3) to determine the probability of getting two heads and one tail after three coin tosses.

In order to generalize the results of the preceding discussion, we will now turn our attention to the problem of finding a formula for the probability of getting k heads and $n - k$ tails after n trials. In analogy to the example of three heads and two tails in (34.2), we first need to examine in how many different ways we can arrange k letters H and $n - k$ letters T in a row of length n. To approach this problem, it is helpful to address the following simpler question first: in how many different ways can we arrange n *distinct* objects in a row of length n? Assume, for example, that we are given three squares labeled 1, 2, and 3. Then, for the first position we can choose either one of the three squares. Having filled the first position, we are left with two choices for the second position, and having filled the first two positions, there is only one choice for the last position. Therefore, the total number of arrangements or *permutations* of three distinct objects is $3 \cdot 2 \cdot 1 = 6$.

1	1	2	2	3	3
2	3	1	3	1	2
3	2	3	1	2	1

In general, for an arbitrary positive integer n, the number of permutations is $n \cdot (n-1) \cdots 3 \cdot 2 \cdot 1$. The product $n \cdot (n-1) \cdots 3 \cdot 2 \cdot 1$ is referred to as n *factorial* and is denoted by $n!$. Thus, we have established the following rule:

> The number of permutations of n distinct objects is $n!$.

Note: for the special case $n = 0$ it is customary to define $0! := 1$.

34.2 Exercise. Find the number of permutations of 6 distinct objects.

In order to apply the rule above to the problem of finding the number of arrangements of k letters H and $n-k$ letters T, we resort to a little trick: considering again the example of three letters H and two letters T with $n=5$, we write the letters H and T on five little squares that are labeled with the numbers from 1 to 5.

$$\boxed{H_1}\boxed{H_2}\boxed{H_3}\boxed{T_4}\boxed{T_5}$$

Then there are $5! = 120$ different permutations in the indices from 1 to 5, but looking only at the letters H and T, some of these permutations are identical. For instance, the permutations

$$\boxed{H_1}\boxed{T_4}\boxed{H_2}\boxed{H_3}\boxed{T_5} \tag{34.4}$$

and

$$\boxed{H_3}\boxed{T_5}\boxed{H_1}\boxed{H_2}\boxed{T_4} \tag{34.5}$$

display the same sequence in the letters H and T, but different permutations in the numbers from 1 to 5. Since there are $3!$ ways to arrange three distinct objects (the squares with an H) and $2!$ ways to arrange two distinct objects (the squares with a T), we may infer that there are $3! \cdot 2!$ permutations of the numbers from 1 to 5 that display the sequence $HTHHT$ as in (34.4) and (34.5) above. Given this observation, it is easy to see that the total number of permutations in the letters H and T is

$$\frac{5!}{3! \cdot 2!} = 10 \quad \text{(see also (34.2))}.$$

In general, the number of permutations of k letters H and $n-k$ letters T is $n!/(k!(n-k)!)$. This quotient is referred to as a *binomial coefficient* (see the ⋆remark⋆ below) and is denoted by $\binom{n}{k}$. In summary:

$$\boxed{\text{The number of permutations of two objects in a row of length } n \text{ with } k \text{ times the first and } n-k \text{ times the second object is } \binom{n}{k} = \frac{n!}{k!(n-k)!}.} \tag{34.6}$$

Remark. Since there is only one permutation of n identical letters (H or T) in a row of length n, it follows that the statement in (34.6) applies also to the special cases $k=0$ and $k=n$. For in setting $0! = 1$ (see the definition above), we obtain $\binom{n}{n} = \binom{n}{0} = 1$.

34.3 Exercise. Find the number of permutations of four letters T and six letters H in a row of length ten.

Let us denote by $P_n(k)$ the probability of receiving k heads and $n-k$ tails after n coin tosses. Since there is a total of 2^n outcomes (each with probability $1/2^n$), the statement in (34.6) implies that

$$\boxed{P_n(k) = \frac{1}{2^n}\binom{n}{k}.} \tag{34.7}$$

34.4 Example. To determine the probability of receiving exactly four heads after tossing a coin nine times we substitute $n=9$ and $k=4$ in (34.7). This yields

$$P_9(4) = \frac{1}{2^9}\binom{9}{4} = \frac{1}{2^9} \cdot \frac{9!}{4!5!} = \frac{126}{512} \approx 0.246.$$

34.5 Exercise. Find the probability of receiving no more than three heads after nine coin tosses.

34.6 Exercise. Use formula (34.7) to show that $P_n(k) = P_n(n-k)$ and give an intuitive explanation for this equation.

Since the sum of the probabilities $P_n(k)$ represents the total probability of all possible outcomes, it follows that this sum is actually equal to 1:

34.7 Theorem. *For any integers n and k with $0 \le k \le n$ we have*

$$1 = \frac{1}{2^n} \sum_{k=0}^{n} \binom{n}{k}.$$

★ *Remark.* Theorem 34.7 is a special case of a more general result known as the *binomial formula*: for any nonnegative integer n and all $a, b \in \mathbb{R}$ we have

$$(a+b)^n = \sum_{k=0}^{n} \binom{n}{k} a^k b^{n-k}.$$

A proof of this formula is given in Appendix E (and also in Exercise 42.31). The statement of Theorem 34.7 follows from the binomial formula by setting $a = b = 1/2$ (think about it!). To reveal an interesting pattern in the set of binomial coefficients, we explicitly write out the binomial formula for a few small values of the exponent n:

$$\begin{aligned}
(a+b)^0 &= 1 \\
(a+b)^1 &= 1a + 1b \\
(a+b)^2 &= 1a^2 + 2ab + 1b^2 \\
(a+b)^3 &= 1a^3 + 3a^2b + 3ab^2 + 1b^3 \\
(a+b)^4 &= 1a^4 + 4a^3b + 6a^2b^2 + 4ab^3 + 1b^4.
\end{aligned}$$

Extracting the binomial coefficients from the terms on the right-hand side, we obtain a triangular array known as the *Pascal Triangle:*

$$\begin{array}{ccccccccc}
 & & & & 1 & & & & \\
 & & & 1 & & 1 & & & \\
 & & 1 & & 2 & & 1 & & \\
 & 1 & & 3 & & 3 & & 1 & \\
1 & & 4 & & 6 & & 4 & & 1 \\
\vdots & & & & \vdots & & & & \vdots
\end{array}$$

The binomial coefficients in each row are built by adding the diagonally adjacent coefficients in the preceding row (see also Exercise 34.12): $2 = 1+1$, $3 = 1+2$, $4 = 1+3$, $6 = 3+3$, etc. ★

The Long Term Expectation for a Coin-Tossing Experiment

To continue our discussion of random experiments we assume that a coin is tossed n times. According to equation (34.1), we expect that for large values of n the ratio of the number of heads $N_n(H)$ over the total number of trials n is approximately equal to $1/2$. In other words, the probability that the ratio $N_n(H)/n$ is about $1/2$ should be close to 1.

34.8 Example. Let us determine the probability that after eight trials we have no more than five heads and no fewer than three heads. So, we require $N_8(H)$ to satisfy the following condition:

$$\frac{3}{8} \le \frac{N_8(H)}{8} \le \frac{5}{8}.$$

This inequality is satisfied for $N_8(H) = 3$ with probability $P_8(3)$, for $N_8(H) = 4$ with probability $P_8(4)$, and for $N_8(H) = 5$ with probability $P_8(5)$. Thus, the total probability is

$$\begin{aligned}
P_8(3) + P_8(4) + P_8(5) &= \frac{1}{2^8}\binom{8}{3} + \frac{1}{2^8}\binom{8}{4} + \frac{1}{2^8}\binom{8}{5} \\
&= \frac{56}{256} + \frac{70}{256} + \frac{56}{256} = \frac{182}{256} \approx 0.71.
\end{aligned}$$

In order to see how this probability depends on the number of coin tosses, we increase the value of n from 8 to 16 while maintaining the same bounds for the ratio of the number of heads to the total number of trials:

$$\frac{3}{8} = \frac{6}{16} \leq \frac{N_{16}(H)}{16} \leq \frac{10}{16} = \frac{5}{8}.$$

The probability for $N_{16}(H)$ to satisfy this inequality is

$$P_{16}(6) + P_{16}(7) + P_{16}(8) + P_{16}(9) + P_{16}(10) = \frac{1}{2^{16}} \sum_{k=6}^{10} \binom{16}{k} \approx 0.79.$$

So for $n = 16$ there is an increase in probability by about 0.08 from 0.71 to 0.79. In choosing successively larger values for n we would find that the corresponding probabilities are converging to 1.

34.9 Exercise. Compute the probability for the inequality $12 \leq N_{32}(H) \leq 20$ to be satisfied after 32 trials.

As a matter of course, the result of Example 34.8 does not depend on the bounds 3/8 and 5/8 that were chosen completely arbitrarily. If we had chosen bounds closer to 1/2, for example 7/16 and 9/16, our observations would have been very similar: the convergence of the probabilities would have been somewhat slower, but the limiting value 1 would have been the same. In fact, no matter how close to 1/2 we choose the bounds, the probability will always approach 1. Based on this observation we will now give a precise mathematical formulation for the long term expectation of a coin-tossing experiment. Let ε be a number greater than 0, but typically very small. Then, for sufficiently large values of n, the probability that the difference between the ratio $N_n(H)/n$ and 1/2 is less than or equal to ε will be close to 1, and in the limit, as n tends to ∞, the probability will actually be equal to 1. To say that the difference between the ratio $N_n(H)/n$ and 1/2 is less than or equal to ε simply means that

$$\frac{1}{2} - \varepsilon \leq \frac{N_n(H)}{n} \leq \frac{1}{2} + \varepsilon,$$

or equivalently,

$$\left| \frac{N_n(H)}{n} - \frac{1}{2} \right| \leq \varepsilon.$$

(Note: the bounds 3/8 and 5/8 in Example 34.8 are obtained by setting $\varepsilon = 1/8$, because $3/8 = 1/2 - 1/8$ and $5/8 = 1/2 + 1/8$.) In summarizing the preceding discussion we arrive at the following theorem, which at this point is still a conjecture:

34.10 Theorem. *For any $\varepsilon > 0$ we have*

$$\lim_{n\to\infty} \frac{1}{2^n} \sum_{\substack{k=0 \\ |k/n-1/2|\leq\varepsilon}}^{n} \binom{n}{k} = 1$$

and

$$\lim_{n\to\infty} \frac{1}{2^n} \sum_{\substack{k=0 \\ |k/n-1/2|>\varepsilon}}^{n} \binom{n}{k} = 0.$$

Note: the notation above is supposed to convey the fact that the sums are respectively taken only over those values of k that satisfy the conditions

$$\left| \frac{k}{n} - \frac{1}{2} \right| \leq \varepsilon \quad \text{and} \quad \left| \frac{k}{n} - \frac{1}{2} \right| > \varepsilon.$$

It is important to understand that the two equations in Theorem 34.10 are actually equivalent. To see this, we make use of Theorem 34.7:

$$1 = \frac{1}{2^n}\sum_{k=0}^{n}\binom{n}{k} = \frac{1}{2^n}\sum_{\substack{k=0\\|k/n-1/2|\leq\varepsilon}}^{n}\binom{n}{k} + \frac{1}{2^n}\sum_{\substack{k=0\\|k/n-1/2|>\varepsilon}}^{n}\binom{n}{k}.$$

Hence

$$1 = \lim_{n\to\infty}\frac{1}{2^n}\sum_{\substack{k=0\\|k/n-1/2|\leq\varepsilon}}^{n}\binom{n}{k} + \lim_{n\to\infty}\frac{1}{2^n}\sum_{\substack{k=0\\|k/n-1/2|>\varepsilon}}^{n}\binom{n}{k}.$$

If the first limit on the right-hand side of this equation is equal to 1, then the second has to be 0; equivalently, if the second is 0 then the first has to be 1. In other words, the first equation in Theorem 34.10 implies the second and vice versa. Given this observation, it will suffice to prove in Chapter 35 only the second equation in Theorem 34.10.

Remark. Theorem 34.10 is a very special case of a far more general result in the theory of probability—the so-called *weak law of large numbers* (see, for example, [Ham]). The proof of this law is less involved but unfortunately considerably more abstract than the proof we will give for Theorem 34.10 in the next chapter. It also requires familiarity with a number of probabilistic concepts that are definitely beyond the scope of this text.

Additional Exercises

34.11. Use a computer to find the probability that after 1,000 coin tosses, the number of heads is less than or equal to 550 and greater than or equal to 450.

34.12. Show that for all positive integers n and all $k \in \{1, \ldots, n\}$ we have $\binom{n+1}{k} = \binom{n}{k-1} + \binom{n}{k}$.

34.13. Assume that a coin is tossed 6 times. Determine the probability that at least one of the first two tosses and exactly two out of the final four tosses result in a head.

34.14. What is the probability that in rolling two dice the numbers add up to 7? What is the probability that the numbers add up to 2?

34.15. ★ Determine the probability that in a class of 23 students, at least two have the same date of birth.

Chapter 35

Integration by Parts and Tossing Coins (Part 2)

Approximating Products by Integrals

With the goal of proving Theorem 34.10 in mind, our first objective in this chapter is to develop estimates for factorials and, by implication, binomial coefficients. Since $n!$ is a *product* of integers, it is appropriate to broaden the scope of our discussion to the study of arbitrary products. The standard *product notation* is very similar to summation notation: for integers m and n with $m \leq n$ and real numbers $a_m, a_{m+1}, \ldots, a_n$, the product $a_m a_{m+1} \cdots a_n$ is denoted by

$$\prod_{k=m}^{n} a_k.$$

35.1 Example.

$$\prod_{k=2}^{5} k^2 = 2^2 \cdot 3^2 \cdot 4^2 \cdot 5^2 = 14400$$

and

$$n! = \prod_{k=1}^{n} k.$$

35.2 Exercise. Find the value of $\prod_{k=1}^{4} 2^k$.

Frequently helpful in manipulating products is the natural logarithm, because it transforms multiplication into summation: for *positive* numbers $a_m, a_{m+1}, \ldots, a_n$, the general property $\ln(ab) = \ln(a)+\ln(b)$ (see Theorem 21.1a) implies that

$$\ln\left(\prod_{k=m}^{n} a_k\right) = \ln(a_m) + \ln(a_{m+1}) + \cdots + \ln(a_n) = \sum_{k=m}^{n} \ln(a_k).$$

Since the natural exponential function is the inverse of the natural logarithm function, we obtain

$$\boxed{\prod_{k=m}^{n} a_k = e^{\sum_{k=m}^{n} \ln(a_k)}.} \tag{35.1}$$

Applying these results to $n!$ for some integer $n \geq 2$ (see Example 35.1), it follows that

$$\ln(n!) = \sum_{k=1}^{n} \ln(k) = \sum_{k=2}^{n} \ln(k) \quad \text{(because } \ln(1) = 0\text{)} \tag{35.2}$$

and

$$n! = e^{\sum_{k=2}^{n} \ln(k)}. \tag{35.3}$$

Thus, the problem of finding an estimate for $n!$ can be reduced to the problem of finding an estimate for the sum $\sum_{k=2}^{n} \ln(k)$. This is a significant advantage, because sums can be interpreted as approximations for integrals (see Chapter 17), and equivalently, *integrals can be regarded as approximations for sums.* So in order to find an estimate for $n!$, we will approximate the sum $\sum_{k=2}^{n} \ln(k)$ by a definite integral of the function $\ln(x)$ and then determine the value of this integral via the fundamental theorem of calculus.

In Chapter 17, we defined definite integrals as limits of approximating Riemann sums of the form $\sum f(x_k)\Delta x_k$. Setting $f(x) := \ln(x)$, $x_k := k$, and $\Delta x_k := x_k - x_{k-1} = k - (k-1) = 1$, it follows that

$$\sum_{k=2}^{n} f(x_k)\Delta x_k = \sum_{k=2}^{n} \ln(k).$$

Consequently, in regarding $\sum_{k=2}^{n} \ln(k)$ as a right sum approximation for $\int_1^n \ln(x)\,dx$ (see Figure 35.1), we may conclude that

$$\int_1^n \ln(x)\,dx \leq \sum_{k=2}^{n} \ln(k) \quad \text{(because } \ln(x) \text{ is increasing).} \tag{35.4}$$

In other words, the integral $\int_1^n \ln(x)\,dx$ gives us a *lower* estimate for the sum $\sum_{k=2}^{n} \ln(k)$.

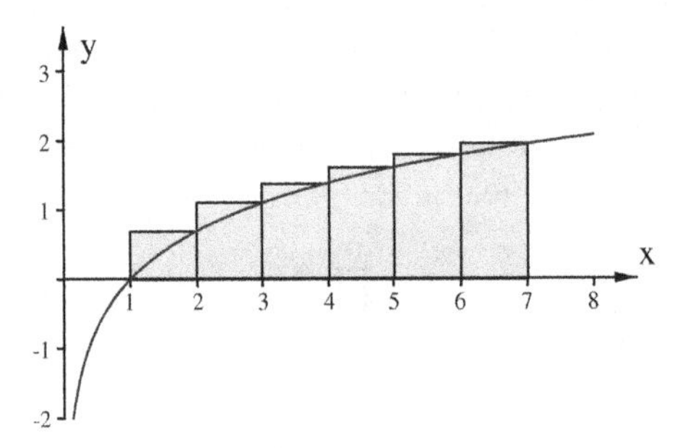

Figure 35.1: lower estimate for $\sum_{k=1}^{n} \ln(k)$ with $n = 7$.

Similarly, in considering $\sum_{k=2}^{n} \ln(k)$ to be a left sum approximation for $\int_2^{n+1} \ln(x)\,dx$ (see Figure 35.2), we find the *upper* estimate

$$\int_2^{n+1} \ln(x)\,dx \geq \sum_{k=2}^{n} \ln(k). \tag{35.5}$$

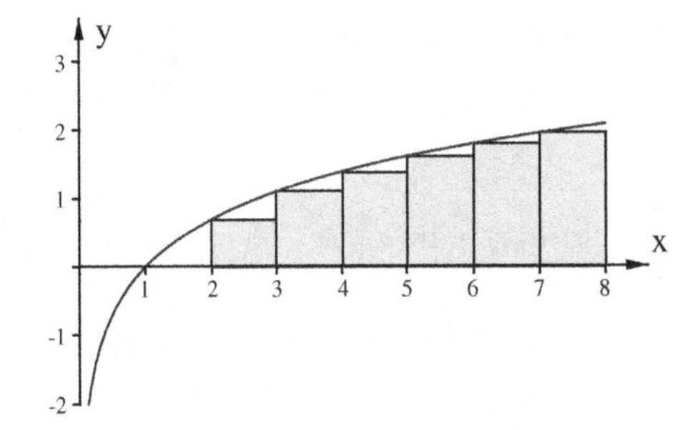

Figure 35.2: upper estimate for $\sum_{k=1}^{n} \ln(k)$ with $n = 7$.

Combining (35.2) with the estimates in (35.4) and (35.5) yields

$$\boxed{\int_1^n \ln(x)\,dx \leq \ln(n!) \leq \int_2^{n+1} \ln(x)\,dx \text{ for all integers } n \geq 1.} \tag{35.6}$$

To generalize this result we assume that for integers m and n with $m < n$, we are given a bounded piecewise continuous *increasing* function $f : [m-1, n+1] \to \mathbb{R}$. (Note: the assumptions of piecewise continuity and boundedness guarantee that the integral of f exists.) Following the same argument as in our derivation of (35.6), it is easy to see that

$$\boxed{\int_{m-1}^{n} f(x)\,dx \leq \sum_{k=m}^{n} f(k) \leq \int_{m}^{n+1} f(x)\,dx.} \tag{35.7}$$

Similarly, if f is *decreasing*, we may infer that

$$\boxed{\int_{m-1}^{n} f(x)\,dx \geq \sum_{k=m}^{n} f(k) \geq \int_{m}^{n+1} f(x)\,dx.} \tag{35.8}$$

35.3 Exercise. Draw appropriate diagrams (analogous to Figures 35.1 and 35.2) to illustrate the inequalities in (35.8).

In order to evaluate the integrals in (35.6), we will now introduce a new method of integration, known as *integration by parts.*

Integration by Parts

Assume that $f : I \to \mathbb{R}$ and $g : I \to \mathbb{R}$ are continuously differentiable functions defined on an open interval I. Then, according to the product rule (see Theorem 8.3), we have $(fg)'(x) = f'(x)g(x) + f(x)g'(x)$ for all $x \in I$, or equivalently,

$$f(x)g'(x) = (fg)'(x) - f'(x)g(x).$$

Integrating both sides of this equation and applying the FTC2 yields

$$\int_a^b f(x)g'(x)\,dx = \int_a^b (fg)'(x)\,dx - \int_a^b f'(x)g(x)\,dx = f(x)g(x)\Big|_a^b - \int_a^b f'(x)g(x)\,dx$$

for all $a, b \in I$. So in order to find the antiderivative of $f(x)g'(x)$ we first integrate only one *part* of this function, namely $g'(x)$, and are then left with the integral of $f'(x)g(x)$. Given this interpretation, it appears plausible to refer to the rule of integration, stated in the following theorem, as *integration by parts*:

35.4 Theorem. *If $f : I \to \mathbb{R}$ and $g : I \to \mathbb{R}$ are continuously differentiable functions defined on an open interval I, then for all $a, b \in I$ we have*

$$\int_a^b f(x)g'(x)\,dx = f(x)g(x)\Big|_a^b - \int_a^b f'(x)g(x)\,dx.$$

The corresponding rule for indefinite integrals is

$$\int f(x)g'(x)\,dx = f(x)g(x) - \int f'(x)g(x)\,dx.$$

Remark. In stating the rule of integration by parts, the letters f and g are frequently replaced with u and v so that

$$\int u(x)v'(x)\,dx = u(x)v(x) - \int u'(x)v(x)\,dx.$$

Also fairly common is the abbreviated form

$$\int u\,dv = uv - \int v\,du$$

that can be justified via the following symbolic calculations with infinitesimals:

$$v'(x)\,dx = \frac{dv}{dx}\,dx = dv \quad \text{and} \quad u'(x)\,dx = \frac{du}{dx}\,dx = du$$

However, the shorthand notation above is somewhat misleading, because it wrongly suggests that, for example, in the integral $\int u\,dv$ we integrate u as a function of v rather than of x.

35.5 Example. We wish to evaluate the integral

$$\int_0^1 xe^x\,dx.$$

Setting $f(x) := x$ and $g'(x) := e^x$, it follows that $f(x)g'(x) = xe^x$, $f'(x) = 1$, and $g(x) = e^x$. Therefore, Theorem 35.4 implies that

$$\int_0^1 \underbrace{xe^x}_{f(x)g'(x)}\,dx = \underbrace{xe^x}_{f(x)g(x)}\Big|_0^1 - \int_0^1 \underbrace{1\cdot e^x}_{f'(x)g(x)}\,dx = xe^x\Big|_0^1 - e^x\Big|_0^1 = 1.$$

In applications of integration by parts, it is usually crucial to make the right choices for $f(x)$ and $g'(x)$. For instance, if in the example above we had mistakenly set $f(x) := e^x$ and $g'(x) := x$ with $f'(x) = e^x$ and $g(x) = x^2/2$, then the corresponding integration

$$\int_0^1 xe^x\,dx = \frac{x^2e^x}{2}\Big|_0^1 - \int_0^1 \frac{x^2e^x}{2}\,dx$$

would have been of little use because the integral $\int_0^1 x^2e^x/2\,dx$ on the right-hand side would have been more complicated than the integral we started with.

35.6 Exercise. Find the values of $\int_0^\pi x\sin(x)\,dx$ and $\int_1^e x\ln(x)\,dx$.

35.7 Example. Let us determine the antiderivative of the function h defined by the equation

$$h(x) := x^2\sin(x).$$

Setting $f(x) := x^2$ and $g'(x) := \sin(x)$, it follows that $f'(x) = 2x$, $g(x) = -\cos(x)$ and

$$\int h(x)\,dx = \int x^2\sin(x)\,dx = -x^2\cos(x) - \int 2x(-\cos(x))\,dx = -x^2\cos(x) + \int 2x\cos(x)\,dx. \quad (35.9)$$

To determine the indefinite integral $\int 2x\cos(x)\,dx$, we again apply integration by parts. Setting this time $f(x) := 2x$ and $g'(x) := \cos(x)$ with $f'(x) = 2$ and $g(x) = \sin(x)$, we obtain

$$\int 2x\cos(x)\,dx = 2x\sin(x) - \int 2\sin(x)\,dx = 2x\sin(x) + 2\cos(x) + C.$$

Combining this result with (35.9) yields

$$\int h(x)\,dx = -x^2\cos(x) + 2x\sin(x) + 2\cos(x) + C.$$

35.8 Exercise. Find the antiderivative of the function h defined by the equation $h(x) := x^2e^x$.

35.9 Example. We wish to evaluate the definite integral

$$\int_0^{\pi/2} \sin(x)\cos(x)\,dx.$$

One way to find the antiderivative of $\sin(x)\cos(x)$ is to use the trigonometric formula $\sin(2x) = 2\sin(x)\cos(x)$ (see Appendix A), but in the current context it is more instructive to apply integration by parts. Setting $f(x) := \sin(x)$ and $g'(x) := \cos(x)$ ($f(x) := \cos(x)$ and $g'(x) := \sin(x)$ would also work), it follows that $f'(x) = \cos(x)$, $g(x) = \sin(x)$ and

$$\int_0^{\pi/2} \sin(x)\cos(x)\,dx = \sin^2(x)\Big|_0^{\pi/2} - \int_0^{\pi/2} \cos(x)\sin(x)\,dx. \tag{35.10}$$

At first sight it may seem that little has been accomplished, because the integral on the right-hand side is the same as the one we started with. A moment's reflection, though, reveals that we only need to add this integral to both sides of (35.10) and divide the resulting equation by 2. This yields

$$\int_0^{\pi/2} \sin(x)\cos(x) = \frac{\sin^2(x)}{2}\Big|_0^{\pi/2} = \frac{1}{2}.$$

35.10 Exercise. Find the value of the definite integral $\int_0^{\pi/2} \sin^2(x)\,dx$. *Hint.* After applying integration by parts once, you will have to solve an integral of $\cos^2(x)$. To do so you may want to apply the formula $\cos^2(x) = 1 - \sin^2(x)$, and then use the same trick as in Example 35.9.

35.11 Example. Let us determine the antiderivative of the function $e^{2x}\sin(3x)$. Setting $f(x) = e^{2x}$ and $g'(x) = \sin(3x)$, it follows that $f'(x) = 2e^{2x}$ and $g(x) = -\cos(3x)/3$. Hence

$$\int e^{2x}\sin(3x)\,dx = -\frac{1}{3}e^{2x}\cos(3x) + \int \frac{2}{3}e^{2x}\cos(3x)\,dx.$$

Using integration by parts a second time, it easily follows that

$$\int e^{2x}\sin(3x)\,dx = -\frac{1}{3}e^{2x}\cos(3x) + \frac{2}{9}e^{2x}\sin(3x) - \int \frac{4}{9}e^{2x}\sin(3x)\,dx.$$

Adding now the integral on the right-hand side to both sides of the equation, we may infer that

$$\left(1 + \frac{4}{9}\right)\int e^{2x}\sin(3x)\,dx = -\frac{1}{3}e^{2x}\cos(3x) + \frac{2}{9}e^{2x}\sin(3x) + C,$$

or equivalently,

$$\int e^{2x}\sin(3x)\,dx = -\frac{3}{13}e^{2x}\cos(3x) + \frac{2}{13}e^{2x}\sin(3x) + C.$$

35.12 Exercise. Evaluate the integral $\int_0^{\pi/2} e^{-x}\sin(2x)\,dx$.

An Estimate for $n!$

With reference to the estimate in (35.6), we will now address the problem of finding the antiderivative of the natural logarithm function via integration by parts. In writing $\ln(x)$ as a product of $\ln(x)$ and 1 (i.e., $\ln(x) = 1 \cdot \ln(x)$), we may set $f(x) := \ln(x)$ and $g'(x) := 1$. This yields $f'(x) = 1/x$, $g(x) = x$, and

$$\int \ln(x)\,dx = x\ln(x) - \int \frac{x}{x}\,dx = x\ln(x) - \int 1\,dx = x\ln(x) - x + C. \tag{35.11}$$

35.13 Exercise. Find the antiderivative of $\arcsin(x)$. *Hint.* First use integration by parts with $f(x) := \arcsin(x)$ and $g'(x) := 1$, and then, in the second step, use integration by substitution.

Using equation (35.11), we see that

$$\int_1^n \ln(x)\,dx = (x\ln(x) - x)\Big|_1^n = n\ln(n) - n + 1$$

and

$$\int_2^{n+1} \ln(x)\,dx = (x\ln(x) - x)\Big|_2^{n+1} = (n+1)\ln(n+1) - (n+1) - 2\ln(2) + 2.$$

Therefore, (35.6) implies that

$$n\ln(n) - n + 1 \le \ln(n!) \le (n+1)\ln(n+1) - n - \ln(4) + 1. \tag{35.12}$$

35.14 Exercise. Show that $\ln\left(\prod_{k=2}^{n} k^k\right) \le (n+1)^2(\ln(n+1) - 1/2)/2 - \ln(4) + 1$ by using (35.7) or (35.8) (whichever applies).

Given that the exponential function is increasing (see Chapter 22), we may use (35.12) to conclude that

$$e^{n\ln(n)-n+1} \le n! \le e^{(n+1)\ln(n+1)-n-\ln(4)+1}. \tag{35.13}$$

According to the definition of the general exponential function (see Chapter 23), we have

$$e^{n\ln(n)} = n^n \quad \text{and} \quad e^{(n+1)\ln(n+1)} = (n+1)^{n+1}.$$

Hence

$$e^{n\ln(n)-n+1} = e^{n\ln(n)}e^{-n+1} = \frac{n^n}{e^{n-1}} \ge \frac{n^n}{e^n},$$

$$e^{(n+1)\ln(n+1)-n-\ln(4)+1} = e^{(n+1)\ln(n+1)}e^{-n+1}e^{-\ln(4)} = \frac{(n+1)^{n+1}}{4e^{n-1}} \le \frac{(n+1)^{n+1}}{e^n}.$$

Combining these results with (35.13), we finally obtain

$$\boxed{\frac{n^n}{e^n} \le n! \le \frac{(n+1)^{n+1}}{e^n}.} \tag{35.14}$$

Notice that this estimate is also valid for $n = 0$ if we set $0^0 := 1$ (which makes sense, because L'Hôpital's rule implies that $\lim_{x\to 0^+} x^x = 1$).

35.15 Exercise. Show that $\prod_{k=2}^{n} k^k \le (1/4)\sqrt{n+1}^{(n+1)^2} e^{1-(n+1)^2/4}$ by using (35.1) and the result of Exercise 35.14.

35.16 Exercise. Explain why the fact that the exponential function is increasing is important in deriving (35.13) from (35.12).

It will be useful for our calculations in the next section to write the term on the right-hand side of (35.14) in a slightly different form:

$$\frac{(n+1)^{n+1}}{e^n} = \frac{n^n}{e^n} \cdot \frac{(n+1)^n(n+1)}{n^n} = \frac{n^n(n+1)}{e^n}\left(1 + \frac{1}{n}\right)^n.$$

With this modification we obtain

$$\frac{n^n}{e^n} \le n! \le \frac{n^n(n+1)}{e^n}\left(1 + \frac{1}{n}\right)^n. \tag{35.15}$$

Tossing Coins (Part 2)

Our goal in this section is to explain the long term behavior of a coin-tossing experiment by giving a proof of the second equation in Theorem 34.10. The first step in this process is to derive an upper estimate for binomial coefficients. For integers n and k with $0 \le k \le n$, the inequalities in (35.15) imply that

$$n! \le \frac{n^n(n+1)}{e^n}\left(1+\frac{1}{n}\right)^n,$$
$$k! \ge \frac{k^k}{e^k}, \quad \text{and}$$
$$(n-k)! \ge \frac{(n-k)^{n-k}}{e^{n-k}}.$$

The last two of these estimates are equivalent to the inequalities

$$\frac{1}{k!} \le \frac{e^k}{k^k} \quad \text{and} \quad \frac{1}{(n-k)!} \le \frac{e^{n-k}}{(n-k)^{n-k}}.$$

Consequently, we may infer that

$$\begin{aligned}\binom{n}{k} &= \frac{n!}{k!(n-k)!} \le (n+1)\left(1+\frac{1}{n}\right)^n \frac{n^n e^k e^{n-k}}{k^k(n-k)^{n-k}e^n} \\ &= (n+1)\left(1+\frac{1}{n}\right)^n \frac{n^k n^{n-k}}{k^k(n-k)^{n-k}} = (n+1)\left(1+\frac{1}{n}\right)^n \frac{1}{(k/n)^k(1-(k/n))^{n-k}} \\ &= (n+1)\left(1+\frac{1}{n}\right)^n \frac{1}{\left((k/n)^{k/n}(1-(k/n))^{1-(k/n)}\right)^n}.\end{aligned} \tag{35.16}$$

To simplify the expression on the right-hand side, we define a function $f : [0,1] \to \mathbb{R}$ via the equation

$$f(x) := \frac{1}{x^x(1-x)^{1-x}} \quad \text{(where } f(0) = f(1) = 1\text{, because } 0^0 := 1\text{)}$$

and set

$$a_n := (n+1)\left(1+\frac{1}{n}\right)^n.$$

Given these definitions, (35.16) can be rewritten in the form

$$\binom{n}{k} \le a_n f\left(\frac{k}{n}\right)^n. \tag{35.17}$$

In order to better understand the properties of f, we take a look at its graph in Figure 35.3. Here we

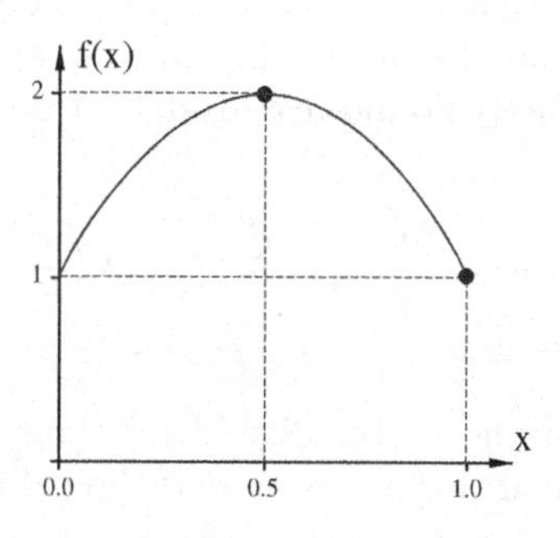

Figure 35.3: graph of $f(x) = 1/(x^x(1-x)^{1-x})$.

see that f has a global maximum at $x = 1/2$ (for a proof see the ⋆remark⋆ below) with

$$f\left(\frac{1}{2}\right) = \frac{1}{(1/2)^{1/2}(1/2)^{1/2}} = 2.$$

Furthermore, f is strictly increasing on $[0, 1/2]$ and strictly decreasing on $[1/2, 1]$.

★ *Remark.* In order to show that f has a global maximum at $x = 1/2$, we first determine the critical points of f. Since

$$\frac{d}{dx}x^x = \frac{d}{dx}e^{x\ln(x)} = e^{x\ln(x)}\frac{d}{dx}(x\ln(x)) = x^x(1+\ln(x)),$$

it follows that

$$\begin{aligned} f'(x) &= -\frac{x^x(1+\ln(x))(1-x)^{1-x} - x^x(1-x)^{1-x}(1+\ln(1-x))}{(x^x(1-x)^{1-x})^2} \\ &= \frac{-(1+\ln(x)) + (1+\ln(1-x))}{x^x(1-x)^{1-x}} \\ &= \frac{\ln(1-x)-\ln(x)}{x^x(1-x)^{1-x}}. \end{aligned}$$

Therefore, x is a critical point of f if and only if

$$0 = \ln(1-x) - \ln(x) = \ln\left(\frac{1-x}{x}\right).$$

In applying the exponential function to both sides of this equation we obtain

$$1 = \frac{1-x}{x},$$

and solving for x yields $x = 1/2$. Since $f(1/2) = 2$ (see above) and $f(0) = f(1) = 1$, it follows that f does indeed assume its global maximum at the critical point $x = 1/2$. ★

To discuss the usefulness of (35.17) in the current context of estimating probabilities of coin-tossing experiments, we will now consider as an example the probability of getting fewer than three heads or more than five heads after eight trials. Given the formula in (34.7), the value of this probability is easily seen to be

$$\frac{1}{2^8}\sum_{k=0}^{2}\binom{8}{k} + \frac{1}{2^8}\sum_{k=6}^{8}\binom{8}{k} \quad \text{(see also Example 34.8).} \tag{35.18}$$

According to (35.17), an upper estimate for this sum is

$$\frac{a_8}{2^8}\sum_{k=0}^{2} f(k/8)^8 + \frac{a_8}{2^8}\sum_{k=6}^{8} f(k/8)^8 = \frac{a_8}{2^8}\sum_{\substack{k=0 \\ |k/8-1/2|>1/8}}^{8} f(k/8)^8 \approx 20.07. \tag{35.19}$$

The alert reader will have noticed that this result is completely ridiculous. For without doing any calculation at all, we can be sure that the probability in (35.18) has to be less than or equal to 1. So an upper bound of 20.07 is of absolutely no interest. In fact, things get even worse as the total number of trials increases. To see this we define

$$p(n) := \frac{a_n}{2^n}\sum_{\substack{k=0 \\ |k/n-1/2|>1/8}}^{n} f\left(\frac{k}{n}\right)^n.$$

Then $p(n)$ is an upper estimate for the probability of receiving fewer than $3n/8$ or more than $5n/8$ heads after n coin tosses. In particular, for $n = 8$ we find the value in (35.19): $p(8) \approx 20.07$. Increasing n in steps of eight yields

$$\begin{aligned} p(16) &\approx 43.09 \\ p(24) &\approx 57.70 \\ p(32) &\approx 64.88 \\ p(40) &\approx 66.58 \\ p(48) &\approx 64.61. \end{aligned}$$

All of these values are even more ridiculous than the one in (35.19), but in the last step from $n = 40$ to $n = 48$ we see for the first time a decrease in the value of $p(n)$. To see what happens for even larger values of n, we increase the step length from 8 to 80:

$$\begin{aligned} p(80) &\approx 42.97592 \\ p(160) &\approx 7.47303 \\ p(240) &\approx 0.92559 \\ p(320) &\approx 0.10037 \\ p(400) &\approx 0.01014 \\ p(480) &\approx 0.00098. \end{aligned}$$

This new sequence of values suggests that $p(n)$ approaches zero as n tends to infinity, which, in light of the second equation in Theorem 34.10, is exactly what we are hoping for. Encouraged by this observation, we will now proceed to complete the proof of Theorem 34.10 by using the estimate in (35.17). More precisely, we will show that

$$\lim_{n\to\infty} \frac{1}{2^n} \sum_{\substack{k=0 \\ |k/n-1/2|>\varepsilon}}^{n} \binom{n}{k} = 0 \text{ for all } \varepsilon > 0.$$

So let us choose an arbitrary $\varepsilon > 0$ (typically small) and let n and k be nonnegative integers with $k \leq n$. If $k/n < (1/2) - \varepsilon$, then the fact that f is strictly increasing on the interval $[0, 1/2]$ implies that

$$f\left(\frac{k}{n}\right) < f\left(\frac{1}{2} - \varepsilon\right) = \frac{1}{(1/2-\varepsilon)^{1/2-\varepsilon}(1/2+\varepsilon)^{1/2+\varepsilon}}. \tag{35.20}$$

If on the other hand $k/n > (1/2) + \varepsilon$, then we have

$$f\left(\frac{k}{n}\right) < f\left(\frac{1}{2} + \varepsilon\right) = \frac{1}{(1/2+\varepsilon)^{1/2+\varepsilon}(1/2-\varepsilon)^{1/2-\varepsilon}}, \tag{35.21}$$

because f is strictly decreasing on the interval $[1/2, 1]$. So in both cases we arrive at the same estimate. To simplify our notation, we define

$$c := \frac{1}{(1/2+\varepsilon)^{1/2+\varepsilon}(1/2-\varepsilon)^{1/2-\varepsilon}}. \tag{35.22}$$

Since f assumes its maximal value 2 at $x = 1/2$, and since $c = f(1/2 - \varepsilon) = f(1/2 + \varepsilon)$, it follows that

$$c < 2. \tag{35.23}$$

Combining now (35.17), (35.20), and (35.21) with (35.22) yields

$$\begin{aligned} \frac{1}{2^n} \sum_{\substack{k=0 \\ |k/n-1/2|>\varepsilon}}^{n} \binom{n}{k} &\leq \frac{a_n}{2^n} \sum_{\substack{k=0 \\ |k/n-1/2|>\varepsilon}}^{n} f\left(\frac{k}{n}\right)^n \leq \frac{a_n}{2^n} \sum_{\substack{k=0 \\ |k/n-1/2|>\varepsilon}}^{n} c^n = \frac{a_n c^n}{2^n} \sum_{\substack{k=0 \\ |k/n-1/2|>\varepsilon}}^{n} 1 \\ &\leq \frac{a_n c^n}{2^n} \sum_{k=0}^{n} 1 = \frac{a_n c^n (n+1)}{2^n}. \end{aligned}$$

Consequently, it is sufficient to show that

$$\lim_{n\to\infty} \frac{a_n c^n (n+1)}{2^n} = 0. \tag{35.24}$$

Using the definition of a_n and the equation $\lim_{n\to\infty}(1+1/n)^n = e$ (see Chapter 29, p.224), we may infer that

$$\lim_{n\to\infty} \frac{a_n c^n (n+1)}{2^n} = \left(\lim_{n\to\infty}\left(1+\frac{1}{n}\right)^n\right)\left(\lim_{n\to\infty}\frac{c^n(n+1)^2}{2^n}\right) = e \lim_{n\to\infty}\frac{(n+1)^2}{(2/c)^n}.$$

Furthermore, according to (35.23) we have $2/c > 1$, and therefore Theorem 29.8 implies that

$$\lim_{n\to\infty}\frac{(n+1)^2}{(2/c)^n} = 0.$$

This shows that equation (35.24) is indeed valid, and our proof of Theorem 34.10 is thus complete. Surprisingly, the simple fact in (35.23) was the key observation that the entire argument depended on.

Additional Exercises

35.17. Solve the following definite and indefinite integrals:

$$\textbf{a)} \int_0^1 \arctan(x)\,dx \quad \textbf{b)} \int x\sinh(x)\,dx \quad \textbf{c)} \int x^3\ln(x)\,dx$$

$$\textbf{d)} \int_0^2 x^3 e^{x^2}\,dx \quad \textbf{e)} \int \ln^2(x)\,dx \quad \textbf{f)} \int x\sqrt{x+1}\,dx$$

35.18. Show that for all $n > 1$ we have

$$\int \frac{1}{(1+x^2)^n}\,dx = \frac{1}{2(n-1)}\left(\frac{x}{(1+x^2)^{n-1}} + (2n-3)\int \frac{1}{(1+x^2)^{n-1}}\,dx\right)$$

and use this observation to determine $\int 1/\cosh^3(x)\,dx$. *Hint.* Apply integration by parts to the integral $\int 1\cdot 1/(1+x^2)^{n-1}\,dx$.

35.19. Use left and right sums to prove the following inequalities:

a) $\dfrac{\sqrt{n+1}^{\ln(n+1)}}{2^{\ln(4)}} \le \prod_{k=4}^{n} k^{1/k} \le \dfrac{\sqrt{n}^{\ln(n)}}{\sqrt{3}^{\ln(3)}}$ for all $n \ge 4$

b) $\dfrac{n^{n\ln(n)}e^{2n}}{e^2 n^{2n}} \le \prod_{k=1}^{n} k^{\ln(k)}$

c) $\prod_{k=1}^{n}(k^2+1) \le \dfrac{(n^2+2n+2)^{n+1}e^{2\arctan(n+1)}}{2e^{2n+\pi/2}} \le \dfrac{5(n^2+2n+2)^{n+1}}{2e^{2n}}$

d) $\ln(n+1) \le \sum_{k=1}^{n}\dfrac{1}{k} \le 1+\ln(n)$

e) $\dfrac{2n}{\pi} \le \sum_{k=1}^{n}\sin\left(\dfrac{\pi k}{2n}\right)$

35.20. Assume that $f: I \to \mathbb{R}$ is an invertible, continuously differentiable function defined on an open interval I such that $f'(x) \ne 0$ for all $x \in I$. Use integration by parts and integration by substitution to show that

$$\int_a^b f^{-1}(x)\,dx = xf^{-1}(x)\Big|_a^b - \int_{f^{-1}(a)}^{f^{-1}(b)} f(t)\,dt \text{ for all } a,b \in I.$$

35.21. Assume that $f : [a, b] \to \mathbb{R}$ is invertible and continuously differentiable such that $f'(x) \neq 0$ for all $x \in [a, b]$. Use the result of Exercise 35.20 to prove the following statements:

a) If $f(a) = a$ and $f(b) = b$, then $\int_a^b f(x)\,dx + \int_a^b f^{-1}(x)\,dx = b^2 - a^2$.

b) If $f(a) = b$ and $f(b) = a$, then $\int_a^b f(x)\,dx = \int_a^b f^{-1}(x)\,dx$.

35.22. Use the result of Exercise 35.20 to find the values of the following definite integrals:

$$\textbf{a)} \int_0^{1/2} \arcsin(x)\,dx \quad \textbf{b)} \int_1^3 \ln(x)\,dx \quad \textbf{c)} \int_0^{\pi/4} \tan(x)\,dx.$$

35.23. The antiderivative of $e^{f(x)} \cos(g(x)) g'(x)$ is...

a) $e^{f(x)} \cos(g(x)) - \int e^{f(x)} f'(x) \sin(g(x))\,dx$.

b) $e^{f(x)} \sin(g(x)) - \int e^{f(x)} \sin(g(x))\,dx$.

c) $e^{f(x)} \sin(g(x)) - \int e^{f(x)} f'(x) \sin(g(x))\,dx$.

Chapter 36

★ Simple Random Experiments

The Definition of a Simple Random Experiment

To generalize our discussion of coin-tossing experiments we now turn our attention to the study of random experiments with two outcomes that are not necessarily equally likely.

36.1 Example. Suppose a die is rolled repeatedly, and each time one of the numbers 1, 2, 3, or 4 appears we write down the letter A. Otherwise, for a 5 or a 6, we write down the letter B. Given these rules, we expect that on average 4 out of 6 trials will result in an A and only 2 out of 6 in a B. Thus, the probability for getting an A is $4/6 = 2/3$, and for a B it is $2/6 = 1/3$. Considering now two successive trials we observe that, on average, two thirds of those trials that resulted in an A after the first roll will result in an A again after the second roll. Therefore, the probability of the outcome AA is $2/3 \cdot 2/3 = 4/9$. Similarly, for the outcome AAB after three trials the corresponding probability is $2/3 \cdot 2/3 \cdot 1/3 = (2/3)^2 1/3$, and for the outcome $AABAB$ after five trials it is $2/3 \cdot 2/3 \cdot 1/3 \cdot 2/3 \cdot 1/3 = (2/3)^3(1/3)^2$.

To generalize the results of Example 36.1 we consider a *simple random experiment* with two possible outcomes A and B that occur with the respective probabilities p and $1-p$ for some $p \in [0,1]$. Then, with $p = 2/3$ and $1-p = 1/3$, the probability for the outcome $AABAB$ in Example 36.1 is $p^3(1-p)^2$. In general, the probability for an outcome that in some order displays k times an A and $n-k$ times a B after n trials is

$$p^k(1-p)^{n-k}. \tag{36.1}$$

(Note: for a coin-tossing experiment we have $p = 1-p = 1/2$ so that all outcomes after n trials occur with *equal* probability $(1/2)^k(1/2)^{n-k} = 1/2^n$.) In order to find the *total* probability of all outcomes that display k times an A and $n-k$ times a B after n trials we only need to multiply the probability in (36.1) by the number of permutations of k letters A and $n-k$ letters B in a row of length n. Since in Chapter 34 we determined this number to be $\binom{n}{k}$, we arrive at the following conclusion:

$$\boxed{\begin{array}{l}\text{The probability for receiving } k \text{ times an } A \text{ and}\\ n-k \text{ times a } B \text{ after } n \text{ trials is } \dbinom{n}{k}p^k(1-p)^{n-k}\end{array}} \tag{36.2}$$

36.2 Exercise. Assume that a simple random experiment is performed with probability $1/4$ for A. What is the probability of receiving twice an A and three times a B after five trials?

36.3 Exercise. Under the same assumptions as in Exercise 36.2, determine the probability of receiving at least two times but no more than four times an A after six trials.

Since the total probability of all possible outcomes is always equal to 1, we may apply (36.2) to obtain the following generalization of Theorem 34.7:

36.4 Theorem. *For any nonnegative integers n and k with $k \leq n$ and any $p \in [0,1]$ we have*

$$1 = \sum_{k=0}^{n} \binom{n}{k} p^k (1-p)^{n-k}.$$

Note: the statement of this theorem also follows from the binomial formula with $a = p$ and $b = 1-p$ (see the ⋆remark⋆ on p.262).

The Long Term Expectation for a Simple Random Experiment

For a simple random experiment with two outcomes A and B that occur with the respective probabilities $p = 2/3$ and $1 - p = 1/3$ (see Example 36.1) we naturally expect the ratio of the number of As over the total number of trials n to approach $2/3$ as n tends to ∞. By analogy, for arbitrary probabilities p and $1 - p$, we expect this ratio to converge to p. Consequently, the long term expectation for a simple random experiment is properly expressed by the following theorem (c.f. Theorem 34.10):

36.5 Theorem. *For any $p \in [0,1]$ and all $\varepsilon > 0$ we have*

$$\lim_{n \to \infty} \sum_{\substack{k=0 \\ |k/n - p| \leq \varepsilon}}^{n} \binom{n}{k} p^k (1-p)^{n-k} = 1$$

and

$$\lim_{n \to \infty} \sum_{\substack{k=0 \\ |k/n - p| > \varepsilon}}^{n} \binom{n}{k} p^k (1-p)^{n-k} = 0.$$

Proof. According to Theorem 36.4, it is sufficient to prove only the second equation in Theorem 36.5. To do so we choose an arbitrary $\varepsilon > 0$ and use (35.17) (with the same definitions for f and a_n as on p.271) to infer that

$$\sum_{\substack{k=0 \\ |k/n - p| > \varepsilon}}^{n} \binom{n}{k} p^k (1-p)^{n-k} \leq a_n \sum_{\substack{k=0 \\ |k/n - p| > \varepsilon}}^{n} f\left(\frac{k}{n}\right)^n p^k (1-p)^{n-k}.$$

Introducing the function

$$g_p(x) := \begin{cases} f(x) p^x (1-p)^{1-x} & \text{if } x \in [0,1], \\ 0 & \text{if } x \in \mathbb{R} \setminus [0,1], \end{cases}$$

we may write the inequality above in the form

$$\sum_{\substack{k=0 \\ |k/n - p| > \varepsilon}}^{n} \binom{n}{k} p^k (1-p)^{n-k} \leq a_n \sum_{\substack{k=0 \\ |k/n - p| > \varepsilon}}^{n} g_p\left(\frac{k}{n}\right)^n.$$

In setting the first derivative of g_p equal to zero, it is easy to show that g_p assumes its global maximum at $x = p$ and that g_p is strictly increasing on $[0,p]$ and strictly decreasing on $[p,1]$ (see Figure 36.1 and Exercise 36.6). Denoting by d the larger of the two values $g_p(p - \varepsilon)$ and $g_p(p + \varepsilon)$ (i.e., $d := \max\{g_p(p-\varepsilon), g_p(p+\varepsilon)\}$) and observing that $g_p(p) = 1$, it therefore follows that

$$d < 1.$$

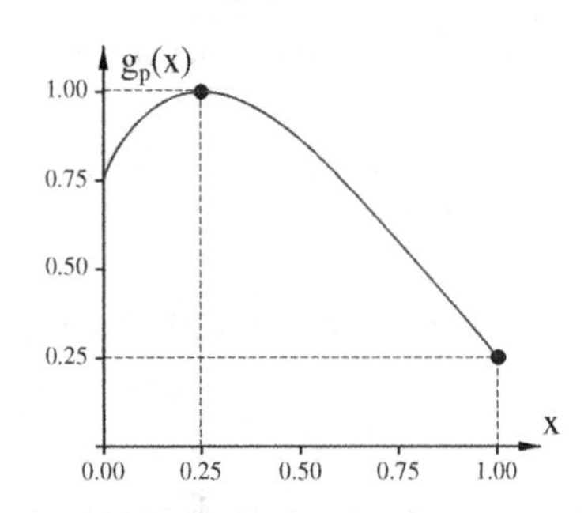

Figure 36.1: graph of $g_p(x)$ for $p = 1/4$.

(Note: this inequality is valid also if $p - \varepsilon \notin [0,1]$ and/or $p + \varepsilon \notin [0,1]$, because $g_p(x) = 0$ for all $x \in \mathbb{R} \setminus [0,1]$.) Since g_p is strictly increasing on $[0,p]$ and strictly decreasing on $[1,p]$ we may infer that $g_p(x) < d$ for all $x \in [0,1]$ with $|x - p| > \varepsilon$. Hence

$$\sum_{\substack{k=0 \\ |k/n-p|>\varepsilon}}^{n} \binom{n}{k} p^k (1-p)^{n-k} \le a_n \sum_{\substack{k=0 \\ |k/n-p|>\varepsilon}}^{n} d^n \le a_n (n+1) d^n.$$

Given that d is strictly less than 1, it is easy to see that $\lim_{n\to\infty} a_n (n+1) d^n = 0$ (see Theorem 29.8). Consequently, we obtain

$$\lim_{n\to\infty} \sum_{\substack{k=0 \\ |k/n-p|>\varepsilon}}^{n} \binom{n}{k} p^k (1-p)^{n-k} = 0,$$

and this completes the proof of Theorem 36.5. □

36.6 Exercise. Show that g_p assumes its global maximum at $x = p$ and that g_p is strictly increasing on $[0,p]$ and strictly decreasing on $[p,1]$.

Chapter 37

Trapezoid Estimates and Stirling's Formula

Good and Bad Estimates

Teacher: In Chapter 35, we found the following estimate for $n!$:

$$\frac{n^n}{e^n} \leq n! \leq \frac{(n+1)^{n+1}}{e^n}. \tag{37.1}$$

For clarification and to learn more about estimation techniques in general, we are now going to show that this estimate is really not very good and then discuss ways to improve it.

Simplicio: What does it mean to say an estimate is "not very good"?

Teacher: There is no definite answer to this question, but in many cases the quality of an estimate can be determined by examining the ratio of the lower and upper bounds: if this ratio is close to 1, the estimate is fairly good, and if it is close to 0, the estimate is poor. According to this criterion, the estimate in (37.1) would have to be classified as poor because

$$\begin{aligned}\lim_{n\to\infty} \frac{\text{lower bound}}{\text{upper bound}} &= \lim_{n\to\infty} \frac{n^n/e^n}{(n+1)^{n+1}/e^n} = \lim_{n\to\infty} \frac{n^n}{(n+1)^{n+1}} = \lim_{n\to\infty} \frac{1}{(1+1/n)^n(n+1)} \\ &= \frac{1}{e} \lim_{n\to\infty} \frac{1}{n+1} = 0 \quad \text{(see Chapter 29, p.224).}\end{aligned}$$

Simplicio: Why exactly is it a problem that the quotient of lower and upper bounds approaches zero?

Teacher: Because it shows that for large values of n, the upper bound is very large compared to the lower bound so that little information is provided regarding the value of $n!$.

Sophie: I see your point, but it doesn't seem to answer Simplicio's question. Even if the estimate is not "good" it is still "good enough." After all, we used it successfully in the proof of Theorem 34.10.

Teacher: I cannot argue with that, but you will see that the problem of improving the estimate in (37.1) is worth our attention all the same, because it will lead us to new insights regarding the approximation of definite integrals by sums.

Trapezoid and Midpoint Approximations

As an alternative to the approximation of areas under a graph by rectangles (left or right sums) we will discuss in this section two new methods of approximation using *trapezoids.*

37.1 Example. Let us consider the function f defined by the equation

$$f(x) := \frac{x^2}{4} + 1.$$

We wish to find an upper estimate for the integral of f over the interval $[-2, 2]$. To begin with, we subdivide $[-2, 2]$ into n subintervals of equal length $\Delta x = 4/n$ with endpoints $x_k = -2 + k\Delta x$ for $0 \le k \le n$. Then we place trapezoids with vertical side lengths $f(x_{k-1})$ and $f(x_k)$ over each of the subintervals $[x_{k-1}, x_k]$ (see Figure 37.1).

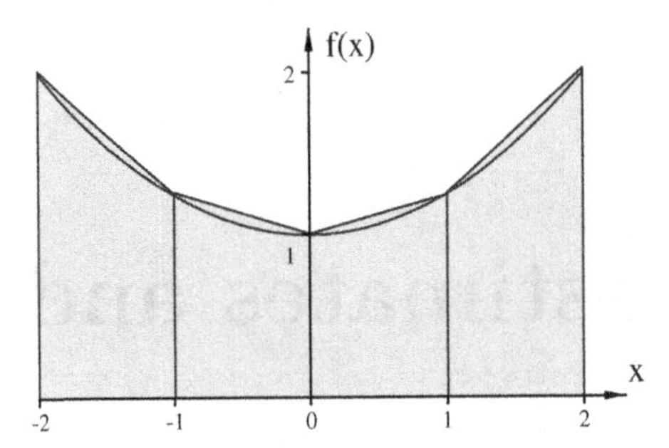

Figure 37.1: upper estimate using trapezoids for $n = 4$.

Since the area of each of these trapezoids is easily seen to be

$$\frac{f(x_{k-1}) + f(x_k)}{2}\Delta x \quad \text{(see Exercise 25.3)},$$

we obtain

$$\int_{-2}^{2} f(x)\,dx \le \sum_{k=1}^{n} \frac{f(x_{k-1}) + f(x_k)}{2}\Delta x.$$

The sum on the right-hand side can be simplified as follows:

$$\begin{aligned}\sum_{k=1}^{n} \frac{f(x_{k-1}) + f(x_k)}{2} &= \sum_{k=1}^{n} \frac{f(x_{k-1})}{2} + \sum_{k=1}^{n} \frac{f(x_k)}{2} = \sum_{k=0}^{n-1} \frac{f(x_k)}{2} + \sum_{k=1}^{n} \frac{f(x_k)}{2} \\ &= \left(-\frac{f(x_n)}{2} + \sum_{k=0}^{n} \frac{f(x_k)}{2}\right) + \left(-\frac{f(x_0)}{2} + \sum_{k=0}^{n} \frac{f(x_k)}{2}\right) \\ &= -\frac{f(-2) + f(2)}{2} + \sum_{k=0}^{n} f(x_k).\end{aligned}$$

Hence

$$\int_{-2}^{2} f(x)\,dx \le -\frac{f(-2) + f(2)}{2}\Delta x + \sum_{k=0}^{n} f(x_k)\Delta x = -2 \cdot \frac{4}{n} + \sum_{k=0}^{n} \left(\frac{1}{4}\left(-2 + \frac{4k}{n}\right)^2 + 1\right)\frac{4}{n}.$$

For $n = 4$, the value of the approximating sum is $11/2 = 5.5$. Comparing this result with the actual value of the integral, which is $16/3 \approx 5.33$, we see that even for relatively small values of n, a trapezoid approximation can be fairly accurate.

To find a lower estimate using trapezoids, we again subdivide the interval $[-2, 2]$ into n subintervals of equal length $4/n$, but this time we insert the approximating trapezoids beneath the tangent lines to the graph of f at the *midpoints* $(x_{k-1} + x_k)/2$ of the subintervals $[x_{k-1}, x_k]$ (see Figure 37.2). Since the area of each of these trapezoids is obviously equal to the area of a rectangle of width $\Delta x = 4/n$ and height $f((x_{k-1} + x_k)/2)$, it follows that a lower estimate for the area under the graph of f over the interval $[-2, 2]$ is given by the inequality

$$\int_{-2}^{2} f(x)\,dx \ge \sum_{k=1}^{n} f\left(\frac{x_{k-1} + x_k}{2}\right)\Delta x = \sum_{k=1}^{n} \left(\frac{1}{4}\left(-2 + \frac{2(2k-1)}{n}\right)^2 + 1\right)\frac{4}{n}.$$

For $n = 4$, the value of the sum on the right is $21/4 = 5.25$, which is even closer to the actual value $16/3$ than is the upper estimate $11/2$.

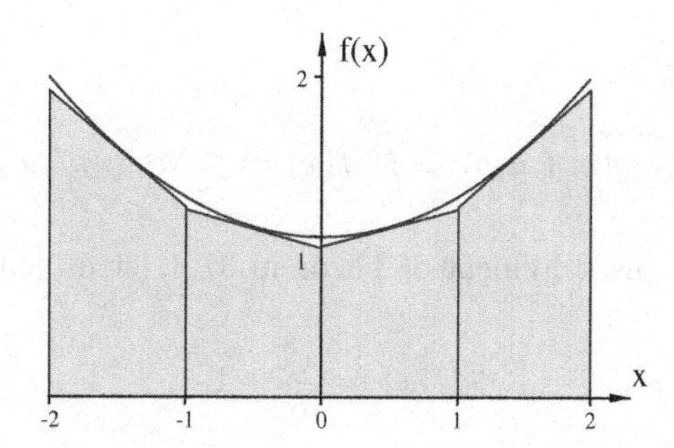

Figure 37.2: lower estimate using trapezoids.

37.2 Exercise. Find upper and lower estimates for $\int_{-1}^{1} e^x\,dx$ using trapezoid and midpoint approximations with $n = 10$.

To generalize the results of Example 37.1, we assume that we are given an arbitrary function $f : [a,b] \to \mathbb{R}$ and n subintervals $[x_{k-1}, x_k]$ with

$$x_k := a + k\Delta x \text{ for } 0 \le k \le n \text{ and } \Delta x := \frac{b-a}{n}. \tag{37.2}$$

Then the *trapezoid sum* of f over $[a,b]$ is

$$\boxed{\mathrm{Trap}(n, f, a, b) := -\frac{f(a)+f(b)}{2}\Delta x + \sum_{k=0}^{n} f(x_k)\Delta x,}$$

and the *midpoint sum* is

$$\boxed{\mathrm{Mid}(n, f, a, b) := \sum_{k=1}^{n} f\left(\frac{x_{k-1}+x_k}{2}\right)\Delta x.}$$

(Note: this formula for the midpoint sum is consistent with the definition given in Chapter 17, p.144.) Referring back to our discussion in Example 37.1, it is not difficult to understand that trapezoid and midpoint approximations are respectively upper and lower estimates if f is concave up. By contrast, if f is concave down, the roles are reversed in the sense that an upper estimate is represented by the midpoint sum and a lower estimate by the trapezoid sum (see Figure 37.3).

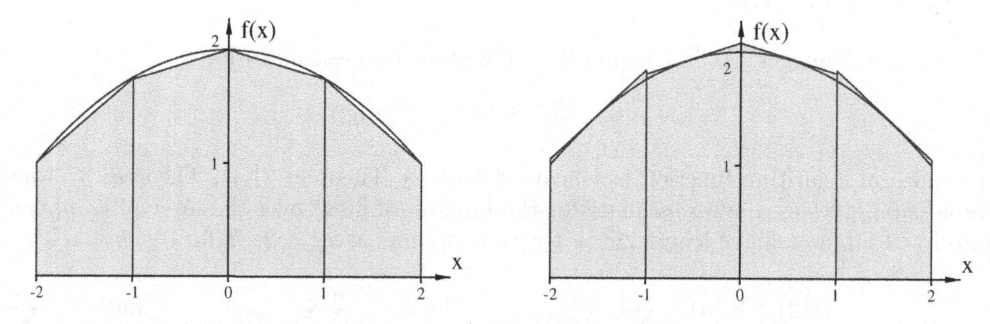

Figure 37.3: upper and lower estimates.

37.3 Theorem. *For a function $f : [a,b] \to \mathbb{R}$ and a positive integer n, we assume that Δx and x_k are defined as in (37.2). If f is concave up, then*

$$\mathrm{Mid}(n, f, a, b) \le \int_a^b f(x)\,dx \le \mathrm{Trap}(n, f, a, b),$$

and if f is concave down, then

$$\text{Trap}(n, f, a, b) \le \int_a^b f(x)\,dx \le \text{Mid}(n, f, a, b).$$

37.4 Example. To illustrate the statement of Theorem 37.3, let us consider the function f defined by the equation

$$f(x) := \frac{x^3}{8} + 1.$$

We wish to find a lower estimate for the integral of f over the interval $[-2, 3]$. Since the second derivative $f''(x) = 3x/4$ is negative on $[-2, 0]$ and positive on $[0, 3]$, we may infer that f is concave down on $[-2, 0]$ and concave up on $[0, 3]$. Therefore, in order to obtain a lower estimate for the integral of f, we need to use a trapezoid approximation on $[-2, 0]$ and a midpoint approximation on $[0, 3]$ (see Figure 37.4). Choosing n_1 subdivisions for $[-2, 0]$ with $\Delta x_1 = 2/n_1$ and n_2 subdivisions for $[0, 3]$ with $\Delta x_2 = 3/n_2$, we may infer that

$$\int_{-2}^3 f(x)\,dx \ge \text{Trap}(n_1, f, -2, 0) + \text{Mid}(n_2, f, 0, 3).$$

37.5 Exercise. Evaluate the terms on the right-hand side of this inequality for $n_1 = 20$ and $n_2 = 30$.

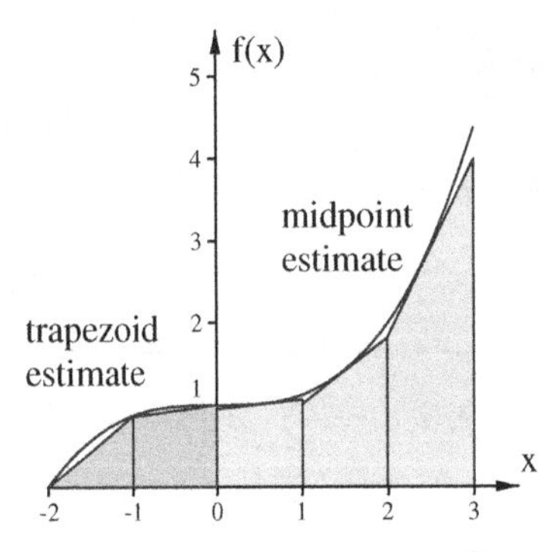

Figure 37.4: lower estimate for $\int_{-2}^3 f(x)\,dx$.

Stirling's Formula

Let n be a positive integer. As in Chapter 35, we wish to find estimates for

$$n! = e^{\sum_{k=2}^n \ln(k)} \quad \text{(see (35.3))}.$$

Since the natural logarithm function is concave down (see Theorem 21.1), Theorem 37.3 implies that a trapezoid sum gives us a lower estimate for the integral of $\ln(x)$ over the interval $[2, n]$. Subdividing $[2, n]$ into $n-2$ subintervals of length $\Delta x = 1$ with endpoints at $x_k = 2+k$ for $0 \le k \le n-2$, we obtain

$$\int_2^n \ln(x)\,dx \ge -\frac{\ln(2)+\ln(n)}{2} + \sum_{k=0}^{n-2} \ln(x_k)\Delta x = -\frac{\ln(2n)}{2} + \sum_{k=0}^{n-2} \ln(2+k) = -\frac{\ln(2n)}{2} + \sum_{k=2}^{n} \ln(k).$$

In the current context of estimating $n!$, this inequality is to be regarded as an upper estimate for the sum $\sum_{k=2}^n \ln(k)$ rather than a lower estimate for the integral. So we write

$$\int_2^n \ln(x)\,dx + \frac{\ln(2n)}{2} \ge \sum_{k=2}^n \ln(k). \tag{37.3}$$

To find a lower estimate for $\sum_{k=2}^{n} \ln(k)$, we set up a midpoint sum for the integral $\int_{3/2}^{n+1/2} \ln(x)\,dx$. We choose $3/2$ and $n+1/2$ as the boundaries of integration in order to guarantee that the integers $k = 2, 3, \ldots, n$ in the sum $\sum_{k=2}^{n} \ln(k)$ are the midpoints of the subintervals in the corresponding midpoint approximation. Setting $\Delta x = 1$ and $x_k = 3/2 + k$ for $0 \le k \le n-1$, it follows that

$$\begin{aligned}\int_{3/2}^{n+1/2} \ln(x)\,dx &\le \sum_{k=1}^{n-1} \ln\left(\frac{x_{k-1}+x_k}{2}\right)\Delta x = \sum_{k=1}^{n-1} \ln\left(\frac{3/2+(k-1)+3/2+k}{2}\right) \\ &= \sum_{k=1}^{n-1} \ln(k+1) = \sum_{k=2}^{n} \ln(k).\end{aligned} \tag{37.4}$$

Combining (37.3) with (37.4), we may infer that

$$\int_{3/2}^{n+1/2} \ln(x)\,dx \le \sum_{k=2}^{n} \ln(k) \le \int_{2}^{n} \ln(x)\,dx + \frac{\ln(2n)}{2}.$$

To evaluate the integrals on the left- and right-hand sides, we recall that $\int \ln(x)\,dx = x\ln(x) - x + C$ (see Chapter 35). This yields

$$\begin{aligned}\int_{3/2}^{n+1/2} \ln(x)\,dx &= \left(n+\frac{1}{2}\right)\ln\left(n+\frac{1}{2}\right) - (n-1) - \frac{3}{2}\ln\left(\frac{3}{2}\right), \\ \int_{2}^{n} \ln(x)\,dx &= n\ln(n) - n - 2\ln(2) + 2.\end{aligned}$$

Using these results and the fact that the exponential function is increasing, we obtain

$$e^{(n+1/2)\ln(n+1/2)-(n-1)-3\ln(3/2)/2} \le e^{\sum_{k=2}^{n}\ln(k)} = n! \le e^{n\ln(n)-n-2\ln(2)+2+\ln(2n)/2}. \tag{37.5}$$

In simplifying the exponential expressions on both sides, we arrive at the final estimate:

$$\boxed{\frac{\sqrt{8}(n+1/2)^n\sqrt{n+1/2}}{\sqrt{27}e^{n-1}} \le n! \le \frac{n^n\sqrt{n}}{2\sqrt{2}e^{n-2}} \quad \text{for all } n \ge 2.} \tag{37.6}$$

37.6 Exercise. Use the definition of the general exponential function as given in Chapter 23 to prove that the estimates in (37.5) and (37.6) are equivalent.

In order to generalize the central ideas of the preceding discussion, we assume that for integers $m < n$ we are given a downward concave function $f : [m-1/2, n+1/2] \to \mathbb{R}$. Then, Theorem 37.3 implies that

$$-\frac{f(m)+f(n)}{2} + \sum_{k=m}^{n} f(k) = \text{Trap}(n-m, f, m, n) \le \int_{m}^{n} f(x)\,dx$$

and

$$\int_{m-1/2}^{n+1/2} f(x)\,dx \le \text{Mid}(n-m+1, f, m-1/2, n+1/2) = \sum_{k=m}^{n} f(k).$$

Hence

$$\boxed{\int_{m-1/2}^{n+1/2} f(x)\,dx \le \sum_{k=m}^{n} f(k) \le \frac{f(m)+f(n)}{2} + \int_{m}^{n} f(x)\,dx.}$$

Similarly, if f is concave up, then

$$\boxed{\int_{m-1/2}^{n+1/2} f(x)\,dx \ge \sum_{k=m}^{n} f(k) \ge \frac{f(m)+f(n)}{2} + \int_{m}^{n} f(x)\,dx.}$$

37.7 Exercise. Use the preceding estimates (whichever applies) and (35.1) to show that the following inequality is valid for all integers $n \geq 2$:

$$\frac{\sqrt{n}^{n^2+n}}{2\sqrt[4]{e}^{n^2-4}} \leq \prod_{k=1}^{n} k^k \leq \frac{\sqrt[8]{2\sqrt{e}}^{9}\sqrt{n+1/2}^{(n+1/2)^2}}{\sqrt[8]{3}^{9}\sqrt[4]{e}^{(n+1/2)^2}}.$$

In order to verify that (37.6) is a better estimate than (37.1), we examine the quotient of lower and upper bounds:

$$\frac{\dfrac{\sqrt{8}(n+1/2)^n\sqrt{n+1/2}}{\sqrt{27}e^{n-1}}}{\dfrac{n^n\sqrt{n}}{2\sqrt{2}e^{n-2}}} = \frac{8(n+1/2)^n\sqrt{n+1/2}}{e\sqrt{27}n^n\sqrt{n}} = \frac{8}{e\sqrt{27}}\left(1+\frac{1}{2n}\right)^n\sqrt{1+\frac{1}{2n}}.$$

Now using formula (29.4), we obtain

$$\begin{aligned} \lim_{n\to\infty}\left(\frac{8}{e\sqrt{27}}\left(1+\frac{1}{2n}\right)^n\sqrt{1+\frac{1}{2n}}\right) &= \frac{8}{e\sqrt{27}}\left(\lim_{n\to\infty}\left(1+\frac{1}{2n}\right)^n\right)\left(\lim_{n\to\infty}\sqrt{1+\frac{1}{2n}}\right) \\ &= \frac{8}{e\sqrt{27}}\cdot e^{1/2}\cdot 1 \approx 0.93. \end{aligned} \tag{37.7}$$

This shows that (37.6) is indeed a better estimate than (37.1), because in the case of (37.6) the quotient of lower and upper bounds does not converge to 0 but rather to a value close to 1. A famous result that gives a *precise* asymptotic representation of $n!$ is attributed to Stirling and is known as *Stirling's formula* (for a proof see [Rud1]):

37.8 Theorem. $\lim_{n\to\infty}\dfrac{n!e^n}{n^n\sqrt{n}} = \sqrt{2\pi}$.

To see how this result compares with the estimate (37.6), we multiply (37.6) by $e^n/(n^n\sqrt{n})$. This yields

$$\frac{e\sqrt{8}(n+1/2)^n\sqrt{n+1/2}}{\sqrt{27}n^n\sqrt{n}} \leq \frac{n!e^n}{n^n\sqrt{n}} \leq \frac{e^2}{2\sqrt{2}}.$$

According to (37.7), the limit of the term on the left-hand side of this inequality converges to $e\sqrt{8e}/\sqrt{27} \approx 2.44$, and the value of $e^2/(2\sqrt{2})$ is approximately 2.61. So while we were not able to prove that the limit of $n!e^n/(n^n\sqrt{n})$ is $\sqrt{2\pi} \approx 2.51$, our result does show that the limit has to be somewhere between 2.44 and 2.61, which is not bad.

Additional Exercises

37.9. Use trapezoid and midpoint estimates to find lower and upper estimates for the following sums:

$$\text{a) } \sum_{k=3}^{n}\arctan(k+1) \qquad \text{b) } \sum_{k=5}^{n}\frac{k}{k^2+1}$$

37.10. Use trapezoid and midpoint sums to find upper and lower estimates for the integral $\int_0^{10} x^3\,dx$. Then find upper and lower estimates using left and right sums and compare the errors made in each type of approximation. Use 10 subdivisions along the interval $[0, 10]$.

37.11. Use trapezoid and midpoint approximations to find lower and upper estimates for the following products:

$$\text{a)}\ \prod_{k=4}^{n} k^{\ln(k)} \qquad \text{b)}\ \prod_{k=1}^{n} e^{e^k}$$

37.12. Use a trapezoid or midpoint estimate to prove that

$$\sum_{k=1}^{n} \frac{1}{k\sqrt{k}} \geq \frac{5}{2} - \frac{1}{\sqrt{n}}\left(2 - \frac{1}{2n}\right).$$

37.13. Use the result of Exercise 37.12 to prove that

$$\prod_{k=1}^{n} e^{1/(k\sqrt{k})} \geq \frac{\sqrt{e}^{5}}{\sqrt{e}^{(4-1/n)/\sqrt{n}}}.$$

37.14. According to the criterion discussed in the first section of this chapter, the estimate $n \leq \sqrt{n^2 + 2n - 1} \leq n + 1$ would have to be classified as...

a) good.

b) bad.

c) immoral.

Chapter 38

Improper Integrals and Infinite Trumpets

Infinite Areas Can Be Finite

In Chapter 17 we introduced the definite integral of bounded piecewise continuous functions over intervals of finite length. For theoretical reasons it is often necessary to consider integrals also over infinite intervals like, for instance, $[1, \infty)$, $(-\infty, 2]$, or $(-\infty, \infty)$.

38.1 Example. Let us examine the area under the graph of the function $f(x) := 2^{-x}$ on the interval $[0, \infty)$ (see Figure 38.1). At first sight it may appear that this area has to be infinite because it is stretched out over a region of infinite length, but we will show that the rapid decrease in the values $f(x) = 2^{-x}$ for large x actually causes it to be finite. For this purpose we place a sequence of rectangles

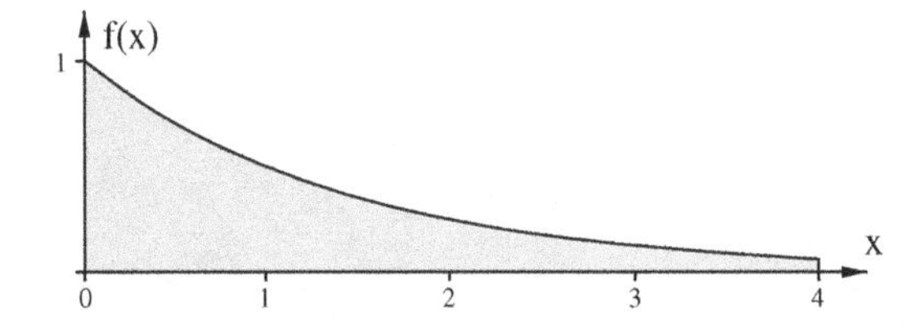

Figure 38.1: area under the graph of $f(x) = 2^{-x}$.

along the graph of f as shown in Figure 38.2. In order to calculate the area of these rectangles, it is

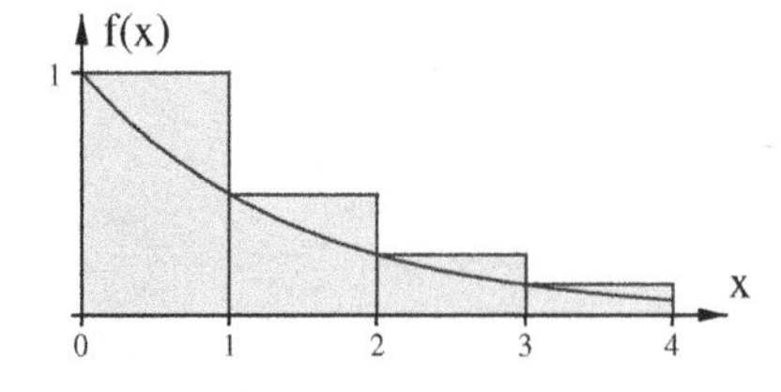

Figure 38.2: upper estimate for the area under the graph of $f(x) = 2^{-x}$.

helpful to stack them up: we place the second rectangle on top of the first, then the third on top of the second, and so forth (see Figure 38.3). The leftmost rectangle over the interval $[0, 1]$ has height 1. After placing the second rectangle on top of it, we have a stack of height $3/2 = 2 - 2^{-1}$. Placing the third and then the fourth rectangle on top of the first two, the height of the stack increases to $7/4 = 2 - 2^{-2}$ and then to $15/8 = 2 - 2^{-3}$. In general, after adding on the nth rectangle, the height of the stack is

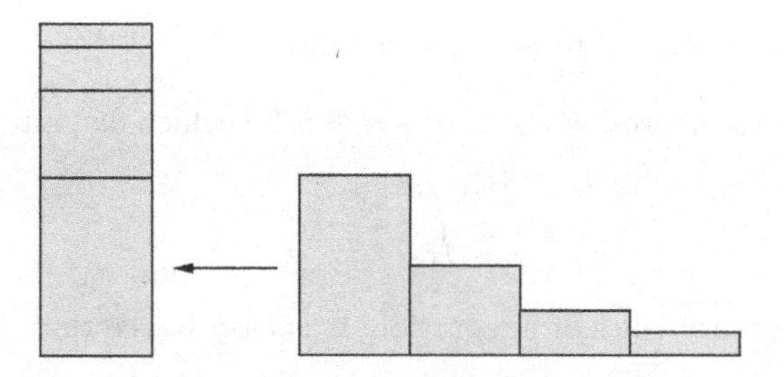

Figure 38.3: stacking up the rectangles.

$2 - 2^{-(n-1)}$, and this is also the value of its area because the width of the stack is 1. Since $2^{-(n-1)}$ approaches zero as n tends to infinity, the total area of the rectangles is

$$\lim_{n \to \infty} (2 - 2^{-(n-1)}) = 2.$$

This shows in particular that the total area of the rectangles is finite, and therefore, *the area between the graph of f and the x-axis over the interval* $[0, \infty)$ *is finite* as well, because it actually is smaller than the area of the rectangles (see Figure 38.2).

Improper Integrals over Infinite Intervals

In Example 38.1, we determined the total area of the rectangles as a limit of the areas of finite stacks. A similar approach is appropriate also in defining *improper integrals* over infinite intervals: we regard an integral of the form $\int_a^\infty f(x)\,dx$ as a limit of integrals $\int_a^b f(x)\,dx$ over finite intervals $[a, b]$ as b tends to ∞. In order to ensure that each integral $\int_a^b f(x)\,dx$ is well defined, we need to assume that f is piecewise continuous and bounded on $[a, b]$ for all $b > a$ (see Chapter 17). This leads us to the following definition:

38.2 Definition. **a)** If $f : [a, \infty) \to \mathbb{R}$ is a function that is piecewise continuous and bounded on every interval $[a, b]$ with $b > a$, then we say that the *improper integral* $\int_a^\infty f(x)\,dx$ is *convergent* if $\lim_{b\to\infty} \int_a^b f(x)\,dx$ exists and is finite. Otherwise we say that $\int_a^\infty f(x)\,dx$ is *divergent.* If $\int_a^\infty f(x)\,dx$ is convergent, then its value is given by the defining equation

$$\int_a^\infty f(x)\,dx := \lim_{b\to\infty} \int_a^b f(x)\,dx.$$

b) If $f : (-\infty, a] \to \mathbb{R}$ is a function that is piecewise continuous and bounded on every interval $[b, a]$ with $b < a$, then we say that the *improper integral* $\int_{-\infty}^a f(x)\,dx$ is *convergent* if $\lim_{b\to-\infty} \int_b^a f(x)\,dx$ exists and is finite. Otherwise we say that $\int_{-\infty}^a f(x)\,dx$ is *divergent.* If $\int_{-\infty}^a f(x)\,dx$ is convergent, then its value is given by the defining equation

$$\int_{-\infty}^a f(x)\,dx := \lim_{b\to-\infty} \int_b^a f(x)\,dx.$$

c) If $f : (-\infty, \infty) \to \mathbb{R}$ is a function that is piecewise continuous and bounded on every interval $[-b, b]$ with $b > 0$, then we say that the *improper integral* $\int_{-\infty}^\infty f(x)\,dx$ is *convergent* if both $\int_0^\infty f(x)\,dx$ and $\int_{-\infty}^0 f(x)\,dx$ are convergent. Otherwise we say that $\int_{-\infty}^\infty f(x)\,dx$ is *divergent.* If $\int_{-\infty}^\infty f(x)\,dx$ is convergent, then we define

$$\int_{-\infty}^\infty f(x)\,dx := \int_0^\infty f(x)\,dx + \int_{-\infty}^0 f(x)\,dx.$$

38.3 Exercise. Use Definition 38.2a to evaluate the integral $\int_0^\infty 2^{-x}\,dx$.

38.4 Example. We wish to determine the values $\alpha > 0$ for which the integral

$$\int_1^\infty \frac{1}{x^\alpha}\,dx$$

is convergent. Using elementary rules of integration, it is easy to see that for all $b > 1$ we have

$$\int_1^b \frac{1}{x^\alpha}\,dx = \begin{cases} \dfrac{b^{1-\alpha}-1}{1-\alpha} & \text{if } \alpha \neq 1, \\ \ln(b) & \text{if } \alpha = 1. \end{cases}$$

Since

$$\lim_{b\to\infty} b^{1-\alpha} = \begin{cases} \infty & \text{if } 0 < \alpha < 1, \\ 0 & \text{if } \alpha > 1, \end{cases}$$

it follows that

$$\lim_{b\to\infty} \frac{b^{1-\alpha}-1}{1-\alpha} = \begin{cases} \infty & \text{if } 0 < \alpha < 1, \\ \dfrac{1}{\alpha-1} & \text{if } \alpha > 1. \end{cases}$$

Using this result and the fact that $\lim_{b\to\infty} \ln(b) = \infty$ (see Theorem 21.1), we conclude that

$$\boxed{\int_1^\infty \frac{1}{x^\alpha}\,dx = \frac{1}{\alpha-1} \text{ if } \alpha > 1 \text{ and } \int_1^\infty \frac{1}{x^\alpha}\,dx \text{ is divergent if } 0 < \alpha \leq 1.} \tag{38.1}$$

Note: for $\alpha \leq 0$, the inequality $1/x^\alpha \geq 1$ for all $x \geq 1$ implies trivially that $\int_1^\infty 1/x^\alpha\,dx$ is divergent (see also Theorem 38.7 below).

38.5 Exercise. Determine for which $\alpha \in \mathbb{R}$ the integral $\displaystyle\int_2^\infty \frac{1}{x\ln^\alpha(x)}\,dx$ is convergent.

38.6 Example. Let us consider the integral

$$\int_{-\infty}^\infty \frac{1}{1+x^2}\,dx.$$

According to Definition 38.2c, this integral is convergent because

$$\int_0^\infty \frac{1}{1+x^2}\,dx = \lim_{b\to\infty} \int_0^b \frac{1}{1+x^2}\,dx = \lim_{b\to\infty} \arctan(b) = \frac{\pi}{2}$$

and

$$\int_{-\infty}^0 \frac{1}{1+x^2}\,dx = \lim_{b\to-\infty} \int_b^0 \frac{1}{1+x^2}\,dx = \lim_{b\to-\infty} -\arctan(b) = \frac{\pi}{2}.$$

These results also show that the value of $\int_{-\infty}^\infty 1/(1+x^2)\,dx$ is π, because

$$\int_{-\infty}^\infty \frac{1}{1+x^2}\,dx = \int_{-\infty}^0 \frac{1}{1+x^2}\,dx + \int_0^\infty \frac{1}{1+x^2}\,dx = \frac{\pi}{2} + \frac{\pi}{2} = \pi.$$

Simplicio: The process of calculating an improper integral of the form $\int_{-\infty}^\infty f(x)\,dx$ by breaking it up into two integrals over the intervals $[0,\infty)$ and $(-\infty, 0]$ seems a little cumbersome. Isn't there an easier way to do it?

Sophie: I was asking myself the same question, and I noticed that the integral in Example 38.6 is more easily evaluated as a single limit:

$$\int_{-\infty}^{\infty} \frac{1}{1+x^2}\,dx = \lim_{b\to\infty} \int_{-b}^{b} \frac{1}{1+x^2}\,dx = \lim_{b\to\infty} 2\arctan(b) = \pi.$$

Teacher: Are you suggesting that it might always be possible to evaluate integrals of the form $\int_{-\infty}^{\infty} f(x)\,dx$ by computing the limit $\lim_{b\to\infty} \int_{-b}^{b} f(x)\,dx$?

Sophie: It does sound reasonable, doesn't it?

Teacher: Consider the integral $\int_{-\infty}^{\infty} \sin(x)\,dx$. Here, $\int_{0}^{\infty} \sin(x)\,dx$ is divergent because for every positive integer n we have

$$\int_0^{2n\pi} \sin(x)\,dx = -\cos(x)\Big|_0^{2n\pi} = 0 \text{ and}$$

$$\int_0^{(2n+1)\pi} \sin(x)\,dx = -\cos(x)\Big|_0^{(2n+1)\pi} = 2.$$

Therefore...

Simplicio: How does that prove that $\int_0^{\infty} \sin(x)\,dx$ is divergent?

Teacher: Because for $b = n\pi$ the values of $\int_0^b \sin(x)\,dx$ alternate between 0 and 2 so that the limit $\lim_{b\to\infty} \int_0^b \sin(x)\,dx$ does not exist.

Sophie: So according to Definition 38.2c, the integral $\int_{-\infty}^{\infty} \sin(x)\,dx$ is divergent as well.

Teacher: Correct. But now consider the limit $\lim_{b\to\infty} \int_{-b}^{b} f(x)\,dx$. This limit does actually exist and is finite, because

$$\lim_{b\to\infty} \int_{-b}^{b} \sin(x)\,dx = \lim_{b\to\infty} -\cos(x)\Big|_{-b}^{b} = \lim_{b\to\infty} (\cos(-b) - \cos(b)) = 0.$$

Consequently, the convergence of an integral of the form $\int_{-\infty}^{\infty} f(x)\,dx$ cannot be deduced from the existence of the limit $\lim_{b\to\infty} \int_{-b}^{b} f(x)\,dx$.

Sophie: So it was just a coincidence that I found the correct value for the integral $\int_{-\infty}^{\infty} 1/(1+x^2)\,dx$ by evaluating the limit $\lim_{b\to\infty} \int_{-b}^{b} 1/(1+x^2)\,dx$?

Teacher: To answer this question, we need to remember our discussion concerning limits of sums and sums of limits at the very beginning of the course. If the integral $\int_{-\infty}^{\infty} f(x)\,dx$ is convergent, then, by definition, the integrals $\int_{-\infty}^{0} f(x)\,dx$ and $\int_0^{\infty} f(x)\,dx$ are convergent as well, and the limits $\lim_{b\to\infty} \int_0^b f(x)\,dx$ and $\lim_{b\to\infty} \int_{-b}^{0} f(x)\,dx$ exist. Consequently, Theorem 2.9 (which applies also to finite limits at infinity) allows us to infer that

$$\begin{aligned}\int_{-\infty}^{\infty} f(x)\,dx &= \lim_{b\to\infty} \int_{-b}^{0} f(x)\,dx + \lim_{b\to\infty} \int_0^b f(x)\,dx = \lim_{b\to\infty} \left(\int_{-b}^{0} f(x)\,dx + \int_0^b f(x)\,dx\right)\\ &= \lim_{b\to\infty} \int_{-b}^{b} f(x)\,dx.\end{aligned}$$

This shows that your result, Sophie, for the integral $\int_{-\infty}^{\infty} 1/(1+x^2)\,dx$ was not a coincidence, because the sum of two limits is always equal to the limit of the sum. So whenever an integral of the form $\int_{-\infty}^{\infty} f(x)\,dx$ is convergent, its value is equal to the limit $\lim_{b\to\infty} \int_{-b}^{b} f(x)\,dx$, but conversely, the existence of this limit does not necessarily imply the convergence of the integral because, in general, the existence of the limit of a sum does not imply the existence of the individual limits of the terms in the sum.

In order to introduce a simple criterion that allows us to decide whether an improper integral is convergent or divergent, we assume that $g : [a, \infty) \to \mathbb{R}$ and $f : [a, \infty) \to \mathbb{R}$ are piecewise continuous and bounded on $[a, b]$ for all $b > a$ such that $|f(x)| \leq g(x)$ for all $x \in [a, \infty)$. If $\int_a^\infty g(x)\,dx$ is convergent, then the area between the graph of g and the x-axis is finite. Therefore, it is intuitively obvious that the assumption $|f| \leq g$ forces the area between the graph of $|f|$ and the x-axis to be finite as well. Following this line of reasoning, it can be shown that the convergence of $\int_a^\infty g(x)\,dx$ implies the convergence of $\int_a^\infty f(x)\,dx$. A similar *comparison test* also holds for divergent integrals: if $0 \leq g(x) \leq f(x)$, then the divergence of $\int_a^\infty g(x)\,dx$ implies the divergence of $\int_a^\infty f(x)\,dx$.

38.7 Theorem. *Assume that $f : [a, \infty) \to \mathbb{R}$ and $g : [a, \infty) \to \mathbb{R}$ are bounded and piecewise continuous on $[a, b]$ for all $b > a$. If $\int_a^\infty g(x)\,dx$ is convergent and $|f(x)| \leq g(x)$ for all $x \in [a, \infty)$, then $\int_a^\infty f(x)\,dx$ is convergent as well. If $\int_a^\infty g(x)\,dx$ is divergent and $f(x) \geq g(x) \geq 0$ for all $x \in [a, \infty)$, then $\int_a^\infty f(x)\,dx$ is divergent as well. Analogous statements are also true for improper integrals of the form $\int_{-\infty}^a f(x)\,dx$ and $\int_{-\infty}^\infty f(x)\,dx$.*

A rigorous proof of this theorem will be omitted because it would require the use of certain theoretical properties of real numbers that are beyond the scope of this text (see [DS]). The diagrams in Figure 38.4 show typical graphs of f and g for the two cases described in Theorem 38.7.

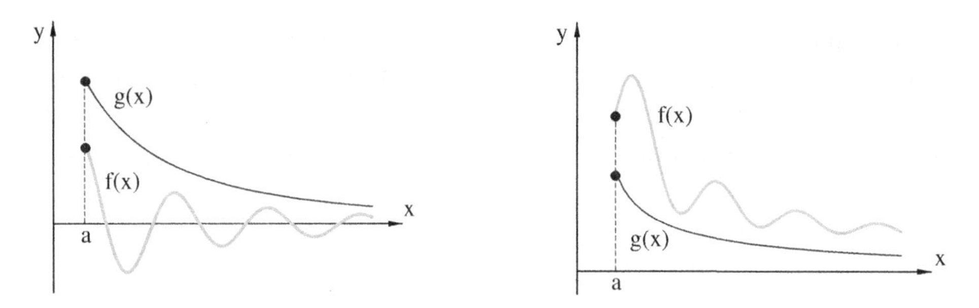

Figure 38.4: convergent and divergent cases.

38.8 Example. Let us consider the integral

$$\int_1^\infty \frac{x}{x^3 + x + 1}\,dx.$$

For any $x \in [1, \infty)$ we have

$$\frac{x}{x^3 + x + 1} \leq \frac{x}{x^3} = \frac{1}{x^2}.$$

Since, according to (38.1), the improper integral $\int_1^\infty 1/x^2\,dx$ is convergent, we may apply Theorem 38.7 with $g(x) = 1/x^2$ and $f(x) = x/(x^3 + x + 1)$ to infer that $\int_1^\infty x/(x^3 + x + 1)\,dx$ is convergent as well.

38.9 Example. We wish to show that the integral

$$\int_1^\infty \frac{\ln(x)}{\sqrt{x}}\,dx$$

is divergent. Since the natural logarithm function is increasing and $\ln(e) = 1$, we may infer that

$$\frac{\ln(x)}{\sqrt{x}} \geq \frac{1}{\sqrt{x}} \quad \text{for all } x \in [e, \infty).$$

Using (38.1) and Theorem 38.7, it follows that $\int_e^\infty \ln(x)/\sqrt{x}\,dx$ is divergent, and therefore

$$\int_1^\infty \frac{\ln(x)}{\sqrt{x}}\,dx = \int_1^e \frac{\ln(x)}{\sqrt{x}}\,dx + \int_e^\infty \frac{\ln(x)}{\sqrt{x}}\,dx$$

is divergent as well.

38.10 Exercise. Use (38.1) and Theorem 38.7 to decide which of the following integrals are convergent: $\int_2^\infty \frac{1}{\sqrt{x-1}}\,dx$, $\int_3^\infty \frac{1}{(x+2)^4}\,dx$, $\int_2^\infty \frac{x}{x^2-1}\,dx$, and $\int_1^\infty \frac{x}{\sqrt{1+x^5}}\,dx$.

Improper Integrals at Singularities

Let us consider the integral $\int_0^1 1/x^\alpha\,dx$ for $\alpha > 0$. In contrast to Example 38.4 where we integrated a bounded function over an infinite interval, the difficulty here is in integrating an unbounded function (because $\lim_{x\to 0^+} 1/x^\alpha = \infty$) over a finite interval. In such a situation, we say that the integrand has a *singularity*. In analogy to our definition of improper integrals over infinite intervals, we define improper integrals at singularities as limits of ordinary definite integrals:

38.11 Definition. **a)** Let $a, b \in \mathbb{R}$ such that $a < b$, and assume that $f : (a, b] \to \mathbb{R}$ is bounded and piecewise continuous on every interval $[c, b]$ with $a < c < b$. Then we say that the *improper integral* $\int_a^b f(x)\,dx$ is *convergent* if $\lim_{c\to a^+} \int_c^b f(x)\,dx$ exists and is finite. Otherwise we say that $\int_a^b f(x)\,dx$ is *divergent.* If $\int_a^b f(x)\,dx$ is convergent, then its value is given by the defining equation

$$\int_a^b f(x)\,dx := \lim_{c\to a^+} \int_c^b f(x)\,dx.$$

b) Let $a, b \in \mathbb{R}$ such that $a < b$ and assume that $f : [a, b) \to \mathbb{R}$ is bounded and piecewise continuous on every interval $[a, c]$ with $a < c < b$. Then we say that the *improper integral* $\int_a^b f(x)\,dx$ is *convergent* if $\lim_{c\to b^-} \int_a^c f(x)\,dx$ exists and is finite. Otherwise we say that $\int_a^b f(x)\,dx$ is *divergent.* If $\int_a^b f(x)\,dx$ is convergent, then its value is given by the defining equation

$$\int_a^b f(x)\,dx := \lim_{c\to b^-} \int_a^c f(x)\,dx.$$

38.12 Example. For $\alpha > 0$, the integral

$$\int_0^1 \frac{1}{x^\alpha}\,dx$$

needs to be evaluated as an improper integral, because $1/x^\alpha$ is undefined at $x = 0$ and $\lim_{x\to 0^+} 1/x^\alpha = \infty$. For $c > 0$ we have

$$\int_c^1 \frac{1}{x^\alpha}\,dx = \begin{cases} \dfrac{1 - c^{1-\alpha}}{1-\alpha} & \text{if } \alpha \neq 1, \\ -\ln(c) & \text{if } \alpha = 1. \end{cases}$$

Since

$$\lim_{c\to 0^+} \frac{1 - c^{1-\alpha}}{1-\alpha} = \begin{cases} \dfrac{1}{1-\alpha} & \text{if } 0 < \alpha < 1, \\ \infty & \text{if } \alpha > 1, \end{cases}$$

and $\lim_{c\to 0^+} -\ln(c) = \infty$, we conclude that

$$\boxed{\int_0^1 \frac{1}{x^\alpha}\,dx = \frac{1}{\alpha - 1} \text{ if } 0 < \alpha < 1 \text{ and } \int_0^1 \frac{1}{x^\alpha}\,dx \text{ is divergent if } \alpha \geq 1.} \tag{38.2}$$

38.13 Example. The integral

$$\int_0^1 \ln(x)\,dx$$

is improper, because $\lim_{x\to 0^+} \ln(x) = -\infty$. Using integration by parts, we obtain

$$\int_c^1 \ln(x)\,dx = (x\ln(x) - x)\Big|_c^1 = c - c\ln(c) - 1 \text{ for all } c > 0.$$

In Chapter 29, p.222, we showed that $\lim_{c\to 0^+} c\ln(c) = 0$. Therefore, $\int_0^1 \ln(x)\,dx$ is convergent because

$$\lim_{c\to 0^+}\int_c^1 \ln(x)\,dx = \lim_{c\to 0^+}(c - c\ln(c) - 1) = -1.$$

38.14 Exercise. Determine whether the following integrals are convergent or divergent and evaluate those that are convergent: $\int_0^{1/2}\frac{1}{x\ln^2(x)}\,dx$, $\int_0^{1/2}\frac{1}{x\sqrt{-\ln(x)}}\,dx$, $\int_0^1\frac{1}{\sqrt{1-x^2}}\,dx$, and $\int_{-2}^2 \ln|x|\,dx$.

The following theorem is completely analogous in meaning and content to Theorem 38.7:

38.15 Theorem. *Let $a, b \in \mathbb{R}$ with $a < b$ and assume that the functions $f : (a, b] \to \mathbb{R}$ and $g : (a, b] \to \mathbb{R}$ are piecewise continuous and bounded on every interval $[c, b]$ with $a < c < b$. If $\int_a^b g(x)\,dx$ is convergent and $|f(x)| \leq g(x)$ for all $x \in (a, b]$, then $\int_a^b f(x)\,dx$ is convergent as well. If $\int_a^b g(x)\,dx$ is divergent and $f(x) \geq g(x) \geq 0$ for all $x \in (a, b]$, then $\int_a^b f(x)\,dx$ is divergent as well. A completely analogous statement is valid also for improper integrals over half-open intervals of the form $[a, b)$ (instead of $(a, b]$).*

38.16 Example. We wish to decide whether the integral

$$\int_0^1 \frac{x^2}{\sqrt{x^7+x^9}}\,dx \tag{38.3}$$

is convergent or divergent. What matters in this example is the behavior of the integrand at zero, because this is where we have a singularity. So let us pick a positive value x close to zero. Then x^9 is very small compared to x^7 and $\sqrt{x^7+x^9} \approx \sqrt{x^7} = x^{7/2}$. Hence

$$\frac{x^2}{\sqrt{x^7+x^9}} \approx \frac{x^2}{x^{7/2}} = \frac{1}{x^{3/2}}.$$

Since, according to (38.2), the integral $\int_0^1 1/x^{3/2}\,dx$ is divergent, we ought to expect the integral in (38.3) to be divergent as well. To give a rigorous proof of this assertion based on Theorem 38.15, we observe that

$$\frac{x^2}{\sqrt{x^7+x^9}} \geq \frac{x^2}{\sqrt{x^7+x^7}} = \frac{x^2}{\sqrt{2x^7}} = \frac{1}{\sqrt{2}}\cdot\frac{1}{x^{3/2}} \quad \text{for all } x \in (0,1).$$

This shows that the integral in (38.3) is indeed divergent, because

$$\frac{1}{\sqrt{2}}\int_0^1 \frac{1}{x^{3/2}}\,dx = \infty.$$

38.17 Exercise. Use (38.2) and Theorem 38.15 to decide which of the following integrals are convergent: $\int_0^1 \frac{x}{\sqrt{x^2+x^3}}\,dx$, $\int_0^1\frac{x^2}{x^4+x^5+x^6}\,dx$, $\int_0^1\frac{1}{\sqrt{x^2+x}}\,dx$, $\int_0^{1/2}\frac{1}{\sqrt{x^3-x^5}}\,dx$, and $\int_0^2\frac{x+1}{\sqrt[3]{x^4+x^6}}\,dx$.

Infinite Trumpets

In Chapter 16 we discussed the problem of computing volumes of solids of revolution obtained by revolving the graph of a function f about the x-axis over a finite interval $[a, b]$. To broaden the scope of this discussion, we will now study solids of revolution over intervals of infinite length.

Teacher: Let us consider the function $f : [1, \infty) \to \mathbb{R}$ defined by the equation $f(x) := 1/x$. We wish to determine the volume of the *infinite trumpet* obtained by revolving the graph of f about the x-axis (see Figure 38.5).

Simplicio: Exactly why should I be interested in the volume of an infinite trumpet?

Teacher: My wife and I have several infinite trumpets at home, and our children have enjoyed them for years.

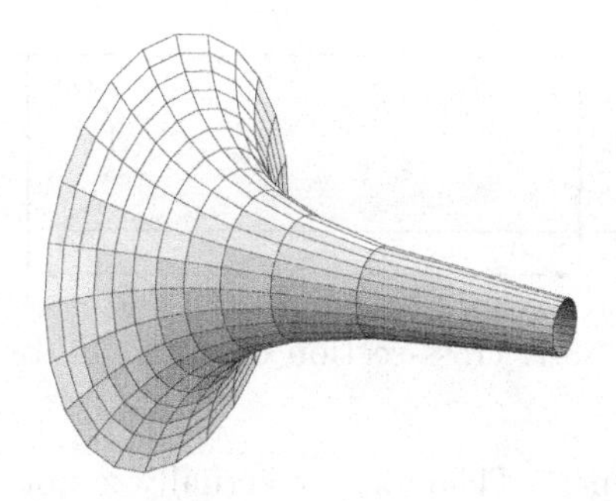

Figure 38.5: solid of revolution generated by $f(x) = 1/x$.

Simplicio: I see.

Teacher: For any $b > 1$, the volume $V(b)$ of the finite segment of the trumpet over the interval $[1, b]$ can be computed from the formula

$$V(b) = \pi \int_1^b f(x)^2\, dx \quad \text{(see Chapter 19, p.158)}.$$

Consequently, the total volume V equals the limit of $V(b)$ as b tends to infinity. In other words, V is equal to $\pi \int_1^\infty f(x)^2\, dx$. Using (38.1) with $\alpha = 2$, we find that

$$V = \pi \int_1^\infty \frac{1}{x^2}\, dx = \pi.$$

In particular, our calculation shows that the volume of the infinite trumpet is *finite*.

Sophie: This is very strange, because according to equation (38.1) with $\alpha = 1$, the cross-sectional area of the trumpet in the xy-plane (i.e., the area between the graphs of the functions $1/x$ and $-1/x$ over the interval $[1, \infty)$) is *infinite*. How is it possible that an infinite area revolved about the x-axis produces a finite volume?

Teacher: In order to understand this phenomenon, let us imagine that we cover the cross-sectional area of the infinite trumpet with a layer of paint of constant thickness h (see Figure 38.6). The amount of paint needed is infinite because the volume of the paint on the cross-section is equal to the infinite cross-sectional area multiplied by the positive constant h.

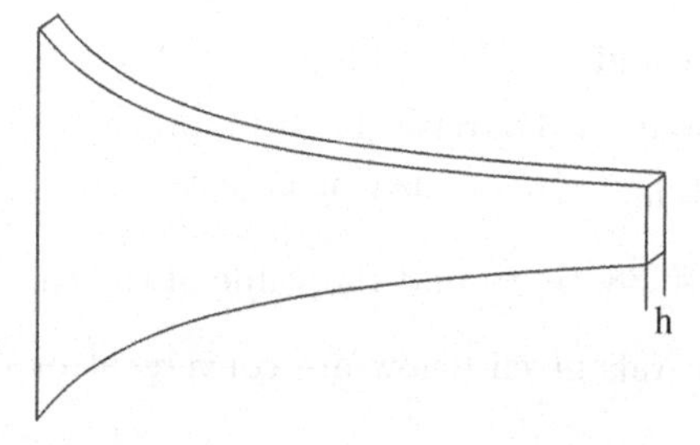

Figure 38.6: painted cross-section.

Simplicio: Are you saying that we can fill up the entire volume of the trumpet with a finite amount of paint, but covering just the cross-section requires an infinite amount?

Teacher: It seems highly implausible, but let us take a closer look at the layer of paint in Figure 38.6 for large values of x. Since the thickness h is constant, we can certainly find an $x > 1$ such that $1/x$ is very small compared to h. Then the circular cross-section of the trumpet at x (perpendicular to the x-axis) is very small compared to the area of the cross-section of the layer of paint of thickness h (see Figure 38.7). So for large x, the volume of the layer of paint increases at a much faster rate

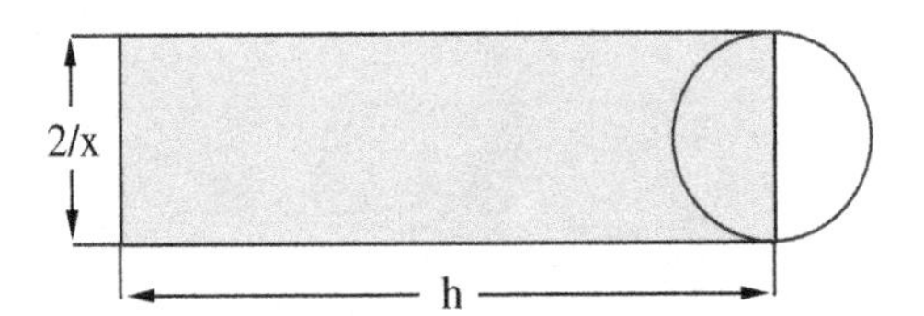

Figure 38.7: cross-section of the paint for large x.

than the volume of the trumpet. Therefore, it actually is not surprising that the trumpet has a finite volume but an infinite cross-sectional area between the graphs of $1/x$ and $-1/x$.

38.18 Exercise. For $\alpha > 0$, we denote by V_α the volume of the infinite trumpet obtained by rotating the graph of the function $f(x) := 1/x^\alpha$ about the x-axis over the interval $[1, \infty)$. For which values of α is V_α finite?

Additional Exercises

38.19. Find the values of the following improper integrals:

$$\textbf{a)} \int_0^\infty e^{-x}\,dx \quad \textbf{b)} \int_0^\infty \sin(x)e^{-x}\,dx \quad \textbf{c)} \int_2^\infty \frac{1}{x^2-1}\,dx \quad \textbf{d)} \int_{-\infty}^\infty \frac{1}{x^2+2x+5}\,dx$$

38.20. ★ Which of the following integrals are convergent and which are divergent? Explain your answers.

$$\textbf{a)} \int_0^1 \frac{\tan(x)}{x^{3/2}}\,dx \quad \textbf{b)} \int_0^1 \frac{\tan(x)}{x^2}\,dx \quad \textbf{c)} \int_0^1 \frac{x}{\sin(x)}\,dx \quad \textbf{d)} \int_1^\infty \frac{\sin(x)}{x}\,dx$$

38.21. Assume that $f : [0, \infty) \to \mathbb{R}$ is a positive, continuous, and strictly decreasing function that satisfies the following conditions: $f(0) = a$, $\int_0^\infty f(x)\,dx$ is convergent and $\lim_{x\to\infty} f(x) = 0$.

- **a)** Show that $bf(b) \le cf(b) + \int_c^b f(x)\,dx$ for all $b, c \in \mathbb{R}$ with $b > c$.
- **b)** Use a) to show that $\lim_{b\to\infty} bf(b) \le \int_c^\infty f(x)\,dx$ for all $c \in \mathbb{R}$.
- **c)** Use b) to show that $\lim_{b\to\infty} bf(b) = 0$.
- **d)** Explain why f is invertible and show that $\lim_{x\to 0^+} f^{-1}(x) = \infty$.
- **e)** Use c) and d) and the result of Exercise 35.20 to prove that $\int_0^a f^{-1}(x)\,dx = \int_0^\infty f(x)\,dx$. In particular, this shows that $\int_0^a f^{-1}(x)\,dx$ is convergent.

38.22. Use the result in Exercise 38.21e to find the value of the the integral $\int_0^1 \ln(x)\,dx$.

38.23. Decide whether the integrals given below are convergent or divergent and evaluate those that are convergent.

$$\textbf{a)} \int_e^\infty \frac{1}{x\ln^2(x)}\,dx \quad \textbf{b)} \int_0^\infty e^{-e^x}e^x\,dx \quad \textbf{c)} \int_0^\infty \frac{1}{1+\sqrt{x}}\,dx \quad \textbf{d)} \int_0^\infty \frac{x+1}{x^2+4x+5}\,dx$$

38.24. For each of the functions listed below, find the volume of the corresponding solid of revolution over the given interval.

- **a)** $f(x) := e^{-x}$ over $[0, \infty)$
- **b)** $f(x) := 1/(x-1)^2$ over $[2, \infty)$
- **c)** $f(x) := x/(1 + x^3)$ over $[0, \infty)$

d) $f(x) := xe^{-x}$ over $[0, \infty)$

38.25. Assume that $f : \mathbb{R} \to (0, \infty)$ is a continuous function such that $\ln(f(x))/x \geq 2.416$ for all $x \geq 1$. Show that $\int_1^\infty 1/f(x)\,dx$ is convergent.

38.26. Assume that $f : \mathbb{R} \to [-10, 10]$ is a continuously differentiable function. Show that $\int_1^\infty f'(x)/x\,dx$ is convergent.

PART V

TAYLOR APPROXIMATION

Chapter 39

Taylor Polynomials

Approximating Functions by Polynomials

Let us consider the seemingly simple problem of estimating the value of the sine function at 1. Since $\sin(1)$ is the solution of the equation $\arcsin(x) - 1 = 0$, it is tempting to think that we might be able to approximate $\sin(1)$ by applying Newton's method (see Chapter 13) to the function $f(x) = \arcsin(x) - 1$. This approach, however, is circular, because it only reformulates the problem of finding approximations for $\sin(1)$ in terms of the equally difficult problem of finding estimates for the values of $\arcsin(x)$ and its derivative. For this reason we will follow a different strategy and derive estimates for $\sin(1)$ via *approximations by polynomials*. The simplest case is the well known tangent line approximation.

In general, if I is an open interval and $f : I \to \mathbb{R}$ is a differentiable function, then the tangent line to the graph of f at the point $(x_0, f(x_0))$ is given by the equation

$$y(x) = f'(x_0)(x - x_0) + f(x_0) \quad \text{(see Chapter 4, p.35).} \tag{39.1}$$

39.1 Example. In order to obtain an estimate for $\sin(1)$ we apply (39.1) to the function $f(x) := \sin(x)$. A convenient choice for x_0 is $\pi/3$, because $\pi/3$ is close to 1, and the values of sine and cosine at $\pi/3$ are known to be $\sqrt{3}/2$ and $1/2$ respectively (see Exercise 6.2). Since the derivative of sine is cosine, we obtain

$$y(x) = \frac{1}{2}\left(x - \frac{\pi}{3}\right) + \frac{\sqrt{3}}{2}.$$

Consequently, for $x = 1$ we have

$$\sin(1) \approx y(1) = \frac{1}{2}\left(1 - \frac{\pi}{3}\right) + \frac{\sqrt{3}}{2} \approx 0.842426. \tag{39.2}$$

This estimate is already fairly precise, because in comparing it with the correct value $\sin(1) \approx 0.841471$, we see that the error is less than 0.001.

For the approximation in Example 39.1 we had to use our prior knowledge of the approximate values of $\pi/3$ and $\sqrt{3}$, and it would be preferable to obtain an estimate for $\sin(1)$ in a more self-contained manner. For instance, we could use $x_0 = 0$ instead of $x_0 = \pi/3$, because then the equation for the tangent line at $(0, 0)$ is

$$y(x) = x, \tag{39.3}$$

and we easily find the approximation

$$\sin(1) \approx y(1) = 1. \tag{39.4}$$

Unfortunately, the error in our approximation is now fairly large and in order to find estimates that are not only self-contained but also reasonably accurate, we need to increase the degree of the approximating polynomial. To formalize this idea, we assume that we are given a function $f : I \to \mathbb{R}$ that is twice differentiable on the open interval I. In analogy to the case of the tangent line approximation, our objective is to find a second degree polynomial of the form

$$y(x) = c(x - x_0)^2 + b(x - x_0) + a$$

such that

$$f(x_0) = y(x_0), \ f'(x_0) = y'(x_0), \text{ and } f''(x_0) = y''(x_0). \tag{39.5}$$

The first two of these equations are already satisfied for a tangent line approximation, but the third is an added requirement that will allow us to determine the coefficient c of the quadratic term $(x - x_0)^2$. Since $y(x_0) = a$, $y'(x_0) = b$ and $y''(x_0) = 2c$, the conditions in (39.5) imply that

$$a = f(x_0), \ b = f'(x_0), \text{ and } c = \frac{f''(x_0)}{2}. \tag{39.6}$$

39.2 Exercise. Verify that with these values for the coefficients a, b, and c the conditions in (39.5) are satisfied.

Given the values in (39.6), we obtain the following second degree approximation for f at $(x_0, f(x_0))$:

$$y(x) = \frac{f''(x_0)}{2}(x - x_0)^2 + f'(x_0)(x - x_0) + f(x_0). \tag{39.7}$$

39.3 Example. For $f(x) = \sin(x)$ and $x_0 = \pi/3$, we have $f(x_0) = \sqrt{3}/2$, $f'(x_0) = 1/2$, and $f''(x_0) = -\sqrt{3}/2$. Therefore, according to (39.7), the corresponding second-order approximation is

$$y(x) = -\frac{\sqrt{3}}{4}\left(x - \frac{\pi}{3}\right)^2 + \frac{1}{2}\left(x - \frac{\pi}{3}\right) + \frac{\sqrt{3}}{2}.$$

Setting $x = 1$, we now obtain the estimate

$$\sin(1) \approx y(1) = -\frac{\sqrt{3}}{4}\left(1 - \frac{\pi}{3}\right)^2 + \frac{1}{2}\left(1 - \frac{\pi}{3}\right) + \frac{\sqrt{3}}{2} \approx 0.841462.$$

Comparing this result with the correct value for sin(1) given in Example 39.1, we see that the error has decreased to less than 0.00001.

39.4 Example. Let us modify Example 39.3 by choosing again $x_0 = 0$ instead of $x_0 = \pi/3$. Then, for $f(x) = \sin(x)$, we have $f(x_0) = \sin(0) = 0$, $f'(x_0) = \cos(0) = 1$, and $f''(x_0) = -\sin(0) = 0$. Substituting these values in (39.7) yields

$$y(x) = x.$$

So in this special case the second-order approximation is equal to the first-order approximation in (39.3), and there is consequently no improvement in our estimate for sin(1). In order to reduce the error also for $x_0 = 0$, we will now discuss third degree approximations.

Let $f : I \to \mathbb{R}$ be a function that is three times differentiable on the open interval I. We wish to find a third degree polynomial of the form

$$y(x) = d(x - x_0)^3 + c(x - x_0)^2 + b(x - x_0) + a$$

such that not only $f(x_0) = y(x_0)$, $f'(x_0) = y'(x_0)$, and $f''(x_0) = y''(x_0)$, but also $f'''(x_0) = y'''(x_0)$. Given these conditions, it is easy to see that

$$a = f(x_0), \ b = f'(x_0), \ c = \frac{f''(x_0)}{2}, \text{ and } d = \frac{f'''(x_0)}{6}.$$

Hence

$$y(x) = \frac{f'''(x_0)}{6}(x - x_0)^3 + \frac{f''(x_0)}{2}(x - x_0)^2 + f'(x_0)(x - x_0) + f(x_0). \tag{39.8}$$

39.5 Exercise. Verify that this polynomial satisfies the conditions $f(x_0) = y(x_0)$, $f'(x_0) = y'(x_0)$, $f''(x_0) = y''(x_0)$, and $f'''(x_0) = y'''(x_0)$.

39.6 Example. For $f(x) = \sin(x)$ and $x_0 = 0$, we have $f(x_0) = 0$, $f'(x_0) = 1$, $f''(x_0) = 0$, and $f'''(x_0) = -1$. Therefore, according to (39.8), the third degree approximation of sine at zero is

$$y(x) = -\frac{1}{6}x^3 + x. \tag{39.9}$$

Setting $x = 1$ yields

$$\sin(1) \approx y(1) = \frac{5}{6} \approx 0.833333.$$

Given the correct value $\sin(1) \approx 0.841471$, we see that the error is now significantly smaller than in the estimate (39.4), but still greater than in (39.2). To find an estimate with $x_0 = 0$, which is better than (39.2), we would have to further increase the degree of the approximating polynomial (see Example 39.9 below).

To generalize the results of our discussion up to this point, we assume that n is a nonnegative integer and that $f : I \to \mathbb{R}$ is an n times differentiable function defined on an open interval I. Then for $x_0 \in I$, we denote by $P_{n,x_0}(x)$ the unique polynomial (see Exercise 39.8) of degree less than or equal to n that satisfies the conditions

$$\begin{aligned} f(x_0) &= P_{n,x_0}(x_0), \\ f'(x_0) &= P'_{n,x_0}(x_0), \\ &\vdots \\ f^{(n)}(x_0) &= P^{(n)}_{n,x_0}(x_0). \end{aligned} \tag{39.10}$$

The polynomial $P_{n,x_0}(x)$ is referred to as the *nth-order Taylor polynomial of f at x_0*. According to (39.1), (39.7), and (39.8), the Taylor polynomials of degrees $n = 1, 2, 3$ are

$$\begin{aligned} P_{1,x_0}(x) &= f'(x_0)(x - x_0) + f(x_0), \\ P_{2,x_0}(x) &= \frac{f''(x_0)}{2}(x - x_0)^2 + f'(x_0)(x - x_0) + f(x_0), \\ P_{3,x_0}(x) &= \frac{f'''(x_0)}{6}(x - x_0)^3 + \frac{f''(x_0)}{2}(x - x_0)^2 + f'(x_0)(x - x_0) + f(x_0). \end{aligned}$$

Continuing this pattern, it is easy to see that

$$\boxed{P_{n,x_0}(x) = \sum_{k=0}^{n} \frac{f^{(k)}(x_0)}{k!}(x - x_0)^k.} \tag{39.11}$$

39.7 Exercise. Verify that $P_{n,x_0}(x)$, as given by this formula, satisfies the conditions in (39.10).

39.8 Exercise. Show that if $p(x)$ is a polynomial of degree less than or equal to n with $p^{(k)}(x_0) = f^{(k)}(x_0)$ for $0 \le k \le n$, then $p(x) = P_{n,x_0}(x)$ for all $x \in \mathbb{R}$. *Hint.* Explain why $q(x) := p(x + x_0)$ is a polynomial and then prove that $p(x) = q(x - x_0) = P_{n,x_0}(x)$.

A simplification of the formula for $P_{n,x_0}(x)$ can be obtained for the special case $x_0 = 0$ (if $0 \in I$): setting $Q_n(x) := P_{n,0}(x)$ yields

$$\boxed{Q_n(x) = \sum_{k=0}^{n} \frac{f^{(k)}(0)}{k!}x^k.} \tag{39.12}$$

$Q_n(x)$ is referred to as the *nth-order MacLaurin polynomial of f*.

39.9 Example. For $n = 5$ and $f(x) = \sin(x)$, we have $f(0) = f''(0) = f^{(4)}(0) = 0$, $f'(0) = f^{(5)}(0) = 1$, and $f'''(0) = -1$. Hence

$$Q_5(x) = \frac{1}{5!}x^5 + \frac{0}{4!}x^4 + \frac{-1}{3!}x^3 + \frac{0}{2!}x^2 + \frac{1}{1!}x^1 + \frac{0}{0!}x^0 = \frac{x^5}{120} - \frac{x^3}{6} + x. \tag{39.13}$$

Thus, for $x = 1$ we obtain the estimate $\sin(1) \approx Q_5(1) = 101/120 \approx 0.841666$. The error is now less than 0.0002 and also smaller than in (39.2). The improvement in the approximation of $\sin(x)$ by the MacLaurin polynomials $Q_1(x)$ (see (39.2)), $Q_3(x)$ (see (39.9)), and $Q_5(x)$ is graphically illustrated in Figure 39.1.

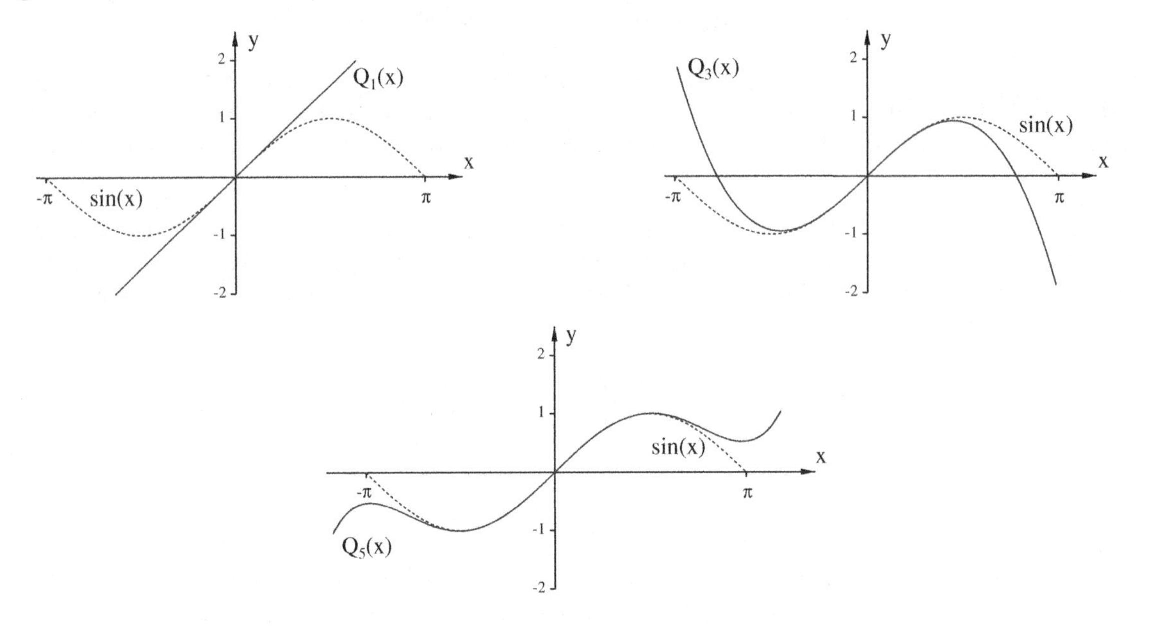

Figure 39.1: 1st, 3rd, and 5th degree approximations of $\sin(x)$.

39.10 Example. To find a general formula for the MacLaurin polynomials of $f(x) = \sin(x)$, let us take a closer look at the polynomial in (39.13). We notice that the coefficients of the even powers x^{2k} are equal to zero while those of the odd powers x^{2k+1} are of the form $\pm 1/(2k+1)!$. In other words, for any nonnegative integer n we have

$$Q_{2n+2}(x) = Q_{2n+1}(x) = \frac{\pm 1}{(2n+1)!}x^{2n+1} + \cdots + \frac{1}{5!}x^5 + \frac{-1}{3!}x^3 + \frac{1}{1!}x.$$

In order to write this result in a more concise form, using summation notation, we observe that the alternating sign in front of the coefficients can be expressed by the factor $(-1)^k$, which equals 1 for even values of k and -1 for odd ones. This yields

$$\boxed{Q_{2n+2}(x) = Q_{2n+1}(x) = \sum_{k=0}^{n} \frac{(-1)^k}{(2k+1)!}x^{2k+1}.}$$

39.11 Exercise. Given a nonnegative integer n, find a formula for the MacLaurin polynomial $Q_{2n}(x)$ for $f(x) = \cos(x)$.

Approximating π and e

39.12 Example. We wish to find an estimate for the value of π using a fifth degree MacLaurin approximation. Given that $\arcsin(1/2) = \pi/6$, we may estimate the value of $\pi/6$ by using the MacLaurin polynomials of the function $f(x) := \arcsin(x)$. Substituting the values of the derivatives of $\arcsin(x)$ (see (27.3)) into equation (39.12), it is easy (but laborious) to verify that

$$\begin{aligned} Q_1(x) &= x, \\ Q_2(x) &= x, \\ Q_3(x) &= x + \frac{1}{6}x^3, \\ Q_4(x) &= x + \frac{1}{6}x^3, \\ Q_5(x) &= x + \frac{1}{6}x^3 + \frac{3}{40}x^5. \end{aligned}$$

The corresponding approximations for the value of $\pi/6$ are $Q_1(1/2) = 1/2$, $Q_3(1/2) = 25/40$, and $Q_5(1/2) = 2009/3840 \approx 0.523177$. In multiplying the last of these values by 6, we obtain

$$\pi \approx 3.139.$$

A comparison with the correct value $\pi \approx 3.1416$ shows that the error is less than 0.003.

39.13 Example. Let us estimate the value of Euler's number e by using again a fifth degree MacLaurin approximation. Since

$$\left.\frac{d^n}{dx^n}e^x\right|_{x=0} = \left.e^x\right|_{x=0} = e^0 = 1$$

for all nonnegative integers n, the Maclaurin polynomials of $f(x) := e^x$ are

$$\begin{aligned} Q_1(x) &= 1 + x, \\ Q_2(x) &= 1 + x + \frac{1}{2}x^2, \\ Q_3(x) &= 1 + x + \frac{1}{2}x^2 + \frac{1}{6}x^3, \\ Q_4(x) &= 1 + x + \frac{1}{2}x^2 + \frac{1}{6}x^3 + \frac{1}{24}x^4, \\ Q_5(x) &= 1 + x + \frac{1}{2}x^2 + \frac{1}{6}x^3 + \frac{1}{24}x^4 + \frac{1}{120}x^5. \end{aligned}$$

This yields the estimate

$$e \approx Q_5(1) = \frac{163}{60} \approx 2.7167.$$

Given the correct value $e \approx 2.7183$, we see that the error in our estimate is less than 0.002.

In Examples 39.9, 39.12, and 39.13 we determined the margin of error by comparing our estimates with the known values for $\sin(1)$, π, and e. Approximating values that are already known is, of course, meaningful only for instructional purposes. For this reason we will discuss in the next chapter a method for estimating errors that is more self-contained and does not rely on comparisons with known results.

Additional Exercises

39.14. Find a general formula for the Taylor polynomial $P_{n,x_0}(x)$ of $f(x) = \ln(x)$ at $x_0 = 1$.

39.15. Use the Taylor polynomial of degree 7 of $f(x) = \ln(x)$ at $x_0 = 1$ to find an approximation for $\ln(1/2)$. Use the same polynomial to find approximations for $\ln(3/2)$, $\ln(2)$, and $\ln(3)$. What do you notice?

39.16. For each of the following functions find the MacLaurin polynomials $Q_1(x), \ldots, Q_n(x)$ up to the indicated degree n and use a computer to plot the graphs of the MacLaurin polynomials on the same set of axes as the graph of f.

a) $f(x) = \cos(x)$, $n = 6$ (plot the graph of $Q_k(x)$ only for even values of k)

b) $f(x) = e^x$, $n = 4$

c) $f(x) = \arctan(x)$, $n = 7$ (plot the graph of $Q_k(x)$ only for odd values of k)

39.17. Find the Taylor polynomial $P_{7,3}(x)$ of $f(x) := x^7 - 3x^5 - 4x^4 + x^3 - 5x^2 + x + 3$. What do you notice? Which general result stated in this chapter could have been used to determine $P_{7,3}(x)$ without doing any calculations?

39.18. Find the Taylor polynomial of degree 4 of each of the functions listed below at the given point x_0.

a) $f(x) = 2^x$, $x_0 = 2$ **b)** $f(x) = xe^x$, $x_0 = 0$

c) $f(x) = \dfrac{1}{x^2}$, $x_0 = 1$ **d)** $f(x) = \dfrac{1}{1-x}$, $x_0 = 0$

39.19. Find a general formula for the Taylor polynomial of degree n of each of the functions in Exercise 39.18.

39.20. Find a general formula for the MacLaurin polynomial $Q_{2n+1}(x)$ of the function $\sin(x)\cos(x)$.

39.21. Find a general formula for the MacLaurin polynomial $Q_{2n}(x)$ of the function $\cos^2(x)$.

Chapter 40

Taylor's Theorem

The Mean Value Theorem

In order to approach the problem of estimating the accuracy of Taylor approximations (see Chapter 39), we will first consider the simplest case of a Taylor approximation of degree 0. If I is an open interval and $f : I \to \mathbb{R}$ is differentiable then, according to (39.11), the Taylor polynomial $P_{0,x_0}(x)$ has the *constant* value $f(x_0)$. Thus, for $x \neq x_0$ the difference between $f(x)$ and $P_{0,x_0}(x)$ is

$$f(x) - P_{0,x_0}(x) = f(x) - f(x_0) = \frac{f(x) - f(x_0)}{x - x_0}(x - x_0). \tag{40.1}$$

It is intuitively not difficult to understand that there must exist at least one point t between x and x_0 at which the slope of the tangent line to the graph of f is equal to the slope of the secant line that connects $(x, f(x))$ with $(x_0, f(x_0))$ (see Figure 40.1). Since the slope of this secant line is $(f(x)-f(x_0))/(x-x_0)$, it follows that there is at least one point t between x and x_0 such that $(f(x) - f(x_0))/(x - x_0) = f'(t)$. This statement is known as the *mean value theorem* of differentiation:

40.1 Theorem. *If $f : [a, b] \to \mathbb{R}$ is a continuous function such that f is differentiable on (a, b), then there is a point $t \in (a, b)$ that satisfies the equation*

$$\frac{f(b) - f(a)}{b - a} = f'(t).$$

For a rigorous proof of this theorem the reader is referred to [Rud1].

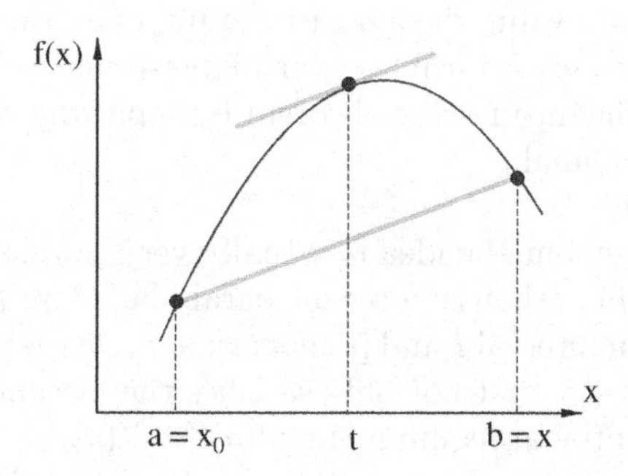

Figure 40.1: the mean value theorem.

Applying the mean value theorem to equation (40.1) (with $a = x_0$ and $b = x$ if $x > x_0$, or $a = x$ and $b = x_0$ if $x < x_0$), we may infer that

$$f(x) - P_{0,x_0}(x) = f(x) - f(x_0) = f'(t)(x - x_0) \text{ for some } t \text{ between } x \text{ and } x_0. \tag{40.2}$$

Since there is no general method for determining the value of t (Theorem 40.1 only asserts its existence), an equation like (40.2) is useful mainly for finding estimates.

40.2 Example. For the function $f(x) := \sin(x)$ we have

$$-1 \leq f'(x) = \cos(x) \leq 1 \text{ for all } x \in \mathbb{R},$$

and therefore, equation (40.2) implies that

$$|f(x) - P_{0,x_0}(x)| = |f'(t)(x - x_0)| = |f'(t)||x - x_0| \leq 1 \cdot |x - x_0| = |x - x_0|.$$

Since $P_{0,x_0}(x) = \sin(x_0)$, it follows that the absolute value of the difference between $\sin(x)$ and $\sin(x_0)$ is always less than or equal to the absolute value of the difference between x and x_0. While not entirely trivial, this statement is still rather crude and we will have to improve our method for estimating errors significantly before we can come up with more useful results.

40.3 Exercise. Use the mean value theorem to show that $|\arctan(x) - \arctan(x_0)| \leq |x - x_0|$ for all $x, x_0 \in \mathbb{R}$.

Remark. In Chapter 8, p.64 we mentioned that a function must be constant if it is defined on an open interval and has a vanishing derivative. Using the mean value theorem, we are now able to give a rigorous proof of this fact. Assume that $f : I \to \mathbb{R}$ is a differentiable function defined on an open interval I such that $f'(x) = 0$ for all $x \in I$. In order to show that f is constant, it suffices to prove that $f(x_1) = f(x_2)$ for all $x_1, x_2 \in I$ with $x_1 \neq x_2$. (Think about it!) If $x_1 \neq x_2$, then according to the mean value theorem, there is a t between x_1 and x_2 such that

$$f(x_2) - f(x_1) = f'(t)(x_2 - x_1).$$

Since by assumption $f'(t) = 0$, it follows that $f(x_2) - f(x_1) = 0$, and therefore $f(x_1) = f(x_2)$ as desired. This result also allows us to rigorously establish Theorem 14.2, because if $F : I \to \mathbb{R}$ and $G : I \to \mathbb{R}$ are differentiable functions defined on an open interval I such that $F' = G'$, then $(F - G)' = 0$ and, by implication, $F - G$ must be constant. In other words, there must exist a constant $C \in \mathbb{R}$ such that $G(x) = F(x) + C$ for all $x \in I$.

40.4 Exercise. Give an example of a function $f : D \subset \mathbb{R} \to \mathbb{R}$ that satisfies the following conditions: f is differentiable, $f'(x) = 0$ for all $x \in D$, and f is *not* constant.

Taylor's Theorem

Teacher: In the previous section we arrived at an estimate for the error in the approximation of $f(x)$ by $P_{0,x_0}(x)$ by applying the mean value theorem to the interval (x_0, x) (or (x, x_0) if $x < x_0$). In order to generalize this approach to error estimates for higher-order approximations of $f(x)$ by $P_{n,x_0}(x)$, we first need to generalize the mean value theorem by replacing the secant line in Figure 40.1 with a higher-order "secant polynomial."

Simplicio: A secant polynomial?

Teacher: This may sound strange, but the idea is actually very simple. To get started we will consider a second-order secant polynomial, that is, a secant parabola. More precisely, for a twice differentiable function f defined on an open interval I and points $a, b \in I$, we wish to find a second-order polynomial $p(x) = \alpha(x - a)^2 + \beta(x - a) + \gamma$ that not only satisfies the (secant line) conditions $p(a) = f(a)$ and $p(b) = f(b)$ but also the additional requirement $p'(a) = f'(a)$.

Sophie: Am I right in assuming that we are going to use these conditions to determine the coefficients α, β, and γ?

Teacher: Yes indeed. From the given equation $p(x) = \alpha(x - a)^2 + \beta(x - a) + \gamma$, we easily deduce that $p(a) = \gamma$ and $p'(a) = \beta$. Consequently, the conditions $p(a) = f(a)$ and $p'(a) = f'(a)$ imply that $\beta = f'(a)$ and $\gamma = f(a)$. Given these observations, the condition $p(b) = f(b)$ allows us to infer that

$$f(b) = \alpha(b - a)^2 + \beta(b - a) + \gamma = \alpha(b - a)^2 + f'(a)(b - a) + f(a),$$

and solving for α yields

$$\alpha = \frac{f(b) - f'(a)(b-a) - f(a)}{(b-a)^2}.$$

So the secant polynomial p is given by the equation

$$p(x) = \frac{f(b) - f'(a)(b-a) - f(a)}{(b-a)^2}(x-a)^2 + f'(a)(x-a) + f(a).$$

Simplicio: I understand how we found the coefficients of p, but I do not see how this is going to lead us to a generalization of the mean value theorem, or how exactly such a generalization should be formulated.

Teacher: Here is how it works: the mean value theorem as stated in Theorem 40.1 says that there is at least one point between a and b at which the derivative of the function f is equal to the slope or, equivalently, the derivative of the secant line. Given that we are now considering a secant parabola instead of a secant line, it may not be altogether unreasonable to expect that there ought to be a point between a and b where instead of equal first derivatives we now have equal *second* derivatives. In other words, we wish to show that

$$f''(t) = p''(t) = \frac{2(f(b) - f'(a)(b-a) - f(a))}{(b-a)^2} \text{ for some } t \text{ between } a \text{ and } b. \tag{40.3}$$

Sophie: The formal analogy between Theorem 40.1 and equation (40.3) is fairly obvious, but intuitively speaking, the assertion in (40.3) appears considerably more obscure than the corresponding assertion of equality of slopes in 40.1.

Teacher: I agree. The result in (40.3) is undoubtedly more abstract than the original mean value theorem, because second derivatives are conceptually less transparent than first derivatives. Since there is unfortunately little we can do to overcome this difficulty, we will have to be content to establish the validity of (40.3) on a purely formal level. To do so we define a function g as the difference of f and p:

$$g(x) := f(x) - p(x).$$

Using the equations $p(a) = f(a)$ and $p(b) = f(b)$, it follows that $g(a) = g(b) = 0$, and therefore, the mean value theorem, applied to g, implies that there is a point t_0 between a and b such that

$$g'(t_0) = \frac{g(b) - g(a)}{b-a} = 0.$$

Since the condition $p'(a) = f'(a)$ implies that $g'(a) = 0$, we may again apply the mean value theorem (this time to g' instead of g) to infer the existence of a point t between a and t_0 that satisfies the following equation:

$$0 = \frac{g'(t_0) - g'(a)}{t_0 - a} = g''(t) = f''(t) - p''(t).$$

Consequently, $f''(t) = p''(t)$ as desired.

In order to apply (40.3) to the problem of estimating the error in the approximation of $f(x)$ by $P_{1,x_0}(x)$, we set $a = x_0$ and $b = x$ to conclude that there exists a t between x_0 and x such that

$$f''(t) = \frac{2(f(x) - f'(x_0)(x-x_0) - f(x_0))}{(x-x_0)^2} = \frac{2(f(x) - P_{1,x_0}(x))}{(x-x_0)^2}.$$

In rearranging the terms, we arrive at the following result:

$$\boxed{f(x) - P_{1,x_0}(x) = \frac{f''(t)}{2}(x-x_0)^2 \text{ for some } t \text{ between } x \text{ and } x_0.} \tag{40.4}$$

40.5 Example. Let us again consider the function $f(x) := \sin(x)$. Since $f''(x) = -\sin(x)$, we may infer that

$$|f''(x)| \leq 1 \text{ for all } x \in \mathbb{R}.$$

Therefore, (40.4) implies that

$$|f(x) - P_{1,x_0}(x)| = \frac{1}{2}|f''(t)|(x - x_0)^2 \leq \frac{1}{2}(x - x_0)^2.$$

Choosing, for example, $x = 1$ and $x_0 = \pi/3$, we obtain

$$|\sin(1) - P_{1,\pi/3}(1)| \leq \frac{1}{2}\left(1 - \frac{\pi}{3}\right)^2 \approx 0.0011.$$

This estimate for the error in the Taylor approximation of $\sin(1)$ by $P_{1,\pi/3}(1)$ is very reasonable, because we saw in Example 39.1 that the actual error is slightly less than 0.001.

40.6 Exercise. Use equation (40.4) to show that $1 - \cos(x) \leq \pi^2/72$ for all $x \in [-\pi/6, \pi/6]$. *Hint.* Determine the MacLaurin polynomial $Q_1(x)$ for $f(x) := \cos(x)$.

A generalization of the equations (40.2) and (40.4) to the case of Taylor polynomials of degree n is given in Theorem 40.7 below. This generalized result is known as *Taylor's theorem*, and its proof follows essentially the same idea as the proof of (40.4).

40.7 Theorem. *Assume that $f : I \to \mathbb{R}$ is an $n+1$ times differentiable function on an open interval I. Then for all $x, x_0 \in I$ with $x \neq x_0$ there exists a t between x and x_0 (i.e., $t \in (x, x_0) \cup (x_0, x)$) such that*

$$f(x) - P_{n,x_0}(x) = \frac{f^{(n+1)}(t)}{(n+1)!}(x - x_0)^{n+1}.$$

Proof. ★ Let us define

$$g(y) := f(y) - P_{n,x_0}(y) - \frac{f(x) - P_{n,x_0}(x)}{(x - x_0)^{n+1}}(y - x_0)^{n+1}.$$

Since $P_{n,x_0}(x_0) = f(x_0)$, it is easy to see that $g(x) = 0 = g(x_0)$. Thus, there exists a t_0 between x and x_0 such that $g'(t_0) = 0$. Furthermore, the derivative of g is

$$g'(y) = f'(y) - P'_{n,x_0}(y) - (n+1)\frac{f(x) - P_{n,x_0}(x)}{(x - x_0)^{n+1}}(y - x_0)^n.$$

Since $P'_{n,x_0}(x_0) = f'(x_0)$, it follows that $g'(t_0) = 0 = g'(x_0)$. Consequently, there exists a t_1 between t_0 and x_0 such that $g''(t_1) = 0$. Using the fact that $P^{(k)}_{n,x_0}(x_0) = f^{(k)}(x_0)$ for all $k \in \{0, \ldots, n\}$, we can continue this process to find values $t_0, \ldots, t_n$ such that t_k is between t_{k-1} and x_0 and $g^{(k+1)}(t_k) = 0$. Then, for $t := t_n$ we obtain

$$0 = g^{(n+1)}(t) = f^{(n+1)}(t) - (n+1)!\frac{f(x) - P_{n,x_0}(x)}{(x - x_0)^{n+1}},$$

because $P^{(n+1)}_{n,x_0}(y) = 0$. (Note: a derivative of order greater than n of a polynomial of degree less than or equal to n is always zero.) Hence

$$f(x) - P_{n,x_0}(x) = \frac{f^{(n+1)}(t)}{(n+1)!}(x - x_0)^{n+1},$$

as desired. ★ □

40.8 Exercise. Use Taylor's theorem to show that the following inequality is valid for all $x \in \mathbb{R}$:

$$|\sin(x) - Q_n(x)| \leq \frac{|x|^{n+1}}{(n+1)!}.$$

An Error Estimate for e

40.9 Example. For $f(x) := e^x$, we wish to derive an estimate for the error in the approximation of $f(1) = e$ by MacLaurin polynomials without assuming any prior knowledge concerning the value of e. To get started, we will derive a rough upper estimate for e from the defining equation

$$\ln(e) = \int_1^e \frac{1}{t}\,dt = 1 \quad \text{(see Chapter 22, p.179).} \tag{40.5}$$

Since the function $g(t) := 1/t$ is concave up on $(0, \infty)$, we may use a midpoint sum (see Theorem 37.3) to obtain the following lower estimate for $\int_1^3 g(t)\,dt$:

$$\int_1^3 \frac{1}{t}\,dt = \int_1^3 g(t)\,dt \geq \mathrm{Mid}(2, g, 1, 3) = g\left(\frac{3}{2}\right) + g\left(\frac{5}{2}\right) = \frac{16}{15} > 1.$$

Since the integral of g over the interval $[1, 3]$ is greater than 1, equation (40.5) implies that $3 > e$. Furthermore, since $f^{(k)}(x) = e^x$ for all integers $k \geq 0$, the nth degree MacLaurin polynomial of f is

$$Q_n(x) = \sum_{k=0}^{n} \frac{f^{(k)}(0)}{k!} x^k = \sum_{k=0}^{n} \frac{e^0}{k!} x^k = \sum_{k=0}^{n} \frac{x^k}{k!} \quad \text{(see (39.12)).}$$

According to Taylor's theorem, we can find a point t between 0 and 1 such that

$$|e - Q_n(1)| = |f(1) - Q_n(1)| = \frac{|f^{(n+1)}(t)|}{(n+1)!} |1 - 0|^{n+1} = \frac{e^t}{(n+1)!}.$$

Using the estimate $e < 3$ as derived above and the fact that the exponential function is increasing, we may infer that $e^t \leq e^1 < 3$ for all $t \leq 1$. Hence

$$|e - Q_n(1)| < \frac{3}{(n+1)!}. \tag{40.6}$$

For $n = 5$ (see Example 39.13) we obtain

$$|e - Q_5(1)| < \frac{3}{720} = \frac{1}{240} \approx 0.0042.$$

So without assuming any prior knowledge concerning the value of e, we are able to deduce that e differs from $Q_5(1) \approx 2.7167$ by less than 0.0042.

As n increases, the upper bound for the error in (40.6) rapidly decreases to zero. For instance, for $n = 10$ we have

$$|e - Q_{10}(1)| < \frac{3}{11!} \approx 7.5 \cdot 10^{-8}.$$

Since $Q_{10}(1) \approx 2.718281801$, we may infer that

$$2.718281725 \approx Q_{10}(1) - 7.5 \cdot 10^{-8} < e < Q_{10}(1) + 7.5 \cdot 10^{-8} \approx 2.718281876.$$

Choosing larger values for n, the estimate can be made even more precise.

Additional Exercises

40.10. Use a Taylor polynomial of degree 5 for $f(x) = \ln(x)$ and $x_0 = 1$ to find an approximation for $\ln(3/2)$. Then apply Taylor's theorem to estimate the accuracy of the approximation.

40.11. Use the MacLaurin polynomial $Q_1(x)$ for $f(x) = \sin(x)$ to show that $\lim_{x\to 0} \sin(x)/x = 1$. *Hint.* Use equation (40.4) with $x_0 = 0$ to replace $\sin(x)$ with $Q_1(x) - \sin(t)x^2/2$.

40.12. Use the MacLaurin polynomial $Q_2(x)$ for $f(x) = \cos(x)$ to show that $\lim_{x\to 0}(1 - \cos(x))/x^2 = 1/2$. *Hint.* Think about it.

40.13. Use Exercise 40.8 and Example 39.10 to find an estimate for $\sin(2)$ with an error less than 10^{-4}.

40.14. Use a MacLaurin polynomial of degree 7 for $f(x) = \arcsin(x)$ to find an estimate for π. Then use Taylor's theorem to obtain an estimate for the error in the approximation. *Hint.* Take a look again at Example 39.12.

40.15. ★ Prove the following generalization of the mean value theorem: assume that $f : [a, b] \to \mathbb{R}$ and $g : [a, b] \to \mathbb{R}$ are continuous functions that are differentiable on (a, b). If $g(b) \neq g(a)$, then there exists a $t \in (a, b)$ such that

$$\boxed{\frac{f(b) - f(a)}{g(b) - g(a)} = \frac{f'(t)}{g'(t)}.}$$

Hint. Apply the ordinary mean value theorem to the function

$$h(x) := f(x) - f(a) - \frac{f(b) - f(a)}{g(b) - g(a)}(g(x) - g(a)).$$

40.16. Prove the mean value theorem for integrals: if $f : [a, b] \to \mathbb{R}$ is a continuous function, then there exists a $t \in (a, b)$ such that

$$\boxed{f(t) = \frac{1}{b - a}\int_a^b f(x)\,dx.}$$

Hint. Apply Theorem 40.1 to the function $g(x) := \int_a^x f(u)\,du$.

40.17. Find a counterexample to each of the following statements:

a) If $f : [a, b] \to \mathbb{R}$ is continuous on (a, b) and differentiable on (a, b), then there is a $t \in (a, b)$ such that $f'(t) = (f(b) - f(a))/(b - a)$.

b) If $f : [a, b] \to \mathbb{R}$ is piecewise continuous, then there is a $t \in [a, b]$ such that $f(t) = \int_a^b f(x)\,dx/(b-a)$.

40.18. For the function $f(x) := x^2$, find the second-order secant polynomial $p(x)$ that satisfies the following conditions: $p(1) = f(1)$, $p(3) = f(3)$, and $p'(1) = f'(1)$. What do you notice? Form a general conjecture and prove it.

40.19. Assume that $f : \mathbb{R} \to \mathbb{R}$ is a differentiable function. Which of the following conditions guarantees the existence of a critical point of f in the interval $(2, 4)$? Explain your answer.

a) $f(2) = 1$ and $f(4) = 3$ b) $f(2) = 1$ and $f(4) = 1$ c) $f(2) = 2$ and $f(4) = 0$

40.20. Prove the statement of Exercise 39.8 by means of Taylor's theorem.

Chapter 41

Infinite Series

From Taylor Polynomials to Taylor Series

In Chapters 39 and 40 we developed techniques for approximating functions by polynomials and for estimating the corresponding errors. In several examples we saw that the accuracy of the approximation increased with the degree of the approximating polynomial. In the present chapter we will discuss how the error can be reduced to zero by replacing the approximating Taylor polynomials P_{n,x_0} with infinite sums of the form

$$\sum_{k=0}^{\infty} \frac{f^{(k)}(x_0)}{k!}(x-x_0)^k. \tag{41.1}$$

In general, infinite sums are referred to as *(infinite) series*, and the particular sum in (41.1) is said to be a *Taylor series*. For $x_0 = 0$, the Taylor series (41.1) simplifies to the *MacLaurin series*

$$\sum_{k=0}^{\infty} \frac{f^{(k)}(0)}{k!}x^k.$$

41.1 Example. Let us consider the exponential function $f(x) := e^x$. In Example 40.9, we saw that the MacLaurin polynomials of f are of the form

$$Q_n(x) = \sum_{k=0}^{n} \frac{x^k}{k!}.$$

Given an arbitrary value $x \neq 0$, we may employ Taylor's theorem to find a t between x and 0 such that

$$|f(x) - Q_n(x)| = \frac{|f^{(n+1)}(t)|}{(n+1)!}|x|^{n+1} = e^t\frac{|x|^{n+1}}{(n+1)!} \leq e^{|x|}\frac{|x|^{n+1}}{(n+1)!}. \tag{41.2}$$

We wish to show that the error $|f(x) - Q_n(x)|$ converges to zero as n tends to infinity. For this purpose it is, according to (41.2), sufficient to prove that

$$\lim_{n\to\infty} e^{|x|}\frac{|x|^{n+1}}{(n+1)!} = 0 \text{ for all } x \in \mathbb{R},$$

or equivalently,

$$\lim_{n\to\infty} \frac{|x|^n}{n!} = 0 \text{ for all } x \in \mathbb{R}. \tag{41.3}$$

For a given $x \in \mathbb{R}$ we choose a positive integer m such that $m > 2|x|$. Then

$$\frac{|x|^n}{n!} = \frac{|x|^m}{m!}\prod_{k=m+1}^{n}\frac{|x|}{k} < \frac{|x|^m}{m!}\prod_{k=m+1}^{n}\frac{1}{2} = \frac{|x|^m}{2^{n-m}m!} \text{ for all } n > m.$$

Since $\lim_{n\to\infty} 1/2^n = 0$, it follows that

$$0 \leq \lim_{n\to\infty} \frac{|x|^n}{n!} \leq \lim_{n\to\infty} \frac{2^m|x|^m}{2^n m!} = \frac{2^m|x|^m}{m!} \lim_{n\to\infty} \frac{1}{2^n} = 0.$$

This proves (41.3) and according to (41.2) it therefore follows that

$$\lim_{n\to\infty} |f(x) - Q_n(x)| = 0$$

or equivalently

$$e^x = f(x) = \lim_{n\to\infty} Q_n(x). \tag{41.4}$$

Introducing the definition

$$\sum_{k=0}^{\infty} \frac{x^k}{k!} := \lim_{n\to\infty} \sum_{k=0}^{n} \frac{x^k}{k!} = \lim_{n\to\infty} Q_n(x),$$

we may restate the result in (41.4) in the following form:

$$\boxed{\sum_{k=0}^{\infty} \frac{x^k}{k!} = e^x \text{ for all } x \in \mathbb{R}.}$$

Thus, we have demonstrated that, at least in the special case of the exponential function, we can reduce the error in the MacLaurin approximation to zero by increasing the degree of the approximating polynomial to infinity. To generalize this result will be the main purpose of this chapter and the next.

41.2 Exercise. Use Taylor's theorem to show that for the MacLaurin polynomials $Q_n(x)$ of $f(x) = \cos(x)$ we have $\cos(x) = \lim_{n\to\infty} Q_n(x)$ for all $x \in \mathbb{R}$.

Infinite Series and Convergence Tests

Before stating a general definition for infinite series we define, for a given integer m, a *sequence* to be an infinite succession of real numbers $a_m, a_{m+1}, a_{m+2}, \ldots$ for which we use the notation $(a_n)_{n=m}^{\infty}$ (see Chapter 3 for a more rigorous but slightly less general definition of sequences).

41.3 Definition. If $(a_n)_{n=m}^{\infty}$ is a sequence, then we call $S_n := \sum_{k=m}^{n} a_k$ a *partial sum* of the *series* $\sum_{k=m}^{\infty} a_k$. If $\lim_{n\to\infty} S_n$ exists and is finite, then the series

$$\sum_{k=m}^{\infty} a_k := \lim_{n\to\infty} S_n$$

is said to be *convergent*. Otherwise, if $\lim_{n\to\infty} S_n$ is either infinite or nonexistent, the series $\sum_{k=m}^{\infty} a_k$ is said to be *divergent*.

41.4 Example. For the series in Example 41.1 we have $m = 0$ and $a_k = x^k/k!$. Furthermore, the series is convergent because we showed the limit of its partial sums $S_n = Q_n(x)$ to be equal to e^x.

The following theorem establishes two simple properties of infinite series that are direct consequences of Theorem 2.9 (or more precisely, of an appropriate generalization of Theorem 2.9 to limits at infinity).

41.5 Theorem. *Let $\lambda \in \mathbb{R}$. If $\sum_{k=m}^{\infty} a_k$ and $\sum_{k=m}^{\infty} b_k$ are convergent series, then $\sum_{k=m}^{\infty}(a_k + b_k)$ and $\sum_{k=m}^{\infty} \lambda a_k$ are convergent as well and*

$$\sum_{k=m}^{\infty} (a_k + b_k) = \sum_{k=m}^{\infty} a_k + \sum_{k=m}^{\infty} b_k,$$

$$\sum_{k=m}^{\infty} \lambda a_k = \lambda \sum_{k=m}^{\infty} a_k.$$

41.6 Exercise. Use Definition 41.3 and Theorem 2.9 to prove Theorem 41.5.

Now we will discuss a number of *convergence tests* that help us decide whether a given series is convergent or divergent. The first of these is the *comparison test*:

41.7 Theorem. *If $(a_n)_{n=1}^{\infty}$ and $(b_n)_{n=1}^{\infty}$ are sequences such that $\sum_{k=m}^{\infty} b_k$ is convergent and $b_n \geq |a_n|$ for all positive integers n, then $\sum_{k=m}^{\infty} a_k$ is convergent as well. Furthermore, if $\sum_{k=m}^{\infty} b_k$ is divergent and $0 \leq b_n \leq a_n$ for all positive integers n, then $\sum_{k=m}^{\infty} a_k$ is divergent as well.*

Proof. ★ Let us define functions $f : [m-1, \infty) \to \mathbb{R}$ and $g : [m-1, \infty) \to \mathbb{R}$ via the equations

$$f(x) := a_n \text{ and } g(x) := b_n \text{ for all integers } n \geq m \text{ and all } x \in [n-1, n).$$

Then the following equations are easily seen to be valid for all integers $n \geq m$:

$$\sum_{k=m}^{n} a_k = \int_{m-1}^{n} f(x)\,dx \quad \text{and} \quad \sum_{k=m}^{n} b_k = \int_{m-1}^{n} g(x)\,dx.$$

Given this observation, the statement of Theorem 41.7 is a direct consequence of Theorem 38.7. ★ □

41.8 Exercise. Assume that $(a_n)_{n=1}^{\infty}$ is a sequence such that $\sum_{k=m}^{\infty} |a_k|$ is convergent. Use the comparison test to show that $\sum_{k=m}^{\infty} a_k$ is convergent as well.

Intuitively, Theorem 41.7 is not difficult to understand: if $b_n \geq |a_n|$, then the partial sums of the series $\sum_{k=m}^{\infty} b_k$ are greater than or equal to the partial sums of the series $\sum_{k=m}^{\infty} |a_k|$. Therefore, it is natural to expect that the convergence of the partial sums of $\sum_{k=m}^{\infty} b_k$ to a finite value implies that the partial sums of $\sum_{k=m}^{\infty} a_k$ converge to a finite value as well. Similarly, if $\sum_{k=m}^{\infty} b_k$ is divergent and $0 \leq b_n \leq a_n$, then we have $\lim_{n\to\infty} \sum_{k=m}^{n} b_k = \infty$ and therefore also $\lim_{n\to\infty} \sum_{k=m}^{n} a_k = \infty$.

Comparison tests are of course only useful if we have some independent means of establishing the convergence or divergence of the comparison series $\sum_{k=m}^{\infty} b_k$. A tool that is frequently very helpful in this context is the *integral test:*

41.9 Theorem. *If, for a given integer m the function $f : [m, \infty) \to \mathbb{R}$ is positive, decreasing, and piecewise continuous on every interval $[m, c]$ for $c > m$, then $\sum_{k=m}^{\infty} f(k)$ is convergent if and only if $\int_m^{\infty} f(x)\,dx$ is convergent.*

Proof. For any integer $k \geq m$ we define $b_k := \int_k^{k+1} f(x)\,dx$. Assuming that $\int_m^{\infty} f(x)\,dx$ is convergent, we obtain

$$\infty > \int_m^{\infty} f(x)\,dx = \lim_{n\to\infty} \int_m^{n+1} f(x)\,dx = \lim_{n\to\infty} \sum_{k=m}^{n} b_k.$$

This shows that $\sum_{k=m}^{\infty} b_k$ is convergent and, according to the comparison test, $\sum_{k=m}^{\infty} f(k)$ is convergent as well because if f is positive and decreasing, then $b_{k-1} \geq f(k) = |f(k)|$ for all $k \geq m+1$. If on the other hand $\int_m^{\infty} f(x)\,dx$ is divergent, then

$$\infty = \int_m^{\infty} f(x)\,dx = \lim_{n\to\infty} \sum_{k=m}^{n} b_k.$$

So in this case $\sum_{k=m}^{\infty} b_k$ is divergent and, referring again to the comparison test, we may conclude that $\sum_{k=m}^{\infty} f(k)$ is divergent as well because $0 \leq b_k \leq f(k)$ for all $k \geq m$. This proves that $\int_m^{\infty} f(x)\,dx$ is convergent if and only if $\sum_{k=m}^{\infty} f(k)$ is convergent. □

To give an intuitive explanation for the integral test we observe that the series $\sum_{k=m+1}^{\infty} f(k)$ is a *lower* estimate for the value of $\int_m^{\infty} f(x)\,dx$ (see Figure 41.1). Consequently, if the value of the integral $\int_m^{\infty} f(x)\,dx$ is finite, the value of the series $\sum_{k=m}^{\infty} f(k)$ must be finite as well. In other words, the series must be convergent. Similarly, the fact that $\sum_{k=m}^{\infty} f(k)$ is an *upper* estimate for $\int_m^{\infty} f(x)\,dx$ (see Figure 41.1) implies that $\sum_{k=m}^{\infty} f(k)$ is divergent whenever $\int_m^{\infty} f(x)\,dx$ is known to be divergent as well.

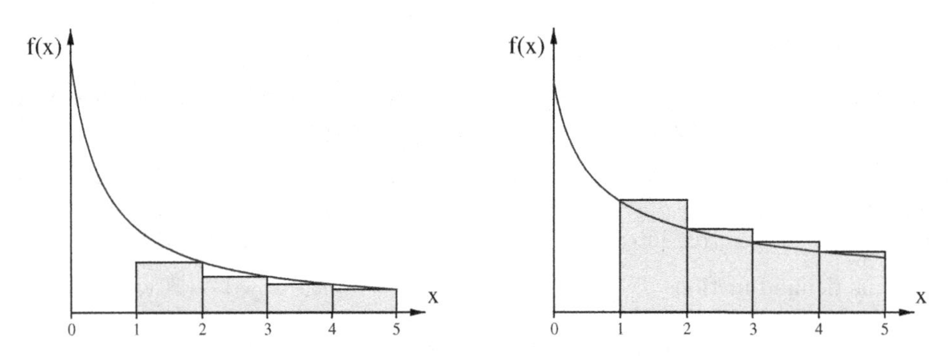

Figure 41.1: integral test for convergent and divergent series.

41.10 Example. Let us consider the *harmonic series*

$$\sum_{k=1}^{\infty} \frac{1}{k}.$$

The function $f(x) := 1/x$ is positive and decreasing on the interval $[1, \infty)$. Furthermore, according to (38.1) we have $\int_1^{\infty} 1/x\, dx = \infty$, and therefore $\sum_{k=1}^{\infty} 1/k = \sum_{k=1}^{\infty} f(k)$ is divergent.

41.11 Example. As another example, let us consider the series

$$\sum_{k=1}^{\infty} \frac{1}{k^2}.$$

The function $f(x) := 1/x^2$ is positive and decreasing on $[1, \infty)$ and, according to (38.1), we have $\int_1^{\infty} 1/x^2\, dx = 1$. Consequently, $\sum_{k=1}^{\infty} 1/k^2 = \sum_{k=1}^{\infty} f(k)$ is convergent.

41.12 Exercise. Use the integral test to decide which of the following series are convergent:

$$\sum_{k=0}^{\infty} e^{-k}, \quad \sum_{k=2}^{\infty} \frac{1}{k \ln(k)}, \quad \text{and} \quad \sum_{k=2}^{\infty} \frac{1}{k \ln^2(k)}.$$

The next theorem is commonly referred to as the *sequence test.*

41.13 Theorem. *If $\sum_{k=m}^{\infty} a_k$ is a convergent series, then the sequence $(a_n)_{n=m}^{\infty}$ converges to zero, i.e., $\lim_{n\to\infty} a_n = 0$.*

Proof. If $\sum_{k=m}^{\infty} a_k$ is convergent, then

$$\sum_{k=m}^{\infty} a_k = \lim_{n\to\infty} \sum_{k=m}^{n} a_k = \lim_{n\to\infty} \sum_{k=m}^{n-1} a_k.$$

Hence

$$0 = \lim_{n\to\infty} \sum_{k=m}^{n} a_k - \lim_{n\to\infty} \sum_{k=m}^{n-1} a_k = \lim_{n\to\infty} \left(\sum_{k=m}^{n} a_k - \sum_{k=m}^{n-1} a_k \right) = \lim_{n\to\infty} a_n$$

as desired. □

Simplicio: It seems to me there is a contradiction between the statement of Theorem 41.13 and the result of Example 41.10.
Teacher: Why is that?
Simplicio: Isn't it true that the sequence $(1/k)_{k=1}^{\infty}$ converges to zero?
Teacher: Certainly.

Simplicio: But then, according to Theorem 41.13, the series $\sum_{k=1}^{\infty} 1/k$ should be convergent rather than divergent as we claimed in Example 41.10.

Sophie: No, Simplicio, Theorem 41.13 only asserts that the convergence of the series implies the convergence of the sequence to zero. Your conclusion goes in the wrong direction—from the convergence of the sequence to zero to the convergence of the series.

Teacher: Sophie is right. Theorem 41.13 is not a statement of equivalence, and it ought to be viewed as a *divergence test* rather than a convergence test: since the series $\sum_{k=m}^{\infty} a_k$ is convergent only if the sequence $(a_n)_{n=m}^{\infty}$ converges to zero, it follows that the series $\sum_{k=m}^{\infty} a_k$ is *divergent* whenever the sequence $(a_n)_{n=m}^{\infty}$ does not converge to zero. For instance, for $a_n := (-1)^n$ the limit $\lim_{n\to\infty} a_n$ does not exist (because the values of a_n alternate between 1 and -1), and therefore the series $\sum_{k=1}^{\infty} a_k = \sum_{k=1}^{\infty}(-1)^k$ is divergent.

41.14 Exercise. Show that the series $\sum_{k=1}^{\infty} \sin(\pi k/2)$ is divergent.

41.15 Example. For a given $q \in \mathbb{R}$ we refer to

$$\sum_{k=0}^{\infty} q^k$$

as a *geometric series.* Since

$$\lim_{n\to\infty} q^n = \begin{cases} 0 & \text{if } -1 < q < 1, \\ 1 & \text{if } q = 1, \\ \infty & \text{if } q > 1, \\ \text{does not exist} & \text{if } q \leq -1, \end{cases} \tag{41.5}$$

the sequence test (Theorem 41.13) implies that a geometric series is divergent whenever $|q| \geq 1$. Furthermore, considering the partial sums $S_n = \sum_{k=0}^{n} q^k$, we see that

$$(1-q)S_n = (1-q)\sum_{k=0}^{n} q^k = \sum_{k=0}^{n} q^k - \sum_{k=0}^{n} q^{k+1} = \sum_{k=0}^{n} q^k - \sum_{k=1}^{n+1} q^k = 1 - q^{n+1},$$

and for $q \neq 1$ it follows that

$$S_n = \frac{1-q^{n+1}}{1-q}. \tag{41.6}$$

Using (41.5), (41.6), and Definition 41.3, we may infer that a geometric series is convergent whenever $|q| < 1$ because

$$\sum_{k=0}^{\infty} q^k = \lim_{n\to\infty} S_n = \lim_{n\to\infty} \frac{1-q^{n+1}}{1-q} = \frac{1-\lim_{n\to\infty} q^{n+1}}{1-q} = \frac{1}{1-q}.$$

We summarize our results in the following theorem:

41.16 Theorem. *For $q \in \mathbb{R}$ the geometric series $\sum_{k=0}^{\infty} q^k$ is divergent if $|q| \geq 1$ and it is convergent if $|q| < 1$. In the latter case we have*

$$\sum_{k=0}^{\infty} q^k = \frac{1}{1-q}.$$

41.17 Exercise. Prove the statement in (41.5).

41.18 Example. We wish to determine the value of the series

$$\sum_{k=1}^{\infty} \frac{1}{4^k}. \tag{41.7}$$

According to Theorem 41.16 with $q = 1/4$, we have

$$\sum_{k=0}^{\infty} \left(\frac{1}{4}\right)^k = \frac{1}{1-1/4} = \frac{4}{3}.$$

The first term in the series on the right-hand side of the equation is $(1/4)^0 = 1$. Since this term is missing in the series (41.7), which starts at $k = 1$, we may conclude that

$$\sum_{k=1}^{\infty} \frac{1}{4^k} = \sum_{k=0}^{\infty} \left(\frac{1}{4}\right)^k - 1 = \frac{4}{3} - 1 = \frac{1}{3}.$$

The same result can also be obtained by factoring out $1/4$ and changing the index of summation:

$$\sum_{k=1}^{\infty} \frac{1}{4^k} = \frac{1}{4}\sum_{k=1}^{\infty} \frac{1}{4^{k-1}} = \frac{1}{4}\sum_{k=0}^{\infty} \frac{1}{4^k} = \frac{1}{4} \cdot \frac{4}{3} = \frac{1}{3}.$$

41.19 Exercise. Find the value of the series $\sum_{k=2}^{\infty} 1/5^k$.

The most important convergence test in the context of representing functions by Taylor series is the *root test*:

41.20 Theorem. *Assume that $(a_n)_{n=m}^{\infty}$ is a sequence for which the limit $L = \lim_{n\to\infty} \sqrt[n]{|a_n|}$ exists. Then*

a) $\sum_{k=m}^{\infty} a_k$ *is convergent if $L < 1$,*
b) $\sum_{k=m}^{\infty} a_k$ *is divergent if $L > 1$, and*
c) *the test is inconclusive if $L = 1$.*

Proof. **a)** If $L < 1$, then for $q := (1 + L)/2$ we have $L < q < 1$. This implies that $\sqrt[n]{|a_n|} < q$ for all sufficiently large n. More precisely, there exists an integer $N \geq m$ such that $\sqrt[n]{|a_n|} < q$ for all $n \geq N$, or equivalently $|a_n| < q^n$ for all $n \geq N$. Since $0 < q < 1$, Theorem 41.16 implies that $\sum_{k=0}^{\infty} q^k$ is convergent and, therefore, the comparison test (Theorem 41.7) allows us to infer that $\sum_{k=m}^{\infty} a_k$ is convergent as well.
b) If $L > 1$, then there is an integer $N \geq m$ such that $\sqrt[n]{|a_n|} > 1$ for all $n \geq N$. Hence $|a_n| > 1$ for all $n \geq N$. This shows that $(a_n)_{n=m}^{\infty}$ does not converge to zero, and according to the sequence test (Theorem 41.13) we may conclude that $\sum_{n=1}^{\infty} a_n$ is divergent.
c) In Examples 41.10 and 41.11, we showed that $\sum_{k=1}^{\infty} 1/k$ is divergent and that $\sum_{k=1}^{\infty} 1/k^2$ is convergent. So in order to demonstrate that the root test is inconclusive for $L = 1$, we only need to prove that $\lim_{n\to\infty} \sqrt[n]{1/n} = \lim_{n\to\infty} \sqrt[n]{1/n^2} = 1$, and for this purpose it clearly suffices to show that $\lim_{n\to\infty} \sqrt[n]{n} = 1$. According to Theorem 29.8, we have $\lim_{x\to\infty} \ln(x)/x = 0$ and therefore,

$$\lim_{n\to\infty} \sqrt[n]{n} = \lim_{x\to\infty} x^{1/x} = \lim_{x\to\infty} e^{\ln(x)/x} = e^0 = 1.$$

This completes the proof of Theorem 41.20. □

41.21 Example. Let us consider the series

$$\sum_{k=1}^{\infty} kq^k$$

for a given $q \in \mathbb{R}$ with $|q| < 1$. Since $\lim_{n\to\infty} \sqrt[n]{n} = 1$ (see the proof of Theorem 41.20c), it follows that

$$\lim_{n\to\infty} \sqrt[n]{|nq^n|} = |q| \lim_{n\to\infty} \sqrt[n]{n} = |q| < 1.$$

Therefore, the root test (Theorem 41.20a) implies that $\sum_{k=1}^{\infty} kq^k$ is convergent whenever $|q| < 1$.

41.22 Exercise. Use the root test to show that the series $\sum_{k=1}^{\infty} q^{k^2/(2k+1)}$ is convergent if $0 < q < 1$, and explain why the series is divergent if $q \geq 1$.

A close relative of the root test is the *ratio test:*

41.23 Theorem. *If $(a_n)_{n=m}^{\infty}$ is a sequence of nonzero terms (i.e., $a_n \neq 0$ for all integers $n \geq m$) such that the limit $L = \lim_{n\to\infty} |a_{n+1}/a_n|$ exists, then*

a) *$\sum_{k=m}^{\infty} a_k$ is convergent if $L < 1$,*

b) *$\sum_{k=m}^{\infty} a_k$ is divergent if $L > 1$, and*

c) *the test is inconclusive if $L = 1$.*

Proof. **a)** If $L < 1$, then for $q := (L+1)/2$ we have $L < q < 1$. This implies that $|a_{n+1}/a_n| < q$ for all sufficiently large n. In other words, there is an integer $N \geq m$ such that $|a_{n+1}/a_n| < q$ for all $n \geq N$. So if $n > N$, then

$$\left|\frac{a_n}{a_N}\right| = \left|\frac{a_{N+1}}{a_N}\right| \cdot \left|\frac{a_{N+2}}{a_{N+1}}\right| \cdots \left|\frac{a_n}{a_{n-1}}\right| < \underbrace{q \cdot q \cdots q}_{n-N \text{ times}} = q^{n-N}$$

and therefore

$$|a_n| < q^{n-N}|a_N| \text{ for all } n > N.$$

Since the series $\sum_{k=N+1}^{\infty} q^{k-N}|a_N| = |a_N| \sum_{k=1}^{\infty} q^k$ is convergent (by Theorem 41.16), the comparison test allows us to conclude that $\sum_{k=m}^{\infty} a_k$ is convergent as well.
b) If $L > 1$, then there is an integer $N \geq m$ such that $|a_{n+1}/a_n| > 1$ for all $n \geq N$. Consequently, if $n > N$, then

$$\left|\frac{a_n}{a_N}\right| = \left|\frac{a_{N+1}}{a_N}\right| \cdot \left|\frac{a_{N+2}}{a_{N+1}}\right| \cdots \left|\frac{a_n}{a_{n-1}}\right| > 1,$$

or equivalently,

$$|a_n| > |a_N| \text{ for all } n > N.$$

Since $|a_N| > 0$, the sequence $(a_n)_{n=m}^{\infty}$ does not converge to zero. Consequently, the series $\sum_{k=m}^{\infty} a_k$ is divergent by the sequence test.
c) As in the proof of Theorem 41.20c, we consider the series $\sum_{k=1}^{\infty} 1/k$ and $\sum_{k=1}^{\infty} 1/k^2$. Since

$$\lim_{n\to\infty} \frac{1/(n+1)}{1/n} = \lim_{n\to\infty} \frac{n}{n+1} = 1 \quad \text{and} \quad \lim_{n\to\infty} \frac{1/(n+1)^2}{1/n^2} = \lim_{n\to\infty} \frac{n^2}{(n+1)^2} = 1,$$

it follows that the ratio test is indeed inconclusive for $L = 1$, because the harmonic series $\sum_{k=1}^{\infty} 1/k$ is divergent, while the series $\sum_{k=1}^{\infty} 1/k^2$ is convergent (see Examples 41.10 and 41.11). □

41.24 Example. In Example 41.1 we saw that the series

$$\sum_{k=0}^{\infty} \frac{x^k}{k!}$$

is convergent for all $x \in \mathbb{R}$. This result can also be easily derived from the ratio test: For $x \in \mathbb{R}$ and $a_n := x^n/n!$ we have

$$\lim_{n\to\infty} \left|\frac{a_{n+1}}{a_n}\right| = \lim_{n\to\infty} \left|\frac{x^{n+1}n!}{x^n(n+1)!}\right| = \lim_{n\to\infty} \frac{|x|}{n+1} = 0 < 1.$$

Thus, Theorem 41.23a implies that the series $\sum_{k=0}^{\infty} x^k/k!$ is convergent.

41.25 Exercise. Use the ratio test to show that the series $\sum_{k=1}^{\infty} k!/k^k$ is convergent.

Additional Exercises

41.26. Use appropriate convergence tests to decide which of the following series are convergent:

a) $\sum_{k=1}^{\infty} \arctan(k)$ b) $\sum_{k=1}^{\infty} \tan\left(\frac{1}{k}\right)$ c) $\sum_{k=1}^{\infty} \frac{1}{\sqrt{k}^3}$ d) $\sum_{k=1}^{\infty} \left(\frac{\arctan(k)}{\pi}\right)^k$

e) $\sum_{k=1}^{\infty} \sin\left(\frac{\pi k}{6}\right)$ f) $\sum_{k=1}^{\infty} \frac{1}{(k+2)^{\sqrt{k}}}$ g) $\sum_{k=1}^{\infty} \frac{1}{\arctan(k)^k}$ h) $\sum_{k=1}^{\infty} \frac{1}{\arctan(k^k)}$

i) $\sum_{k=1}^{\infty} \frac{k^k}{(k!)^2}$ j) $\sum_{k=2}^{\infty} \frac{1}{\ln(k)^{\ln(k)}}$ k) $\sum_{k=2}^{\infty} \frac{1}{\ln(k)\sqrt{\ln(k)}}$ l) $\sum_{k=1}^{\infty} \sin\left(\frac{1}{k}\right)$

m) $\sum_{k=1}^{\infty} \frac{k^2+3k}{k^2+1}$ n) $\sum_{k=1}^{\infty} 0.9^{k^2}$ o) $\sum_{k=1}^{\infty} \left(1-\frac{1}{k}\right)^{k^2}$ p) $\sum_{k=1}^{\infty} \left(1+\frac{2}{k}\right)^{k^2}$

Hint. For j) and k) you may want to use the integral test with the substitution $u = \ln(x)$.

41.27. Find a general formula for the value of the series $\sum_{k=m}^{\infty} q^k$ for $q \in (-1, 1)$. *Hint.* Factor out q^m.

41.28. Figure 41.2 illustrates how an infinite succession of circles $C_0, C_1, C_2, \ldots$ is inscribed within the upper right-hand corner of a unit square. Find the total area and perimeter of the figure composed of these infinitely many circles.

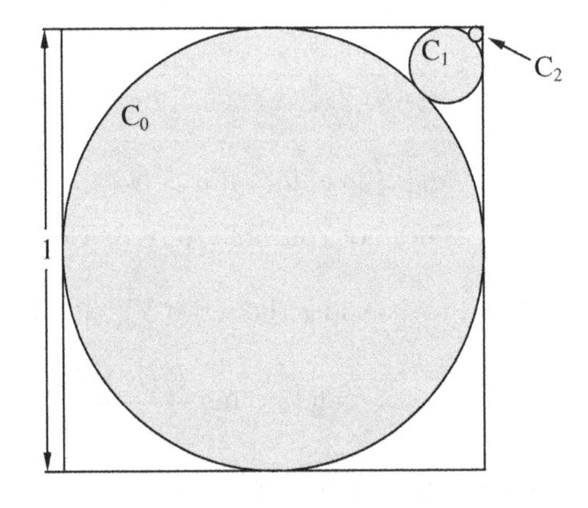

Figure 41.2: circles inscribed within a square.

41.29. Assume that an equilateral triangle of side length 1 is given, from which we cut out an infinite succession of increasingly smaller equilateral triangles in the manner indicated in Figure 41.3. What is the area of the remaining geometric figure?

41.30. The so-called *Koch snowflake* is constructed by adjoining an infinite succession of equilateral triangles to a given equilateral triangle of side length 1 in the manner indicated in Figure 41.4 (one equilateral triangle is always attached to each of the line segments of which the outline of the figure at a given stage in the construction is composed). Find the total area and perimeter of the resulting geometric figure.

41.31. Find the values of the following series:

a) $\sum_{k=0}^{\infty} \frac{1}{6^k}$ b) $\sum_{k=2}^{\infty} 0.1^k$ c) $\sum_{k=1}^{\infty} \frac{1}{k(k+1)}$ d) $\sum_{k=1}^{\infty} e^{-3k}$ e) $\sum_{k=2}^{\infty} 3^{-2k+3}$

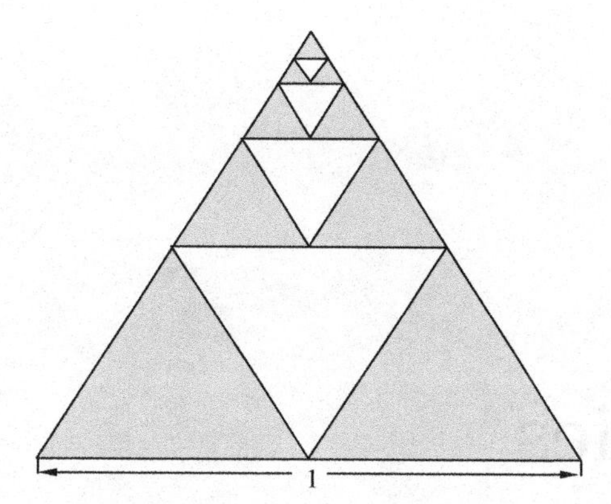

Figure 41.3: cutting out equilateral triangles.

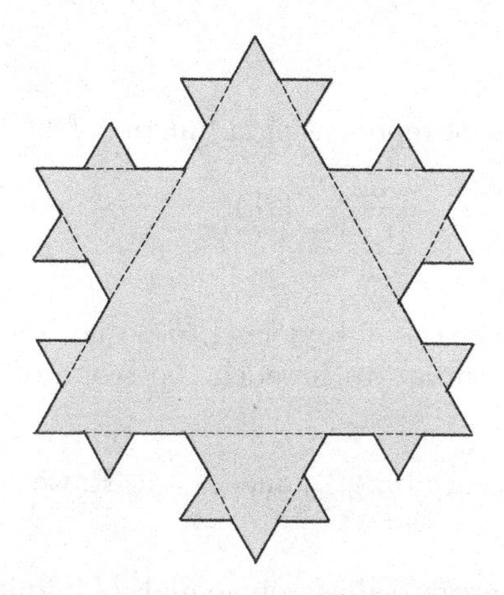

Figure 41.4: the Koch snowflake.

Hint. For c) consider partial sums and use the equation $1/(k(k+1)) = 1/k - 1/(k+1)$.

41.32. Use geometric series to find ratios of integers that are equal to each of the following decimals with periodic expansions: $0.\overline{51} = 0.515151...$, $0.\overline{142857}$, and $0.103\overline{275}$.

41.33. Which of the following conditions guarantee(s) that the series $\sum_{k=0}^{\infty} a_k$ is convergent? Explain your answer.

a) $\lim_{n\to\infty} a_n = 0$ **b)** $\lim_{n\to\infty} n^2 a_n = 0$ **c)** $\lim_{n\to\infty} a_n = 1$ **d)** $\lim_{n\to\infty} \frac{a_{n+1}}{a_n} = 0$

Chapter 42

Taylor Series

Power Series

In Chapter 41 we outlined the idea of representing a function f by its Taylor series

$$\sum_{k=0}^{\infty} \frac{f^{(k)}(x_0)}{k!}(x-x_0)^k.$$

Since many of the important properties of Taylor series do not depend on the particular form of the coefficients $f^{(k)}(x_0)/k!$, it is appropriate to broaden the scope of our discussion by introducing the following definition:

42.1 Definition. For a given sequence $(a_n)_{n=0}^{\infty}$ and x, $x_0 \in \mathbb{R}$, we call $\sum_{k=0}^{\infty} a_k(x-x_0)^k$ a *power series about* x_0.

Intuitively, power series can be regarded as polynomials of infinite degree, but whereas polynomials can always be defined on the entire set of real numbers, the domain of a power series $\sum_{k=0}^{\infty} a_k(x-x_0)^k$ has to be restricted to those values x for which the series is convergent. In this context the following theorem is very important:

42.2 Theorem. *Let $x_0 \in \mathbb{R}$ and assume that $(a_n)_{n=0}^{\infty}$ is a sequence such that $\lambda = \lim_{n\to\infty} \sqrt[n]{|a_n|}$ exists.*

a) *If $0 < \lambda < \infty$, then the series $\sum_{k=0}^{\infty} a_k(x-x_0)^k$ is convergent for all $x \in \mathbb{R}$ that satisfy the inequality $|x-x_0| < 1/\lambda$. The series is divergent if $|x-x_0| > 1/\lambda$ and no general conclusion is possible in the case where $|x-x_0| = 1/\lambda$.*

b) *If $\lambda = 0$, then the series $\sum_{k=0}^{\infty} a_k(x-x_0)^k$ is convergent for all $x \in \mathbb{R}$.*

c) *If $\lambda = \infty$, then the series $\sum_{k=0}^{\infty} a_k(x-x_0)^k$ is convergent only for $x = x_0$.*

Proof. To begin with, we observe that

$$\lim_{n\to\infty} \sqrt[n]{|a_n(x-x_0)^n|} = |x-x_0| \lim_{n\to\infty} \sqrt[n]{|a_n|} = \lambda|x-x_0|. \tag{42.1}$$

So if $\lambda > 0$, then the condition $|x-x_0| < 1/\lambda$ is equivalent to the inequality

$$\lim_{n\to\infty} \sqrt[n]{|a_n(x-x_0)^n|} < 1.$$

Therefore, a) is a direct consequence of Theorem 41.20 with $L = \lim_{n\to\infty} \sqrt[n]{|a_n(x-x_0)^n|}$. If $\lambda = 0$, then (42.1) implies that $\lim_{n\to\infty} \sqrt[n]{|a_n(x-x_0)^n|} = 0 < 1$ for all $x \in \mathbb{R}$, and the convergence of $\sum_{k=0}^{\infty} a_k(x-x_0)^k$ follows from Theorem 41.20a. The proof of c) is left as an exercise. □

42.3 Exercise. Prove c) of Theorem 42.2.

If the limit $\lambda = \lim_{n\to\infty} \sqrt[n]{|a_n|}$ exists, then $r := 1/\lambda$ (with the conventions $r := \infty$ for $\lambda = 0$ and $r := 0$ for $\lambda = \infty$) is said to be the *radius of convergence* of the power series $\sum_{k=0}^{\infty} a_k(x-x_0)^k$. According to Theorem 42.2, we may draw the following conclusion:

The power series $\sum_{k=0}^{\infty} a_k(x-x_0)^k$ is convergent for all $x \in (x_0 - r, x_0 + r)$ and divergent for all $x \in \mathbb{R} \setminus [x_0 - r, x_0 + r]$.

The interval $(x_0 - r, x_0 + r)$ is referred to as the *interval of convergence.* Note: in the case where $r = \infty$, the interval of convergence is $(-\infty, \infty) = \mathbb{R}$, and if $r = 0$, then the series is convergent only at $x = x_0$.

Remark. It can be shown that every power series has a well defined radius of convergence even if $\lim_{n\to\infty} \sqrt[n]{|a_n|}$ does not exist (see [Ros]). So for every power series $\sum_{k=0}^{\infty} a_k(x-x_0)^k$ there exists an $r \in \mathbb{R}$ such that the series is convergent for all $x \in (x_0 - r, x_0 + r)$ and divergent for all $x \in \mathbb{R} \setminus [x_0 - r, x_0 + r]$. To determine the convergence behavior at the endpoints of the interval of convergence (i.e., at $x_0 - r$ and $x_0 + r$) is of little practical interest, and no general rule exists (see Exercise 42.25). We will therefore always be content to work with the interval of convergence as an open interval, not including $x_0 - r$ or $x_0 + r$.

42.4 Example. Let us consider the power series

$$\sum_{k=1}^{\infty} \frac{1}{k^2}(x-2)^k.$$

In this case we have $a_n = 1/n^2$ and

$$\lambda = \lim_{n\to\infty} \sqrt[n]{|a_n|} = \lim_{n\to\infty} \frac{1}{\sqrt[n]{n}^{\,2}} = 1 \quad \text{(see the proof of Theorem 41.20c).}$$

Consequently, the radius of convergence is $r = 1/\lambda = 1$ and the interval of convergence is $(1, 3)$.

42.5 Exercise. Determine the radius and the interval of convergence of the series $\sum_{k=0}^{\infty} 3^k(x-1)^k$.

42.6 Theorem. *Assume that $(a_n)_{n=0}^{\infty}$ is a sequence of nonzero terms (i.e., $a_n \neq 0$ for all integers $n \geq 0$) such that the limit $\lim_{n\to\infty} |a_{n+1}/a_n|$ exists. Then the radius of convergence of the power series $\sum_{k=0}^{\infty} a_k(x-x_0)^k$ is*

$$r = \frac{1}{\lim_{n\to\infty} |a_{n+1}/a_n|} = \lim_{n\to\infty} \left|\frac{a_n}{a_{n+1}}\right|.$$

42.7 Exercise. Prove Theorem 42.6. *Hint.* Use Theorem 41.23 to show that for $\lambda := \lim_{n\to\infty} |a_{n+1}/a_n|$, the power series $\sum_{k=0}^{\infty} a_k(x-x_0)^k$ is convergent for all $x \in (x_0 - 1/\lambda, x_0 + 1/\lambda)$ and divergent for all $x \in \mathbb{R} \setminus [x_0 - 1/\lambda, x_0 + 1/\lambda]$. (The argument will be completely analogous to the derivation of Theorem 42.2 from Theorem 41.20.)

42.8 Example. For the power series

$$\sum_{k=0}^{\infty} \frac{(k!)^2}{(2k)!} x^k$$

we have

$$\lim_{n\to\infty} \left|\frac{a_{n+1}}{a_n}\right| = \lim_{n\to\infty} \frac{((n+1)!)^2(2n)!}{(n!)^2(2(n+1))!} = \lim_{n\to\infty} \frac{(n+1)^2}{(2n+2)(2n+1)} = \frac{1}{4}.$$

Therefore, according to Theorem 42.6, the radius of convergence is $r = 4$ and the interval of convergence is $(-4, 4)$.

42.9 Exercise. Find the radius and the interval of convergence of the series $\sum_{k=0}^{\infty} k!(x+2)^k/k^k$.

For a given power series $\sum_{k=0}^{\infty} a_k(x-x_0)^k$ with radius of convergence r, we define a function f on the interval $(x_0 - r, x_0 + r)$ via the equation

$$f(x) := \sum_{k=0}^{\infty} a_k(x-x_0)^k.$$

Simple formulae for the derivative and antiderivative of f are given in the following theorem:

42.10 Theorem. *Let $x_0 \in \mathbb{R}$ and let $\sum_{k=0}^{\infty} a_k(x-x_0)^k$ be a power series with radius of convergence r. Then the function $f(x) := \sum_{k=0}^{\infty} a_k(x-x_0)^k$, defined on the interval $(x_0 - r, x_0 + r)$, is differentiable, and the derivative and antiderivative of f are given as follows:*

$$f'(x) = \sum_{k=1}^{\infty} k a_k (x-x_0)^{k-1} \text{ and}$$

$$\int f(x)\,dx = \sum_{k=0}^{\infty} \frac{a_k(x-x_0)^{k+1}}{k+1} + C.$$

Furthermore, the radii of convergence of the series $\sum_{k=1}^{\infty} k a_k(x-x_0)^{k-1}$ and $\sum_{k=0}^{\infty} a_k(x-x_0)^{k+1}/(k+1)$ are both equal to r.

The proof of Theorem 42.10 requires analytical techniques that are beyond the scope of this text (see [Ros]). For a partial proof of the latter part of Theorem 42.10 the reader is encouraged to work the following exercise:

42.11 Exercise. Assume that $(a_n)_{n=0}^{\infty}$ is a sequence such that $\lim_{n\to\infty} \sqrt[n]{|a_n|}$ exists. Show that the series $\sum_{k=0}^{\infty} a_k(x-x_0)^k$ and $\sum_{k=1}^{\infty} k a_k(x-x_0)^{k-1}$ have the same radius of convergence.

Intuitively speaking, the statement of Theorem 42.10 simply says the following:

The rules for the differentiation and integration of polynomials are applicable to power series as well.

42.12 Example. To demonstrate the usefulness of power series representations in general and Theorem 42.10 in particular, we will now discuss the problem of finding the antiderivative of the function e^{x^2}. To determine the antiderivative of this function in a closed form using standard integration techniques, such as integration by substitution, turns out to be impossible, but a power series representation can be easily obtained: Since

$$e^x = \sum_{k=0}^{\infty} \frac{x^k}{k!} \text{ for all } x \in \mathbb{R} \text{ (see Example 41.1)},$$

we may substitute x^2 for x to infer that

$$e^{x^2} = \sum_{k=0}^{\infty} \frac{(x^2)^k}{k!} = \sum_{k=0}^{\infty} \frac{x^{2k}}{k!}.$$

According to Theorem 42.10, it therefore follows that

$$\int e^{x^2}\,dx = \sum_{k=0}^{\infty} \frac{x^{2k+1}}{(2k+1)k!} + C.$$

Taylor Series

Let us assume that I is an open interval on which a given function $f : I \to \mathbb{R}$ is infinitely many times differentiable (i.e., $f^{(n)}(x)$ exists for all positive integers n and all $x \in I$). If n is a positive integer and $x_0 \in I$, then Taylor's theorem (Theorem 40.7) implies that for every $x \in I$ with $x \neq x_0$ there exists a t between x and x_0 such that

$$f(x) = P_{n,x_0}(x) + R_{n,x_0}(x), \text{ where } R_{n,x_0}(x) := \frac{f^{(n+1)}(t)}{(n+1)!}(x - x_0)^{n+1}.$$

If $\lim_{n\to\infty} R_{n,x_0}(x) = 0$, then

$$f(x) = \lim_{n\to\infty}(P_{n,x_0}(x) + R_{n,x_0}(x)) = \lim_{n\to\infty} P_{n,x_0}(x) = \sum_{k=0}^{\infty} \frac{f^{(k)}(x_0)}{k!}(x - x_0)^k. \tag{42.2}$$

That is, if $\lim_{n\to\infty} R_{n,x_0}(x) = 0$, then the function f is equal to its Taylor series at x.

42.13 Theorem. *Assume that $f : I \to \mathbb{R}$ is infinitely many times differentiable on the open interval I and let $x_0 \in I$. Then*

$$f(x) = \sum_{k=0}^{\infty} \frac{f^{(k)}(x_0)}{k!}(x - x_0)^k$$

for all $x \in I$ that satisfy the condition $\lim_{n\to\infty} R_{n,x_0}(x) = 0$. Furthermore, the power series representation of f is unique: If $\sum_{k=0}^{\infty} a_k(x - x_0)^k$ is a power series for which there exists an open interval J containing x_0 such that $f(x) = \sum_{k=0}^{\infty} a_k(x - x_0)^k$ for all $x \in J$, then

$$a_n = \frac{f^{(n)}(x_0)}{n!} \quad \textit{for all integers } n \geq 0.$$

Proof. The first part of the theorem has already been proven in (42.2). So it remains to be shown that the power series representation of f is unique. Since by assumption $f(x) = \sum_{k=0}^{\infty} a_k(x - x_0)^k$ for all $x \in J$, and since $x_0 \in J$, it follows that

$$f(x_0) = \sum_{k=0}^{\infty} a_k(x_0 - x_0)^k = a_0.$$

Using Theorem 42.10, we obtain

$$f'(x_0) = \sum_{k=1}^{\infty} k a_k(x_0 - x_0)^{k-1} = a_1.$$

Taking the second derivative of f at x_0 yields

$$f''(x_0) = \sum_{k=2}^{\infty} k(k-1)a_k(x_0 - x_0)^{k-2} = 2a_2,$$

or equivalently

$$\frac{f''(x_0)}{2} = a_2.$$

Continuing in this way by taking successively higher-order derivatives of f, we may infer that

$$a_n = \frac{f^{(n)}(x_0)}{n!}$$

as desired. ∎

Remark. Theorem 42.13 not only asserts the Taylor series of f to be convergent under the given assumption $\lim_{n\to\infty} R_{n,x_0}(x) = 0$, but also specifies the value that it converges to, namely $f(x)$. In general, though, the convergence of the Taylor series of a function f at a given point x does not necessarily imply its identity with $f(x)$. To illustrate this fact the following ⋆remark⋆ provides an example of a function whose MacLaurin series converges for all $x \in \mathbb{R}$, but equals $f(x)$ only at $x = 0$.

★ *Remark.* Let us define a function $f : \mathbb{R} \to \mathbb{R}$ via the equation

$$f(x) := \begin{cases} e^{-1/x^2} & \text{for } x \neq 0, \\ 0 & \text{for } x = 0. \end{cases}$$

It can be shown that f is infinitely many times differentiable on $\mathbb{R}$ and that $f^{(n)}(0) = 0$ for all nonnegative integers n (think about it). Thus, the MacLaurin series of f is

$$\sum_{k=0}^{\infty} \frac{f^{(k)}(0)}{k!} x^k = \sum_{k=0}^{\infty} 0 \cdot x^k = 0.$$

Consequently, the MacLaurin series of f is convergent for all $x \in \mathbb{R}$, but equal to $f(x)$ only at $x = 0$. ★

42.14 Example. Given the result in Example 39.10 concerning the MacLaurin polynomials of $f(x) := \sin(x)$, we may infer that the corresponding MacLaurin series is

$$\sum_{k=0}^{\infty} \frac{(-1)^k}{(2k+1)!} x^{2k+1} \quad \text{(see also Example 39.10).}$$

Since all higher-order derivatives of sine are equal to either $\sin(x)$, $\cos(x)$, $-\sin(x)$, or $-\cos(x)$, it follows that $|f^{(n)}(t)| \leq 1$ for all $t \in \mathbb{R}$. Hence

$$|R_{n,0}(x)| = \left| \frac{f^{(n+1)}(t)}{(n+1)!} x^{n+1} \right| \leq \frac{|x|^{n+1}}{(n+1)!}.$$

Since $\lim_{n\to\infty} |x|^n/n! = 0$ for all $x \in \mathbb{R}$ (see Example 41.1), we may infer that $\lim_{n\to\infty} R_{n,0}(x) = 0$ for all $x \in \mathbb{R}$. Consequently, Theorem 42.13 implies that

$$\sin(x) = \sum_{k=0}^{\infty} \frac{(-1)^k}{(2k+1)!} x^{2k+1} \text{ for all } x \in \mathbb{R}. \tag{42.3}$$

42.15 Example. In order to find a power series representation for $\cos(x)$, we apply Theorem 42.10 and differentiate both sides of equation (42.3). This yields

$$\cos(x) = \sum_{k=0}^{\infty} \frac{(-1)^k}{(2k)!} x^{2k} \text{ for all } x \in \mathbb{R}. \tag{42.4}$$

Note: according to the uniqueness statement of Theorem 42.13, the power series on the right-hand side of this equation is equal to the MacLaurin series of $\cos(x)$.

42.16 Exercise. Verify that Theorem 42.10 allows us to derive (42.4) from (42.3).

The derivation of the MacLaurin series for $\cos(x)$ from the MacLaurin series for $\sin(x)$ is an example that illustrates the following useful idea: the power series representation of a function f can frequently be determined by manipulating the known representation of another function that is in some way related to f. The standard techniques for manipulating power series are *differentiation* (as in Example 42.15), *integration*, and *substitution*. The following example provides a further illustration:

42.17 Example. We wish to determine the MacLaurin series of $f(x) := \arctan(x)$. According to Theorem 41.16, we know that

$$\frac{1}{1-x} = \sum_{k=0}^{\infty} x^k \text{ for all } x \in (-1,1). \tag{42.5}$$

Since $-x^2 \in (-1,1)$ for all $x \in (-1,1)$, we may *substitute* $-x^2$ for x in (42.5) to infer that

$$\frac{1}{1+x^2} = \sum_{k=0}^{\infty} (-1)^k x^{2k} \text{ for all } x \in (-1,1).$$

Integrating both sides of this equation (see Theorem 42.10) yields

$$\arctan(x) + C = \int \frac{1}{1+x^2}\,dx = \sum_{k=0}^{\infty} \frac{(-1)^k}{2k+1} x^{2k+1}.$$

The constant C can be determined by setting $x = 0$:

$$C = \arctan(0) + C = \sum_{k=0}^{\infty} \frac{(-1)^k}{2k+1} \cdot 0 = 0.$$

Hence

$$\arctan(x) = \sum_{k=0}^{\infty} \frac{(-1)^k}{2k+1} x^{2k+1} \text{ for all } x \in (-1,1).$$

42.18 Exercise. Find the MacLaurin series of the functions $\sin(x^2)$ and $-2x/(1+x^2)^2$. Determine in each case for which $x \in \mathbb{R}$ the MacLaurin series is equal to the function.

42.19 Example. We wish to determine the Taylor series of $f(x) := \ln(x)$ at $x_0 = 1$. *Substituting* $1-x$ for x in (42.5), we obtain

$$\frac{1}{x} = \sum_{k=0}^{\infty} (1-x)^k = \sum_{k=0}^{\infty} (-1)^k (x-1)^k. \tag{42.6}$$

This equation is valid for all $x \in (0,2)$ because $1-x \in (-1,1)$ if and only if $x \in (0,2)$. *Integrating* both sides of equation (42.6) yields

$$\ln(x) + C = \sum_{k=0}^{\infty} \frac{(-1)^k}{k+1} (x-1)^{k+1} = \sum_{k=1}^{\infty} \frac{(-1)^{k+1}}{k} (x-1)^k \text{ for all } x \in (0,2).$$

Setting $x = 1$, it follows that

$$C = \ln(1) + C = \sum_{k=1}^{\infty} \frac{(-1)^{k+1}}{k} \cdot 0 = 0,$$

and therefore,

$$\ln(x) = \sum_{k=1}^{\infty} \frac{(-1)^{k+1}}{k} (x-1)^k \text{ for all } x \in (0,2).$$

42.20 Exercise. Find the Taylor series of $x\ln(x) - x$ at $x_0 = 1$ and determine the set of all $x \in \mathbb{R}$ for which the Taylor series is equal to the function.

The following table summarizes the power series representations that we have derived in this and the preceding chapter:

power series representation	valid for
$e^x = \sum_{k=0}^{\infty} \frac{x^k}{k!}$	$x \in \mathbb{R}$
$\sin(x) = \sum_{k=0}^{\infty} \frac{(-1)^k}{(2k+1)!} x^{2k+1}$	$x \in \mathbb{R}$
$\cos(x) = \sum_{k=0}^{\infty} \frac{(-1)^k}{(2k)!} x^{2k}$	$x \in \mathbb{R}$
$\frac{1}{1-x} = \sum_{k=0}^{\infty} x^k$	$x \in (-1,1)$
$\arctan(x) = \sum_{k=0}^{\infty} \frac{(-1)^k}{2k+1} x^{2k+1}$	$x \in (-1,1)$
$\ln(x) = \sum_{k=1}^{\infty} \frac{(-1)^{k+1}}{k} (x-1)^k$	$x \in (0,2)$

Additional Exercises

42.21. Which of the following series are power series?

a) $\sum_{k=0}^{\infty} (1-x)^{k^2}$ **b)** $\sum_{k=0}^{\infty} k(\sqrt{x}-1)^{2k}$ **c)** $\sum_{k=0}^{\infty} x^{2k^2+3k+1}$ **d)** $\sum_{k=0}^{\infty} (x+2)^{\sqrt{k}}$

42.22. Determine x_0 and the sequence $(a_n)_{n=0}^{\infty}$ in the power series representation $\sum_{k=0}^{\infty} a_k(x-x_0)^k$ for each of the series that you identified as power series in Exercise 42.21.

42.23. Find the MacLaurin series of each of the following functions and determine in each case for which values of x the series is equal to the function.

a) $f(x) = x^3 - 3x^2 + 5$ **b)** $f(x) = \sinh(x)$ **c)** $f(x) = x^3 \sin(x^2)$ **d)** $f(x) = \sin(x)\cos(x)$

e) $f(x) = \cos^2(x) - \sin^2(x)$ **f)** $f(x) = \frac{1}{1+8x^3}$ **g)** $f(x) = \frac{1}{4-2x^2}$ **h)** $f(x) = \frac{1}{(1-x)^4}$

i) $f(x) = \sinh(x)\cosh(x)$ **j)** $f(x) = \frac{x^2}{1+x}$ **k)** $f(x) = \frac{\sin(x^3)}{x}$ **l)** $f(x) = \frac{1}{1+2x}$

42.24. Find the Taylor series of $f(x) := \sqrt{x}$ at $x_0 = 1$ and $x_0 = 2$. Determine in each case the radius of convergence.

42.25. Consider the power series $\sum_{k=1}^{\infty} x^k/k$, $\sum_{k=1}^{\infty} x^k/k^2$, and $\sum_{k=0}^{\infty} x^k$.

a) Show that the interval of convergence for each of the series above is $(-1,1)$.

b) ★ Explain why the series $\sum_{k=1}^{\infty} x^k/k$ is convergent for $x = -1$ (this is the hard part) and divergent for $x = 1$.

c) Show that the series $\sum_{k=1}^{\infty} x^k/k^2$ is convergent for both $x = -1$ and $x = 1$.

d) Explain why the series $\sum_{k=1}^{\infty} x^k$ is divergent for both $x = -1$ and $x = 1$.

42.26. Find the Taylor series for $f(x) := e^x$ at $x_0 = 1$ and explain why this Taylor series is equal to e^x for all $x \in \mathbb{R}$.

42.27. Use MacLaurin series to find the antiderivatives of the following functions:

$$\text{a) } f(x) = e^{-x^3} \quad \text{b) } f(x) = \frac{e^x}{x} \quad \text{c) } f(x) = \frac{\arctan(x^2)}{x}$$

42.28. Find the interval of convergence of each of the following power series:

$$\text{a) } \sum_{k=1}^{\infty} \frac{kx^k}{2^k} \quad \text{b) } \sum_{k=1}^{\infty} \frac{(-4)^k (x-1)^k}{\sqrt{2k+1}} \quad \text{c) } \sum_{k=1}^{\infty} \frac{k^{10} x^k}{10^k} \quad \text{d) } \sum_{k=1}^{\infty} \frac{1 \cdot 3 \cdot 5 \cdots (2k+1)}{k!} (x+3)^k$$

42.29. Find the numerical value of each of the following series:

$$\text{a) } \sum_{k=1}^{\infty} \frac{k}{2^k} \quad \text{b) } \sum_{k=1}^{\infty} \frac{k^2}{3^k}$$

Hint. Use Theorems 41.16 and 42.10.

42.30. Use termwise differentiation of the MacLaurin series of $\sin(x)$ (see Theorem 42.10) to show that the second derivative of $\sin(x)$ is $-\sin(x)$.

42.31. ★ For $\alpha \in \mathbb{R}$ we define a function $f : (-1, 1) \to \mathbb{R}$ via the equation

$$f(x) := (1 + x)^{\alpha}.$$

The sequence of exercises below establishes Newton's *binomial theorem*, and it also gives in e) an alternative proof of the binomial formula as stated in the ⋆remark⋆ on p.262 (see also Appendix E).

a) Verify that the MacLaurin series of f is

$$g(x) := 1 + \sum_{k=1}^{\infty} \frac{\alpha(\alpha - 1) \cdots (\alpha - k + 1)}{k!} x^k.$$

Note: at this point it is not yet clear that $g = f$, but we will give a proof of this equation in d).

b) Show that the radius of convergence of the series in a) is equal to 1.

c) Use Theorem 42.10 to show that

$$\frac{g'(x)}{g(x)} = \frac{\alpha}{1 + x} \quad \text{for all } x \in (-1, 1).$$

d) Apply integration by substitution to the equation in c) in order to prove that $f(x) = g(x)$, or equivalently

$$\boxed{(1 + x)^{\alpha} = 1 + \sum_{k=1}^{\infty} \frac{\alpha(\alpha - 1) \cdots (\alpha - k + 1)}{k!} x^k \text{ for all } x \in (-1, 1).}$$

Note: you will have to use the fact that $g(0) = 1$.

e) Use d) to show that $(x + y)^n = \sum_{k=0}^{n} \binom{n}{k} x^k y^{n-k}$ for all nonnegative integers n and all $x, y \in \mathbb{R}$.

42.32. ★

a) Use the binomial theorem as stated in Exercise 42.31d to find a power series representation of the function $1/\sqrt{1 + x}$ for $x \in (-1, 1)$.

b) Use a) to determine a power series representation of the function $1/\sqrt{1 - x^2}$ for $x \in (-1, 1)$.

c) Use b) and Theorem 42.10 to find a power series representation for the function $\arcsin(x)$ on $(-1, 1)$.

d) For any integer $n \geq 0$ we set

$$I_n := \frac{1}{2}\int_0^1 \frac{u^n}{\sqrt{1-u}}\,du.$$

Use integration by parts to show that $I_n = 2nI_{n-1}/(2n+1)$ for all $n \geq 1$. Furthermore, verify that $I_0 = 1$.

e) Use integration by substitution to show that $I_n = \displaystyle\int_0^1 \frac{x^{2n+1}}{\sqrt{1-x^2}}\,dx.$

f) Use d) and e) to infer that $\displaystyle\int_0^1 \frac{x^{2n+1}}{\sqrt{1-x^2}}\,dx = \frac{2\cdot 4\cdot 6\cdots(2n)}{1\cdot 3\cdot 5\cdots(2n+1)} = \frac{2^{2n}(n!)^2}{(2n+1)!}.$

g) In the integral $\int_0^1 \arcsin(x)/\sqrt{1-x^2}\,dx$, replace $\arcsin(x)$ with its MacLaurin series as given in c), and then use f) to prove that

$$\int_0^1 \frac{\arcsin(x)}{\sqrt{1-x^2}}\,dx = \sum_{k=0}^{\infty} \frac{1}{(2k+1)^2}.$$

h) Use integration by substitution to deduce that $\displaystyle\int_0^1 \frac{\arcsin(x)}{\sqrt{1-x^2}}\,dx = \frac{\pi^2}{8}.$

i) Explain why

$$\sum_{k=1}^{\infty} \frac{1}{k^2} = \sum_{k=0}^{\infty} \frac{1}{(2k+1)^2} + \sum_{k=1}^{\infty} \frac{1}{(2k)^2}.$$

Hint. Write down the first eight terms of the sum on the left-hand side and the first four terms of each of the sums on the right-hand side.

j) Use i) to show that $\displaystyle\frac{3}{4}\sum_{k=0}^{\infty} \frac{1}{k^2} = \sum_{k=0}^{\infty} \frac{1}{(2k+1)^2}.$

k) Use g), h), and j) to conclude that

$$\boxed{\sum_{k=1}^{\infty} \frac{1}{k^2} = \frac{\pi^2}{6}.}$$

PART VI

DIFFERENTIAL EQUATIONS

Chapter 43

Separable and Homogeneous Differential Equations

Some Applications Revisited

In order to motivate our general discussion of separable differential equations in the next section, we begin this chapter with a review of some applications that we studied earlier in the text.

43.1 Example. As explained in Chapter 21, Newton's law of cooling is described by the equation

$$\frac{T'(t)}{H - T(t)} = k, \tag{43.1}$$

where we denote by $T(t)$ the temperature of an object at time t, by H the (constant) surrounding temperature and by k a constant of proportionality which depends on the object's material characteristics. Integrating both sides of equation (43.1) and using the substitution $u = T(t)$ with $du = T'(t)\,dt$ on the left-hand side, we obtain

$$\int \frac{1}{H-u}\,du = kt + C.$$

Hence

$$-\ln|H - T(t)| = -\ln|H-u| = kt + C. \tag{43.2}$$

Applying the exponential function and doing a bit of algebra, it easily follows that

$$T(t) = H \pm De^{-kt} \text{ with } D := e^{-C}.$$

43.2 Example. Let us recall that the velocity $v(t)$ of a sky diver (see Chapter 33) is given by the equation

$$\frac{v'(t)}{v_{max}^2 - v(t)^2} = \frac{g}{v_{max}^2} \tag{43.3}$$

where v_{max} denotes the terminal velocity and g the gravitational acceleration. In analogy to our calculation in Example 43.1 we first integrate both sides of equation (43.3) and then substitute $u = v(t)$ with $du = v'(t)\,dt$. This yields

$$\int \frac{1}{v_{max}^2 - u^2}\,du = \frac{g}{v_{max}^2}t + C.$$

Using the techniques for the integration of rational functions, as discussed in Chapter 33, and substituting $v(t)$ back in for u, we obtain

$$-\frac{1}{2v_{max}}\ln\left(\frac{v_{max} - v(t)}{v_{max} + v(t)}\right) = \frac{g}{v_{max}^2}t + C. \tag{43.4}$$

Finally, we solve for $v(t)$ to conclude that

$$v(t) = v_{max}\frac{1 - De^{-2gt/v_{max}}}{1 + De^{-2gt/v_{max}}} \text{ with } D := e^{-2v_{max}C}. \tag{43.5}$$

Separable Differential Equations

Let I be an open interval, and assume that $p : \mathbb{R} \to \mathbb{R}$ and $q : I \to \mathbb{R}$ are continuous functions. Then

$$p(f(x))f'(x) = q(x) \tag{43.6}$$

is said to be a *separable differential equation.*

43.3 Example. With $p(x) := 1/(H - x)$ and $q(x) := k$, it is easy to see that (43.1) is equivalent to the equation

$$p(T(t))T'(t) = q(t). \tag{43.7}$$

This shows that (43.1) is a separable differential equation. Note: the variable t in (43.7) corresponds to x in (43.6), and $T(t)$ corresponds to $f(x)$.

43.4 Example. To see that (43.3) is a separable differential equation as well, we define $p(x) := 1/(v_{max}^2 - x^2)$ and $q(x) := -g/v_{max}^2$. Then (43.3) is equivalent to the equation

$$p(v(t))v'(t) = q(t).$$

In this case, $v(t)$ corresponds to $f(x)$ in (43.6) and t corresponds again to x.

Traditionally, the letter f in (43.6) is replaced with y, and we write

$$\boxed{p(y(x))y'(x) = q(x)} \tag{43.8}$$

or in abbreviated form

$$\boxed{p(y)y' = q(x).} \tag{43.9}$$

The reason for referring to equations of this form as "separable" is that the variables x and y are separated—the variable y appears in (43.9) only on the left-hand side, and x appears only on the right-hand side.

In order to solve equation (43.8) we proceed as in Examples 43.1 and 43.2: we integrate both sides and use the substitution $u = y(x)$ with $du = y'(x)\,dx$ in the integral on the left-hand side. This yields

$$\int p(u)\,du = \int p(y(x))y'(x)\,dx = \int q(x)\,dx. \tag{43.10}$$

If P and Q are the respective antiderivatives of p and q, then (43.10) can be written in the form

$$P(u) = Q(x) + C.$$

Substituting $y(x)$ back in for u, we obtain

$$\boxed{P(y(x)) = Q(x) + C.} \tag{43.11}$$

To solve this equation for $y(x)$ requires that we apply the inverse function of P (if P is invertible) to both sides of the equation. This final step is analogous to solving (43.2) and (43.4) for $T(t)$ and $v(t)$ respectively.

43.5 Example. Let us consider the separable differential equation

$$\frac{y'(x)}{1+y(x)^2} = x^3.$$

Setting $p(x) := 1/(1+x^2)$ and $q(x) := x^3$, it follows that $P(x) = \arctan(x)$ and $Q(x) = x^4/4$. Therefore, equation (43.11) allows us to infer that

$$\arctan(y(x)) = \frac{x^4}{4} + C.$$

To solve this equation for $y(x)$ we apply the tangent function to both sides and obtain

$$y(x) = \tan\left(\frac{x^4}{4} + C\right).$$

43.6 Exercise. Solve the following separable differential equations:

$$\ln(y(x))y'(x) = e^x, \quad \frac{y'(x)}{y(x)^2 + 2y(x) + 2} = x^2, \quad \text{and} \quad y(x)e^{y(x)}y'(x) = \sin(x).$$

If possible, find an explicit equation for $y(x)$ otherwise leave the equation in the form of (43.11).

Occasionally we encounter equations that cannot be immediately recognized as separable. The following example illustrates how in such cases, an initial algebraic transformation may be helpful.

43.7 Example. Let us consider the equation

$$\frac{y'(x) - x}{y'(x) + x} = y(x). \tag{43.12}$$

Multiplying both sides by $y'(x) + x$ and rearranging the terms yields

$$y'(x)(1 - y(x)) = x(1 + y(x)).$$

Consequently, (43.12) is equivalent to the following separable differential equation:

$$\frac{1 - y(x)}{1 + y(x)} y'(x) = x. \tag{43.13}$$

43.8 Exercise. Solve equation (43.13).

43.9 Exercise. Solve the differential equations $y(x)^2y'(x) - 2xy(x)^2y'(x) - 1 = 0$ and $y'(x) = xy(x) - xy'(x)$.

Initial Value Problems

In physical applications it is usually necessary to determine specific solutions of differential equations that satisfy certain *initial conditions.* For example, in the case of a sky diver we may be interested in finding the velocity at any point in time t given that the initial velocity at time $t = 0$ is zero. Given this assumption, the result in (43.5) allows us to infer that

$$0 = v(0) = v_{max}\frac{1 - D}{1 + D},$$

and therefore $D = 1$. Substituting for D in (43.5), we obtain

$$v(t) = v_{max}\frac{1 - e^{-2gt/v_{max}}}{1 + e^{-2gt/v_{max}}}.$$

This specific solution, which satisfies the condition $v(0) = 0$, is said to be a solution of the *initial value problem*

$$\frac{v'(t)}{v_{max}^2 - v(t)^2} = \frac{g}{v_{max}^2}, \quad v(0) = 0.$$

43.10 Example. Let us consider the following initial value problem:

$$\frac{y'(x)}{1 + y(x)^2} = x^3, \quad y(2) = 1. \tag{43.14}$$

We already saw in Example 43.5 that the *general solution* is

$$y(x) = \tan\left(\frac{x^4}{4} + C\right).$$

Using the initial condition $y(2) = 1$, we obtain

$$1 = y(2) = \tan(4 + C),$$

and therefore

$$C = \arctan(1) - 4 = \frac{\pi}{4} - 4.$$

Thus, the solution of the initial value problem (43.14) is

$$y(x) = \tan\left(\frac{x^4}{4} + \frac{\pi}{4} - 4\right).$$

43.11 Exercise. Find the solutions of the following initial value problems: $y(x)^2 y'(x) = x^3$, $y(1) = 0$ and $\sin(y(x))y'(x) = \cos(x)$, $y(\pi/6) = \pi$.

Homogeneous Differential Equations

In Chapter 6 we saw that the cross-sectional shape of a radar antenna is described by the graph of a function f that satisfies the equation

$$f'(x) = \frac{f(x)}{x} \pm \sqrt{\frac{f(x)^2}{x^2} + 1}. \tag{43.15}$$

Introducing the substitution $h(x) := f(x)/x$, we demonstrated in Chapter 31 that (43.15) is equivalent to the following separable differential equation:

$$\frac{h'(x)}{\sqrt{h(x)^2 + 1}} = \pm\frac{1}{x}. \tag{43.16}$$

To generalize the process of transforming (43.15) into (43.16), we will now consider a so-called *homogeneous differential equation* of the form

$$\boxed{y'(x) = F\left(\frac{y(x)}{x}\right)} \tag{43.17}$$

where F is a given continuous function.

Remark. In setting $F(x) := x \pm \sqrt{x^2 + 1}$ and $y(x) := f(x)$, (43.15) is readily seen to be equivalent to the equation $y'(x) = F(y(x)/x)$. In other words, (43.15) is a special case of a homogeneous differential equation.

In order to transform the general homogeneous equation (43.17) into a separable differential equation, we set $z(x) := y(x)/x$. This yields

$$y'(x) = \frac{d}{dx}(x\,z(x)) = z(x) + x\,z'(x).$$

Now replacing $y'(x)$ in (43.17) with $z(x) + x\,z'(x)$ and $y(x)/x$ with $z(x)$, we obtain

$$z(x) + x\,z'(x) = F(z(x)),$$

or equivalently

$$\boxed{\frac{z'(x)}{F(z(x)) - z(x)} = \frac{1}{x}.} \tag{43.18}$$

43.12 Example. We wish to solve the initial value problem

$$y'(x) = \frac{y(x)^2}{x^2} - \frac{y(x)}{x}, \quad y(1) = 4.$$

Setting $F(z) := z^2 - z$, it follows that $y'(x) = F(y(x)/x)$, and therefore equation (43.18) implies that

$$\frac{z'(x)}{z(x)^2 - 2z(x)} = \frac{1}{x}. \tag{43.19}$$

The general solution of this separable differential equation is

$$z(x) = \frac{2}{1 - Cx^2}. \tag{43.20}$$

43.13 Exercise. Verify that (43.20) is the general solution of (43.19).

Since $z(x) = y(x)/x$, we may infer that

$$y(x) = \frac{2x}{1 - Cx^2}.$$

Using the initial condition $y(1) = 4$, we obtain

$$4 = \frac{2}{1 - C},$$

and therefore

$$C = \frac{1}{2}.$$

Thus, the solution of the initial value problem is

$$y(x) = \frac{2x}{1 - x^2/2} = \frac{4x}{2 - x^2}.$$

43.14 Exercise. Find the general solution of the differential equation $y'(x) = \dfrac{y(x)}{x + y(x)}$.

Additional Exercises

43.15. Find the general solutions of the differential equations given below.

$$\text{a) } y' = \frac{xy - 2y^2}{x^2} \qquad \text{b) } y' = \frac{3xe^{(y^2)}}{y}$$

43.16. Solve the following initial value problems:

a) $y' = e^x\sqrt{1-y^2}$, $y(0) = 1/2$

b) $e^x y' = \sin(y)$, $y(0) = \pi$ (Here the solution can be guessed without any calculation.)

c) $xy' = \dfrac{y}{\ln(y) - \ln(x)} + y$, $y(1) = e$

d) $y' = \dfrac{xy^2 + 2y^2}{x}$, $y(1) = 2$

43.17. Find the general solution of the equation $axy' + by = 0$ where $a \neq 0$.

43.18. Show that each of the differential equations given below is separable.

$$\text{a) } \frac{y'(x) - 2}{1 + 3y'(x)} = y(x) - 2 \qquad \text{b) } y'(x) = \sin(y(x) - x) + \cos(y(x))\sin(x)$$

$$\text{c) } \frac{\ln(y'(x)) - 1}{\ln(y'(x)) + 1} = 2y(x) \qquad \text{d) } y'(x) = \ln(x)\ln(y(x)) + \ln(y(x)) + \ln(x) + 1$$

$$\text{e) } e^{y'(x) - y(x)} = y(x)^2 \qquad \text{f) } y'(x) = xy(x) - 2x + y(x) - 2$$

43.19. Find the general solutions of the differential equations in Exercise 43.18a, b, f. *Hint.* To solve b), you may want to take a look again at Exercise 30.20.

43.20. Show that each of the differential equations given below is homogeneous.

$$\text{a) } y'(x) = \ln(y(x)^2) - 2\ln(x) \quad \text{b) } y'(x) = \frac{y(x) - x}{y(x) + x} + \frac{y(x)}{x} \quad \text{c) } x^2y'(x) = y(x)^2 + xy(x) + x^2$$

43.21. Find the general solutions of the differential equations in Exercise 43.20b, c.

Chapter 44

First-Order Linear Differential Equations and Electric Circuits (Part 1)

R-C Circuits

Teacher: Let us consider a simple electric circuit consisting of a capacitor, a resistor, and a switch as shown in Figure 44.1. We assume that initially there is a positive charge Q_0 (measured in Coulombs)

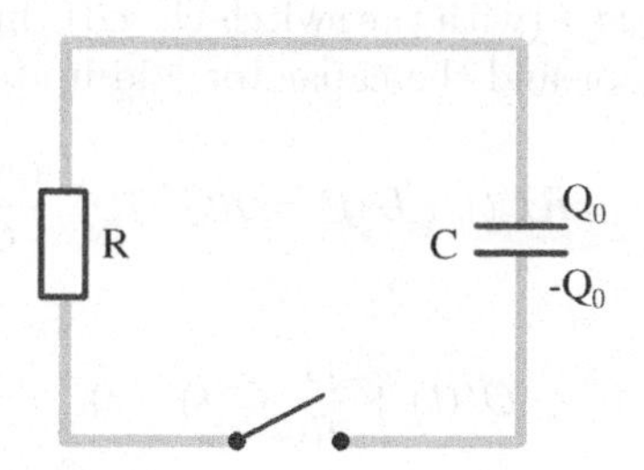

Figure 44.1: R-C circuit.

on one side of the capacitor and a negative charge $-Q_0$ on the other. At time $t = 0$ the switch is closed and an electric current $I(t)$ (measured in Ampères) begins to flow, resulting in a reduction of the charge $Q(t)$ on the capacitor. The instantaneous rate of change of $Q(t)$ is (by definition) equal to the current $I(t)$, that is

$$I(t) = Q'(t). \tag{44.1}$$

By $U(t)$ we denote the difference in electric potential (measured in Volts) between the two sides of the capacitor. If $U(t)$ is not too large, then $Q(t)$ is proportional to $U(t)$. In other words, there is a constant of proportionality C, referred to as the capacitance (measured in Farads), such that

$$Q(t) = CU(t). \tag{44.2}$$

Sophie: Do we simply have to accept this equation as an empirical fact or is there a way to derive it from some underlying laws?

Teacher: It is possible to derive equation (44.2) from Coulomb's law, which quantifies the force between two point charges, but this requires multivariable integration and will have to wait until Chapter 74. However, even in absence of a rigorous derivation, we can still gain some intuitive understanding of the physics behind equation (44.2) by examining the role played by the constant of proportionality

C. How, for example, do we justify that C is placed on the right-hand side as a factor in front of $U(t)$ rather than on the left-hand side in front of $Q(t)$?

Simplicio: That's a good question.

Teacher: And it's not difficult to answer. Just think of the relative magnitude of $Q(t)$ and C for a given value of $U(t)$.

Simplicio: Well, if the word "capacitance" refers to the "capacity" for taking in charge, then one should expect the charge $Q(t)$ to be large if the capacitance is large and small if the capacitance is small.

Teacher: Exactly. So it wouldn't make sense to place C as a factor in front of $Q(t)$ on the left-hand side of equation (44.2), because then, contrary to our intuitive expectation, the charge $Q(t)$ would have to be small if C is large and vice versa.

Simplicio: I get it.

Teacher: Good. To continue our discussion we need to take a closer look at the resistor with resistance R (measured in Ohms). The current $I(t)$ flowing through this resistor creates a difference in electric potential which, according to Ohm's law, is equal to $RI(t)$. In other words, the resistance R in Ohm's law serves as a constant of proportionality that links the potential difference at the resistor to the current flowing through it. Furthermore, the placement of R as a factor in front of $I(t)$ expresses our common sense understanding that, for a given value of $U(t)$, a large resistance should correspond to a small current and vice versa.

Sophie: I can see that, but how are we going to apply all this information to describe the process of discharging the capacitor that you described at the beginning of this lesson?

Teacher: We need to invoke the following general fact, which is commonly referred to as *Kirchoff's first law:*

The sum of the differences in electric potential in a closed circuit equals zero.

For the circuit shown in Figure 44.1 (with the switch closed) this law simply says that the differences in electric potential at the resistor and the capacitor add up to zero. Hence

$$0 = RI(t) + U(t) = RQ'(t) + \frac{Q(t)}{C}, \tag{44.3}$$

or equivalently,

$$Q'(t) + \frac{1}{RC}Q(t) = 0. \tag{44.4}$$

Dividing this equation by $Q(t)$ and rearranging the terms, we obtain the following separable differential equation:

$$\frac{Q'(t)}{Q(t)} = -\frac{1}{RC}.$$

Using the initial condition $Q(0) = Q_0$, it is easy to see that the solution of this differential equation is

$$\boxed{Q(t) = Q_0 e^{-t/RC}.} \tag{44.5}$$

With $U_0 := U(0)$ we may also write

$$\boxed{Q(t) = CU_0 e^{-t/RC}.}$$

44.1 Exercise. Show that the solution of (44.4) is given by equation (44.5).

Equation (44.5) shows that $Q(t)$ decreases to zero as t tends to ∞, because $\lim_{t\to\infty} e^{-t/RC} = 0$. Furthermore, if the resistance or the capacitance are large, the term t/RC is small, and therefore the rate of decrease of $Q(t)$ is small as well. This observation is intuitively not surprising, because if the resistance R is large, then the current $I(t)$ will be small and it will thus take a relatively longer time for a certain amount of charge to flow from one side of the capacitor to the other. Similarly, if the capacitance is large, then the difference in potential at the capacitor (and the resistor) is small, and

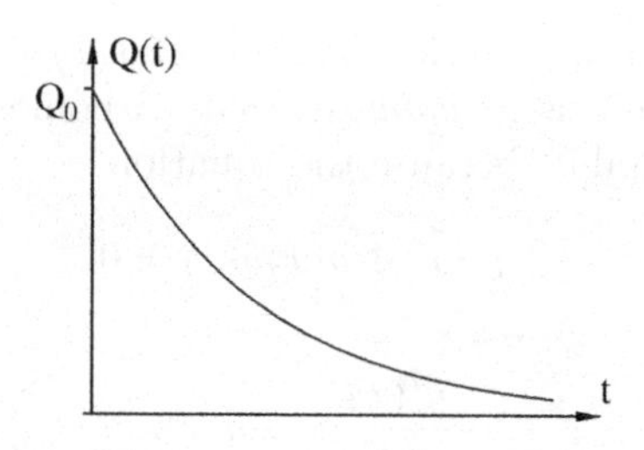

Figure 44.2: graph of $Q(t)$.

according to Ohm's law it follows that in this case the current $I(t)$ through the resistor is small as well. A typical graph of $Q(t)$ is shown in Figure 44.2.

In order to find an equation for $I(t)$, we may combine (44.1) with (44.5) to infer that

$$\boxed{I(t) = -\frac{Q_0}{RC}e^{-t/RC},}$$

or equivalently,

$$\boxed{I(t) = -\frac{U_0}{R}e^{-t/RC}.}$$

As a variation of the circuit in Figure 44.1, we will now consider an R-C circuit with an external voltage source generating a difference in electric potential $E(t)$ as shown in Figure 44.3. Due to this

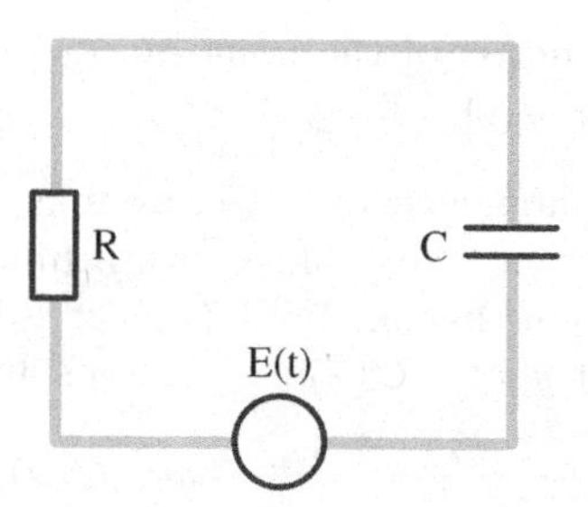

Figure 44.3: R-C circuit with external voltage source.

modification, the differences in electric potential at the resistor and the capacitor add up to $E(t)$ rather than zero. So in place of equation (44.3) we have

$$E(t) = RI(t) + U(t) = RQ'(t) + \frac{Q(t)}{C},$$

or equivalently,*

$$Q'(t) + \frac{1}{RC}Q(t) = \frac{1}{R}E(t). \tag{44.6}$$

This differential equation for $Q(t)$ is in general not separable, and the techniques necessary for solving it will be developed in the next section.

First-Order Linear Differential Equations

Let us assume that $a : I \to \mathbb{R}$ and $f : I \to \mathbb{R}$ are continuous functions defined on an open interval I. Then the equation

$$y'(x) + a(x)y(x) = f(x) \tag{44.7}$$

*Due to certain properties of Maxwell's fundamental equations of electrodynamics, equation (44.6) does not apply if the external voltage source generates an alternating current of very high frequency.

is referred to as a *first-order linear differential equation.* If $f(x) = 0$ for all $x \in I$, the equation is said to be *homogeneous*† and otherwise it is *nonhomogeneous.* A homogeneous first-order linear differential equation is in particular also separable, because the equation

$$y'(x) + a(x)y(x) = 0 \tag{44.8}$$

can be written in the form

$$\frac{y'(x)}{y(x)} = -a(x).$$

If $A : I \to \mathbb{R}$ is an antiderivative of a, then the general solution of this equation is easily seen to be

$$\boxed{y(x) = Ce^{-A(x)}.} \tag{44.9}$$

Furthermore, the solution of the initial value problem

$$y'(x) + a(x)y(x) = 0, \;\; y(x_0) = y_0 \tag{44.10}$$

is

$$\boxed{y(x) = y_0 e^{A(x_0)-A(x)} = y_0 e^{-\int_{x_0}^{x} a(t)\,dt}.} \tag{44.11}$$

Note: an example of a homogeneous first-order linear differential equation was given in (44.4).

44.2 Exercise. Verify that (44.9) is the general solution of equation (44.8).

44.3 Exercise. Show that (44.11) is the solution of the initial value problem (44.10).

44.4 Exercise. Find the general solution of the homogeneous first-order linear differential equations $y'(x) + \sin(x)y(x) = 0$ and $y'(x) - x^2y(x) = 0$.

In order to solve (44.7) in the nonhomogeneous case, we make use of a technique known as *variation of the constant:* we replace the constant C in (44.9) with a function $C(x)$ and then derive a defining equation for $C(x)$ from the assumption that $y(x) = C(x)e^{-A(x)}$ is a solution of (44.7). So let A be an antiderivative of a, and suppose that $y(x) = C(x)e^{-A(x)}$ is a solution of (44.7). It then follows that

$$y'(x) = C'(x)e^{-A(x)} - a(x)C(x)e^{-A(x)}$$

and

$$f(x) = y'(x) + a(x)y(x) = C'(x)e^{-A(x)} - a(x)C(x)e^{-A(x)} + a(x)C(x)e^{-A(x)} = C'(x)e^{-A(x)}.$$

Hence

$$C'(x) = f(x)e^{A(x)}.$$

Integrating both sides, we may infer that there is a constant $D \in \mathbb{R}$ such that

$$C(x) = \int f(x)e^{A(x)}\,dx + D.$$

(Note: since the constant D plays an important role in the process of solving initial value problems, we chose to visibly include it as part of the solution instead of implicitly assuming its presence in the indefinite integral $\int f(x)e^{A(x)}\,dx$.) Substituting for $C(x)$ in the equation $y(x) = C(x)e^{-A(x)}$, we obtain the following general solution of equation (44.7):

$$\boxed{y(x) = \left(\int f(x)e^{A(x)}\,dx + D\right) e^{-A(x)}.} \tag{44.12}$$

†Our use of the word "homogeneous" is here unfortunately not free of ambiguity. In Chapter 43 the same word was used in an entirely different context as a label for differential equations of the form $y' = F(y/x)$.

To solve the initial value problem

$$y'(x) + a(x)y(x) = f(x), \; y(x_0) = y_0 \tag{44.13}$$

we replace the indefinite integral in (44.12) with the specific antiderivative $\int_{x_0}^{x} f(t)e^{A(t)}\,dt$. This yields

$$y(x) = \left(\int_{x_0}^{x} f(t)e^{A(t)}\,dt + D\right)e^{-A(x)}.$$

Setting $x = x_0$, we obtain

$$y_0 = y(x_0) = \left(\int_{x_0}^{x_0} f(t)e^{A(t)}\,dt + D\right)e^{-A(x_0)} = De^{-A(x_0)},$$

or equivalently,

$$D = y_0 e^{A(x_0)}.$$

Consequently, the solution of (44.13) is

$$\boxed{y(x) = \left(\int_{x_0}^{x} f(t)e^{A(t)}\,dt + y_0 e^{A(x_0)}\right)e^{-A(x)}.} \tag{44.14}$$

44.5 Exercise. Verify that $y(x)$, as given in (44.12), is indeed a solution of (44.7).

44.6 Example. We wish to solve the following initial value problem:

$$y'(x) + 3x^2 y(x) = x^5, \; y(0) = 3.$$

Given the general form of a first-order differential equation in (44.7), it follows that $f(x) = x^5$, $a(x) = 3x^2$ and, by implication, $A(x) = x^3$. According to (44.14) with $x_0 = 0$ and $y_0 = 3$, we may thus infer that

$$y(x) = \left(\int_0^x t^5 e^{t^3}\,dt + 3e^{0^3}\right)e^{-x^3}.$$

To evaluate the integral on the right, we use integration by parts and integration by substitution with $u = t^3$. This yields

$$\int_0^x t^5 e^{t^3}\,dt = \int_0^{x^3} \frac{1}{3}ue^u\,du = \frac{1}{3}(u-1)e^u\Big|_0^{x^3} = \frac{1}{3}((x^3-1)e^{x^3}+1).$$

Therefore, the solution of the initial value problem is

$$y(x) = \left(\frac{1}{3}((x^3-1)e^{x^3}+1)+3\right)e^{-x^3} = \frac{x^3}{3} - \frac{1}{3} + \frac{10e^{-x^3}}{3}.$$

44.7 Exercise. Use (44.14) to find the solution of the initial value problem $y'(x)+xy(x) = x$, $y(1) = 2$. Then rewrite the equation $y'(x) + xy(x) = x$ in separable form and use the techniques developed in Chapter 43 to verify your answer.

44.8 Example. We wish to apply the results of our general discussion above to solve equation (44.6). For simplification, we assume that $Q(0) = Q_0 = 0$. Using (44.14) with $A(t) = t/RC$, we obtain

$$Q(t) = \left(\int_0^t \frac{1}{R}E(\tau)e^{\tau/RC}\,d\tau\right)e^{-t/RC}.$$

As a special case, we consider an external voltage source generating a sinusoidal difference in electric potential $E(t) = E_0\sin(\omega t)$ so that

$$Q(t) = \left(\int_0^t \frac{E_0}{R}\sin(\omega\tau)e^{\tau/RC}\,d\tau\right)e^{-t/RC}. \tag{44.15}$$

Using integration by parts, it is easy to see that

$$\begin{aligned}\int_0^t \frac{E_0}{R}\sin(\omega\tau)e^{\tau/RC}\,d\tau &= \frac{E_0Ce^{\tau/RC}}{1+\omega^2R^2C^2}(\sin(\omega\tau)-\omega RC\cos(\omega\tau))\Big|_0^t \\ &= \frac{E_0C}{1+\omega^2R^2C^2}\left(e^{t/RC}(\sin(\omega t)-\omega RC\cos(\omega t))+\omega RC\right).\end{aligned} \tag{44.16}$$

44.9 Exercise. Verify equation (44.16). *Hint.* Take a look at Example 35.11.

Combining (44.15) with (44.16) yields

$$Q(t) = \frac{E_0C}{1+\omega^2R^2C^2}\left(\sin(\omega t)-\omega RC\cos(\omega t)+\omega RCe^{-t/RC}\right).$$

With $\alpha := \arcsin\left(1/\sqrt{1+\omega^2R^2C^2}\right)$, this equation can also be written in the form

$$Q(t) = E_0C\left(-\frac{\cos(\omega t+\alpha)}{\sqrt{1+\omega^2R^2C^2}}+\frac{\omega RCe^{-t/RC}}{1+\omega^2R^2C^2}\right) \tag{44.17}$$

because the addition law for cosine (see Appendix A) allows us to infer that

$$\begin{aligned}-\frac{\cos(\omega t+\alpha)}{\sqrt{1+\omega^2R^2C^2}} &= -\frac{\cos(\omega t)\cos(\alpha)-\sin(\omega t)\sin(\alpha)}{\sqrt{1+\omega^2R^2C^2}} \\ &= -\frac{\cos(\omega t)\sqrt{1-\sin^2\left(\arcsin\left(\frac{1}{\sqrt{1+\omega^2R^2C^2}}\right)\right)}-\sin(\omega t)\sin\left(\arcsin\left(\frac{1}{\sqrt{1+\omega^2R^2C^2}}\right)\right)}{\sqrt{1+\omega^2R^2C^2}} \\ &= \frac{\sin(\omega t)-\omega RC\cos(\omega t)}{1+\omega^2R^2C^2}.\end{aligned}$$

Differentiating both sides of (44.17), we obtain the following equation for the current $I(t)$:

$$I(t) = \omega E_0C\left(\frac{\sin(\omega t+\alpha)}{\sqrt{1+\omega^2R^2C^2}}-\frac{e^{-t/RC}}{1+\omega^2R^2C^2}\right).$$

A typical graph of $I(t)$ is shown in Figure 44.4. Since $\lim_{t\to\infty} e^{-t/RC} = 0$, it follows that for large

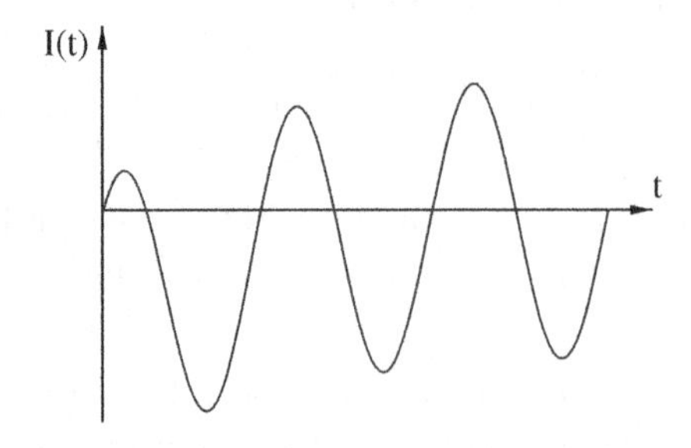

Figure 44.4: $I(t)$ with sinusoidal excitation.

values of t we have

$$I(t) \approx \frac{\omega E_0C\sin(\omega t+\alpha)}{\sqrt{1+\omega^2R^2C^2}} =: I_s(t).$$

The current $I_s(t)$ is the so-called *steady state solution*, and the term $\omega E_0Ce^{-t/RC}/(1+\omega^2R^2C^2)$, which converges to zero, is said to be the *transient part* of the solution.

R-L Circuits

Let us consider an electric circuit consisting of a resistor with resistance R, an inductor, and an external voltage source generating a difference in electric potential $E(t)$ as shown in Figure 44.5. According to Faraday's law, the difference in electric potential between the two ends of the inductor is proportional to the rate of change of the current $I(t)$. In other words, the difference in electric potential is $LI'(t)$, where L is a constant of proportionality known as inductance (measured in Henry). Since the difference in electric potential between the two ends of the resistor is $RI(t)$, we obtain the following nonhomogeneous

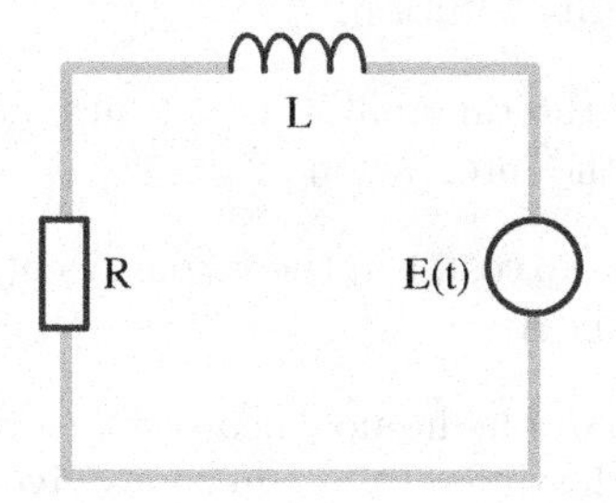

Figure 44.5: R-L circuit with external voltage source.

first-order linear differential equation in $I(t)$:

$$LI'(t) + RI(t) = E(t)$$

or, equivalently,

$$I'(t) + \frac{R}{L}I(t) = \frac{1}{L}E(t).$$

Using equation (44.14) with the initial condition $I(0) = I_0 = 0$, we find the solution

$$I(t) = \left(\int_0^t \frac{1}{L}E(\tau)e^{R\tau/L}\,d\tau\right)e^{-Rt/L}. \tag{44.18}$$

44.10 Exercise. Use equation (44.18) to determine $I(t)$ for $E(t) = E_0\cos(\omega t)$.

Additional Exercises

44.11. Find the general solutions of the following differential equations:

a) $xy' + (x-2)y = 3x^3e^{-x}$

b) $x\ln(x)y' + y = 2\ln(x)$ for $x > 1$

c) $xy' + y = x\sin(x)$ for $x > 0$

44.12. Find the solutions of the following initial value problems:

a) $y' + \cos(x)y = \cos(x)$, $y(\pi) = 0$

b) $(x^2+1)y' - 2xy = x^2 + 1$, $y(1) = \pi$

c) $y' - 2xy = 1$, $y(a) = b$ (Use MacLaurin series for the integration.)

44.13. Use the substitution $u = 1/y$ to find the general solution of the differential equation $y' + 3y/x = x^2y^2$ for $x > 0$.

44.14. Find the current $I(t)$ flowing in an R-C circuit with $R = 500\,\Omega$, $C = 10^{-3}\,F$, and an external voltage source generating a difference in potential $E(t) = 2e^{-t}$. You may assume that $Q(0) = 0$.

44.15. **a)** Find the charge $Q(t)$ in an RC-circuit under the following assumptions: $R = \sqrt{35}\,\Omega$, $C = 0.1\,F$, $Q(0) = 0\,C$, and $E(t) = 5\cos(\omega t)$ with $\omega = 10\,s^{-1}$.

b) Determine the phase shift by writing the solution you found in a) in the form $A\sin(\omega t + \alpha) + \ldots$ (the phase shift is α/ω).

c) Use the result in b) to determine the current $I(t)$ and the steady state solution.

d) Plot the graph of the steady state solution.

44.16. Find the general solution of the differential equation $y'(x) - y(x)/x = \sin(x)$ for $x > 0$. *Hint.* Use Taylor or MacLaurin series for the integration.

44.17. Write the expression $2\sin(t) - 3\cos(t)$ in the forms $A\sin(t + \alpha)$ and $B\cos(t + \beta)$. *Hint.* Take a look at the addition laws in Appendix A.

44.18. Determine the type (i.e., separable, homogeneous, first-order linear homogeneous, or first-order linear nonhomogeneous) of each of the differential equations given below. Note: if an equation fits the description of more than one type, you are to list all types that apply.

a) $y'(x) = 1$ **b)** $y'(x) - x = y(x)y'(x) + 3xy(x)$ **c)** $y'(x) = y(x)$ **d)** $y'(x) = y(x)^2$

e) $y'(x) = \dfrac{y(x)^2}{x^2}$ **f)** $y'(x) = \tan\left(\dfrac{y(x)}{x}\right) + \dfrac{y(x)}{x}$ **g)** $y'(x) = y(x) + e^{2x}$ **h)** $y'(x) = x$

44.19. Find the general solution of each of the differential equations given in Exercise 44.18.

Chapter 45

Complex Numbers

Polynomial Equations

In Chapter 44 we learned that first-order linear differential equations can be used to describe simple electric circuits of type R-C or R-L. In the next chapter we will show that in order to analyze C-L or R-C-L circuits, we need to be able to solve second-order linear differential equations of the form

$$y''(x) + ay'(x) + by(x) = 0 \tag{45.1}$$

where a and b are given constants. It will turn out that the problem of solving this equation can be reduced to the problem of solving the quadratic equation

$$x^2 + ax + b = 0. \tag{45.2}$$

(The reader should not expect to understand the relation between (45.1) and (45.2) at this point, but rather accept it as a motivation for our discussion of complex numbers and polynomials in the present chapter.) More generally, we will show in Chapter 64 that for given coefficients $a_{n-1}, \ldots, a_1, a_0$ the problem of solving the nth-order linear differential equation

$$y^{(n)}(x) + a_{n-1}y^{(n-1)}(x) + \cdots + a_1 y'(x) + a_0 y(x) = 0$$

can be reduced to the problem of solving the polynomial equation

$$x^n + a_{n-1}x^{n-1} + \cdots + a_1 x + a_0 = 0. \tag{45.3}$$

Unfortunately, equations such as (45.2) or in general (45.3) are not always solvable in the real numbers. The following dialogue explains how this difficulty can be dealt with.

Teacher: For $a = 0$ and $b = 1$ equation (45.2) simplifies to

$$x^2 + 1 = 0. \tag{45.4}$$

What are the solutions of this equation?

Simplicio: There aren't any. If x were a solution of (45.4), then x^2 would have to be equal to -1, but this is impossible because the square of any number is always greater than or equal to zero.

Teacher: You are quite right. There aren't any solutions in the real numbers $\mathbb{R}$. For this reason we will now introduce an *imaginary number* i which by definition satisfies the equation

$$\boxed{i^2 = -1 \text{ (or equivalently } i = \sqrt{-1}).} \tag{45.5}$$

Simplicio: How can we possibly introduce a number that simply doesn't exist?

Teacher: Intuitively, we may be reluctant to accept the existence of a number the square of which is negative, but from a strictly mathematical point of view the definition of i as a solution of (45.4) does not lead to any logical inconsistencies.

Sophie: So as long as something is not logically inconsistent we may consider it a legitimate object of study?

Teacher: Why not? We do it all the time. We introduce new definitions and build theorems and entire theories around them. This is a very familiar process, even at the most fundamental level. For instance, in analogy to the definition of i as a solution of (45.4) we may regard -1 to be defined as the solution of the equation $x + 1 = 0$. Of course, we are so used to working with negative numbers that we would never stop to think how the existence of -1 is to be justified, but intuitively speaking a negative number hardly appears more "real" than an imaginary one. After all, I can easily imagine to have one apple or two apples or no apples, but negative one apple? How can I have less than nothing?

Sophie: Are you saying that we ought to be content to regard negative numbers as a logically consistent extension of the natural counting numbers, and not worry about any intrinsic reality that we may want to attach to them?

Teacher: Yes indeed. We work with negative numbers mainly because they are useful, not because they are in some sense "real." In the same way, we will now introduce the *complex numbers* as an extension of the real numbers purely for the sake of utility to gain greater algebraic flexibility. A *complex number* is by definition a sum of the form

$$a + ib \text{ with } a, b \in \mathbb{R}.$$

The set of all complex numbers is denoted by $\mathbb{C}$. The *real part* of a complex number $z = a + ib$ is defined to be a and the *imaginary part* is b. In other words, in denoting the real and imaginary parts respectively by $\mathcal{R}e(z)$ and $\mathcal{I}m(z)$, we define

$$\mathcal{R}e(z) := a,$$
$$\mathcal{I}m(z) := b.$$

If $\mathcal{I}m(z) = 0$, then $z = a$ is a real number. Therefore, $\mathbb{C}$ is an *extension* of $\mathbb{R}$ because every real number can be regarded as a complex number with imaginary part zero. We will see later in this chapter that having extended the set of available numbers from $\mathbb{R}$ to $\mathbb{C}$, we are (in principle) able to find a solution in $\mathbb{C}$ for every equation of the form given in (45.3).

Properties of Complex Numbers

Since every complex number is completely determined by a pair of real numbers—the real and imaginary parts—it is natural to represent complex numbers as points in a Cartesian coordinate system called

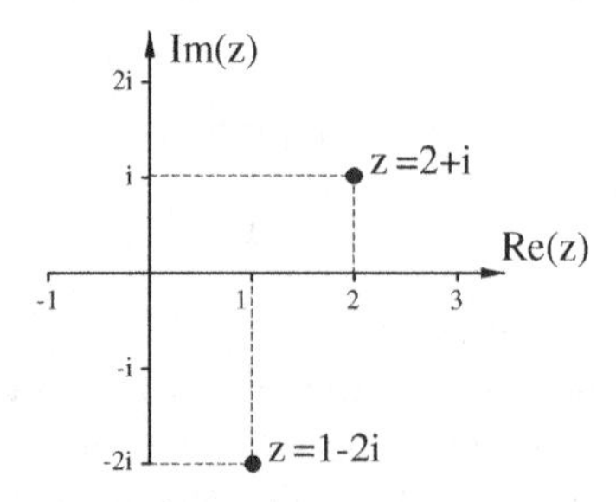

Figure 45.1: the complex plane.

the *complex plane*. The real part is represented by the x-coordinate and the imaginary part by the y-coordinate (see Figure 45.1). The x- and y-coordinate axes are also referred to as the *real* and *imaginary axes*.

If $z_1 = a_1 + ib_1$ and $z_2 = a_2 + ib_2$ are complex numbers, then the *sum* of z_1 and z_2 is defined as

$$z_1 + z_2 := (a_1 + a_2) + i(b_1 + b_2).$$

Graphically, the sum $z_1 + z_2$ is represented by the endpoint of the diagonal passing through the origin in the parallelogram spanned by the lines connecting z_1 and z_2 with the origin (see Figure 45.2). The

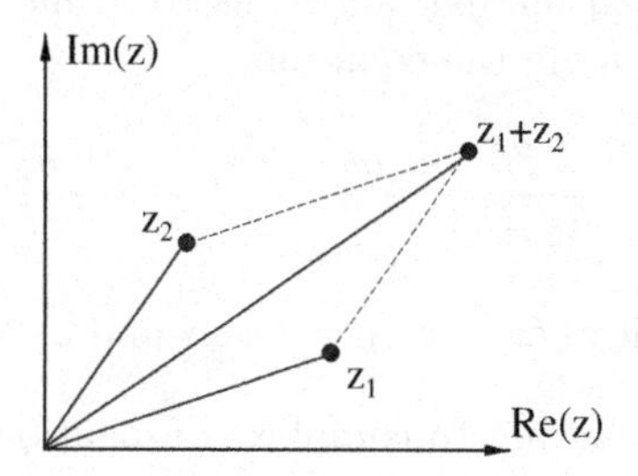

Figure 45.2: addition in the complex plane.

product of z_1 and z_2 is defined by the equation

$$z_1 z_2 := (a_1 a_2 - b_1 b_2) + i(b_1 a_2 + a_1 b_2). \tag{45.6}$$

The rationale behind this definition is simply that an evaluation of the product $(a_1 + ib_1)(a_2 + ib_2)$, using (45.5) and the same distributive laws as are familiar from the multiplication of real numbers, yields the term on the right-hand side of equation (45.6). We wish to emphasize, though, that this explanation does not constitute a proof for it obviously is logically impossible to apply the distributive laws to the multiplication of complex numbers before such a multiplication has actually been defined.

Given the definitions of addition and multiplication of complex numbers, it is natural to define the *difference* of two complex numbers z_1 and z_2 as the sum of z_1 and $(-1)z_2$. So we set

$$z_1 - z_2 := z_1 + (-1)z_2 = (a_1 - a_2) + i(b_1 - b_2).$$

45.1 Exercise. For $z_1 = 2 - 3i$ and $z_2 = 2 + i$, find $z_1 + z_2$, $z_1 z_2$ and $z_1 - z_2$.

45.2 Exercise. Show that for all numbers $w, z_1, z_2 \in \mathbb{C}$ we have $w(z_1 + z_2) = wz_1 + wz_2$.

For any complex number $z = a + ib$, we say that

$$\bar{z} := a - ib$$

is the *complex conjugate* of z. Graphically, complex conjugation is equivalent to a reflection across the real axis of the complex plane (see Figure 45.3).

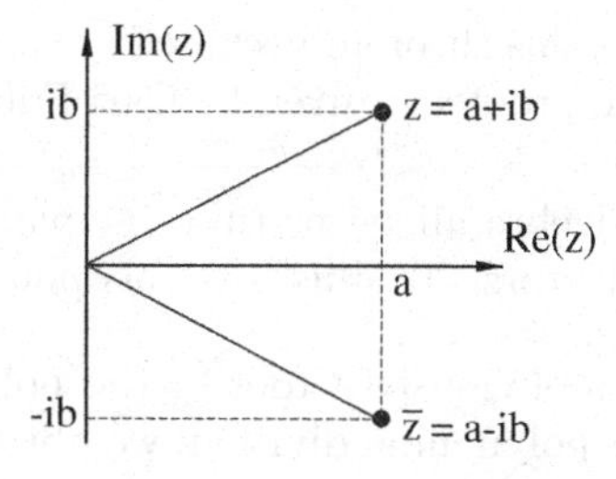

Figure 45.3: complex conjugation.

The *absolute value* of a complex number $z = a + ib$ is defined to be the distance from z to the origin in the complex plane, i.e.,

$$|z| := \sqrt{a^2 + b^2}.$$

Since the product of z and $\bar{z}$ is easily seen to be equal to $a^2 + b^2 = |z|^2$, it follows that

$$|z| = \sqrt{z\bar{z}}.$$

45.3 Exercise. Verify this equation.

The fact that $z\bar{z} = |z|^2$ is a real number allows us to define the *quotient* of two complex numbers $z_1 = a_1 + ib_1$ and $z_2 = a_2 + ib_2 \neq 0$ via the equation

$$\frac{z_1}{z_2} := \frac{z_1\bar{z}_2}{z_2\bar{z}_2} = \frac{a_1a_2 + b_1b_2}{a_2^2 + b_2^2} + \frac{i(b_1a_2 - a_1b_2)}{a_2^2 + b_2^2}.$$

45.4 Exercise. Find the quotient z_1/z_2 for $z_1 = 2 + 4i$ and $z_2 = 1 - 3i$.

Several important properties relating to complex conjugation and absolute values are summarized in the following theorem:

45.5 Theorem. *For all $z, z_1, z_2 \in \mathbb{C}$ we have*

a) $\overline{z_1 + z_2} = \bar{z}_1 + \bar{z}_2$, $\overline{z_1z_2} = \bar{z}_1\bar{z}_2$, and $\overline{\left(\frac{z_1}{z_2}\right)} = \frac{\bar{z}_1}{\bar{z}_2}$ *(if $z_2 \neq 0$),*

b) $\mathcal{R}e(z) = \frac{z + \bar{z}}{2}$ *and* $\mathcal{I}m(z) = \frac{z - \bar{z}}{2i}$,

c) $|z_1z_2| = |z_1||z_2|$ *and* $\left|\frac{z_1}{z_2}\right| = \frac{|z_1|}{|z_2|}$ *(if $z_2 \neq 0$).*

45.6 Exercise. Prove Theorem 45.5.

The Fundamental Theorem of Algebra

In the introduction to this chapter we mentioned that there is a close relationship between the problem of finding solutions of certain linear differential equations and the problem of finding solutions of polynomial equations of the form (45.3). The following theorem, which is known as the *fundamental theorem of algebra,* asserts that equation (45.3) always has a solution in $\mathbb{C}$:

45.7 Theorem. *Let $a_0, a_1, \ldots, a_n \in \mathbb{C}$ and let $p(z) := a_nz^n + \cdots + a_1z + a_0$ be a nonconstant complex polynomial, then there is a $z_0 \in \mathbb{C}$ such that $p(z_0) = 0$. In other words, every nonconstant complex polynomial has a root! (Note: A polynomial is nonconstant if its degree is greater than zero. So $p(z)$ is nonconstant if $n \geq 1$ and $a_n \neq 0$.)*

We cannot discuss the proof for this theorem because it requires analytic techniques that are beyond the scope of this text. The original proof was given by Carl Friedrich Gauss in his doctoral dissertation at the age of nineteen.

The fundamental theorem of algebra allows us (in principle) to find for every complex polynomial a factorization into complex linear factors. To illustrate this point, we consider an example.

45.8 Example. Since $z_1 = -1$ is obviously a root of the polynomial $p(z) := z^3 + 1$, it follows that $z^3 + 1$ is divisible by $z + 1$. Using polynomial division, we obtain

$$\frac{p(z)}{z + 1} = z^2 - z + 1,$$

or equivalently,

$$p(z) = (z + 1)(z^2 - z + 1).$$

Given the quadratic formula in Appendix D, the roots of the polynomial $q(z) := z^2 - z + 1$ are easily seen to be $1/2 \pm (\sqrt{3}/2)i$. Hence

$$q(z) = \left(z - \frac{1}{2} - \frac{\sqrt{3}}{2}i\right)\left(z - \frac{1}{2} + \frac{\sqrt{3}}{2}i\right),$$

and the complete factorization of $p(z)$ is therefore

$$p(z) = (z+1)\left(z - \frac{1}{2} - \frac{\sqrt{3}}{2}i\right)\left(z - \frac{1}{2} + \frac{\sqrt{3}}{2}i\right).$$

In general, if $p(z) = a_n z^n + \cdots + a_1 z + a_0$ is a nonconstant complex polynomial of degree n, and z_1 is a complex root of $p(z)$, then we can use polynomial division to find a polynomial $q_1(z)$ of degree $n-1$ such that

$$p(z) = (z - z_1)q_1(z).$$

If $n - 1 \geq 1$, then the fundamental theorem of algebra implies that $q_1(z)$ has a root z_2, and again using polynomial division we can find a polynomial $q_2(z)$ of degree $n-2$ such that $q_1(z) = (z - z_2)q_2(z)$. Hence

$$p(z) = (z - z_1)(z - z_2)q_2(z).$$

Continuing in this way, we can in general find for any $k \in \{1, \ldots, n\}$ a polynomial $q_k(z)$ of degree $n-k$ and roots $z_1, \ldots, z_k$ such that

$$p(z) = (z - z_1)(z - z_2)\cdots(z - z_k)q_k(z).$$

For $k = n$, the degree of $q_n(z)$ is zero, so that $q_n(z)$ is actually constant. Since this constant—$q_n(z)$—is the coefficient of z^n in the polynomial $(z - z_1)(z - z_2)\cdots(z - z_n)q_n(z)$, and since the coefficient of z^n in $p(z)$ is by definition equal to a_n, it follows that $q_n(z) = a_n$ for all $z \in \mathbb{C}$. This proves that

$$p(z) = a_n(z - z_1)\cdots(z - z_n).$$

As a matter of course, the roots $z_1, \ldots, z_n$ are not necessarily always distinct. So let us assume that only the first k roots are pairwise distinct, and that each of the remaining roots $z_{k+1}, \ldots, z_n$ is equal to one of the roots $z_1, \ldots, z_k$. Then, there exist positive integers $m_1, \ldots, m_k$ such that $m_1 + \cdots + m_k = n$ and

$$p(z) = a_n(z - z_1)^{m_1}\cdots(z - z_k)^{m_k}.$$

Each integer m_j is referred to as the *multiplicity* of the root z_j. The following theorem summarizes these results:

45.9 Theorem. *Let $p(z) = a_n z^n + \cdots + a_1 z + a_0$ be a complex polynomial and assume that $z_1, \ldots, z_k$ are the distinct roots of $p(z)$ with multiplicities $m_1, \ldots, m_k$. Then $m_1 + \cdots + m_k = n$ and*

$$p(z) = a_n(z - z_1)^{m_1}\cdots(z - z_k)^{m_k}.$$

As a corollary to this theorem, we record the following observation:

Any complex polynomial of degree $n \geq 1$ has exactly n roots counting multiplicities.

45.10 Example. The polynomial

$$p(z) = (z-2)(z+3i)(z-1)(z-2)(z-2)(z+3i)$$

has three distinct roots $z_1 = 2$, $z_2 = -3i$, and $z_3 = 1$ with the corresponding multiplicities $m_1 = 3$, $m_2 = 2$, and $m_3 = 1$. Thus, $p(z) = (z-2)^3(z+3i)^2(z-1)$.

Note: in practical applications it may be difficult or impossible to find the exact values of the roots $z_1, \ldots, z_k$, and we will therefore frequently have to be content with computer-generated approximations.

45.11 Exercise. Find a factorization of the form given in Theorem 45.9 for the polynomial $p(z) := z^4 + 2z^2 + 1$.

As a special case, we now consider a polynomial $p(z) = a_n z^n + \cdots + a_1 z + a_0$ with *real* coefficients $a_0, a_1, \ldots, a_n \in \mathbb{R}$. If $z_0 \in \mathbb{C}$ is a root of $p(z)$ (i.e., $p(z_0) = 0$), then $\bar{z}_0$ is a root of $p(z)$ as well. For if the coefficients a_i are real, then $a_i = \bar{a}_i$, and Theorem 45.5a therefore implies that

$$p(\bar{z}_0) = a_n \bar{z}_0^n + \cdots + a_1 \bar{z}_0 + a_0 = \overline{a_n z_0^n} + \cdots + \overline{a_1 z_0} + \overline{a_0} = \overline{p(z_0)} = 0.$$

This proves that all complex roots (with nonzero imaginary parts) of a polynomial with real coefficients appear in pairs z_0, $\bar{z}_0$. Consequently, if z_0 is a complex root of $p(z)$, then $p(z)$ is divisible by the quadratic factor $(z - z_0)(z - \bar{z}_0)$. Since

$$(z - z_0)(z - \bar{z}_0) = z^2 - (z_0 + \bar{z}_0)z + z_0\bar{z}_0 = z^2 - 2\mathcal{R}e(z_0)z + |z_0|^2 \quad \text{(see Theorem 45.5b)},$$

it follows that $(z - z_0)(z - \bar{z}_0)$ is a polynomial with real coefficients. Therefore, the polynomial $q(z)$, obtained by dividing $p(z)$ by $(z - z_0)(z - \bar{z}_0)$, has real coefficients and its complex roots thus appear in pairs of complex conjugates as well. Given this observation, it is easy to see that the multiplicity of the complex root z_0 of $p(z)$ is equal to the multiplicity of the root $\bar{z}_0$. So if we denote by $d_1, \ldots, d_j$ the distinct real roots of $p(z)$ with multiplicities $p_1, \ldots, p_j$, and by $z_1, \bar{z}_1, \ldots, z_l, \bar{z}_l$ the distinct complex roots with multiplicities $q_1, \ldots, q_l$, then Theorem 45.9 allows us to infer that $p_1 + \cdots + p_j + 2q_1 + \cdots + 2q_l = n$ and

$$p(z) = a_n(z - d_1)^{p_1} \cdots (z - d_j)^{p_j} (z - z_1)^{q_1} (z - \bar{z}_1)^{q_1} \cdots (z - z_l)^{q_l} (z - \bar{z}_l)^{q_l}.$$

Thus we have established the following theorem:

45.12 Theorem. *Let $p(z) = a_n z^n + \cdots + a_1 z + a_0$ be a polynomial with real coefficients, let $d_1, \ldots, d_j$ be the distinct real roots of $p(z)$ with multiplicities $p_1, \ldots, p_j$, and let $z_1, \bar{z}_1, \ldots, z_l, \bar{z}_l$ be the distinct complex roots of $p(z)$ with multiplicities $q_1, \ldots, q_l$. Then $p_1 + \cdots + p_j + 2q_1 + \cdots + 2q_l = n$ and*

$$p(z) = a_n(z - d_1)^{p_1} \cdots (z - d_j)^{p_j} (z - z_1)^{q_1} (z - \bar{z}_1)^{q_1} \cdots (z - z_l)^{q_l} (z - \bar{z}_l)^{q_l}.$$

Furthermore, setting $b_k := 2\mathcal{R}e(z_k)$ and $c_k := |z_k|^2$ for all $k \in \{1, \ldots, l\}$, we obtain the following factorization of $p(z)$ into real linear and real quadratic factors:

$$p(z) = a_n(z - d_1)^{p_1} \cdots (z - d_j)^{p_j} (z^2 + b_1 z + c_1)^{q_1} \cdots (z^2 + b_l z + c_l)^{q_l}.$$

Note: an illustration of how to apply this theorem will be given in Example 45.20 below.

The Complex Exponential Function

In Chapter 42 we saw that

$$e^x = \sum_{k=0}^{\infty} \frac{x^k}{k!} \quad \text{for all } x \in \mathbb{R}.$$

In order to establish this result, we showed that the limit of the partial sums $\sum_{k=0}^{n} x^k/k!$ exists and is equal to e^x for all $x \in \mathbb{R}$. In a very similar fashion, it can be shown that the limit of these partial sums also exists if x is replaced with a complex number z (for a definition of the limit of a sequence of complex numbers see the ⋆remark⋆ below). This observation allows us to define the *complex exponential function* via the equation

$$\boxed{e^z := \sum_{k=0}^{\infty} \frac{z^k}{k!} \quad \text{for } z \in \mathbb{C}.} \tag{45.7}$$

Note: as in the real case we use the convention $z^0 := 1$ for all $z \in \mathbb{C}$.

★ *Remark.* In order to adequately define the limit of a sequence of complex numbers, let us recall that in Chapter 3 we gave a precise definition for the limit of a sequence of real numbers. Given Definition 3.2, it is not difficult to see that a sequence $(a_n)_{n=0}^{\infty}$ of real numbers converges to a number $L \in \mathbb{R}$ (i.e., $\lim_{n\to\infty} a_n = L$) if and only if

$$\lim_{n\to\infty} |a_n - L| = 0.$$

Thus it appears natural to define the limit of a sequence of complex numbers as follows: if $(z_n)_{n=1}^{\infty}$ is a sequence of complex numbers, then we say that $(z_n)_{n=1}^{\infty}$ converges to a number $L \in \mathbb{C}$ if

$$\lim_{n\to\infty} |z_n - L| = 0,$$

and in this case we write

$$L = \lim_{n\to\infty} z_n.$$

Note: The symbol $\lim_{n\to\infty}$ in the expression $\lim_{n\to\infty} |z_n - L|$ signifies the taking of the limit of a sequence of real numbers in the sense of Definition 3.2 (because $|z_n - L|$ is a real number). By contrast, the same symbol in the expression $\lim_{n\to\infty} z_n$ refers to the newly defined limit of a sequence of *complex* numbers. So if we wanted to be very precise in our use of notation, we should perhaps introduce the symbols $\lim^{\mathbb{R}}_{n\to\infty}$ and $\lim^{\mathbb{C}}_{n\to\infty}$ to distinguish between real and complex limits, but for simplicity's sake we will follow conventional usage and denote both types of limits by the standard symbol $\lim_{n\to\infty}$.

Having defined the notion of a limit for sequences of complex numbers, it now makes sense to say that an infinite series of complex numbers is convergent if the limit of the sequence of its partial sums exists. More precisely, the series $\sum_{k=1}^{\infty} z_k$ is said to be convergent if $\lim_{n\to\infty} \sum_{k=1}^{n} z_k$ exists, and in this case we write

$$\sum_{k=1}^{\infty} z_k = \lim_{n\to\infty} \sum_{k=1}^{n} z_k.$$

Consequently, underlying the defining equation (45.7) is the assertion that the limit of the sequence of the partial sums $\sum_{k=0}^{n} z^k/k!$ exists for all $z \in \mathbb{C}$. For a proof of this claim, based on a generalization of the ratio test to complex series, the reader is referred to [Rud1]. ★

45.13 Theorem. *For all $z_1, z_2, z \in \mathbb{C}$ and all positive integers n we have*

a) $e^{z_1+z_2} = e^{z_1} e^{z_2}$ *and*

b) $(e^z)^n = e^{nz}$.

Proof. ★ **a)** Using the definition of the complex exponential function and the binomial formula* (see Chapter 34, p.262), we obtain

$$\begin{aligned}
e^{z_1} e^{z_2} &= \left(\sum_{k=0}^{\infty} \frac{z_1^k}{k!}\right)\left(\sum_{k=0}^{\infty} \frac{z_2^k}{k!}\right) = \left(1 + z_1 + \frac{z_1^2}{2} + \frac{z_1^3}{6} + \ldots\right)\left(1 + z_2 + \frac{z_2^2}{2} + \frac{z_2^3}{6} + \ldots\right) \\
&= 1 + (z_1 + z_2) + \left(\frac{z_1^2}{2} + z_1 z_2 + \frac{z_2^2}{2}\right) + \left(\frac{z_1^3}{6} + \frac{z_1^2 z_2}{2} + \frac{z_1 z_2^2}{2} + \frac{z_2^3}{6}\right) + \ldots \\
&= 1 + (z_1 + z_2) + \frac{1}{2}(z_1^2 + 2z_1 z_2 + z_2^2) + \frac{1}{6}(z_1^3 + 3z_1^2 z_2 + 3z_1 z_2^2 + z_2^3) + \ldots \\
&= 1 + \sum_{k=0}^{1} \binom{1}{k} z_1^{1-k} z_2^k + \frac{1}{2}\sum_{k=0}^{2} \binom{2}{k} z_1^{2-k} z_2^k + \frac{1}{6}\sum_{k=0}^{3} \binom{3}{k} z_1^{3-k} z_2^k + \ldots \\
&= \sum_{n=0}^{\infty} \frac{1}{n!} \sum_{k=0}^{n} \binom{n}{k} z_1^{n-k} z_2^k = \sum_{n=0}^{\infty} \frac{(z_1 + z_2)^n}{n!} \\
&= e^{z_1+z_2}.
\end{aligned}$$

*The binomial formula for complex numbers is the same as for real numbers and the proofs are identical as well (see Appendix D). So the equation $(a+b)^n = \sum_{k=0}^{n} \binom{n}{k} a^{n-k} b^k$ is valid regardless of whether $a, b \in \mathbb{R}$ or $a, b \in \mathbb{C}$.

Note: This proof is not entirely rigorous, because the termwise multiplication of the series $\sum_{k=0}^{\infty} z_1^k/k!$ with $\sum_{k=0}^{\infty} z_2^k/k!$ that we used in the second step needs some careful justification. For a more detailed discussion the reader is referred to [Rud1]. ★

b) Using the result in a), it follows that

$$(e^z)^n = \underbrace{e^z \cdot \cdots \cdot e^z}_{n \text{ times}} = e^{z+\cdots+z} = e^{nz}$$

as desired. □

If $z = a + ib \in \mathbb{C}$, then according to Theorem 45.13 we have $e^z = e^a e^{ib}$. To evaluate the term e^{ib}, we apply the defining equation (45.7). This yields

$$e^{ib} = \sum_{k=0}^{\infty} \frac{i^k b^k}{k!}.$$

Since $i^4 = 1$, it is easy to see that for all nonnegative integers m we have

$$\begin{aligned} i^{4m} &= 1, \\ i^{4m+1} &= i, \\ i^{4m+2} &= -1, \\ i^{4m+3} &= -i. \end{aligned}$$

Hence

$$\begin{aligned} e^{ib} &= \sum_{m=0}^{\infty} \frac{i^{4m} b^{4m}}{(4m)!} + \sum_{m=0}^{\infty} \frac{i^{4m+1} b^{4m+1}}{(4m+1)!} + \sum_{m=0}^{\infty} \frac{i^{4m+2} b^{4m+2}}{(4m+2)!} + \sum_{m=0}^{\infty} \frac{i^{4m+3} b^{4m+3}}{(4m+3)!} \\ &= \sum_{m=0}^{\infty} \frac{b^{4m}}{(4m)!} + \sum_{m=0}^{\infty} \frac{i b^{4m+1}}{(4m+1)!} + \sum_{m=0}^{\infty} \frac{-b^{4m+2}}{(4m+2)!} + \sum_{m=0}^{\infty} \frac{-i b^{4m+3}}{(4m+3)!}. \end{aligned} \tag{45.8}$$

The first and third terms in (45.8) add up to

$$\sum_{k=0}^{\infty} \frac{(-1)^k}{(2k)!} b^{2k} = \cos(b)$$

and the second and fourth terms add up to

$$\sum_{k=0}^{\infty} \frac{(-1)^k i}{(2k+1)!} b^{2k+1} = i \sin(b) \quad \text{(see Chapter 42, p.326).}$$

Thus, we have established the following important result, known as *Euler's formula*:

$$\boxed{e^z = e^a(\cos(b) + i\sin(b)).} \tag{45.9}$$

In particular, for $z = i\pi$ we obtain the famous formula

$$\boxed{e^{i\pi} = -1.} \tag{45.10}$$

45.14 Exercise. Use (45.9) to show that for all $z \in \mathbb{C}$ the complex conjugate of e^z is $e^{\bar{z}}$.

A frequently useful application of Euler's formula is the representation of complex numbers in *polar form* (see also Appendix A for a closely related discussion of polar coordinates). Given a complex number $z = a + ib$, we denote by θ the angle between the line connecting z with the origin and the

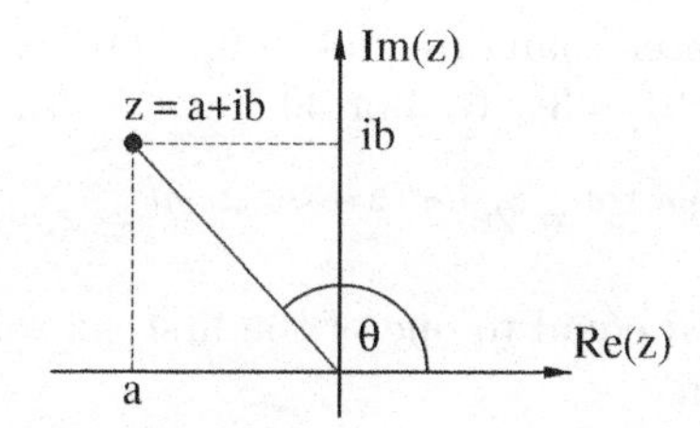

Figure 45.4: polar representation of complex numbers.

positive real axis (see Figure 45.4). Since the distance of z from the origin is equal to $|z| = \sqrt{a^2+b^2}$, it is obvious that

$$a = |z|\cos(\theta) \quad \text{and} \quad b = |z|\sin(\theta).$$

This shows that $z = |z|(\cos(\theta) + i\sin(\theta))$, and therefore, Euler's formula implies that

$$\boxed{z = |z|e^{i\theta}.}$$

45.15 Example. We wish to find the polar representation of $z = -2-5i$. Given that $-2-5i$ is located in the third quadrant of the complex plane, it follows that $\theta = \pi + \arctan(5/2)$. Since $|z| = \sqrt{2^2+5^2} = \sqrt{29}$, we may thus infer that

$$z = \sqrt{29}e^{i(\pi+\arctan(5/2))}.$$

45.16 Exercise. Find the polar representation of $-2+2i$.

45.17 Example. To give a simple illustration of the usefulness of polar representations we will now discuss the problem of finding higher-order roots. More specifically, we wish to determine the sixth roots of -64 in $\mathbb{C}$. In other words, our task is to find all $z \in \mathbb{C}$ that satisfy the equation

$$z^6 = -64. \tag{45.11}$$

The fact that -64 is located on the negative real axis implies that its corresponding polar angle is $\theta = \pi$, and therefore

$$-64 = |-64|e^{i\pi} = 64e^{i\pi} = 2^6e^{i\pi} \quad \text{(see also (45.10))}.$$

Given this polar representation of -64, one obvious solution of (45.11) is $z_0 = 2e^{i\pi/6}$ because, according to Theorem 45.13b, we have

$$z_0^6 = 2^6\left(e^{i\pi/6}\right)^6 = 64e^{i\pi} = -64.$$

Furthermore, since $e^{2\pi i} = \cos(2\pi) + i\sin(2\pi) = 1$, it follows that for all positive integers k the number $z_k = 2e^{i(\pi+2\pi k)/6}$ is a solution of (45.11) as well, because

$$z_k^6 = 2^6\left(e^{i(\pi+2\pi k)/6}\right)^6 = 2^6e^{i(\pi+2\pi k)} = 64e^{i\pi}e^{2\pi i} = -64.$$

Since the index k runs through the entire set of nonnegative integers, it may be tempting to think that (45.11) has infinitely many solutions, but this is actually not the case because for values $k \geq 6$ we observe a periodic recurrence of z_k to one of the first six solutions $z_0, \ldots, z_5$. For example,

$$z_6 = 2e^{i(\pi+2\pi\cdot 6)/6} = 2e^{i\pi/6}e^{2\pi i} = z_0$$

and

$$z_7 = 2e^{i(\pi+2\pi\cdot 7)/6} = 2e^{i(\pi+2\pi)/6}e^{2\pi i} = z_1$$

In general, if k is an integer greater than or equal to 6, there exist integers m and n such that $m \geq 1$, $0 \leq n \leq 5$ and $k = n + 6m$ (e.g. $21 = 3 + 6 \cdot 3$ or $30 = 0 + 6 \cdot 5$), and therefore,

$$z_k = 2e^{i(\pi+2\pi(n+6m))/6} = 2e^{i(\pi+2\pi n)/6}e^{2\pi mi} = z_n \left(e^{2\pi i}\right)^m = z_n \cdot 1^m = z_n.$$

This shows that each z_k is indeed equal to one of the first six solutions $z_0, \ldots, z_5$. In other words, the six distinct sixth roots of -64 are

$$\begin{aligned} z_0 &= 2e^{i\pi/6}, \\ z_1 &= 2e^{i(\pi+2\pi)/6} = 2e^{i\pi/2}, \\ z_2 &= 2e^{i(\pi+2\pi\cdot 2)/6} = 2e^{i5\pi/6}, \\ z_3 &= 2e^{i(\pi+2\pi\cdot 3)/6} = 2e^{i7\pi/6}, \\ z_4 &= 2e^{i(\pi+2\pi\cdot 4)/6} = 2e^{i3\pi/2}, \\ z_5 &= 2e^{i(\pi+2\pi\cdot 5)/6} = 2e^{i11\pi/6}. \end{aligned} \tag{45.12}$$

The perfectly symmetric distribution of these roots in the complex plane is shown in Figure 45.5.

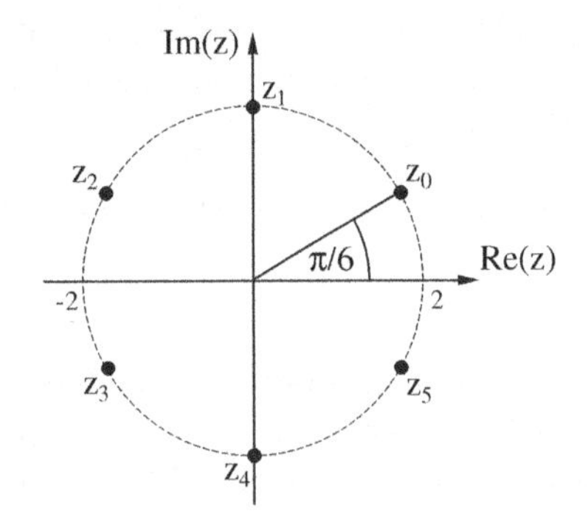

Figure 45.5: the sixth roots of -64.

45.18 Exercise. Find all fourth roots of -1.

It is no coincidence that in Example 45.17 the number of roots of -64 turned out to be equal to the order of the roots. The following exercise will show that, in general, the number of nth roots of a given complex number is equal to n.

45.19 Exercise. Explain why any complex number $w = |w|e^{i\theta}$ has exactly n distinct roots of order n that are given by the formula

$$\boxed{z_k = \sqrt[n]{|w|}e^{i(\theta+2\pi k)/n} \text{ for } k \in \{0, \ldots, n-1\}.}$$

Furthermore, describe the distribution of these roots in the complex plane.

45.20 Example. Expanding on the results in Example 45.17, we will now discuss the problem of finding a factorization of the polynomial $p(z) = z^6 + 64$ into three real quadratic factors (see Theorem 45.12). To begin with, we observe that the complex roots of $p(z)$ are the solutions of equation (45.11) as listed in (45.12). Consequently, $p(z)$ can be factored into six complex linear factors as follows:

$$p(z) = (z - z_0)(z - z_1)(z - z_2)(z - z_3)(z - z_4)(z - z_5).$$

Given the values in (45.12), and given also the distribution of the roots in Figure 45.5, it is easy to see that $z_0 = \bar{z}_5$, $z_1 = \bar{z}_4$ and $z_2 = \bar{z}_3$. Thus, the three quadratic factors that make up the factorization of

$p(z)$ are

$$\begin{aligned}(z-z_0)(z-z_5) &= z^2 - 2\mathcal{R}e(z_0)z + |z_0|^2 = z^2 - 2\mathcal{R}e\left(2e^{i\pi/6}\right)z + \left|2e^{i\pi/6}\right|^2 \\ &= z^2 - 2\cdot 2\cos\left(\frac{\pi}{6}\right)z + 2^2 = z^2 - 2\sqrt{3}z + 4, \\ (z-z_1)(z-z_4) &= z^2 - 2\mathcal{R}e\left(2e^{i\pi/2}\right)z + \left|2e^{i\pi/2}\right|^2 = z^2 + 4, \\ (z-z_2)(z-z_3) &= z^2 - 2\mathcal{R}e\left(2e^{i5\pi/6}\right)z + \left|2e^{i5\pi/6}\right|^2 = z^2 + 2\sqrt{3}z + 4.\end{aligned}$$

Hence

$$p(z) = (z^2 - 2\sqrt{3}z + 4)(z^2 + 4)(z^2 + 2\sqrt{3}z + 4).$$

45.21 Exercise. Find a factorization of $p(z) = z^4 + 1$ into two real quadratic factors.

Differentiation of Complex-Valued Functions

Assume that $u : I \to \mathbb{R}$ and $v : I \to \mathbb{R}$ are differentiable functions defined on an open interval I. Then

$$f(x) := u(x) + iv(x)$$

is the defining equation for a *complex-valued function* $f : I \to \mathbb{C}$. The derivative of this function is obtained by separately differentiating the real and imaginary parts u and v, i.e.,

$$f'(x) := u'(x) + iv'(x) \text{ for all } x \in I.$$

As an important special case, we consider for a given $\lambda = a + ib \in \mathbb{C}$ the function

$$f(x) := e^{\lambda x} \text{ for } x \in \mathbb{R}.$$

Using Euler's formula, it follows that

$$f(x) = e^{ax+ibx} = e^{ax}(\cos(bx) + i\sin(bx)).$$

Consequently, the real and imaginary parts of f are

$$u(x) := e^{ax}\cos(bx) \quad \text{and} \quad v(x) := e^{ax}\sin(bx).$$

Hence

$$\begin{aligned}f'(x) = u'(x) + iv'(x) &= ae^{ax}\cos(bx) - be^{ax}\sin(bx) + i(ae^{ax}\sin(bx) + be^{ax}\cos(bx)) \\ &= (a+ib)e^{ax}(\cos(bx) + i\sin(bx)) = \lambda e^{\lambda x}.\end{aligned}$$

This result shows that the differentiation of the exponential function $e^{\lambda x}$ follows the same rule regardless of whether λ is real or complex.

45.22 Theorem. *For all* $\lambda \in \mathbb{C}$ *we have*

$$\boxed{\frac{d}{dx}e^{\lambda x} = \lambda e^{\lambda x}.}$$

Additional Exercises

45.23. Find the real and imaginary parts of the complex number $\dfrac{2+3i}{4-i}$.

45.24. Find the absolute value of the complex number in Exercise 45.23.

45.25. Find a factorization of the polynomial $p(x) = x^5 - 32$ into one real linear factor and two real quadratic factors.

45.26. Find a factorization of the polynomial $p(z) = z^8 - 16$ into eight complex linear factors.

45.27. Find a factorization of the polynomial $p(x) = x^4 + x^2 + 1$ into two real quadratic factors.

45.28. Find a factorization of the polynomial $p(z) = z^3 + z^2 - 2$ into three complex linear factors. *Hint.* One root can be guessed easily.

45.29. Find all fifth roots of $\dfrac{1}{\sqrt{2}} - \dfrac{i}{\sqrt{2}}$.

45.30. For $\lambda = 1/5 + i$, plot the range of the function $f(x) = e^{\lambda x}$ in the complex plane. *Hint.* Use Euler's formula.

45.31. Find $\theta, r \in \mathbb{R}$ such that $-4 - 3i = re^{i\theta}$

45.32. Write each of the numbers given below in the form $re^{i\theta}$.

$$\text{a) } \sqrt{3}+i \quad \text{b) } \sqrt{3}-i \quad \text{c) } -\sqrt{3}+i \quad \text{d) } -\sqrt{3}-i$$

45.33. Find the real and imaginary parts of $(1-i/\sqrt{3})^{10}$. *Hint.* Determine r and θ such that $1-i/\sqrt{3} = re^{i\theta}$.

45.34. Prove the addition laws for sine and cosine:

$$\sin(s+t) = \sin(s)\cos(t) + \sin(t)\cos(s),$$
$$\cos(s+t) = \cos(s)\cos(t) - \sin(s)\sin(t).$$

Hint. Use Euler's formula to show that $\sin(x) = (e^{ix} - e^{-ix})/2i$, and find an analogous formula for $\cos(x)$ (see also Theorem 45.5b).

45.35. For $f_1(x) = u_1(x) + iv_1(x)$ and $f_2(x) = u_2(x) + iv_2(x)$ show that $(f_1 f_2)' = f_1' f_2 + f_1 f_2'$.

45.36. In each of the statements below there is a mistake. Identify the mistake and explain your reasoning.

a) a, b, and c are *real* numbers, and the polynomial $p(z) = z^3 + az^2 + bz + c$ has three roots z_1, z_2, and z_3 such that $\mathcal{R}e(z_1) = 1$, $\mathcal{R}e(z_2) = 2$, $\mathcal{R}e(z_3) = 3$, and $\mathcal{I}m(z_1) \neq 0$.

b) $p(z)$ is a nonconstant complex polynomial such that $|p(z)| \geq 1$ for all $z \in \mathbb{C}$.

Chapter 46

Second-Order Linear Differential Equations and Electric Circuits (Part 2)

R-C-L Circuits

Having discussed the simplest types of electric circuits in Chapter 44, we now move on to a more complex example by combining in one circuit a resistor with resistance R, a capacitor with capacitance C, an inductor with inductance L, and a switch (see Figure 46.1). We assume that, as long as the

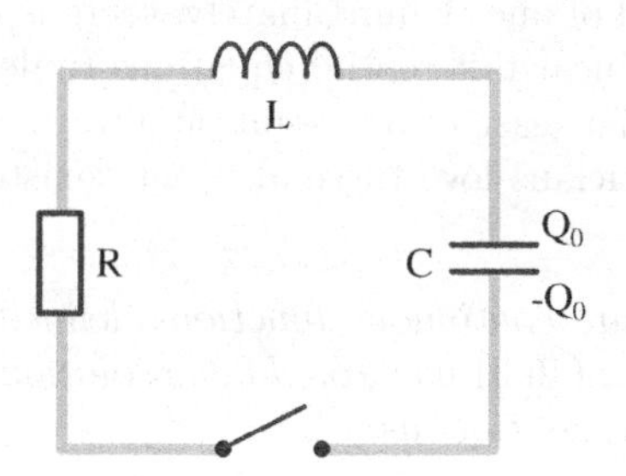

Figure 46.1: R-C-L circuit

switch is open, there is a charge Q_0 on one side of the capacitor and a charge $-Q_0$ on the other. At time $t = 0$ the switch is closed and a current $I(t)$ starts to flow. According to Kirchoff's first law, the differences in electric potential at the resistor, the capacitor, and the inductor add up to zero (see Chapter 44, p.338). Hence

$$0 = RI(t) + \frac{Q(t)}{C} + LI'(t) \text{ for } t \geq 0.$$

In differentiating both sides and substituting $I(t)$ for $Q'(t)$ (see (44.1)), we obtain

$$0 = RI'(t) + \frac{Q'(t)}{C} + LI''(t) = RI'(t) + \frac{I(t)}{C} + LI''(t),$$

or equivalently,

$$I''(t) + \frac{R}{L}I'(t) + \frac{1}{LC}I(t) = 0. \tag{46.1}$$

This equation is a special case of a so-called *homogeneous second-order linear differential equation with constant coefficients.*

Second-Order Linear Differential Equations

Let us assume that $a : I \to \mathbb{R}$, $b : I \to \mathbb{R}$, and $f : I \to \mathbb{R}$ are continuous functions defined on an open interval I. Then we say that

$$y''(x) + a(x)y'(x) + b(x)y(x) = f(x) \tag{46.2}$$

is a *second-order linear differential equation.* The equation is *homogeneous* if $f(x) = 0$ for all $x \in I$, and otherwise it is *nonhomogeneous.* Furthermore, if a and b are constant, the equation is said to have *constant coefficients.*

46.1 Example. **a)** $y''(x) + x^2y'(x) + e^x y(x) = \sin(x)$—this equation is second-order linear with $a(x) = x^2$, $b(x) = e^x$, and $f(x) = \sin(x)$.

b) $y''(x)+2y(x) = 0$—this equation is second-order linear as well with $a(x) = f(x) = 0$ and $b(x) = 2$.

c) $y''(x) + x^2y'(x) + 2y(x)^2 = \sin(x)$—this equation is second-order, but *not* linear because of the term $y(x)^2$.

d) $y'''(x)^2 + x^2y'(x) + 2y(x)y'(x) = \sin(x)$—this equation is *not* second-order because it involves a third derivative of y, and due to the square and product terms $y'''(x)^2$ and $y(x)y'(x)$ it is also not linear.

An *initial value problem* corresponding to (46.2) is

$$\begin{aligned} &y''(x) + a(x)y'(x) + b(x)y(x) = f(x) \\ &y(x_0) = z_1, \; y'(x_0) = z_2, \end{aligned} \tag{46.3}$$

where $x_0 \in I$ and $z_1, z_2 \in \mathbb{R}$. Notice that in contrast to first-order linear differential equations, we now have two initial conditions instead of one. Unfortunately, there is no general strategy for solving initial value problems for second-order linear differential equations (unless the coefficients are constant), but the following theorem guarantees at least that a solution always exists even if it may be impossible to write down explicitly. (Mathematicians love these abstract "existence statements," but physicists and engineers hate them.)

46.2 Theorem. *If a, b, and f are continuous functions defined on an open interval I, then for all $z_1, z_2 \in \mathbb{R}$ the initial value problem* (46.3) *has exactly one solution $y : I \to \mathbb{R}$ that is twice continuously differentiable (i.e., y'' exists and is continuous).*

The proof of this theorem usually involves the theory of infinite-dimensional vector spaces and is definitely beyond the scope of this text (see [WB] or [Wa]).

Teacher: Here is an example that will test your comprehension of Theorem 46.2: for continuous functions $a, b : \mathbb{R} \to \mathbb{R}$ let us consider the *homogeneous* second-order linear differential equation

$$y''(x) + a(x)y'(x) + b(x)y(x) = 0, \tag{46.4}$$

and let us assume that $y_0 : \mathbb{R} \to \mathbb{R}$ is twice differentiable such that for some $x_0 \in \mathbb{R}$ we have

$$y_0(x) = 0 \text{ for all } x \le x_0 \text{ and } y_0(x) > 0 \text{ for all } x > x_0 \text{ (see Figure 46.2).}$$

Is it possible that a function with these properties is a solution of equation (46.4)? In other words, is it possible that $y_0''(x) + a(x)y_0'(x) + b(x)y_0(x) = 0$ for all $x \in \mathbb{R}$?

Simplicio: In absence of an explicit definition for the coefficient functions a and b this question seems impossible to answer.

Teacher: There really is not a lot of information to work with—all we have are the stated properties of y_0 and Theorem 46.2.

Simplicio: If I am not mistaken, Theorem 46.2 only asserts the existence of solutions to equation (46.4), but it does not in any way specify what these solutions might look like. So what is the point?

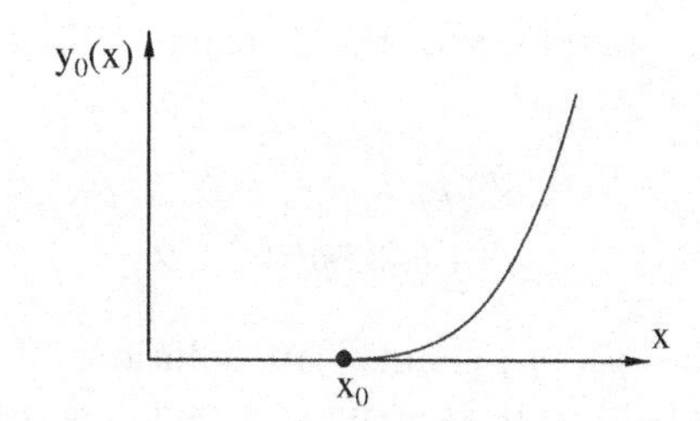

Figure 46.2: graph of y_0.

Teacher: It is not quite true that Theorem 46.2 is only an existence statement because there is also a claim regarding the *uniqueness* of solutions!

Sophie: As far as I can tell, the assertion of uniqueness is made in reference only to solutions of initial value problems. How is that going to help us if all we are given is a differential equation without any initial values?

Teacher: A good question, Sophie, but consider the assumption $y_0(x) = 0$ for all $x \leq x_0$. Given this property of y_0, what do you think are the values $y_0(x_0)$ and $y_0'(x_0)$?

Sophie: They would both have to be equal to zero.

Teacher: Correct.

Sophie: So are you suggesting that we ought to decide whether y_0, as shown in Figure 46.2, can be the unique solution of the initial value problem

$$\begin{aligned} &y''(x) + a(x)y'(x) + b(x)y(x) = 0 \\ &y(x_0) = 0, \ y'(x_0) = 0? \end{aligned} \tag{46.5}$$

Teacher: That, indeed, is the right approach, because the initial value problem (46.5) is very easy to solve.

Simplicio: One possible solution is obviously $y_1(x) := 0$, because if $y_1(x) = 0$ for all $x \in \mathbb{R}$, then $y_1'(x)$ and $y_1''(x)$ are equal to zero as well and all equations in (46.5) are trivially satisfied, but...

Teacher: No "but"—the riddle is solved.

Simplicio: How? I don't see it.

Teacher: Since, as you said, y_1 is a solution of (46.5), Theorem 46.2 implies that y_1 *is the only solution of* (46.5). Therefore, y_0 cannot be a solution of (46.5) as well, because the condition $y_0(x) > 0$ for all $x > x_0$ obviously implies that $y_0 \neq y_1$. Combining this observation with the fact that y_0 satisfies the initial conditions in (46.5), we are thus forced to conclude that y_0 is *not* a solution of the differential equation in (46.5) (which is identical with (46.4)). This conclusion is intuitively not surprising, because it is highly implausible to imagine that, for example, the circuit in Figure 46.1 (with the switch closed) could be entirely "flat" for $t \leq x_0$ (no current in the circuit and no charge on the capacitor) and then suddenly generate a positive current out of nowhere for $t > x_0$. Such a behavior can occur only if an external voltage source is present and activated at time $t = x_0$. We will see, however, that the differential equation describing an R-C-L circuit with an external voltage source is nonhomogeneous, thus causing the argument outlined above to fail because $y_1 = 0$ is obviously not a solution of a nonhomogeneous equation.

The Set of Solutions of a Homogeneous Equation

In working with electric circuits, and in many other applications involving linear differential equations, it is very important to be aware of the so-called *principle of linearity (or superposition)*, which asserts that sums and scalar multiples of solutions of homogeneous linear differential equations are solutions as well (see Theorem 46.4 and Exercise 46.5 below).

46.3 Example. To illustrate the usefulness of the principle of linearity, we consider the initial value problem

$$\begin{aligned} y''(x) + y(x) &= 0 \\ y(0) = 3,\ y'(0) &= 2. \end{aligned} \tag{46.6}$$

The functions $y_1(x) := \cos(x)$ and $y_2(x) := \sin(x)$ are solutions of the differential equation in (46.6) because $y_1'' + y_1 = -\cos(x) + \cos(x) = 0 = -\sin(x) + \sin(x) = y_2'' + y_2$. The principle of linearity applied to y_1 and y_2 says that for all constants $C_1, C_2 \in \mathbb{R}$, the function $y := C_1y_1 + C_2y_2$ is a solution of the differential equation in (46.6) as well. This statement is easily verified from elementary rules of differentiation as the following calculation shows:

$$\begin{aligned} y'' + y &= (C_1y_1 + C_2y_2)'' + C_1y_1 + C_2y_2 = C_1y_1'' + C_2y_2'' + C_1y_1 + C_2y_2 \\ &= C_1(y_1'' + y_1) + C_2(y_2'' + y_2) = C_1 \cdot 0 + C_2 \cdot 0 = 0. \end{aligned}$$

Based on this observation, it is now possible to solve the initial value problem (46.6) by properly adjusting the constants C_1 and C_2: if the function $y = C_1y_1 + C_2y_2$ is to satisfy the initial conditions in (46.6), then it must be the case that

$$\begin{aligned} 3 &= y(0) = C_1y_1(0) + C_2y_2(0) = C_1\cos(0) + C_2\sin(0) = C_1, \\ 2 &= y'(0) = C_1y_1'(0) + C_2y_2'(0) = -C_1\sin(0) + C_2\cos(0) = C_2. \end{aligned}$$

Thus, the solution of (46.6) is

$$y(x) = 3y_1(x) + 2y_2(x) = 3\cos(x) + 2\sin(x).$$

The fact that in this example the constants C_1 and C_2 are equal to the initial values 3 and 2 respectively is coincidental. It will not always be so simple.

46.4 Theorem. *If y, y_1, and y_2 are solutions of the homogeneous equation*

$$y''(x) + a(x)y'(x) + b(x)y(x) = 0, \tag{46.7}$$

then for all $C \in \mathbb{R}$ the functions Cy and $y_1 + y_2$ are solutions as well.

Proof. If y_1 and y_2 are solutions of (46.7), then

$$\begin{aligned} y_1'' + a(x)y_1' + b(x)y_1 &= 0, \\ y_2'' + a(x)y_2' + b(x)y_2 &= 0. \end{aligned}$$

Hence

$$(y_1 + y_2)'' + a(x)(y_1 + y_2)' + b(x)(y_1 + y_2) = y_1'' + a(x)y_1' + b(x)y_1 + y_2'' + a(x)y_2' + b(x)y_2 = 0.$$

This shows that $y_1 + y_2$ is a solution of (46.7). Similarly, if y is a solution of (46.7), then

$$(Cy)'' + a(x)(Cy)' + b(x)(Cy) = C(y'' + a(x)y' + b(x)y) = 0,$$

and therefore, Cy is a solution of (46.7) as well. □

46.5 Exercise. Assume that y_1 and y_2 are solutions of the differential equation (46.7). Use Theorem 46.4 to show that for all $C_1, C_2 \in \mathbb{R}$ the function $C_1y_1 + C_2y_2$ is a solution of (46.7) as well.

In order to generalize the results of Example 46.3 we consider the initial value problem

$$\begin{aligned} y''(x) + a(x)y'(x) + b(x)y(x) &= 0 \\ y(x_0) = z_1,\ y'(x_0) &= z_2. \end{aligned} \tag{46.8}$$

Assuming that y_1 and y_2 are solutions of the differential equation in (46.8), the principle of linearity (as stated in Theorem 46.4 and Exercise 46.5) implies that for all $C_1, C_2 \in \mathbb{R}$ the function

$$y(x) = C_1 y_1(x) + C_2 y_2(x)$$

is a solution of the differential equation in (46.8) as well. In order for this function to satisfy the initial conditions in (46.8), we need to choose C_1 and C_2 in such a way that

$$C_1 y_1(x_0) + C_2 y_2(x_0) = y(x_0) = z_1,$$
$$C_1 y_1'(x_0) + C_2 y_2'(x_0) = y'(x_0) = z_2.$$

Multiplying the first equation by $y_2'(x_0)$, the second by $y_2(x_0)$, and then subtracting the second from the first yields

$$C_1(y_1(x_0)y_2'(x_0) - y_1'(x_0)y_2(x_0)) = z_1 y_2'(x_0) - z_2 y_2(x_0).$$

Introducing the additional *assumption*

$$y_1(x_0)y_2'(x_0) - y_1'(x_0)y_2(x_0) \neq 0 \tag{46.9}$$

it follows that

$$C_1 = \frac{z_1 y_2'(x_0) - z_2 y_2(x_0)}{y_1(x_0)y_2'(x_0) - y_1'(x_0)y_2(x_0)}. \tag{46.10}$$

A completely analogous calculation also shows that

$$C_2 = \frac{z_2 y_1(x_0) - z_1 y_1'(x_0)}{y_1(x_0)y_2'(x_0) - y_1'(x_0)y_2(x_0)}. \tag{46.11}$$

46.6 Exercise. Verify equation (46.11).

In summary, a solution of (46.8) of the form $y = C_1 y_1 + C_2 y_2$ exists whenever the solutions y_1 and y_2 satisfy condition (46.9).

Remark. In the preceding discussion we essentially showed that a system of two linear equations

$$ax + by = e$$
$$cx + dy = f$$

has exactly one solution if $ad - bc \neq 0$, and that in this case the solution (x, y) is given by the equations

$$x = \frac{ed - fb}{ad - bc} \quad \text{and} \quad y = \frac{fa - ce}{ad - bc}. \tag{46.12}$$

Here x corresponds to C_1, y to C_2, e to z_1, f to z_2, and the coefficients a, b, c, and d correspond to $y_1(x_0)$, $y_2(x_0)$, $y_1'(x_0)$, and $y_2'(x_0)$ respectively.

46.7 Exercise. Use (46.12) to solve the system given below, and verify that your solution is correct by substituting it back into the equations.

$$2x - 3y = 1$$
$$x + 5y = 2.$$

In light of condition (46.9) it is helpful to introduce the following definition:

46.8 Definition. If $f : I \to \mathbb{R}$ and $g : I \to \mathbb{R}$ are differentiable functions defined on an open interval I, then

$$W_{f,g}(x) := f(x)g'(x) - f'(x)g(x) \text{ for all } x \in I.$$

is said to be the *Wronski determinant* of f and g.

46.9 Theorem. *Let a and b be continuous functions defined on an open interval I and assume that y_1 and y_2 are solutions of the homogeneous equation (46.7). If there exists an $x_1 \in I$ such that $W_{y_1,y_2}(x_1) \neq 0$ then*

a) *$W_{y_1,y_2}(x) \neq 0$ for all $x \in I$, and*

b) *the unique solution y of (46.8) is given by the equation*

$$y(x) = \frac{z_1 y_2'(x_0) - z_2 y_2(x_0)}{W_{y_1,y_2}(x_0)} y_1(x) + \frac{z_2 y_1(x_0) - z_1 y_1'(x_0)}{W_{y_1,y_2}(x_0)} y_2(x) \text{ for all } x \in I.$$

In particular, for every solution y of (46.7) there are constants $C_1, C_2 \in \mathbb{R}$ such that $y = C_1 y_1 + C_2 y_2$. The term $C_1 y_1 + C_2 y_2$ is also referred to as the general solution of equation (46.7).

Proof. **a)** Under the assumption that $W_{y_1,y_2}(x_1) \neq 0$, we need to show that $W_{y_1,y_2}(x_2) \neq 0$ for all $x_2 \in I$. According to Theorem 46.2, the initial value problem

$$\begin{aligned} &y''(x) + a(x)y'(x) + b(x)y(x) = 0, \\ &y(x_2) = 1, \ y'(x_2) = 0 \end{aligned}$$

has exactly one solution which we will choose to call f. Furthermore, by g we denote the unique solution of the initial value problem

$$\begin{aligned} &y''(x) + a(x)y'(x) + b(x)y(x) = 0, \\ &y(x_2) = 0, \ y'(x_2) = 1. \end{aligned}$$

According to Exercise 46.5, the function $y_1(x_2)f + y_1'(x_2)g$ is a solution of (46.7). Using the definitions of f and g as solutions of the initial value problems above, it thus follows that

$$\begin{aligned} y_1(x_2)f(x_2) + y_1'(x_2)g(x_2) &= y_1(x_2), \\ y_1(x_2)f'(x_2) + y_1'(x_2)g'(x_2) &= y_1'(x_2). \end{aligned}$$

Therefore, $y_1(x_2)f + y_1'(x_2)g$ is a solution of the initial value problem

$$\begin{aligned} &y''(x) + a(x)y'(x) + b(x)y(x) = 0, \\ &y(x_2) = y_1(x_2), \ y'(x_2) = y_1'(x_2). \end{aligned}$$

Since y_1 is obviously a solution of this initial value problem as well, and since solutions of initial value problems are unique (Theorem 46.2), we may infer that $y_1 = y_1(x_2)f + y_1'(x_2)g$. With a completely analogous argument it can also be shown that $y_2 = y_2(x_2)f + y_2'(x_2)g$. Given these observations, we obtain

$$\begin{aligned} W_{y_1,y_2}(x_1) &= y_1(x_1)y_2'(x_1) - y_1'(x_1)y_2(x_1) \\ &= (y_1(x_2)f(x_1) + y_1'(x_2)g(x_1))(y_2(x_2)f'(x_1) + y_2'(x_2)g'(x_1)) \\ &\quad - (y_1(x_2)f'(x_1) + y_1'(x_2)g'(x_1))(y_2(x_2)f(x_1) + y_2'(x_2)g(x_1)) \\ &= (f(x_1)g'(x_1) - f'(x_1)g(x_1))(y_1(x_2)y_2'(x_2) - y_1'(x_2)y_2(x_2)) \\ &= W_{f,g}(x_1)W_{y_1,y_2}(x_2). \end{aligned}$$

This proves that $W_{y_1,y_2}(x_2) \neq 0$, because otherwise we would have $W_{y_1,y_2}(x_1) = W_{f,g}(x_1) \cdot 0 = 0$ in contradiction to the assumption $W_{y_1,y_2}(x_1) \neq 0$.
b) The statement in b) follows from a), (46.10), and (46.11) by observing that $0 \neq W_{y_1,y_2}(x_0) = y_1(x_0)y_2'(x_0) - y_1'(x_0)y_2(x_0)$. □

The importance of Theorem 46.9 can be understood as follows: in order to find a solution of (46.8), we only need to find two solutions y_1 and y_2 of (46.7) with $W_{y_1,y_2} \neq 0$, and then determine coefficients

$C_1, C_2 \in \mathbb{R}$ (using (46.10) and (46.11) or Theorem 46.9b) such that $C_1y_1 + C_2y_2$ satisfies the initial conditions in (46.8). (In fact, the ⋆remark⋆ below shows that it is actually sufficient to find only one nonzero solution y_1 of (46.7), because the second solution y_2 can always be constructed from the first in such a way that the condition $W_{y_1,y_2} \neq 0$ is satisfied. This observation is of particular interest for equations with nonconstant coefficients for which a general method for finding solutions does not exist, but for which a single solution can sometimes be guessed.) Furthermore, it is always sufficient to check whether W_{y_1,y_2} is different from zero at a single point in I, because in that case Theorem 46.9a implies that W_{y_1,y_2} is different from zero at all points in I.

★ *Remark.* Let us assume that we are given a solution $y_1 \neq 0$ of (46.7). We wish to use y_1 to generate a second solution y_2 such that $W_{y_1,y_2} \neq 0$. For this purpose, we will use the method of the variation of the constant that we already employed in Chapter 44 (see p.340). More precisely, we will determine a function $C(x)$ such that

$$y_2(x) = C(x)y_1(x)$$

is a solution of (46.7). As in Chapter 44, the logic of the argument is to derive a defining equation for $C(x)$ from the assumption that $C(x)y_1(x)$ is a solution of (46.7). Using the product rule, we obtain

$$\begin{aligned} y_2'(x) &= C(x)y_1'(x) + C'(x)y_1(x), \\ y_2''(x) &= C(x)y_1''(x) + 2C'(x)y_1'(x) + C''(x)y_1(x). \end{aligned}$$

If y_2 is a solution of (46.7), then

$$\begin{aligned} 0 &= y_2''(x) + a(x)y_2(x) + b(x)y_2(x) \\ &= C(x)y_1''(x) + 2C'(x)y_1'(x) + C''(x)y_1(x) + a(x)(C(x)y_1'(x) + C'(x)y_1(x)) + b(x)C(x)y_1(x) \\ &= y_1(x)C''(x) + (2y_1'(x) + a(x)y_1(x))C'(x) + (y_1''(x) + a(x)y_1'(x) + b(x)y_1(x))C(x) \\ &= y_1(x)C''(x) + (2y_1'(x) + a(x)y_1(x))C'(x) \quad \text{(because } y_1 \text{ is a solution of (46.7)),} \end{aligned}$$

and therefore,

$$C''(x) + \frac{2y_1'(x) + a(x)y_1(x)}{y_1(x)}C'(x) = 0.$$

This shows that in order for y_2 to be a solution of (46.7), $C'(x)$ has to be a solution of the homogeneous first-order linear differential equation

$$y'(x) + \frac{2y_1'(x) + a(x)y_1(x)}{y_1(x)}y(x) = 0.$$

Given our results in Chapter 44, the solution of this differential equation is of the form $De^{-F(x)}$, where F is an antiderivative of $(2y_1'(x) + a(x)y_1(x))/y_1(x)$, and $D \in \mathbb{R}$ is a constant (see (44.9)). Which value we pick for D is essentially irrelevant, but $D = 1$ turns out to be the simplest choice. Hence

$$C'(x) = e^{-F(x)}.$$

Using integration by substitution and denoting the antiderivative of $a(x)$ by $A(x)$, it is easy to see that

$$F(x) = \int \left(\frac{2y_1'(x)}{y_1(x)} + a(x) \right) dx = \ln(y_1(x)^2) + A(x) + E.$$

To choose a particular antiderivative, we set the constant E equal to zero. This yields

$$C'(x) = e^{-\ln(y_1(x)^2) - A(x)} = \frac{e^{-A(x)}}{y_1(x)^2}.$$

Since this result shows $C(x)$ to be an antiderivative of $e^{-A(x)}/y_1(x)^2$, we may infer that

$$\boxed{y_2(x) = y_1(x) \int \frac{e^{-A(x)}}{y_1(x)^2}\, dx.}$$

To verify that y_1 and y_2 satisfy the condition $W_{y_1,y_2} \neq 0$, we observe that

$$\begin{aligned} W_{y_1,y_2}(x) &= y_1(x)y_2'(x) - y_1'(x)y_2(x) \\ &= y_1(x)\left(y_1'(x)\int \frac{e^{A(x)}}{y_1(x)^2}\,dx + y_1(x)\frac{e^{-A(x)}}{y_1(x)^2}\right) - y_1'(x)y_1(x)\int \frac{e^{-A(x)}}{y_1(x)^2}\,dx \\ &= e^{-A(x)} \neq 0 \quad (\text{because } e^y \neq 0 \text{ for all } y \in \mathbb{R}). \end{aligned} \tag{46.13}$$

The method of generating y_2 from y_1, as outlined above, is known as *reduction of the order*, because $C'(x)$ is found as a solution of a first-order linear differential equation, while the original differential equation (46.7) is of second order.

As an example, let us consider the initial value problem

$$\begin{aligned} &y''(x) + xy'(x) - y(x) = 0 \\ &y(1) = 1, \; y'(1) = -1. \end{aligned} \tag{46.14}$$

In this case we have $a(x) = x$ and by implication $A(x) = x^2/2$. Since $y_1(x) := x$ is easily seen to be a nonzero solution of the differential equation in (46.14), we may infer that

$$y_2(x) = x\int_1^x \frac{e^{-t^2/2}}{t^2}\,dt.$$

To choose a definite integral with lower boundary 1 for the antiderivative of $e^{-t^2/2}/t^2$ is convenient because the initial values are given at $x = 1$. According to Theorem 46.9b, the solution of the initial value problem above is

$$y(x) = \frac{y_2'(1) + y_2(1)}{W_{y_1,y_2}(1)}y_1(x) + \frac{-y_1(1) - y_1'(1)}{W_{y_1,y_2}(1)}y_2(x).$$

Using the result in (46.13), we find $W_{y_1,y_2}(1)$ to be equal to $e^{-A(1)} = e^{-1/2}$ and, given the defining equations for y_1 and y_2, it follows that $y_1(1) = y_1'(1) = 1$ and $y_2(1) = 0$ (see Theorem 17.10d). To determine $y_2'(1)$ we apply the product rule and the FTC1:

$$y_2'(1) = \frac{d}{dx}\left(x\int_1^x \frac{e^{-t^2/2}}{t^2}\,dt\right)\Bigg|_{x=1} = \left(\int_1^x \frac{e^{-t^2/2}}{t^2}\,dt + x\frac{e^{-x^2/2}}{x^2}\right)\Bigg|_{x=1} = e^{-1/2}.$$

Hence

$$y(x) = \frac{e^{-1/2}}{e^{-1/2}}y_1(x) - \frac{2}{e^{-1/2}}y_2(x) = x - 2\sqrt{e}x\int_1^x \frac{e^{-t^2/2}}{t^2}\,dt.$$

In absence of an explicit formula for the antiderivative of $e^{-t^2/2}/t^2$, we may regard this result to be our final answer. ★

Homogeneous Equations with Constant Coefficients

As mentioned above, there is no general method for solving second-order linear differential equations unless the coefficient functions a and b are constant. It is this special case that we turn to now: for $a, b \in \mathbb{R}$, we consider the *homogeneous second-order linear differential equation with constant coefficients*

$$y''(x) + ay'(x) + by(x) = 0. \tag{46.15}$$

The variable x is understood to range over the entire set of real numbers. An intitial value problem corresponding to (46.15) is

$$\begin{aligned} &y''(x) + ay'(x) + by(x) = 0 \\ &y(x_0) = z_1, \; y'(x_0) = z_2. \end{aligned} \tag{46.16}$$

Given the result in (44.9), we know that every solution of a homogeneous first-order linear differential equation with constant coefficients ($y'(x) + ay(x) = 0$) is of the form $y(x) = Ce^{-ax}$. Motivated by this observation, we will attempt to find (possibly complex) solutions of (46.15) of the form $y(x) = e^{\lambda x}$ for some $\lambda \in \mathbb{C}$ (see Chapter 45). Given the defining equation $y(x) = e^{\lambda x}$, we may apply Theorem 45.22 to infer that

$$y''(x) + ay'(x) + by(x) = \lambda^2 e^{\lambda x} + a\lambda e^{\lambda x} + be^{\lambda x} = (\lambda^2 + a\lambda + b)y(x).$$

This shows that

$$\boxed{y(x) = e^{\lambda x} \text{ is a solution of (46.15) if and only if } \lambda \text{ is a root of the polynomial } x^2 + ax + b.}$$

Note: the solution $e^{\lambda x}$ is real- or complex-valued depending on whether the root λ is real or complex.

46.10 Definition. The *characteristic polynomial* of equation (46.15) is defined to be

$$P(x) := x^2 + ax + b.$$

Since finding solutions of the form $y(x) = e^{\lambda x}$ requires that we determine the roots of the characteristic polynomial, it is appropriate to consider three cases depending on the number of real roots of $P(x)$.

Case 1: $P(x)$ has two distinct real roots λ_1 and λ_2, i.e., $P(x) = (x - \lambda_1)(x - \lambda_2)$ with $\lambda_1 \neq \lambda_2$.

In this case $y_1(x) := e^{\lambda_1 x}$ and $y_2(x) := e^{\lambda_2 x}$ are two distinct real-valued solutions of equation (46.15). Using Definition 46.8, it is easy to see that

$$W_{y_1,y_2}(x) = (\lambda_2 - \lambda_1)e^{(\lambda_1+\lambda_2)x} \tag{46.17}$$

and therefore,

$$W_{y_1,y_2}(0) = \lambda_2 - \lambda_1 \neq 0.$$

(Remember that, according to Theorem 46.9a, it is sufficient to check whether the Wronski determinant of y_1 and y_2 is different from zero at one particular point—in this case at zero.) Given the statement in Theorem 46.9b, we may thus infer that the general solution of equation (46.15) is

$$\boxed{y(x) = C_1 e^{\lambda_1 x} + C_2 e^{\lambda_2 x}.} \tag{46.18}$$

46.11 Example. We wish to solve the initial value problem

$$\begin{aligned} &y''(x) + 2y'(x) - 8y(x) = 0 \\ &y(0) = 1, \ y'(0) = 3. \end{aligned} \tag{46.19}$$

The roots of the characteristic polynomial $P(x) = x^2 + 2x - 8$ are $\lambda_1 = 2$ and $\lambda_2 = -4$. Thus, according to (46.18), the general solution of the differential equation in (46.19) is

$$y(x) = C_1 e^{2x} + C_2 e^{-4x}.$$

Using the given initial values to determine C_1 and C_2, we obtain

$$\begin{aligned} C_1 + C_2 &= y(0) = 1, \\ 2C_1 - 4C_2 &= y'(0) = 3. \end{aligned}$$

Solving for C_1 and C_2 yields $C_1 = 7/6$ and $C_2 = -1/6$. Consequently, the solution of the initial value problem (46.19) is

$$y(x) = \frac{7e^{2x}}{6} - \frac{e^{-4x}}{6}.$$

Note: alternatively, we could have determined the coefficients C_1 and C_2 also from the formulae (46.10) and (46.11) or from Theorem 46.9b.

Case 2: $P(x)$ has only one real root λ, i.e., $P(x) = (x - \lambda)^2$.

In this case one real-valued solution of (46.15) is $y_1(x) := e^{\lambda x}$ and, as the following calculation will show, a second solution is $y_2(x) := xy_1(x) = xe^{\lambda x}$:

$$\begin{aligned} y_2''(x) + ay_2'(x) + by_2(x) &= (\lambda^2 x + 2\lambda)e^{\lambda x} + a(\lambda x + 1)e^{\lambda x} + bxe^{\lambda x} \\ &= (P(\lambda)x + a + 2\lambda)e^{\lambda x} = (a + 2\lambda)e^{\lambda x}. \end{aligned}$$

Since $x^2 + ax + b = P(x) = (x - \lambda)^2 = x^2 - 2\lambda x + \lambda^2$, we may infer that $\lambda = -a/2$ and therefore,

$$y_2''(x) + ay_2'(x) + by_2(x) = (a - a)e^{\lambda x} = 0$$

as desired. Moreover, the Wronski determinant is easily seen to be

$$W_{y_1,y_2}(x) = e^{2\lambda x}. \tag{46.20}$$

Hence

$$W_{y_1,y_2}(0) = 1 \neq 0.$$

Referring again to Theorem 46.9b, the general solution of equation (46.15) is

$$\boxed{y(x) = C_1 e^{\lambda x} + C_2 x e^{\lambda x}.} \tag{46.21}$$

46.12 Example. Let us consider the initial value problem

$$\begin{aligned} y''(x) - 4y'(x) + 4y(x) &= 0 \\ y(2) = 1,\ y'(2) &= -1. \end{aligned} \tag{46.22}$$

The only root of the characteristic polynomial $P(x) = x^2 - 4x + 4$ is $\lambda = 2$. Thus, according to (46.21), the general solution of the differential equation in (46.22) is

$$y(x) = C_1 e^{2x} + C_2 x e^{2x}.$$

Using the initial conditions in (46.22), we obtain

$$\begin{aligned} C_1 e^4 + 2C_2 e^4 &= y(2) = 1, \\ 2C_1 e^4 + C_2(e^4 + 4e^4) &= y'(2) = -1. \end{aligned}$$

Solving for C_1 and C_2 yields $C_1 = 7e^{-4}$ and $C_2 = -3e^{-4}$. Consequently, the solution of the initial value problem (46.22) is

$$y(x) = 7e^{2(x-2)} - 3xe^{2(x-2)}.$$

Case 3: $P(x)$ has two complex roots $\lambda_1 + i\lambda_2$ and $\lambda_1 - i\lambda_2$ with $\lambda_2 \neq 0$, i.e.,

$$P(x) = (x - (\lambda_1 + i\lambda_2))(x - (\lambda_1 - i\lambda_2)) = x^2 - 2\lambda_1 x + \lambda_1^2 + \lambda_2^2.$$

(Remember that complex roots of polynomials with real coefficients always appear in pairs z_0, $\bar{z}_0$. This was explained in Chapter 45, p.350.)

In this case $z(x) := e^{(\lambda_1 + i\lambda_2)x}$ is a complex-valued solution of (46.15) which, according to Euler's formula in (45.9), can also be written in the form $z(x) = e^{\lambda_1 x}(\cos(\lambda_2 x) + i\sin(\lambda_2 x))$. In respectively setting $y_1(x)$ and $y_2(x)$ equal to the real and imaginary parts of $z(x)$, we obtain $y_1(x) := e^{\lambda_1 x}\cos(\lambda_2 x)$ and $y_2(x) := e^{\lambda_1 x}\sin(\lambda_2 x))$. Since z is a solution of (46.15), it follows that

$$0 = z''(x) + az'(x) + bz(x) = (y_1''(x) + ay_1'(x) + by_1(x)) + i(y_2''(x) + ay_2'(x) + by_2(x))$$

and therefore,

$$y_1''(x) + ay_1'(x) + by_1(x) = 0,$$
$$y_2''(x) + ay_2'(x) + by_2(x) = 0.$$

This shows that y_1 and y_2 are *real-valued* solutions of (46.15). Furthermore, the corresponding Wronski determinant is easily seen to be

$$W_{y_1,y_2}(x) = \lambda_2 e^{2\lambda_1 x} \tag{46.23}$$

with

$$W_{y_1,y_2}(0) = \lambda_2 \neq 0.$$

Therefore, Theorem 46.9b implies that the general solution of (46.15) is

$$\boxed{y(x) = C_1 e^{\lambda_1 x}\cos(\lambda_2 x) + C_2 e^{\lambda_1 x}\sin(\lambda_2 x).} \tag{46.24}$$

46.13 Example. We wish to solve the initial value problem

$$\begin{aligned} &y''(x) - 4y'(x) + 13y(x) = 0 \\ &y(0) = 1,\ y'(0) = 4. \end{aligned} \tag{46.25}$$

The roots of the characteristic polynomial $P(x) = x^2 - 4x + 13$ are $2 \pm 3i$, so that $\lambda_1 = 2$ and $\lambda_2 = 3$. Thus, according to (46.24), the general solution of the differential equation in (46.25) is

$$y(x) = C_1 e^{2x}\cos(3x) + C_2 e^{2x}\sin(3x).$$

Using the initial conditions in (46.25), we obtain

$$C_1 = y(0) = 1,$$
$$2C_1 + 3C_2 = y'(0) = 4.$$

Solving for C_1 and C_2 yields $C_1 = 1$, $C_2 = 2/3$. Therefore, the solution of the initial value problem (46.25) is

$$y(x) = e^{2x}\cos(3x) + \frac{2}{3}e^{2x}\sin(3x).$$

46.14 Exercise. Solve the following initial value problems:

a) $y''(x) - 2y'(x) + y(x) = 0,\ y(1) = 2,\ y'(1) = -1$

b) $y''(x) + 4y'(x) + 5y(x) = 0,\ y(0) = 1,\ y'(0) = 0$

c) $y''(x) - 2y'(x) - 3y(x) = 0,\ y(2) = 1,\ y'(2) = 3$

Nonhomogeneous Equations

Let us assume that y_1 and y_2 are solutions of the homogeneous equation (46.7) such that $W_{y_1,y_2} \neq 0$. How can we use y_1 and y_2 to find solutions of the nonhomogeneous equation (46.2)? The general solution of equation (46.7) is of the form

$$C_1 y_1(x) + C_2 y_2(x),$$

and its derivative is

$$C_1 y_1'(x) + C_2 y_2'(x).$$

Motivated by these observations, we will now employ the method of the variation of the constants (which we already used in Chapter 44 and in the ⋆remark⋆ on p.363) to find a solution of (46.2).

More precisely, we will try to find functions $C_1(x)$ and $C_2(x)$ such that for a specific solution y_0 of the nonhomogeneous equation (46.2) the following equations are satisfied:

$$y_0(x) = C_1(x)y_1(x) + C_2(x)y_2(x), \tag{46.26}$$
$$y_0'(x) = C_1(x)y_1'(x) + C_2(x)y_2'(x). \tag{46.27}$$

In analogy to our discussion in Chapter 44, the logic of our argument will be to derive defining equations for $C_1(x)$ and $C_2(x)$ from the assumptions that (46.26) and (46.27) are satisfied and that y_0 is a solution of (46.2). In applying the product rule to (46.26) and then substituting y_0' for $C_1y_1'+C_2y_2'$ (see (46.27)), we obtain

$$y_0'(x) = C_1(x)y_1'(x) + C_2(x)y_2'(x) + C_1'(x)y_1(x) + C_2'(x)y_2(x) = y_0'(x) + C_1'(x)y_1(x) + C_2'(x)y_2(x).$$

Hence

$$0 = C_1'(x)y_1(x) + C_2'(x)y_2(x). \tag{46.28}$$

Furthermore, (46.27) also implies that

$$y_0''(x) = C_1(x)y_1''(x) + C_2(x)y_2''(x) + C_1'(x)y_1'(x) + C_2'(x)y_2'(x). \tag{46.29}$$

Using (46.26), (46.27), and (46.29) in conjunction with the assumption that y_0 is a solution of (46.2), we may infer that

$$\begin{aligned} f(x) &= y_0''(x) + a(x)y_0'(x) + b(x)y_0(x) \\ &= C_1(x)y_1''(x) + C_2(x)y_2''(x) + C_1'(x)y_1'(x) + C_2'(x)y_2'(x) \\ &\quad + a(x)(C_1(x)y_1'(x) + C_2(x)y_2'(x)) + b(x)(C_1(x)y_1(x) + C_2(x)y_2(x)) \\ &= C_1(x)(y_1''(x) + a(x)y_1'(x) + b(x)y_1(x)) + C_2(x)(y_2''(x) + a(x)y_2'(x) + b(x)y_2(x)) \\ &\quad + C_1'(x)y_1'(x) + C_2'(x)y_2'(x) \\ &= C_1'(x)y_1'(x) + C_2'(x)y_2'(x) \quad \text{(because } y_1 \text{ and } y_2 \text{ are solutions of (46.7))}. \end{aligned} \tag{46.30}$$

Summarizing the results in (46.28) and (46.30), we record that

$$C_1'(x)y_1(x) + C_2'(x)y_2(x) = 0,$$
$$C_1'(x)y_1'(x) + C_2'(x)y_2'(x) = f(x).$$

Thus, according to (46.12), we have

$$C_1'(x) = \frac{-f(x)y_2(x)}{W_{y_1,y_2}(x)},$$
$$C_2'(x) = \frac{f(x)y_1(x)}{W_{y_1,y_2}(x)}.$$

These equations determine the antiderivatives $C_1(x)$ and $C_2(x)$ of $C_1'(x)$ and $C_2'(x)$ only up to additive constants, but in light of the fact that the initial values in (46.3) are given at x_0, the most appropriate choices for $C_1(x)$ and $C_2(x)$ are

$$C_1(x) := \int_{x_0}^{x} \frac{-f(t)y_2(t)}{W_{y_1,y_2}(t)}\, dt,$$
$$C_2(x) := \int_{x_0}^{x} \frac{f(t)y_1(t)}{W_{y_1,y_2}(t)}\, dt.$$

Hence

$$\boxed{y_0(x) = C_1(x)y_1(x) + C_2(x)y_2(x) = \left(\int_{x_0}^{x} \frac{-f(t)y_2(t)}{W_{y_1,y_2}(t)}\, dt\right) y_1(x) + \left(\int_{x_0}^{x} \frac{f(t)y_1(t)}{W_{y_1,y_2}(t)}\, dt\right) y_2(x).} \tag{46.31}$$

The function y_0, defined by this equation, is a specific solution of (46.2) that satisfies the initial conditions

$$\boxed{y_0(x_0) = y_0'(x_0) = 0.} \tag{46.32}$$

46.15 Exercise. Verify the equations in (46.32). *Hint.* Use the FTC1 for the differentiation of the integrals in (46.31).

In order to find not just one but all solutions of (46.2), we observe that if y is an arbitrary solution of (46.2), then $y - y_0$ is a solution of (46.7) (see Exercise 46.16). Thus, according to Theorem 46.9b, there are constants $D_1, D_2 \in \mathbb{R}$ such that $y - y_0 = D_1y_1 + D_2y_2$. Since $y = (y - y_0) + y_0$, we may conclude that *the general solution of equation* (46.2) *is*

$$\boxed{y(x) = D_1y_1(x) + D_2y_2(x) + \left(\int_{x_0}^{x} \frac{-f(t)y_2(t)}{W_{y_1,y_2}(t)}\,dt\right) y_1(x) + \left(\int_{x_0}^{x} \frac{f(t)y_1(t)}{W_{y_1,y_2}(t)}\,dt\right) y_2(x).} \tag{46.33}$$

46.16 Exercise. Show that if y is an arbitrary solution of (46.2), then $y - y_0$ is a solution of (46.7).

To determine the solution of the initial value problem (46.3), we only need to find values for D_1 and D_2 such that the initial conditions in (46.3) are satisfied. In this context it is very convenient that the solution y_0 satisfies the trivial initial conditions in (46.32), because this implies that the formulae for the constants D_1 and D_2 are the same as in (46.10) and (46.11) or Theorem 46.9b. (Think about it!). The following example provides an illustration:

46.17 Example. Let us consider the initial value problem

$$\begin{aligned} &y''(x) + y'(x) - 2y(x) = x \\ &y(1) = 0,\ y'(1) = 1. \end{aligned}$$

The characteristic polynomial $P(x) = x^2 + x - 2$ has two distinct roots $\lambda_1 = 1$ and $\lambda_2 = -2$. Thus, the solutions y_1 and y_2 of the corresponding homogeneous equation $(y''(x) + y'(x) - 2y(x) = 0)$ are

$$y_1(x) := e^x \quad \text{and} \quad y_2(x) := e^{-2x} \quad \text{(see Case 1, p.365).}$$

Since $W_{y_1,y_2}(x) = -3e^{-x}$ (see (46.17)), we may apply (46.33) to determine the general solution as follows:

$$\begin{aligned} y(x) &= D_1e^x + D_2e^{-2x} + \left(\int_1^x \frac{1}{3}te^{-t}\,dt\right) e^x + \left(\int_1^x -\frac{1}{3}te^{2t}\,dt\right) e^{-2x} \\ &= D_1e^x + D_2e^{-2x} - \frac{1}{3}(t+1)e^{-t}\Big|_1^x e^x - \frac{1}{6}\left(t - \frac{1}{2}\right)e^{2t}\Big|_1^x e^{-2x} \\ &= D_1e^x + D_2e^{-2x} - \frac{1}{3}(x+1) + \frac{2e^{x-1}}{3} - \frac{1}{6}\left(x - \frac{1}{2}\right) + \frac{e^{-2(x-1)}}{12} \\ &= D_1e^x + D_2e^{-2x} - \frac{x}{2} - \frac{1}{4} + \frac{2e^{x-1}}{3} + \frac{e^{-2(x-1)}}{12}. \end{aligned}$$

This calculation shows in particular that

$$y_0(x) = -\frac{x}{2} - \frac{1}{4} + \frac{2e^{x-1}}{3} + \frac{e^{-2(x-1)}}{12}.$$

To determine the coefficients D_1 and D_2 from the given initial conditions, we use (46.32) to infer that

$$\begin{aligned} 0 &= y(1) = D_1e + D_2e^{-2}, \\ 1 &= y'(1) = D_1e - 2D_2e^{-2}. \end{aligned}$$

Solving these equations for D_1 and D_2 yields $D_1 = e^{-1}/3$ and $D_2 = -e^2/3$. As we already mentioned, the coefficients D_1 and D_2 could also have been determined using (46.10) and (46.11) or Theorem 46.9b, and the result would have been the same (the reader may want to verify this). Thus, the solution of our initial value problem is

$$y(x) = \frac{e^{x-1}}{3} - \frac{e^{-2(x-1)}}{3} - \frac{x}{2} - \frac{1}{4} + \frac{2e^{x-1}}{3} + \frac{e^{-2(x-1)}}{12}.$$

46.18 Exercise. Find the solution of the initial value problem

$$y''(x) + 2y'(x) + 2y(x) = \sin(2x)$$
$$y(0) = -1, \; y'(0) = 1.$$

The Solution of an R-C-L Circuit

Having discussed second-order linear differential equations in general, we are now ready to solve equation (46.1). As an example, we choose the initial values

$$I(0) = 0, \; I'(0) = J_0. \tag{46.34}$$

The characteristic polynomial of (46.1) is

$$P(x) = x^2 + \frac{R}{L}x + \frac{1}{LC},$$

and the roots of this polynomial are

$$\frac{-RC \pm \sqrt{R^2C^2 - 4LC}}{2LC}.$$

To proceed, we need to distinguish three cases depending on the number of real roots of $P(x)$:

Case 1: $R^2C^2 - 4LC > 0$ (overdamped case).

Setting

$$\alpha := \frac{R}{2L} \quad \text{and} \quad \beta := \frac{\sqrt{R^2C^2 - 4LC}}{2LC}$$

we have two distinct real roots $\lambda_1 = -\alpha + \beta$ and $\lambda_2 = -\alpha - \beta$. Thus, according to (46.18), the general solution is $C_1 e^{(-\alpha-\beta)t} + C_2 e^{(-\alpha+\beta)t}$. Using the initial conditions in (46.34), it is not difficult to show that

$$\boxed{I(t) = \frac{J_0}{2\beta} e^{-\alpha t}(e^{\beta t} - e^{-\beta t}) = \frac{J_0}{\beta} e^{-\alpha t} \sinh(\beta t).}$$

A typical graph of $I(t)$ is shown in Figure 46.3.

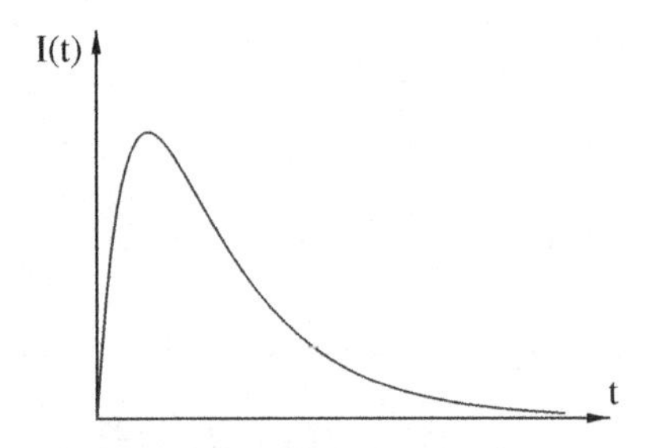

Figure 46.3: graph of $I(t)$ in the overdamped case.

Case 2: $R^2C^2 - 4LC = 0$ *(critically damped case).*

In this case the only root is $\lambda = -R/2L = -\alpha$ so that, according to (46.21), the general solution is $C_1e^{-\alpha t} + C_2te^{-\alpha t}$. Using (46.34), it follows that

$$\boxed{I(t) = J_0te^{-\alpha t}.}$$

The graph of $I(t)$ in the critically damped case is similar to the graph in the overdamped case (see Figure 46.4).

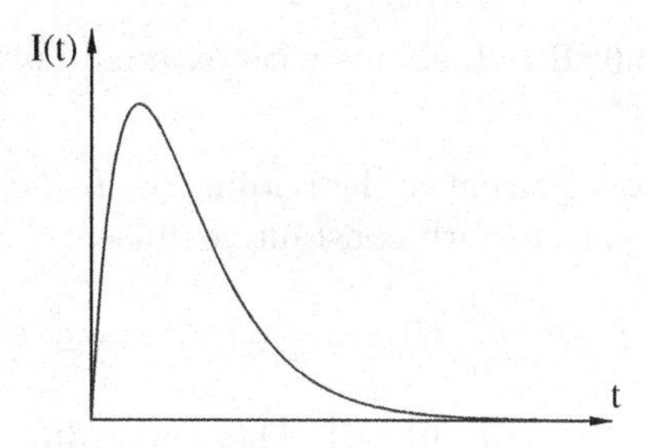

Figure 46.4: graph of $I(t)$ in the critically damped case.

Case 3: $R^2C^2 - 4LC < 0$ *(underdamped or oscillatory case).*

Setting

$$\omega_0 := \frac{\sqrt{4LC - R^2C^2}}{2LC}$$

we find in this case the two complex roots $-\alpha \pm i\omega_0$. The real and imaginary parts are $\lambda_1 = -\alpha$ and $\lambda_2 = \omega_0$. Thus, according to (46.24), the general solution is $C_1e^{-\alpha t}\cos(\omega_0 t) + C_2e^{-\alpha t}\sin(\omega_0 t)$, and using (46.34), we obtain

$$\boxed{I(t) = \frac{J_0}{\omega_0}e^{-\alpha t}\sin(\omega_0 t).}$$

Figure 46.5 shows that the graph of $I(t)$ is oscillating between positive and negative values with a constant period and an exponentially decreasing amplitude.

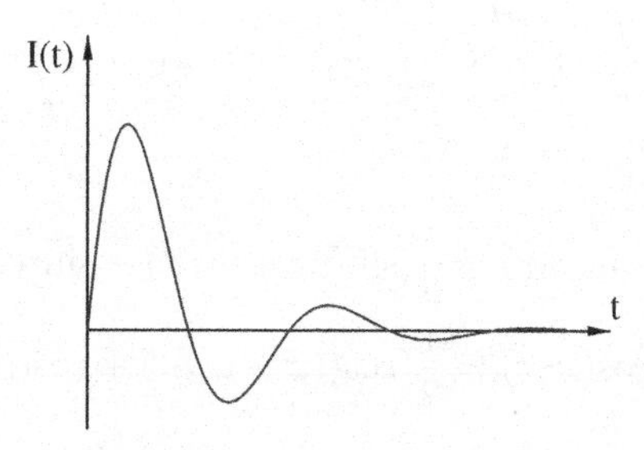

Figure 46.5: graph of $I(t)$ in the underdamped case.

46.19 Example. Let us discuss an R-C-L circuit with an external voltage source generating a difference in electric potential $E(t)$ as shown in Figure 46.6. In this case, the differences in electric potential at the resistor, the capacitor, and the inductor add up to $E(t)$, and therefore

$$RI(t) + \frac{Q(t)}{C} + LI'(t) = E(t).$$

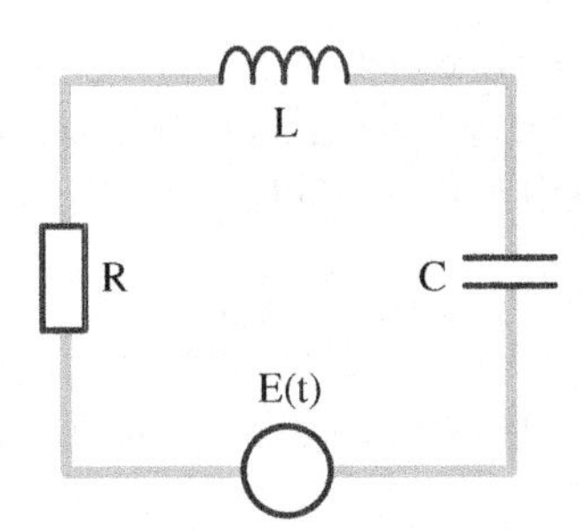

Figure 46.6: R-C-L circuit with external voltage source.

Differentiating both sides of this equation and dividing by L yields the following nonhomogeneous second-order linear differential equation with constant coefficients:

$$I''(t) + \frac{R}{L}I'(t) + \frac{1}{LC}I(t) = \frac{1}{L}E'(t). \tag{46.35}$$

For simplification we assume that $I(0) = I'(0) = 0$. Then, according to (46.31) and (46.32), the solution is

$$I(t) = \left(\int_0^t \frac{-E'(\tau)y_2(\tau)}{LW_{y_1,y_2}(\tau)}\,d\tau\right) y_1(t) + \left(\int_0^t \frac{E'(\tau)y_1(\tau)}{LW_{y_1,y_2}(\tau)}\,d\tau\right) y_2(t), \tag{46.36}$$

where y_1 and y_2 are understood to be solutions of (46.1) that satisfy the condition $W_{y_1,y_2} \neq 0$. To continue, we assume that the circuit is underdamped with $R^2C^2 - 4LC < 0$, because this is the most interesting case. Then, with the same definitions for α and ω_0 as in Case 3, p.371, the solutions y_1 and y_2 of the homogeneous equation (46.1) may be chosen as

$$\begin{aligned} y_1(t) &:= e^{-\alpha t}\cos(\omega_0 t), \\ y_2(t) &:= e^{-\alpha t}\sin(\omega_0 t). \end{aligned}$$

To give an example, we consider an external voltage source generating a sinusoidal difference in electric potential $E(t) = E_0 \sin(\omega t)$ so that $E'(t) = \omega E_0 \cos(\omega t)$. According to (46.23), we have $W_{y_1,y_2}(t) = \omega_0 e^{-2\alpha t}$, and therefore (46.36) implies that

$$\begin{aligned} I(t) = & -\frac{\omega E_0}{\omega_0 L}\left(\int_0^t e^{\alpha\tau}\cos(\omega\tau)\sin(\omega_0\tau)\,d\tau\right) e^{-\alpha t}\cos(\omega_0 t) \\ & + \frac{\omega E_0}{\omega_0 L}\left(\int_0^t e^{\alpha\tau}\cos(\omega\tau)\cos(\omega_0\tau)\,d\tau\right) e^{-\alpha t}\sin(\omega_0 t). \end{aligned}$$

Given the trigonometric identities

$$\begin{aligned} \cos(\omega\tau)\sin(\omega_0\tau) &= \frac{1}{2}(\sin((\omega+\omega_0)t) - \sin((\omega-\omega_0)t)), \\ \cos(\omega\tau)\cos(\omega_0\tau) &= \frac{1}{2}(\cos((\omega-\omega_0)t) + \cos((\omega+\omega_0)t)), \end{aligned} \tag{46.37}$$

we may use integration by parts to conclude (after some simple but laborious calculations) that

$$I(t) = \frac{\omega E_0}{2\omega_0 L}\left((C-D)\cos(\omega t) - (A-B)\sin(\omega t) - ((C-D)\cos(\omega_0 t) + (A+B)\sin(\omega_0 t))e^{-\alpha t}\right),$$

where

$$\begin{aligned} A &:= \frac{\alpha}{\alpha^2 + (\omega+\omega_0)^2}, & B &:= \frac{\alpha}{\alpha^2 + (\omega-\omega_0)^2}, \\ C &:= \frac{\omega+\omega_0}{\alpha^2 + (\omega+\omega_0)^2}, & D &:= \frac{\omega-\omega_0}{\alpha^2 + (\omega-\omega_0)^2}. \end{aligned}$$

Since $\lim_{t\to\infty} e^{-\alpha t} = 0$, the steady state solution $I_s(t)$ (see also Chapter 44, p.342) is

$$I_s(t) = \frac{\omega E_0}{2\omega_0 L}\left((C-D)\cos(\omega t) - (A-B)\sin(\omega t)\right).$$

We notice that $I_s(t)$ is a trigonometric function of period $2\pi/\omega$. In particular, $I_s(t)$ has the same period as $E(t)$. Furthermore, the transient part of the solution is

$$-\frac{\omega E_0}{2\omega_0 L}((C-D)\cos(\omega_0 t) + (A+B)\sin(\omega t))e^{-\alpha t}.$$

46.20 Example. As a final example, let us consider the C-L circuit shown in Figure 46.7 that is composed of a capacitor, an inductor, and an external voltage source but does not involve a resistor.

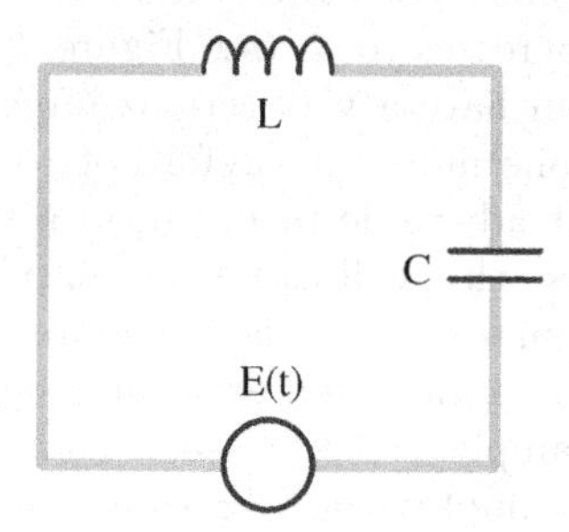

Figure 46.7: C-L circuit with external voltage source.

Then $R = 0$, and the circuit is thus described by the equation

$$I''(t) + \frac{1}{LC}I(t) = \frac{1}{L}E'(t) \quad \text{(see (46.35))}.$$

As above, we assume that $E(t) = E_0\sin(\omega t)$ and that I satisfies the initial conditions $I(0) = I'(0) = 0$. Then the solution is again given by equation (46.36), but this time y_1 and y_2 are solutions of the homogeneous equation

$$I''(t) + \frac{1}{LC}I(t) = 0$$

satisfying the condition $W_{y_1,y_2} \neq 0$. Setting $\omega_0 := 1/\sqrt{LC}$, the characteristic polynomial of this equation is $P(x) = x^2 + \omega_0^2$. The roots of P are $\pm i\omega_0$, and therefore,

$$y_1(t) := \cos(\omega_0 t),$$
$$y_2(t) := \sin(\omega_0 t).$$

Using (46.36) and the fact that $W_{y_1,y_2}(t) = \omega_0$ (see (46.23)), we obtain

$$I(t) = \frac{\omega E_0}{\omega_0 L}\left(-\left(\int_0^t \cos(\omega\tau)\sin(\omega_0\tau)\,d\tau\right)\cos(\omega_0 t) + \left(\int_0^t \cos(\omega\tau)\cos(\omega_0\tau)\,d\tau\right)\sin(\omega_0 t)\right).$$

If $\omega \neq \pm\omega_0$, then we may apply integration by parts or the trigonometric formulae in (46.37) to infer that

$$I(t) = \frac{\omega E_0(\cos(\omega_0 t) - \cos(\omega t))}{L(\omega^2 - \omega_0^2)}.$$

By contrast, if $\omega = \pm\omega_0$, then

$$I(t) = \frac{\omega E_0}{\omega_0 L}\left(-\left(\int_0^t \cos(\omega_0\tau)\sin(\omega_0\tau)\,d\tau\right)\cos(\omega_0 t) + \left(\int_0^t \cos^2(\omega_0\tau)\,d\tau\right)\sin(\omega_0 t)\right) = \frac{E_0 t\sin(\omega_0 t)}{2L}.$$

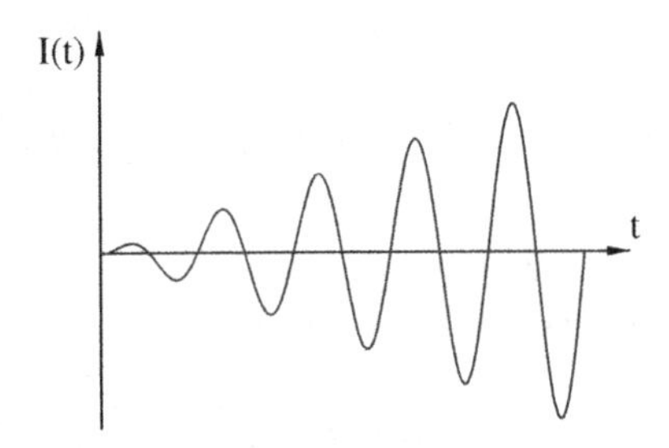

Figure 46.8: graph of $I(t)$ in the resonance case.

The latter case $\omega = \pm\omega_0$ is the so-called *resonance case.* Here the amplitude of the solution $I(t) = E_0 t \sin(\omega_0)/2L$ increases to ∞ as t increases to ∞ (see Figure 46.8).

The phenomenon of resonance can have a very serious impact on oscillatory systems encountered in engineering and technology, and one must be careful not to excite any such system at its natural frequency ω_0. For instance, it is not advisable to install in a C-L circuit an external voltage source with frequency $\omega_0 = 1/\sqrt{LC}$, because this will cause the amplitude of $I(t)$ to increase without limit and eventually burn the circuit. Resonance phenomena can also be observed in oscillatory mechanical systems. In fact, it has happened that bridges and even huge suspension bridges have collapsed due to resonance. An army regiment, for example, is ill advised to cross a bridge in a steady marching rhythm, for the bridge structure may tune in, amplify the vibrations, and eventually collapse.

Additional Exercises

46.21. Find the general solution of each of the following differential equations:

a) $y'' + 10y' + 25y = 0$

b) $y'' - 6y' + 20y = 0$

c) $y'' + 8y' - 2y = 0$

46.22. Find the solutions of the following initial value problems:

a) $y'' - 5y' + 6y = \cosh(x)$, $y(0) = 0$, $y'(0) = 1$

b) $y'' - 2y' + y = xe^x - e^x$, $y(1) = 1$, $y'(1) = -1$

c) $y'' + 2y' + y = \cos^2(x)$, $y(-1) = 1$, $y'(-1) = 2$

d) $y'' - 2y' + 5y = e^x \tan(2x)$, $y(0) = 1$, $y'(0) = 1$

Hint. For the Wronski determinants that appear in (46.33) you may want to use the formulae (46.17), (46.20), and (46.23). Furthermore, the double angle formula for cosine will be helpful for solving the integrals in c).

The next two problems require familiarity with the ⋆remark⋆ on p.363.

46.23. ★ Using the one given solution, find the general solutions of the following differential equations:

a) $y'' - 4xy' + 2(2x^2 - 1)y = 0$, $y_1(x) = e^{(x^2)}$

b) $(2x + 1)y'' - (4x + 4)y' + 4y = 0$, $y_1(x) = e^{2x}$

46.24. ★ Use your result of Exercise 46.23a to solve the initial value problem $y'' - 4xy' + 2(2x^2 - 1)y = x$, $y(0) = 1$, $y'(0) = 2$.

46.25. Assume that $y_1, y_2 : I \to \mathbb{R}$ are solutions of equation (46.2) such that for some $x_0 \in I$ we have $y_1(x_0) = y_2(x_0)$ and $y_1'(x_0) = y_2'(x_0)$. Explain why these assumptions imply that $y_1(x) = y_2(x)$ for all $x \in I$.

46.26. Assume that $y_1, y_2 : I \to \mathbb{R}$ are solutions of equation (46.2) such that for some $x_0, x_1 \in I$ we have $y_1(x_0) = y_2(x_0)$ and $y_1'(x_1) = y_2'(x_1)$. Do these assumptions imply that $y_1(x) = y_2(x)$ for all $x \in I$? Explain your answer.

46.27. Assume that $y : \mathbb{R} \to \mathbb{R}$ is twice continuously differentiable such that $y(0) = y'(0) = 0$.

a) Show that if y is a solution of the homogeneous equation (46.7), then $y(1) = y'(2) = 0$.

b) Is the same conclusion as in a) possible under the assumption that y is a solution of the nonhomogeneous equation (46.2)?

46.28. Find the solution of the R-C-L circuit in Figure 46.6 under the assumption that $R = 400\,\Omega$, $L = 0.2\,H$, $C = 10^{-6}\,F$, $E(t) = 2\cos(1000t)$, $I(0) = 1$, and $I'(0) = 0$.

46.29. In a circuit consisting of a capacitor, an inductor, and a switch, a charge $Q_0 = 3\,C$ has been placed on the capacitor. The capacitance is $C = 0.5\,F$ and the inductance is $L = 0.2\,H$. Find the current $I(t)$ under the assumption that the switch is closed at time $t = 0$. *Hint.* Instead of working with a differential equation for $I(t)$, you should use a differential equation for $Q(t)$ because this will allow you to make use of the initial condition $Q_0 = 3\,C$.

46.30. Assume that $y_1, y_2 : I \to \mathbb{R}$ are two solutions of the homogeneous equation (46.7). Show that for all $x_0 \in I$ we have

$$\boxed{W_{y_1,y_2}(x) = W_{y_1,y_2}(x_0)e^{-\int_{x_0}^{x} a(t)\,dt}}$$

Hint. Differentiate $W_{y_1,y_2}(x)$ and use the fact that y_1 and y_2 are solutions of (46.7) to show that $W_{y_1,y_2}(x)$ is a solution of the first-order linear differential equation $y'(x) + a(x)y(x) = 0$. Solving this equation will yield the desired result.

46.31. Explain how the result of Exercise 46.30 can be used to prove Theorem 46.9a.

Chapter 47

★ Difference Equations and Fibonacci Numbers

Fibonacci Numbers in Nature

Whether we study the structure of a honeycomb, the colors of a rainbow, or the motion of a falling body, everywhere in the natural world we encounter mathematical order and astonishing numerical patterns. A beautiful and indeed amazing example of this sort is hidden in the following succession of numbers, which is known as the *Fibonacci sequence:*

$$0, 1, 1, 2, 3, 5, 8, 13, 21, 34, 55, 89, 144 \ldots \tag{47.1}$$

Taking a closer look, we quickly discover the rule by which this sequence is built: each number is the sum of its two immediate predecessors (i.e., $0+1 = 1$, $1+1 = 2$, $1+2 = 3$, etc.). It may appear that generating numbers in this way is just a meaningless mathematical game but, surprisingly, Fibonacci numbers are frequently encountered in patterns of organic growth. For instance, in examining a pineapple, we notice that the segments on its surface are arranged in three different spiral patterns as illustrated in Figure 47.1. The numbers of spirals of type A, B, and C invariably are the Fibonacci numbers 8, 13, and

Figure 47.1: spiral patterns on a pineapple.

21 respectively. Another example are seed arrangements in flower heads. In Figure 47.2 we are shown two different spiral patterns in the same flower. The numbers of spirals are the Fibonacci numbers 13 and 21. A similar structure can also be found in pine cones (see Figure 47.3) where we count 8 spirals

Figure 47.2: spiral patterns in a flower head.

in one direction and 13 spirals in the other. Furthermore, we will see later in this chapter that the limit of the ratios of successive Fibonacci numbers converges to the so-called *golden section* which played an important role in Greek and Roman architecture and also underlies the spiral growth of sea shells.

  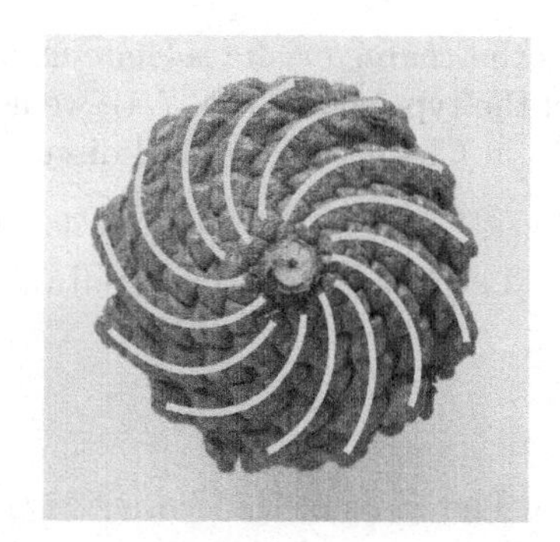

Figure 47.3: spiral patterns on a pinecone.

Our goal in this chapter is to find a formula for the Fibonacci numbers that allows us to calculate the value of the nth Fibonacci number without having to calculate first all of the preceding numbers in the sequence. For this purpose we denote by F_n the nth Fibonacci number in (47.1), i.e., $F_0 = 0$, $F_1 = 1$, $F_2 = 1$ etc. Given the rule for constructing Fibonacci numbers via summation of immediate predecessors, we may infer that $F_{n+2} = F_{n+1} + F_n$, or equivalently,

$$F_{n+2} - F_{n+1} - F_n = 0 \tag{47.2}$$

for all integers $n \geq 0$. This equation is a special case of what in general is referred to as a *difference equation*, and the numbers

$$F_0 = 0 \quad \text{and} \quad F_1 = 1 \tag{47.3}$$

are the corresponding *initial values*. In the next section we will learn that the obvious resemblance of (47.2) and (47.3) to initial value problems for second-order linear differential equations, as discussed in Chapter 46, is not merely formal but indicative of a striking similarity in the corresponding methods of solution as well.

Difference Equations

For constants $a, b \in \mathbb{R}$ we refer to the equation

$$y_{n+2} + ay_{n+1} + by_n = 0 \tag{47.4}$$

as a *homogeneous second-order linear difference equation with constant coefficients.* (Note: a more extensive discussion of the subject, which also covers nonhomogeneous difference equations, can be found in [WB] or [El].) The corresponding *initial value problem* is

$$\begin{aligned} y_{n+2} + ay_{n+1} + by_n &= 0 \\ y_0 = w_0, \ y_1 &= w_1. \end{aligned} \tag{47.5}$$

In analogy to our discussion in Chapter 46, we will try to find a solution of the form $y_n = e^{\lambda n}$ for some $\lambda \in \mathbb{C}$. However, in the present context it will be more convenient to write z instead of e^{λ} so that $y_n := z^n$ for some $z \in \mathbb{C}$. Given this definition for y_n, we may infer that

$$y_{n+2} + ay_{n+1} + by_n = z^{n+2} + az^{n+1} + bz^n = (z^2 + az + b)z^n.$$

Thus, $y_n = z^n$ is a possibly complex (nonzero) solution of (47.4) if and only if z is a solution of the quadratic equation

$$x^2 + ax + b = 0.$$

Motivated by this observation, we refer to

$$P(x) := x^2 + ax + b$$

as the *characteristic polynomial* of the difference equation (47.4). In order to solve difference equations of the type shown in (47.4), we need to consider three cases depending on the number of roots of $P(x)$. As in Chapter 46, we will discuss one example for each of the three cases:

Case 1: $P(x)$ has two distinct real roots.

Let us consider the initial value problem

$$\begin{aligned} y_{n+2} + 2y_{n+1} - 3y_n &= 0 \\ y_0 = 1, \ y_1 &= 2. \end{aligned} \tag{47.6}$$

The roots of the characteristic polynomial $P(x) = x^2 + 2x - 3$ are $z_1 = 1$ and $z_2 = -3$. Thus we have two solutions $y_{1,n} := 1^n = 1$ and $y_{2,n} := (-3)^n$. In order to determine the solution y_n of the initial value problem, we need to find coefficients $C_1, C_2 \in \mathbb{R}$ such that $y_n = C_1 y_{1,n} + C_2 y_{2,n}$ satisfies the initial conditions in (47.6) (here we appeal again to the principle of linearity, which is easily seen to be valid also for homogeneous linear difference equations). Using the initial values and the definitions of $y_{1,n}$ and $y_{2,n}$, we obtain

$$\begin{aligned} 1 = y_0 &= C_1 y_{1,0} + C_2 y_{2,0} = C_1 + C_2, \\ 2 = y_1 &= C_1 y_{1,1} + C_2 y_{2,1} = C_1 - 3C_2. \end{aligned}$$

Solving these equations for C_1 and C_2 yields $C_1 = 5/4$ and $C_2 = -1/4$. Consequently, the solution of the initial value problem is

$$y_n = \frac{5}{4} - \frac{1}{4}(-3)^n. \tag{47.7}$$

47.1 Exercise. Verify that the solution of the initial value problem (47.6) is given by (47.7).

47.2 Exercise. State and prove the principle of linearity for homogeneous second-order linear difference equations (with constant coefficients).

Remark. In analogy to the results established in Chapter 46 concerning the condition in (46.9), it can be shown that any initial value problem of the form given in (47.5) has a unique solution $y_n = C_1 y_{1,n} + C_2 y_{2,n}$ whenever $y_{1,n}$ and $y_{2,n}$ are solutions of (47.4) that satisfy the condition

$$y_{1,n} y_{2,n+1} - y_{1,n+1} y_{2,n} \neq 0.$$

Furthermore, just as in Theorem 46.9a, it can also be shown that this condition is satisfied for all n if only it is satisfied for one particular n. Proofs of these statements are given in [WB].

Case 2: $P(x)$ has one real root.

Let us consider the initial value problem

$$\begin{aligned} y_{n+2} - 4y_{n+1} + 4y_n &= 0 \\ y_0 = 0,\ y_1 &= 3. \end{aligned} \tag{47.8}$$

The only root of the characteristic polynomial $P(x) = x^2 - 4x + 4$ is $z = 2$. Therefore, one solution of the difference equation in (47.8) is $y_{1,n} := 2^n$. Another solution is $y_{2,n} := ny_{1,n} = n2^n$, because

$$\begin{aligned} y_{2,n+2} - 4y_{2,n+1} + 4y_{2,n} &= (n+2)2^{n+2} - 4(n+1)2^{n+1} + 4n2^n \\ &= 2^n(4(n+2) - 8(n+1) + 4n) = 0. \end{aligned}$$

Now we need to determine coefficients C_1 and C_2 such that $y_n = C_1 y_{1,n} + C_2 y_{2,n}$ is a solution of the initial value problem (47.8). Using the initial values, we obtain

$$\begin{aligned} 0 &= y_0 = C_1, \\ 3 &= y_1 = 2C_1 + 2C_2, \end{aligned}$$

and solving for C_1 and C_2 yields $C_1 = 0$ and $C_2 = 3/2$. Consequently, the solution of the initial value problem (47.8) is

$$y_n = \frac{3}{2}n2^n. \tag{47.9}$$

Case 3: $P(x)$ has no real roots.

Let us consider the initial value problem

$$\begin{aligned} y_{n+2} - 2y_{n+1} + 2y_n &= 0 \\ y_0 = 1,\ y_1 &= -1. \end{aligned} \tag{47.10}$$

The characteristic polynomial $P(x) = x^2 - 2x + 2$ has no real roots, but two complex roots at $1 \pm i$. The polar representation (see Chapter 45) of $z := 1 + i$ is easily seen to be $\sqrt{2}e^{i\pi/4}$. Hence

$$z^n = \sqrt{2}^n e^{in\pi/4} = \sqrt{2}^n \left(\cos\left(\frac{\pi}{4}n\right) + i\sin\left(\frac{\pi}{4}n\right)\right). \tag{47.11}$$

Motivated by this observation, we define $y_{1,n}$ and $y_{2,n}$ to be the real and imaginary parts of z^n respectively, i.e.,

$$y_{1,n} := \sqrt{2}^n \cos\left(\frac{\pi}{4}n\right) \text{ and } y_{2,n} := \sqrt{2}^n \sin\left(\frac{\pi}{4}n\right).$$

We wish to verify that $y_{1,n}$ and $y_{2,n}$ are indeed solutions of the difference equation in (47.10): since z^n is a complex solution of this difference equation, it follows that

$$\begin{aligned} 0 &= z^{n+2} - 2z^{n+1} + 2z^n \\ &= \sqrt{2}^{n+2}\left(\cos\left(\frac{\pi}{4}(n+2)\right) + i\sin\left(\frac{\pi}{4}(n+2)\right)\right) - 2\sqrt{2}^{n+1}\left(\cos\left(\frac{\pi}{4}(n+1)\right) + i\sin\left(\frac{\pi}{4}(n+1)\right)\right) \\ &\quad + 2\sqrt{2}^n\left(\cos\left(\frac{\pi}{4}n\right) + i\sin\left(\frac{\pi}{4}n\right)\right) \quad \text{(by (47.11))} \\ &= y_{1,n+2} - 2y_{1,n+1} + 2y_{1,n} + i(y_{2,n+2} - 2y_{2,n+1} + 2y_{2,n}), \end{aligned}$$

and therefore,

$$\begin{aligned} y_{1,n+2} - 2y_{1,n+1} + 2y_{1,n} &= 0, \\ y_{2,n+2} - 2y_{2,n+1} + 2y_{2,n} &= 0 \end{aligned}$$

as desired. To determine coefficients C_1 and C_2 such that $y_n = C_1 y_{1,n} + C_2 y_{2,n}$ is a solution of (47.10), we fill in the initial values to conclude that

$$1 = y_0 = C_1,$$
$$-1 = y_1 = C_1 + C_2.$$

Solving for C_1 and C_2 yields $C_1 = 1$ and $C_2 = -2$. Thus, the solution of the initial value problem (47.10) is

$$y_n = \sqrt{2}^n \cos\left(\frac{\pi}{4}n\right) - 2\sqrt{2}^n \sin\left(\frac{\pi}{4}n\right). \tag{47.12}$$

47.3 Exercise. Solve the following initial value problems:

a) $y_{n+2} - 4y_{n+1} - 5y_n = 0,\ y_0 = 2,\ y_1 = -1.$

b) $y_{n+2} - y_{n+1} + y_n/4 = 0,\ y_0 = -2,\ y_1 = 0.$

c) $y_{n+2} + 4y_{n+1} + 5y_n = 0,\ y_0 = 1,\ y_1 = 1.$

A Formula for Fibonacci Numbers

In this section we will derive a formula for Fibonacci numbers that was first discovered by Leonhard Euler. To this end we need to solve the initial value problem given by (47.2) and (47.3). The characteristic polynomial of equation (47.2) is $P(x) = x^2 - x - 1$, and the corresponding roots are $z_1 = (1+\sqrt{5})/2$ and $z_2 = (1-\sqrt{5})/2$. Thus, we define

$$y_{1,n} := \left(\frac{1+\sqrt{5}}{2}\right)^n \quad \text{and} \quad y_{2,n} := \left(\frac{1-\sqrt{5}}{2}\right)^n.$$

In order to find constants $C_1, C_2 \in \mathbb{R}$ such that $F_n = C_1 y_{1,n} + C_2 y_{2,n}$ (remember that F_n is the nth Fibonacci number), we use the initial values in (47.3) to infer that

$$0 = F_0 = C_1 + C_2,$$
$$1 = F_1 = \left(\frac{1+\sqrt{5}}{2}\right) C_1 + \left(\frac{1-\sqrt{5}}{2}\right) C_2.$$

Solving for C_1 and C_2 yields $C_1 = 1/\sqrt{5}$ and $C_2 = -1/\sqrt{5}$. Thus,

$$\boxed{F_n = \frac{1}{\sqrt{5}}\left(\left(\frac{1+\sqrt{5}}{2}\right)^n - \left(\frac{1-\sqrt{5}}{2}\right)^n\right).} \tag{47.13}$$

To add another twist to this problem—just for fun—we will now try to find a function $F : \mathbb{R} \to \mathbb{R}$ that satisfies the same initial value problem as F_n but with an arbitrary real number x in place of n. In other words, F is to satisfy the following conditions:

$$\begin{aligned} &F(x+2) - F(x+1) - F(x) = 0 \text{ for all } x \in \mathbb{R} \text{ and} \\ &F(0) = 0, F(1) = 1. \end{aligned} \tag{47.14}$$

These conditions do not determine F uniquely (see Exercise 47.4), but they guarantee that $F(n) = F_n$ for all nonnegative integers n. A natural candidate for F seems to be the function $\sqrt{5}^{-1}((1+\sqrt{5})/2)^x + \sqrt{5}^{-1}((1-\sqrt{5})/2)^x$, but unfortunately, the term $((1-\sqrt{5})/2)^x$ is in general undefined because $1-\sqrt{5}$ is a negative number. We can try to get around this problem by observing that $((1-\sqrt{5})/2)^n = (-1)^n((\sqrt{5}-1)/2)^n$, but in replacing n again by x we only find the expression $((\sqrt{5}-1)/2)^x$ to be well defined for all $x \in \mathbb{R}$, while $(-1)^x$ is still undefined. However, since $\cos(\pi n) = (-1)^n$ for all integers n, we may attempt to set

$$F(x) := \frac{1}{\sqrt{5}}\left(\frac{\sqrt{5}+1}{2}\right)^x - \frac{\cos(\pi x)}{\sqrt{5}}\left(\frac{\sqrt{5}-1}{2}\right)^x.$$

Given this definition, we observe that

$$\begin{aligned}
F(x+2) - F(x+1) - F(x) &= \frac{1}{\sqrt{5}}\left(\frac{\sqrt{5}+1}{2}\right)^{x+2} - \frac{\cos(\pi x + 2\pi)}{\sqrt{5}}\left(\frac{\sqrt{5}-1}{2}\right)^{x+2} \\
&\quad - \frac{1}{\sqrt{5}}\left(\frac{\sqrt{5}+1}{2}\right)^{x+1} + \frac{\cos(\pi x + \pi)}{\sqrt{5}}\left(\frac{\sqrt{5}-1}{2}\right)^{x+1} \\
&\quad - \frac{1}{\sqrt{5}}\left(\frac{\sqrt{5}+1}{2}\right)^{x} + \frac{\cos(\pi x)}{\sqrt{5}}\left(\frac{\sqrt{5}-1}{2}\right)^{x} \\
&= \frac{1}{\sqrt{5}}\left(\frac{\sqrt{5}+1}{2}\right)^{x}\left(\left(\frac{1+\sqrt{5}}{2}\right)^2 - \frac{1+\sqrt{5}}{2} - 1\right) \\
&\quad - \frac{\cos(\pi x)}{\sqrt{5}}\left(\frac{\sqrt{5}-1}{2}\right)^{x}\left(\left(\frac{1-\sqrt{5}}{2}\right)^2 - \frac{1-\sqrt{5}}{2} - 1\right) \\
&= 0 \quad \text{(because } (1 \pm \sqrt{5})/2 \text{ is a solution of } x^2 - x - 1 = 0\text{)}.
\end{aligned}$$

Since trivially $F(0) = 0$ and $F(1) = 1$, it follows that F does indeed satisfy the conditions in (47.14). The graph of F is shown in Figure 47.4.

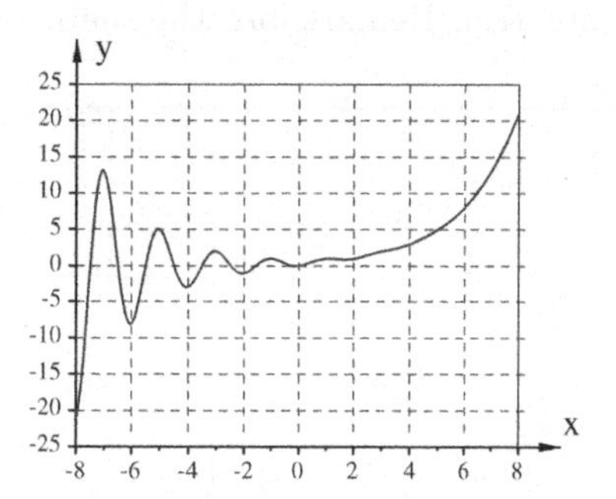

Figure 47.4: graph of the Fibonacci function.

47.4 Exercise. Prove that the Fibonacci function F is not uniquely determined by the conditions in (47.14) by demonstrating that for all integers m these conditions are also satisfied for the function

$$G_m(x) := \frac{1}{\sqrt{5}}\left(\frac{\sqrt{5}+1}{2}\right)^x - \frac{\cos^{2m+1}(\pi x)}{\sqrt{5}}\left(\frac{\sqrt{5}-1}{2}\right)^x.$$

Our final objective is to examine the ratios of successive Fibonacci numbers F_n and F_{n+1}. Given the formula in (47.13), we obtain

$$\frac{F_{n+1}}{F_n} = \frac{\left((1+\sqrt{5})/2\right)^{n+1} - \left((1-\sqrt{5})/2\right)^{n+1}}{\left((1+\sqrt{5})/2\right)^{n} - \left((1-\sqrt{5})/2\right)^{n}} = \left(\frac{1+\sqrt{5}}{2}\right)\frac{1 - \left((1-\sqrt{5})/(1+\sqrt{5})\right)^{n+1}}{1 - \left((1-\sqrt{5})/(1+\sqrt{5})\right)^{n}}.$$

Since $\left|(1-\sqrt{5})/(1+\sqrt{5})\right| < 1$, it follows that $\lim_{n\to\infty}\left((1-\sqrt{5})/(1+\sqrt{5})\right)^n = 0$, and therefore

$$\lim_{n\to\infty}\frac{F_{n+1}}{F_n} = \frac{1+\sqrt{5}}{2}.$$

The limiting value $(1+\sqrt{5})/2$ is the so-called *golden section.*

In geometry, a rectangle is said to be a *golden rectangle* if the ratio of its side lengths is equal to the golden section. Interestingly, in extending a golden rectangle by a square as shown in Figure 47.5, we again obtain a golden rectangle. For if the sides a and b of a given rectangle satisfy the condition

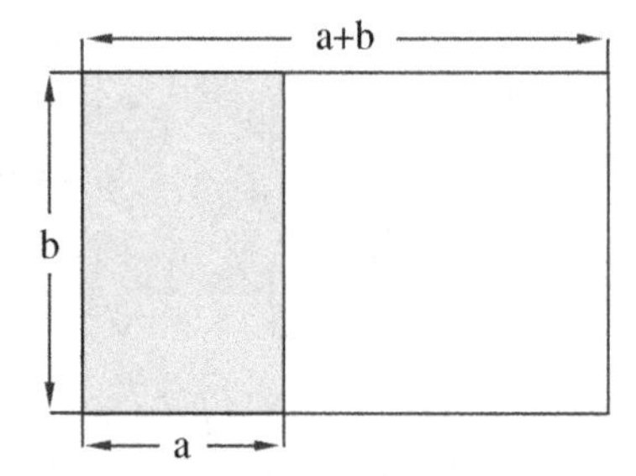

Figure 47.5: extension of a golden rectangle by a square.

$b/a = (1+\sqrt{5})/2$, then b/a is a solution of the equation $x^2 - x - 1 = 0$, or equivalently $1 + x = x^2$ and, by implication, the ratio of the sides $a+b$ and b of the extended rectangle is equal to the golden section as well:

$$\frac{a+b}{b} = \frac{1+b/a}{b/a} = \frac{(b/a)^2}{b/a} = \frac{b}{a} = \frac{1+\sqrt{5}}{2}.$$

In repeatedly extending a golden rectangle by a square we can generate a sequence of outward spiraling golden rectangles as shown in Figure 47.6. Remarkably, the shape of the spiral drawn in Figure 47.6 is

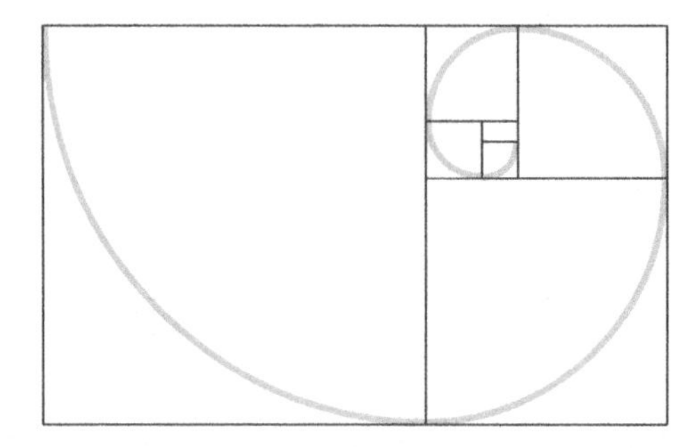

Figure 47.6: a sequence of golden rectangles spiraling outward.

very similar to the cross-sectional shape of the shell of a nautilus (see Figure 47.7).

Additional Exercises

47.5. Decide which of the following equations defines the general solution of the difference equation $y_{n+2} + 2y_{n+1} - 3y_n = 0$:

a) $y_n = C_1 2^n + C_2 3^n$

b) $y_n = C_1 2^n \cos(3n) + C_2 2^n \sin(3n)$

Figure 47.7: shell of a nautilus.

c) $y_n = C_1 + C_2(-1)^n 3^n$

d) $y_n = C_1 3^n + C_2 n 3^n$

47.6. Repeat Exercise 47.5 for the difference equation $y_{n+2} - 6y_{n+1} + 9y_n = 0$.

47.7. In Exercises 47.5 and 47.6 you picked two solutions y_n from the list a)–d) corresponding to the given difference equations. For each of the remaining two solutions y_n (those that you didn't pick), find a difference equation for which y_n is the general solution.

47.8. Given the initial values $F_0 = 0$ and $F_1 = 1$, a sequence of integers F_n is built from the recursive equation

$$F_{n+2} = -4F_{n+1} - 8F_n.$$

Write down the first 6 terms of this sequence (i.e., $F_0, \ldots, F_5$) and then find a general formula for F_n.

47.9. Solve the initial value problem $y_{n+2} + 4y_{n+1} + 4y_n = 0$, $y_0 = 3$, $y_1 = -1$.

47.10. Use Euler's formula (and elementary trigonometry) to find values θ and r such that $-3 + 2i = re^{i\theta}$. Then use your result to solve the initial value problem $y_{n+2} + 6y_{n+1} + 13y_n = 0$, $y_0 = 3$, $y_1 = -1$.

Chapter 48

The Laplace Transform

Definition of the Laplace Transform

In this chapter we will study an important tool for solving differential equations—the *Laplace transform*. A fully satisfactory treatment of this subject is beyond the scope of this text, for it would require a parallel study of the so-called *Fourier transform* (see [Kö1]). We will therefore limit our discussion to those properties and applications of the Laplace transform that are easily accessible.

48.1 Definition. Let us assume that for a given function $f : [0, \infty) \to \mathbb{R}$ and a value $s \in \mathbb{R}$, the integral $\int_0^\infty f(t)e^{-st}\,dt$ is convergent. Then the *Laplace transform of f at s* is defined to be

$$\mathcal{L}[f](s) := \int_0^\infty f(t)e^{-st}\,dt.$$

According to this definition, the Laplace transform creates from a given real-valued function f another function $\mathcal{L}[f]$ whose value at a point s is the improper integral $\int_0^\infty f(t)e^{-st}\,dt$. Since this integral is by no means always convergent (see Exercise 48.2), our first objective is to identify a class of functions for which the Laplace transform is well defined (see Definition 48.3).

48.2 Exercise. Show that for $f(t) := e^{t^2}$ the integral $\int_0^\infty f(t)e^{-st}\,dt$ is divergent for all $s \in \mathbb{R}$. *Hint.* Use a comparison test as stated in Theorem 38.7.

48.3 Definition. For any $s \in \mathbb{R}$, we define the function $e_s : [0, \infty) \to \mathbb{R}$ via the equation $e_s(t) := e^{-st}$, and for a given $a \in \mathbb{R}$ we denote by $\mathcal{E}_a$ the set of all functions $f : [0, \infty) \to \mathbb{R}$ that satisfy the following conditions:

- **a)** f is piecewise continuous on $[0, b]$ for all $b > 0$, and
- **b)** $e_s f$ is bounded on $[0, \infty)$ for all $s > a$ (for the definition of "bounded" see p.145).

If $f \in \mathcal{E}_a$ for some $a \in \mathbb{R}$, then f is said to be of *exponential order*.

The following theorem shows that the assumptions of piecewise continuity in a) and boundedness in b) guarantee that the integral $\int_0^\infty f(t)e^{-st}\,dt$ is convergent for all $s > a$.

48.4 Theorem. *Let $a \in \mathbb{R}$ and let f be a real-valued function defined on $[0, \infty)$.*

- **a)** *If $e_a f$ is bounded then $e_s f$ is bounded for all $s > a$.*
- **b)** *If $e_s f$ is bounded for all $s > a$, then $\lim_{t\to\infty} e^{-st} f(t) = 0$ for all $s > a$.*
- **c)** *If $f \in \mathcal{E}_a$, then the integral $\int_0^\infty |f(t)e^{-st}|\,dt$ is convergent for all $s > a$.*

Proof. **a)** This is trivial, because if $s > a$, then $|e_s f| \leq |e_a f|$ (on the domain $[0, \infty)$).
b) Let $s > a$ and choose an $r \in \mathbb{R}$ such that $s > r > a$. Since $e_s f$ is assumed to be bounded for all $s > a$, it follows that in particular $e_r f$ is bounded. Consequently, there is a constant $M \in \mathbb{R}$ such that $|e^{-rt} f(t)| \leq M$ for all $t \in [0, \infty)$. Thus,

$$|e^{-st} f(t)| = |e^{-rt} f(t)| e^{-(s-r)t} \leq M e^{-(s-r)t}, \tag{48.1}$$

and by implication,

$$0 \leq \lim_{t\to\infty} |e^{-st} f(t)| \leq \lim_{t\to\infty} M e^{-(s-r)t} = 0 \quad \text{(because } s - r > 0\text{)}.$$

This shows that $\lim_{t\to\infty} |e^{-st} f(t)| = \lim_{t\to\infty} e^{-st} f(t) = 0$.
c) Let $f \in \mathcal{E}_a$ and $s > a$. As in the proof of b), the boundedness of $e_r f$ for any value r strictly between s and a (i.e., $s > r > a$) implies that we can find a constant $M \in \mathbb{R}$ such that $|f(t)e^{-rt}| \leq M$ for all $t \in [0, \infty)$. Since $s - r > 0$, the integral $\int_0^\infty M e^{-(s-r)t}\, dt$ is convergent and, according to (48.1) and Theorem 38.7, we may thus infer that $\int_0^\infty |f(t)e^{-st}|\, dt$ is convergent as well. Note: the assumption of piecewise continuity as stated in Definition 48.3a entered the proof only indirectly in that it allowed us to apply Theorem 38.7. □

As an important corollary to Theorem 48.4c we record the following statement:

If $f \in \mathcal{E}_a$ for some $a \in \mathbb{R}$, then the Laplace transform $\mathcal{L}[f](s)$ exists for all $s > a$.

48.5 Example. Let us determine the Laplace transform of $f(t) := \cos(at)$ for $a \in \mathbb{R}$. According to Theorem 48.4a, f is in $\mathcal{E}_0$ because f is continuous and $|f(t)| \leq 1$ for all $t \in \mathbb{R}$. Assuming therefore s to be a value strictly greater than zero, we may use integration by parts to infer that

$$\int_0^\infty f(t)e^{-st}\, dt = \lim_{b\to\infty} \int_0^b \cos(at)e^{-st}\, dt = \lim_{b\to\infty} \left(\frac{e^{-st}}{s^2 + a^2} (a \sin(at) - s\cos(at)) \right) \Bigg|_0^b = \frac{s}{s^2 + a^2}.$$

Hence

$$\mathcal{L}[f](s) = \frac{s}{s^2 + a^2} \tag{48.2}$$

for all $s > 0$.

48.6 Exercise. Show that the constant function $f(t) := 1$ is an element of $\mathcal{E}_0$ and that $\mathcal{L}[f](s) = 1/s$ for all $s > 0$.

Properties of the Laplace Transform

Theorem 48.8 below gives a list of properties of the Laplace transform that are going to be of central importance for everything that follows in this chapter and the next. The proof of this theorem relies in part on the following fundamental result concerning Riemann integrals:

48.7 Theorem. *If $f : I \to \mathbb{R}$ is bounded and piecewise continuous on an interval I, then for all $a \in I$ the function $F_a : I \to \mathbb{R}$ defined by the equation $F_a(x) := \int_a^x f(t)\, dt$ is continuous. (Note: if f were assumed to be continuous instead of only piecewise continuous, then the FTC1 would allow us to infer that F_a is differentiable, and the asserted continuity of F_a would thus be a trivial consequence of Theorem 7.7.)*

Proof. Since f is bounded, there is a constant $M \in \mathbb{R}$ such that $|f(t)| \leq M$ for all $t \in I$. Therefore, according to Definition 17.9 and Theorem 17.10c, e, it is the case that

$$\begin{aligned} |F_a(x) - F_a(y)| &= \left| \int_a^x f(t)\, dt - \int_a^y f(t)\, dt \right| = \left| \int_a^x f(t)\, dt + \int_y^a f(t)\, dt \right| = \left| \int_y^x f(t)\, dt \right| \\ &\leq \left| \int_y^x |f(t)|\, dt \right| \leq \left| \int_y^x M\, dt \right| = M|x - y| \end{aligned}$$

for all $x, y \in I$. Thus, $\lim_{y\to x} |F_a(x)-F_a(y)| \le \lim_{y\to x} M|x-y| = 0$ and, by implication, $\lim_{y\to x} F_a(y) = F_a(x)$. Consequently, F_a is indeed continuous at all points $x \in I$. $\square$

48.8 Theorem. *Let $a \in \mathbb{R}$ and let $f, g \in \mathcal{E}_a$.*

a) *For all $\lambda, \mu \in \mathbb{R}$ we have $\lambda f + \mu g \in \mathcal{E}_a$ and*

$$\mathcal{L}[\lambda f + \mu g](s) = \lambda \mathcal{L}[f](s) + \mu \mathcal{L}[g](s) \text{ for all } s > a.$$

b) *If we set $h(t) := e^{ct} f(t)$ for some $c \in \mathbb{R}$, then $h \in \mathcal{E}_{a+c}$ and*

$$\mathcal{L}[h](s) = \mathcal{L}[f](s-c) \text{ for all } s > a + c.$$

c) *Let $F(t) := \int_0^t f(\tau)\, d\tau$ and $c := \max\{a, 0\}$ (i.e., c is the larger of the two values a and 0). Then $F \in \mathcal{E}_c$ and*

$$\mathcal{L}[F](s) = \frac{1}{s}\mathcal{L}[f](s) \text{ for all } s > c.$$

d) *If n is a positive integer and $p(t) := t^n f(t)$, then $p \in \mathcal{E}_a$ and*

$$\mathcal{L}[p](s) = (-1)^n \frac{d^n}{ds^n}\mathcal{L}[f](s) \text{ for all } s > a.$$

e) *For $c > 0$ and $q(t) := \begin{cases} 0 & \text{if } 0 \le t < c, \\ f(t-c) & \text{if } t \ge c \end{cases}$ we have $q \in \mathcal{E}_a$ and*

$$\mathcal{L}[q](s) = e^{-cs}\mathcal{L}[f](s) \text{ for all } s > a.$$

f) *Suppose that f is $n-1$ times differentiable and that $f^{(n-1)}$ is continuous and piecewise differentiable (i.e., $f^{(n-1)}$ is differentiable on every interval $[0, b]$ except possibly at finitely many points). Furthermore, assume that $f^{(n)}$ is piecewise continuous on every interval $[0, b]$. (Note: at those finitely many points in $[0, b]$ where $f^{(n-1)}$ is possibly nondifferentiable, the function $f^{(n)}$ is per se undefined, but this is irrelevant because we can assign arbitrary values to $f^{(n)}$ at these finitely may points without losing the property of piecewise continuity on $[0, b]$.) If $f^{(k)} \in \mathcal{E}_a$ for all $k \in \{1, \ldots, n-1\}$, then $\mathcal{L}[f^{(n)}](s)$ exists for all $s > a$ and*

$$\mathcal{L}[f^{(n)}](s) = s^n \mathcal{L}[f](s) - (s^{n-1}f(0) + s^{n-2}f'(0) + \cdots + f^{(n-1)}(0)) \text{ for all } s > a.$$

Proof. **a)** Since $|e_s(\lambda f + \mu g)| \le |\lambda||e_s f| + |\mu||e_s g|$, it is obvious that $e_s(\lambda f + \mu g)$ is bounded for all $s > a$ because $e_s f$ and $e_s g$ are bounded for all $s > a$ by assumption. Furthermore, the piecewise continuity of $\lambda f + \mu g$ on $[0, b]$ is a trivial consequence of the piecewise continuity of f and g on any interval $[0, b]$. Therefore, $\lambda f + \mu g \in \mathcal{E}_a$, and

$$\mathcal{L}[\lambda f + \mu g](s) = \int_0^\infty (\lambda f(t) + \mu g(t))e^{-st}\, dt = \lambda \int_0^\infty f(t)e^{-st}\, dt + \mu \int_0^\infty g(t)e^{-st}\, dt = \lambda\mathcal{L}[f](s) + \mu\mathcal{L}[g](s)$$

for all $s > a$.

b) Since $e_{s+c}h = e_s f$, it follows that $e_{s+c}h$ is bounded for all $s > a$, and therefore h is an element of $\mathcal{E}_{a+c}$ (the piecewise continuity of h on any interval $[0, b]$ is obvious). If $s > a + c$, then

$$\mathcal{L}[h](s) = \int_0^\infty h(t)e^{-st}\, dt = \int_0^\infty f(t)e^{-(s-c)t}\, dt = \mathcal{L}[f](s-c).$$

★ **c)** In order to show that F is in $\mathcal{E}_c$, it is sufficient to prove that $e_s F$ is bounded for all $s > c$ because, according to Theorem 48.7, F is continuous and therefore condition a) of Definition 48.3 is satisfied.

So let s be a number strictly greater than c, and r a number strictly between s and c (i.e., $s > r > c$). Since f is contained in $\mathcal{E}_a$, it follows that $e_r f$ is bounded, and by implication there exists an $M \in \mathbb{R}$ such that $|f(t)e^{-rt}| \leq M$ for all $t \in [0, \infty)$ or, equivalently, $|f(t)| \leq Me^{rt}$ for all $t \in [0, \infty)$. Using Theorem 17.10e and the fact that $s > r > c = \max\{a, 0\} \geq 0$, we observe that

$$\begin{aligned} |e^{-st}F(t)| &= \left| e^{-st} \int_0^t f(\tau)\, d\tau \right| \leq e^{-st} \int_0^t |f(\tau)|\, d\tau \leq e^{-st} \int_0^t Me^{r\tau}\, d\tau \\ &= \frac{Me^{-st}}{r}(e^{rt} - 1) \leq \frac{Me^{-(s-r)t}}{r} \leq \frac{M}{r} \end{aligned}$$

for all $t \geq 0$. This proves that $e_s F$ is bounded for all $s > c$ as desired. Now let $b > 0$ and let $t_1 < \cdots < t_m$ be finitely many points in $(0, b)$ such that f is continuous on $(0, b) \setminus \{t_1, \ldots, t_m\}$. In other words, f is continuous on each of the intervals $(0, t_1), (t_1, t_2), \ldots, (t_{m-1}, t_m), (t_m, b)$. Then the FTC1 implies that F is differentiable on each of these intervals, and since F is also continuous on $[0, b]$, we may use integration by parts and the FTC2 to conclude that

$$\begin{aligned} \int_0^b F(t)e^{-st}\, dt &= \int_0^{t_1} F(t)e^{-st}\, dt + \sum_{k=1}^{m-1} \int_{t_k}^{t_{k+1}} F(t)e^{-st}\, dt + \int_{t_m}^b F(t)e^{-st}\, dt \\ &= -\frac{1}{s}F(t)e^{-st}\Big|_0^{t_1} + \frac{1}{s}\int_0^{t_1} f(t)e^{-st}\, dt + \sum_{k=1}^{m-1}\left(-\frac{1}{s}F(t)e^{-st}\Big|_{t_k}^{t_{k+1}} + \frac{1}{s}\int_{t_k}^{t_{k+1}} f(t)e^{-st}\, dt \right) \\ &\quad - \frac{1}{s}F(t)e^{-st}\Big|_{t_m}^{b} + \frac{1}{s}\int_{t_m}^b f(t)e^{-st}\, dt \\ &= -\frac{1}{s}F(b)e^{-sb} + \frac{1}{s}\int_0^b f(t)e^{-st}\, dt \quad \text{(because } F(0) = 0\text{)}. \end{aligned}$$

Applying now Theorem 48.4b, we see that $\lim_{b\to\infty} F(b)e^{-sb} = 0$ for all $s > c$, and therefore

$$\begin{aligned} \mathcal{L}[F](s) &= \int_0^\infty F(t)e^{-st}\, dt = \lim_{b\to\infty} \int_0^b F(t)e^{-st}\, dt \\ &= \lim_{b\to\infty}\left(-\frac{1}{s}F(b)e^{-sb} + \frac{1}{s}\int_0^b f(t)e^{-st}\, dt \right) = \frac{1}{s}\mathcal{L}[f](s). \ \bigstar \end{aligned}$$

★ **d)** Let $s > a$ and choose $r \in \mathbb{R}$ such that $s > r > a$. Then $|e^{-st}p(t)| = |e^{-(s-r)t}t^n||e^{-rt}f(t)|$. Furthermore, according to Theorem 29.8, we know that $\lim_{t\to\infty} e^{-(s-a)t}t^n = 0$, and since $e_r f$ is bounded on $[0, \infty)$, it follows that $e_s p$ is bounded on $[0, \infty)$ as well. Hence $p \in \mathcal{E}_a$. Given a value $s > a$ we now use the definition of the derivative to infer that

$$\frac{d}{ds}\mathcal{L}[f](s) = \lim_{\Delta s\to 0} \frac{\int_0^\infty f(t)e^{-(s+\Delta s)t}\, dt - \int_0^\infty f(t)e^{-st}\, dt}{\Delta s} = \lim_{\Delta s\to 0} \int_0^\infty f(t)\frac{e^{-(s+\Delta s)t} - e^{-st}}{\Delta s}\, dt.$$

Since it can be shown (see [Kö1]) that

$$\lim_{\Delta s\to 0} \int_0^\infty f(t)\frac{e^{-(s+\Delta s)t} - e^{-st}}{\Delta s}\, dt = \int_0^\infty f(t) \lim_{\Delta s\to 0} \frac{e^{-(s+\Delta s)t} - e^{-st}}{\Delta s}\, dt,$$

and since

$$\lim_{\Delta s\to 0} \frac{e^{-(s+\Delta s)t} - e^{-st}}{\Delta s} = \frac{d}{ds}e^{-st} = -te^{-st},$$

we may conclude that

$$\frac{d}{ds}\mathcal{L}[f](s) = -\int_0^\infty tf(t)e^{-st}\, dt.$$

Repeating the same calculation with $tf(t)$ in place of $f(t)$, we obtain

$$\frac{d^2}{ds^2}\mathcal{L}[f](s) = (-1)^2 \int_0^\infty t^2 f(t) e^{-st}\, dt,$$

and continuing this way we readily infer that

$$\frac{d^n}{ds^n}\mathcal{L}[f](s) = (-1)^n \int_0^\infty t^n f(t) e^{-st}\, dt = (-1)^n \mathcal{L}[p](s). \bigstar$$

e) First we need to show that q is in $\mathcal{E}_a$. If $t \geq c$ and $s > a$, then $e_s f$ is bounded and $e^{-st}q(t) = e^{-sc}e^{-s(t-c)}f(t-c)$. Thus, $e_s q$ is bounded as well. Since the assumption $f \in \mathcal{E}_a$ clearly implies that q is piecewise continuous on every interval $[0, b]$, it follows that q is indeed an element of $\mathcal{E}_a$ as desired. Choosing now an arbitrary value $s > a$, we obtain

$$\mathcal{L}[q](s) = \int_0^\infty q(t)e^{-st}\, dt = \int_c^\infty f(t-c)e^{-st}\, dt = \int_0^\infty f(u)e^{-s(u+c)}\, du = e^{-cs}\mathcal{L}[f](s).$$

$\bigstar$ **f)** Let $s > a$. Since

$$\mathcal{L}[f^{(n)}](s) = \int_0^\infty f^{(n)}(t)e^{-st}\, dt = \lim_{b\to\infty} \int_0^b f^{(n)}(t)e^{-st}\, dt \tag{48.3}$$

we need to evaluate $\int_0^b f^{(n)}(t)e^{-st}\, dt$. Given that $f^{(n)}$ is assumed to be piecewise continuous on $[0, b]$, we may again choose points $t_1 < \cdots < t_m$ in $(0, b)$ such that $f^{(n)}$ is continuous on $(0, b) \setminus \{t_1, \ldots, t_m\}$. This yields

$$\int_0^b f^{(n)}(t)e^{-st}\, dt = \int_0^{t_1} f^{(n)}(t)e^{-st}\, dt + \sum_{k=1}^{m-1} \int_{t_k}^{t_{k+1}} f^{(n)}(t)e^{-st}\, dt + \int_{t_m}^b f^{(n)}(t)e^{-st}\, dt. \tag{48.4}$$

Since $f^{(n)}$ is not assumed to be bounded, the integrals over the intervals $(0, t_1), (t_1, t_2), \ldots, (t_m, b)$ are potentially improper and thus require computations using limits. Considering an interval (t_k, t_{k+1}) and choosing an arbitrary point $x \in (t_k, t_{k+1})$, we obtain

$$\begin{aligned} \int_{t_k}^{t_{k+1}} f^{(n)}(t)e^{-st}\, dt &= \int_{t_k}^{x} f^{(n)}(t)e^{-st}\, dt + \int_{x}^{t_{k+1}} f^{(n)}(t)e^{-st}\, dt \\ &= \lim_{\varepsilon\to 0+} \int_{t_k+\varepsilon}^{x} f^{(n)}(t)e^{-st}\, dt + \lim_{\varepsilon\to 0+} \int_{x}^{t_{k+1}-\varepsilon} f^{(n)}(t)e^{-st}\, dt. \end{aligned} \tag{48.5}$$

Since the functions $f, f', \ldots, f^{(n-1)}$ are by assumption continuous (see also Theorem 7.7) and contained in $\mathcal{E}_a$, we may conclude that

$$\begin{aligned} \lim_{\varepsilon\to 0+} \int_{t_k+\varepsilon}^{x} f^{(n)}(t)e^{-st}\, dt &= \lim_{\varepsilon\to 0+} \left(f^{(n-1)}(t)e^{-st}\Big|_{t_k+\varepsilon}^{x} + s\int_{t_k+\varepsilon}^{x} f^{(n-1)}(t)e^{-st}\, dt \right) \\ &= f^{(n-1)}(t)e^{-st}\Big|_{t_k}^{x} + s\int_{t_k}^{x} f^{(n-1)}(t)e^{-st}\, dt \\ &= f^{(n-1)}(t)e^{-st}\Big|_{t_k}^{x} + s\left(f^{(n-2)}(t)e^{-st}\Big|_{t_k}^{x} + s\int_{t_k}^{x} f^{(n-2)}(t)e^{-st}\, dt \right) \\ &= f^{(n-1)}(t)e^{-st}\Big|_{t_k}^{x} + sf^{(n-2)}(t)e^{-st}\Big|_{t_k}^{x} + s^2\int_{t_k}^{x} f^{(n-2)}(t)e^{-st}\, dt \\ &\ \ \vdots \\ &= \sum_{i=0}^{n-1} s^i f^{(n-1-i)}(t)e^{-st}\Big|_{t_k}^{x} + s^n \int_{t_k}^{x} f(t)e^{-st}\, dt. \end{aligned}$$

Similarly, it can be shown that

$$\lim_{\varepsilon\to 0^+}\int_x^{t_{k+1}-\varepsilon} f^{(n)}(t)e^{-st}\,dt = \sum_{i=0}^{n-1} s^i f^{(n-1-i)}(t)e^{-st}\Big|_x^{t_{k+1}} + s^n\int_x^{t_{k+1}} f(t)e^{-st}\,dt.$$

Therefore, (48.5) implies that

$$\int_{t_k}^{t_{k+1}} f^{(n)}(t)e^{-st}\,dt = \sum_{i=0}^{n-1} s^i f^{(n-1-i)}(t)e^{-st}\Big|_{t_k}^{t_{k+1}} + s^n\int_{t_k}^{t_{k+1}} f(t)e^{-st}\,dt$$

for all $k \in \{1,\dots,m\}$. Performing analogous calculations for the integrals $\int_0^{t_1} f^{(n)}(t)e^{-st}\,dt$ and $\int_{t_m}^b f^{(n)}(t)e^{-st}\,dt$ in (48.4), we easily arrive at the following conclusion:

$$\int_0^b f(t)e^{-st}\,dt = \sum_{i=0}^{n-1} s^i f^{(n-1-i)}(t)e^{-st}\Big|_0^b + s^n\int_0^b f(t)e^{-st}\,dt. \tag{48.6}$$

Given that $f, f', \dots, f^{(n-1)} \in \mathcal{E}_a$, we may apply Theorem 48.4b to infer that

$$\lim_{t\to\infty} f^{(k)}(t)e^{-st} = 0 \text{ for all } k \in \{0,\dots,n-1\}.$$

Combining this observation with (48.3) and (48.6) yields

$$\begin{aligned}\mathcal{L}[f^{(n)}](s) &= \lim_{b\to\infty}\left(\sum_{i=0}^{n-1} s^i f^{(n-1-i)}(t)e^{-st}\Big|_0^b + s^n\int_0^b f(t)e^{-st}\,dt\right)\\ &= -\sum_{i=0}^{n-1} s^i f^{(n-1-i)}(0) + s^n\int_0^\infty f(t)e^{-st}\,dt\\ &= -(f^{(n-1)}(0) + sf^{(n-2)}(0) + \cdots + s^{n-2}f'(0) + s^{n-1}f(0)) + s^n\mathcal{L}[f](s).\end{aligned}$$

In particular, our argument shows that $\mathcal{L}[f^{(n)}](s)$ exists for all $s > a$. ★ □

In order to demonstrate the usefulness of Theorem 48.8 in computations of Laplace transforms, we will now consider some examples:

48.9 Example. Let $a \in \mathbb{R}$ and $f(t) := e^{at}$. According to Theorem 48.8b and the result in Exercise 48.6, we have

$$\mathcal{L}[f](s) = \mathcal{L}[1](s-a) = \frac{1}{s-a} \tag{48.7}$$

for all $s > a$.

48.10 Example. Let us determine the Laplace transform of $f(t) := \cosh(at) = (e^{at} + e^{-at})/2$ for some $a \in \mathbb{R}$. Using (48.7) and Theorem 48.8a, we find that

$$\mathcal{L}[f](s) = \frac{1}{2}\left(\mathcal{L}[e_{-a}](s) + \mathcal{L}[e_a](s)\right) = \frac{1}{2}\left(\frac{1}{s-a} + \frac{1}{s+a}\right) = \frac{s}{s^2-a^2} \tag{48.8}$$

for all $s > |a|$.

48.11 Example. Let $f(t) := \sin(at)$ for some $a \in \mathbb{R}$ with $a \neq 0$. Since

$$\sin(at) = -\frac{1}{a}\frac{d}{dt}\cos(at),$$

we may apply (48.2) and Theorem 48.8a, f to conclude that

$$\begin{aligned}\mathcal{L}[f](s) &= -\frac{1}{a}\mathcal{L}\left[\frac{d}{dt}\cos(at)\right](s) = -\frac{1}{a}(-\cos(0) + s\mathcal{L}[\cos(at)](s)) \\ &= -\frac{1}{a}\left(-1 + \frac{s^2}{s^2+a^2}\right) = \frac{a}{s^2+a^2}\end{aligned} \tag{48.9}$$

for all $s > 0$.

48.12 Example. For our final example, we determine the Laplace transform of $f(t) := t^n$ for some nonnegative integer n. According to Theorem 48.8d and Exercise 48.6, we have

$$\mathcal{L}[f](s) = (-1)^n \frac{d^n}{ds^n}\mathcal{L}[1](s) = (-1)^n \frac{d^n}{ds^n}\frac{1}{s} = \frac{n!}{s^{n+1}} \tag{48.10}$$

for all $s > 0$.

48.13 Exercise. Find the Laplace transforms of the functions listed below and decide in each case for which $s \in \mathbb{R}$ the transform is well defined.

a) $f(t) := e^{-t}\sin(2t)$

b) $f(t) := \begin{cases} 0 & \text{if } t < 2, \\ (t-2)^4 & \text{if } t \geq 2. \end{cases}$

c) $f(t) := \sinh(at)$ for some $a \in \mathbb{R}$.

The Gamma Function

In (48.10) we showed that for all positive integers n and all $s > 0$ the Laplace transform of $f(t) := t^n$ is $\mathcal{L}[f](s) = n!/s^{n+1}$, but what happens if we replace the integer n with a real number α? To answer this question we first notice that for $\alpha \leq -1$ the integral $\int_0^1 t^\alpha\,dt$ is divergent (see (38.2)). Therefore, using Theorem 38.15 and the (easily verifiable) inequality $t^\alpha e^{-st} \geq e^{-|s|}t^\alpha$ for all $t \in (0,1]$, it follows that $\int_0^1 t^\alpha e^{-st}\,dt$ is divergent as well for all $\alpha \leq -1$ and all $s \in \mathbb{R}$. Thus, $\mathcal{L}[t^\alpha](s)$ is not well defined for $\alpha \leq -1$. If, however, $\alpha > -1$, then the integral $\int_0^1 t^\alpha e^{-st}\,dt$ is convergent for all $s \in \mathbb{R}$, and since the integral $\int_1^\infty t^\alpha e^{-st}\,dt$ is easily seen to be convergent as well for all $s > 0$, it follows that the Laplace transform of $f(t) := t^\alpha$ exists for all $\alpha > -1$ and all $s > 0$. Using integration by substitution with $u = st$ and $du = s\,dt$, we obtain

$$\mathcal{L}[f](s) = \int_0^\infty t^\alpha e^{-st}\,dt = \frac{1}{s}\int_0^\infty \left(\frac{u}{s}\right)^\alpha e^{-u}\,du = \frac{1}{s^{\alpha+1}}\int_0^\infty u^\alpha e^{-u}\,du. \tag{48.11}$$

This observation motivates the following definition:

48.14 Definition. For $x \in (0,\infty)$ we set

$$\Gamma(x) := \int_0^\infty t^{x-1}e^{-t}\,dt.$$

The function $\Gamma : (0,\infty) \to \mathbb{R}$ defined by this equation is referred to as the *gamma function*.

Given this definition, the result in (48.11) can be stated as follows:

$$\mathcal{L}[f](s) = \frac{\Gamma(\alpha+1)}{s^{\alpha+1}} \quad \text{for all } \alpha > -1 \text{ and all } s > 0. \tag{48.12}$$

The next theorem shows that this formula yields the same result as in (48.10) whenever α is a nonnegative integer.

48.15 Theorem. *The following equations are valid for all $x > 0$ and all integers $n \geq 0$:*

a) $\Gamma(x+1) = x\Gamma(x)$,

b) $\Gamma(n+1) = n!$.

Proof. **a)** Using integration by parts, we obtain

$$\Gamma(x+1) = \int_0^\infty t^x e^{-t}\,dt = \lim_{b\to\infty}\left(-t^x e^{-t}\Big|_0^b + x\int_0^b t^{x-1}e^{-t}\,dt\right) = x\int_0^\infty t^{x-1}e^{-t}\,dt = x\Gamma(x).$$

b) For $n = 0$, we have $\Gamma(1) = \int_0^\infty e^{-t}\,dt = 1 = 0!$. Therefore, we may apply a) for $x = 1$ to infer that $\Gamma(2) = 1\cdot\Gamma(1) = 1 = 1!$. Similarly, for $x = 2$ and $x = 3$ we obtain $\Gamma(3) = 2\Gamma(2) = 2\cdot 1 = 2!$ and $\Gamma(4) = 3\Gamma(3) = 3\cdot 2! = 3!$. Continuing in this manner, we arrive at the desired conclusion: $\Gamma(n+1) = n!$ for all integers $n \geq 0$. □

★ *Remark.* To take a slightly different look at the gamma function, we set $f(n) := (n-1)!$ for all integers $n \geq 1$. Since $n! = n\cdot(n-1)!$, it follows that f satisfies the conditions

$$f(n+1) = n\,f(n) \quad\text{and}\quad f(1) = 0! = 1. \tag{48.13}$$

Given this observation, it is natural and interesting to ask whether it is possible to *extend the domain* of f to all positive real numbers in such a way that the extended function satisfies (48.13) for all *real numbers* $x > 0$. The answer is of course "yes," because Theorem 48.15 shows that the gamma function is indeed an extension of this sort.

The idea of extending the domain of a function from the integers to the real numbers is actually not new—we already encountered it in Chapter 23 with regard to the general exponential function. An exponential a^n is naturally defined as the n-fold product of a with itself only for positive integers n. In analogy to (48.13), the function $g(n) := a^n$ satisfies the conditions $g(n+1) = a\,g(n)$ and $g(1) = a$. The same conditions with an arbitrary real number x in place of n are also satisfied by the exponential function $a^x = e^{x\ln(a)}$ which therefore is properly regarded as the extension of g from the set of positive integers to $\mathbb{R}$. A similar example we also discussed in Chapter 47 where we introduced the Fibonacci function $F : \mathbb{R} \to \mathbb{R}$ as an extension of the sequence of Fibonacci numbers F_n starting from the assumption that F satisfies the same difference equation as F_n. ★

Periodic Functions

In technical applications and in the natural world we encounter oscillatory phenomena in a wide variety of circumstances. Examples are daily fluctuations in temperature, the propagation of waves in water, the motion of a pendulum, the periodic movements of the planets, cyclic changes of population densities in ecosystems, or oscillations in electromagnetic fields. A particular oscillatory system that we have already analyzed is an electric circuit with a sinusoidal external voltage source $E(t) = E_0\sin(\omega t)$ (see Chapter 46). Intuitively, we say here that the function $E(t)$ is periodic with period $2\pi/|\omega|$ because the graph of $E(t)$ over any given interval of length $2\pi/|\omega|$ is identical with the graph of $E(t)$ over any adjoining interval of the same length, and because $2\pi/|\omega|$ is the smallest positive number that satisfies this property. In other words, $T := 2\pi/|\omega|$ is the smallest positive number for which $E(t+T) = E(t)$ for all $t \in \mathbb{R}$. (Notice that for $E(t) = \sin(\omega t)$, the general equation $E(t+T) = E(t)$ is equivalent to the equation $\sin(\omega(t + 2\pi/|\omega|)) = \sin(\omega t \pm 2\pi) = \sin(\omega t)$, which is quite familiar from elementary trigonometry.)

48.16 Definition. A function $f : D \subset \mathbb{R} \to \mathbb{R}$ is said to be *periodic* if there is a real number $T > 0$ such that for all $t \in D$ we have $t + T \in D$ and $f(t+T) = f(t)$. If T is the smallest positive number that satisfies these conditions, then T is said to be the *period* of f.

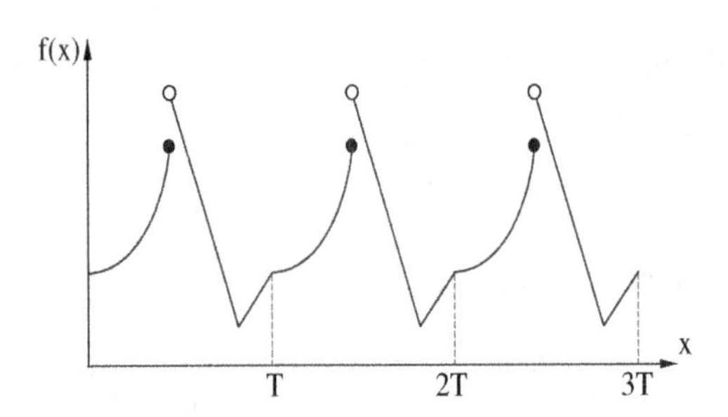

Figure 48.1: a periodic function with period T.

A typical graph of a periodic function of period T is shown in Figure 48.1. (The diagram actually shows only a finite segment of the graph. The complete picture would have to extend to positive infinity.)

Remark. If $f : \mathbb{R} \to \mathbb{R}$ is a periodic function of period T, then the equation $f(t+T) = f(t)$ implies that

$$f(t + kT) = f(t) \tag{48.14}$$

for all nonnegative integers k. Furthermore, it should be noted that not every periodic function has a well-defined period. For example, if $f : \mathbb{R} \to \mathbb{R}$ is a constant function, then the equation $f(t+T) = f(t)$ is satisfied for all $T > 0$ so that a smallest positive T with this property does not exist (see also Exercise 48.36).

48.17 Exercise. Justify equation (48.14).

To determine the Laplace transform of a periodic function, we assume that $f : [0, \infty) \to \mathbb{R}$ is periodic with period T and that f is bounded and piecewise continuous on the interval $[0, T)$. Then it clearly follows that f is bounded on all of $\mathbb{R}$ (because f is periodic), and therefore Theorem 48.4a implies that f is in $\mathcal{E}_0$. Furthermore,

$$\mathcal{L}[f](s) = \int_0^\infty f(t)e^{-st}\,dt = \lim_{n\to\infty} \int_0^{nT} f(t)e^{-st}\,dt = \lim_{n\to\infty} \sum_{k=0}^{n-1} \int_{kT}^{(k+1)T} f(t)e^{-st}\,dt$$

for all $s > 0$. Using integration by substitution with $u = t - kT$ and (48.14), we obtain

$$\int_{kT}^{(k+1)T} f(t)e^{-st}\,dt = \int_0^T f(u + kT)e^{-s(u+kT)}\,du = e^{-skT} \int_0^T f(u)e^{-su}\,du.$$

Hence

$$\mathcal{L}[f](s) = \left(\lim_{n\to\infty} \sum_{k=0}^{n-1} e^{-skT} \right) \int_0^T f(u)e^{-su}\,du.$$

Note: the integral $\int_0^T f(u)e^{-su}\,du$ can be factored out of the sum and the limit, because its value does not depend on either k or n. Furthermore, since s and T are strictly greater than zero, we may infer that $|e^{-sT}| = e^{-sT} < 1$, and according to Theorem 41.16 it therefore follows that

$$\lim_{n\to\infty} \sum_{k=0}^{n-1} e^{-skT} = \sum_{k=0}^{\infty} (e^{-sT})^k = \frac{1}{1 - e^{-sT}}.$$

Thus,

$$\mathcal{L}[f](s) = \frac{1}{1 - e^{-sT}} \int_0^T f(u)e^{-su}\,du.$$

To summarize our discussion, we state the following theorem:

48.18 Theorem. *If $f : [0, \infty) \to \mathbb{R}$ is a periodic function of period T such that f is piecewise continuous and bounded on $[0, T)$, then f is in $\mathcal{E}_0$, and*

$$\mathcal{L}[f](s) = \frac{1}{1 - e^{-sT}} \int_0^T f(t)e^{-st}\, dt$$

for all $s > 0$.

48.19 Example. Let us apply Theorem 48.18 to verify equation (48.9). For $a \neq 0$ the period of the function $f(t) := \cos(at)$ is $T = 2\pi/|a|$. Thus,

$$\mathcal{L}[f](s) = \frac{1}{1 - e^{-2\pi s/|a|}} \int_0^{2\pi/|a|} \cos(at)e^{-st}\, dt$$

for all $s > 0$. Using integration by parts, it is easy to see that

$$\int_0^{2\pi/|a|} \cos(at)e^{-st}\, dt = \frac{s(1 - e^{-2\pi s/|a|})}{s^2 + a^2},$$

and therefore,

$$\mathcal{L}[f](s) = \frac{s(1 - e^{-2\pi s/|a|})}{(1 - e^{-2\pi s/|a|})(s^2 + a^2)} = \frac{s}{s^2 + a^2}$$

as desired.

Functions with Discontinuities

In electric circuits, the function representing the difference in potential generated by the external voltage source is not necessarily always a nice continuous or differentiable function as, for instance, $\sin(\omega t)$. Instead we frequently encounter functions that are discontinuous at some points, or at least very nearly so.

48.20 Example. A typical example of a function with discontinuities is the so-called *Morse-dot function* f_α. The graph in Figure 48.2 shows that f_α is a periodic function of period $T = 2\alpha$ with values that

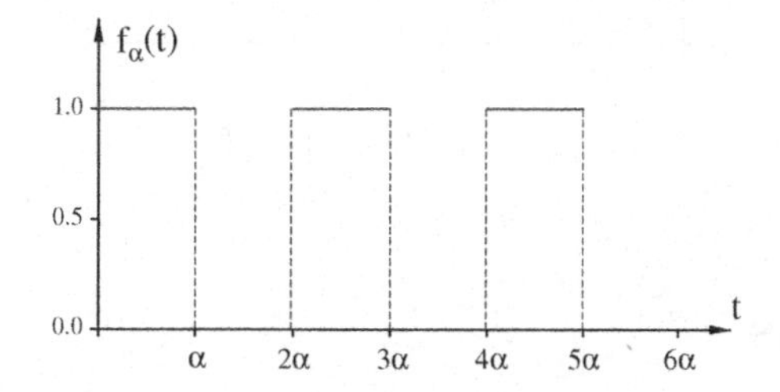

Figure 48.2: Morse-dot function.

alternate between 1 and 0. Since f_α is bounded and piecewise continuous, the restriction of f_α to $[0, \infty)$ is in $\mathcal{E}_0$. Therefore, Theorem 48.18 implies that

$$\mathcal{L}[f_\alpha](s) = \frac{1}{1 - e^{-2\alpha s}} \int_0^{2\alpha} f_\alpha(t)e^{-st}\, dt = \frac{1}{1 - e^{-2\alpha s}} \int_0^{\alpha} e^{-st}\, dt = \frac{1 - e^{-\alpha s}}{s(1 - e^{-2\alpha s})}$$

for all $s > 0$. Simplifying the term on the right-hand side of this equation yields

$$\mathcal{L}[f_\alpha](s) = \frac{1}{s(1 + e^{-\alpha s})}. \tag{48.15}$$

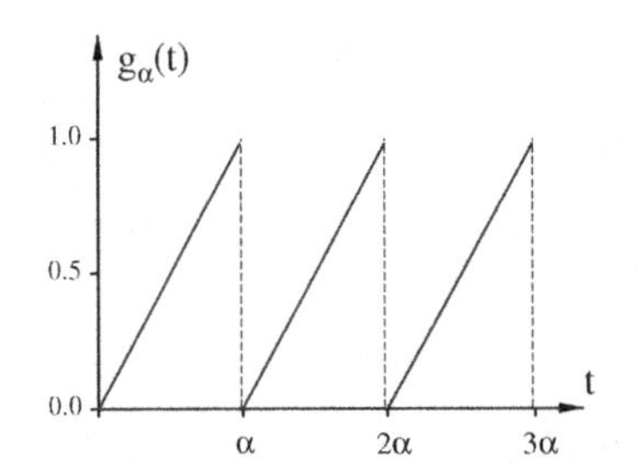

Figure 48.3: saw-tooth function.

48.21 Example. Let us consider the *saw-tooth function* g_α as shown in Figure 48.3. From the graph, we see that g_α is a periodic function of period $T = \alpha$ and that $g_\alpha(t) = t/\alpha$ for all $t \in [0, \alpha)$. Since g_α is bounded and piecewise continuous, the restriction of g_α to $[0, \infty)$ is in $\mathcal{E}_0$. Therefore,

$$\mathcal{L}[g_\alpha](s) = \frac{1}{1 - e^{-\alpha s}} \int_0^\alpha \frac{t}{\alpha} e^{-st}\, dt$$

for all $s > 0$. Using integration by parts, it is easy to see that

$$\int_0^\alpha \frac{t}{\alpha} e^{-st}\, dt = \frac{(1 + \alpha s)(1 - e^{-\alpha s})}{\alpha s^2} - \frac{1}{s}.$$

Hence

$$\mathcal{L}[g_\alpha](s) = \frac{1 + \alpha s}{\alpha s^2} - \frac{1}{s(1 - e^{-\alpha s})}. \tag{48.16}$$

48.22 Exercise. Find the Laplace transform of the periodic function f shown in Figure 48.4.

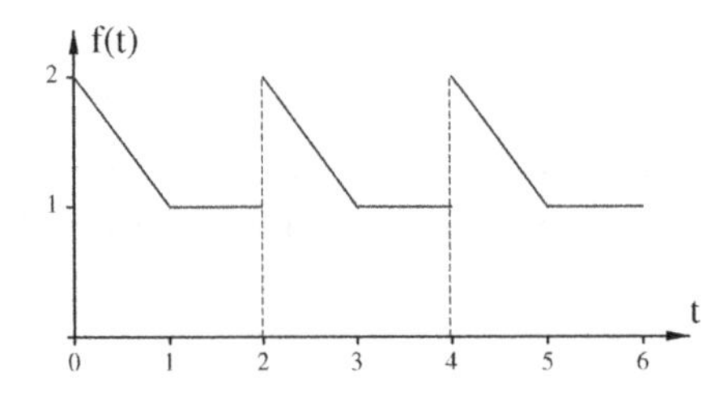

Figure 48.4: graph of f.

Inverse Transforms

In Chapter 49 we will see that in typical applications of the Laplace transform it is usually necessary not only to find the Laplace transform $\mathcal{L}[f]$ of a given function f, but also to find for a given Laplace transform F a function f such that $\mathcal{L}[f] = F$. For instance, we may have to determine a function f such that $\mathcal{L}[f](s) = F(s)$ for $F(s) := s/(s^2 + 1)$. In this case, the function we are looking for is $f(t) = \cos(t)$, because according to (48.2) the Laplace transform of $\cos(t)$ is $s/(s^2+1)$. Thus, we say that $f(t) = \cos(t)$ is the *inverse transform* of $F(s) = s/(s^2 + 1)$. In general, the inverse transform of a function F will be denoted by $\mathcal{L}^{-1}[F]$, and there are three natural questions concerning inverse transforms that need to be answered:

a) Does there exist an $a \in \mathbb{R}$ and an inverse transform $f \in \mathcal{E}_a$ for every function F?

b) If there exists an inverse transform, is it possible that there is more than one? In other words, is the inverse transform uniquely determined by the transform?

c) Is there a general formula for calculating an inverse transform in those cases where it does exist?

The answer to the first question is given by the following theorem:

48.23 Theorem. *If $f \in \mathcal{E}_a$ for some $a \in \mathbb{R}$, then*

$$\lim_{s\to\infty} \mathcal{L}[f](s) = 0.$$

Proof. Let $s > r > a$. Then $e_r f$ is bounded, and therefore we can find an $M \in \mathbb{R}$ such that $|f(t)| \le Me^{rt}$ for all $t \in [0,\infty)$. Using Theorem 17.10e, we obtain

$$|\mathcal{L}[f](s)| = \lim_{b\to\infty}\left|\int_0^b f(t)e^{-st}\,dt\right| \le \lim_{b\to\infty}\int_0^b |f(t)|e^{-st}\,dt \le \int_0^\infty Me^{-(s-r)t}\,dt = \frac{M}{s-r}.$$

Since $\lim_{s\to\infty} M/(s-r) = 0$, it follows that $\lim_{s\to\infty}\mathcal{L}[f](s) = 0$ as well. □

To see how this theorem answers question a) above, let us consider, for example, the function $F(s) := s/(s+1)$. If there did exist a function f of exponential order such that $\mathcal{L}[f] = F$, then Theorem 48.23 would imply that $\lim_{s\to\infty} F(s) = \lim_{s\to\infty}\mathcal{L}[f](s) = 0$ in contradiction to the obvious fact that $\lim_{s\to\infty} F(s) = \lim_{s\to\infty} s/(s+1) = 1$. Consequently, there does not exist a function f of exponential order whose Laplace transform is F, and the answer to question a) is therefore "no." Regarding question b) the following theorem, stated without proof, provides a positive answer.

48.24 Theorem. *Let $a \in \mathbb{R}$ and $f, g \in \mathcal{E}_a$. If there is a number $c > a$ such that $\mathcal{L}[f](s) = \mathcal{L}[g](s)$ for all $s > c$, then $f(t) = g(t)$ at all points $t \in [0,\infty)$ where f and g are both continuous.*

This theorem says that two functions with identical Laplace transforms (for $s > c$) are equal at all points $t \in \mathbb{R}$ except possibly at those points where either one of them is discontinuous.

★ *Remark.* There is a stronger version of Theorem 48.24 called Lerch's theorem (see [Köl]) which says the following: if for given functions $f, g \in \mathcal{E}_a$ there are constants $c, d \in \mathbb{R}$ such that $c > a$, $d > 0$ and $\mathcal{L}[f](c+nd) = \mathcal{L}[g](c+nd)$ for all integers $n \ge 0$, then $f(t) = g(t)$ at all points $t \in \mathbb{R}$ where f and g are both continuous. Lerch's theorem is an improvement over Theorem 48.24, because it only requires equality of the Laplace transforms of f and g on a relatively "small" subset of $\mathbb{R}$ (the set of all points of the form $c+nd$), whereas in Theorem 48.24 equality was required for all $s > c$. ★

Finally, the answer to the third question is also positive, but we will not be able to go into any details because this would require that we consider Laplace transforms defined on $\mathbb{C}$ and use certain fundamental results of Fourier analysis that are beyond the scope of this text. Readers with an interest in this subject are referred to [Köl].

In absence of an easily-accessible inversion formula, we will have to determine inverse transforms in the same way as above where we found the inverse transform of $F(s) = s/(s^2+1)$ simply by using the known fact that $\mathcal{L}[\cos(t)](s) = s/(s^2+1)$. The uniqueness statement of Theorem 48.24 guarantees that $\cos(t)$ is the only continuous function defined on $[0,\infty)$ whose Laplace transform is equal to F.

48.25 Example. We wish to find the inverse transform of

$$F(s) := \frac{2s}{s^2+2s+5}.$$

Completing the square in the denominator yields

$$F(s) = \frac{2s}{(s+1)^2+4} = \frac{2(s+1)}{(s+1)^2+4} - \frac{2}{(s+1)^2+4}.$$

Setting $f(t) := \cos(2t)$ and $g(t) := \sin(2t)$, we may apply (48.2) and (48.9) to infer that $\mathcal{L}[f](s) = s/(s^2+4)$ and $\mathcal{L}[g](s) = 2/(s^2+4)$. Thus, Theorem 48.8a, b implies that the Laplace transform of the function $h(t) := 2e^{-t}f(t) - e^{-t}g(t)$ is

$$\mathcal{L}[h](s) = 2\mathcal{L}[f](s+1) - \mathcal{L}[g](s+1) = \frac{2(s+1)}{(s+1)^2+4} - \frac{2}{(s+1)^2+4}$$

for all $s > -1$. Hence

$$\mathcal{L}^{-1}[F](t) = h(t) = e^{-t}(2\cos(2t) - \sin(2t)).$$

48.26 Example. Let us determine the inverse transform of

$$F(s) := \frac{s}{s^2 - 3s + 2}.$$

Using partial fractions decomposition (see Chapter 33), we obtain

$$F(s) = \frac{2}{s-2} - \frac{1}{s-1}.$$

For $f(t) := e^{2t}$ and $g(t) := e^t$, equation (48.7) implies that $\mathcal{L}[f](s) = 1/(s-2)$ and $\mathcal{L}[g](s) = 1/(s-1)$. Therefore, Theorem 48.8a applied to the function $h(t) := 2f(t) - g(t)$ yields

$$\mathcal{L}[h](s) = \frac{2}{s-2} - \frac{1}{s-1}$$

for all $s > 2$. Hence

$$\mathcal{L}^{-1}[F](t) = 2e^{2t} - e^t.$$

48.27 Example. We wish to determine the inverse transform of

$$F(s) := \frac{e^{-2s}}{s^2}.$$

According to (48.10), we have $\mathcal{L}[t](s) = 1/s^2$, and terefore, Theorem 48.8e implies that

$$\mathcal{L}^{-1}[F](t) = \begin{cases} 0 & \text{if } 0 \le t < 2, \\ t-2 & \text{if } t \ge 2. \end{cases}$$

48.28 Exercise. Find the inverse transforms of $F(s) := \dfrac{e^{-2s}}{s^2 + 2s + 1}$ and $G(s) := \dfrac{s}{s^2 - 8s - 9}$.

A Table of Transforms

Below you are given a list of frequently occurring Laplace transforms. Some of these transforms have been calculated in (48.2), (48.7), (48.8), (48.9), (48.10), (48.12), and Exercise 48.13. The verification of the remaining ones is left as an exercise.

$f(t)$	$\mathcal{L}[f](s)$	$f(t)$	$\mathcal{L}[f](s)$
1	$\frac{1}{s}$	$\sinh(at)$	$\frac{a}{s^2 - a^2}$
t^α	$\frac{\Gamma(\alpha+1)}{s^{\alpha+1}}$	$\cosh(at)$	$\frac{s}{s^2 - a^2}$
e^{at}	$\frac{1}{s-a}$	$t\sin(at)$	$\frac{2as}{(s^2+a^2)^2}$
$t^n e^{at}$	$\frac{n!}{(s-a)^{n+1}}$	$t\cos(at)$	$\frac{s^2-a^2}{(s^2+a^2)^2}$
$\sin(at)$	$\frac{a}{s^2+a^2}$	$t\sinh(at)$	$\frac{2as}{(s^2-a^2)^2}$
$\cos(at)$	$\frac{s}{s^2+a^2}$	$t\cosh(at)$	$\frac{s^2+a^2}{(s^2-a^2)^2}$

Additional Exercises

48.29. Explain why for any functions f and g of exponential order the product fg is of exponential order as well.

48.30. Show that $f(t) := \sin(e^{t^2})$ is of exponential order, but that f' is not.

48.31. Find the Laplace transform of the periodic function shown in Figure 48.5.

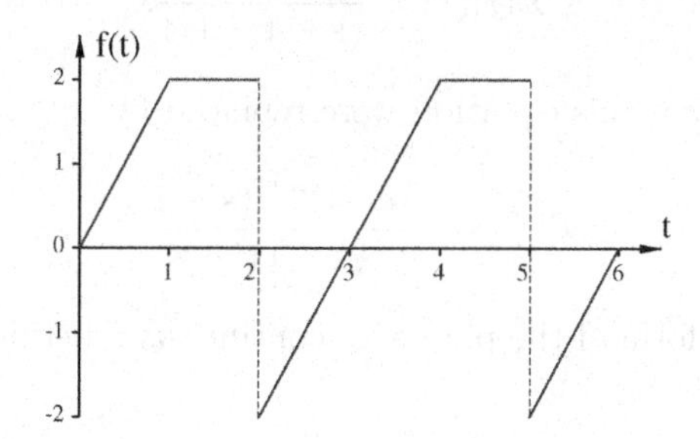

Figure 48.5: graph of a periodic function.

48.32. Use Definition 48.1 to find the Laplace transform of the function

$$f(t) = \begin{cases} 1 & \text{if } 0 \leq t \leq 1, \\ 2 & \text{if } 1 < t \leq 3, \\ 3 & \text{if } 3 < t \leq 4, \\ 0 & \text{if } t > 4. \end{cases}$$

48.33. Use Theorem 48.8 to find the Laplace transforms of the following functions:

a) $f(t) = t^3 \sin(2t)$ **b)** $f(t) = t^2 e^{-3t} \cos(2t)$ **c)** $f(t) = \int_0^t \tau^2 e^{-3\tau} \cos(2\tau)\, d\tau$

d) $f(t) = e^{2t}\sqrt{t}$ **e)** $f(t) = \begin{cases} 0 & \text{if } 0 \leq t < 2, \\ e^t \sin(t-2) & \text{if } t \geq 2. \end{cases}$ **f)** $f(t) = \begin{cases} 0 & \text{if } 0 \leq t < 2, \\ 1 & \text{if } t \geq 2. \end{cases}$

48.34. Let $c > 0$ and assume that for some $a \in \mathbb{R}$ we are given a function $f \in \mathcal{E}_a$. Show that the function $g(t) := f(ct)$ is in $\mathcal{E}_{ca}$ and that $\mathcal{L}[g](s) = \mathcal{L}[f](s/c)/c$ for all $s > ca$.

48.35. Find the Laplace inverse transform of each of the functions given below or explain why the inverse transform does not exist.

a) $F(s) = \dfrac{1}{s^2 + 2s + 1}$ **b)** $F(s) = \dfrac{1}{s^2 - 4s - 5}$ **c)** $F(s) = \dfrac{e^s}{s^4}$

d) $F(s) = \dfrac{s-1}{s^2 - 2s + 10}$ **e)** $F(s) = \dfrac{1}{s(s^2+1)}$ **f)** $F(s) = \dfrac{s^2 - 1}{s^2 + 4}$

g) $F(s) = \sin(s)$ **h)** $F(s) = \dfrac{e^s(s-1)}{s(s^2 - 2s + 2)}$ **i)** $F(s) = \dfrac{e^{-2s}(s-1)}{s(s^2 - 2s + 2)}$

48.36. ★ Find a *nonconstant* function $f : \mathbb{R} \to \mathbb{R}$ that satisfies the following properties:

a) There exists a value $T > 0$ such that $f(t+T) = f(t)$ for all $t \in \mathbb{R}$.

b) For every $T > 0$ that satisfies a) there exists a value $S \in (0, T)$ that satisfies a) as well.

48.37. Let $a \in \mathbb{R}$ and $f \in \mathcal{E}_a$. Use Theorem 48.8 to show that

$$\frac{\mathcal{L}[f](s)}{s-a} = \mathcal{L}\left[e^{at} \int_0^t e^{-a\tau} f(\tau)\, d\tau\right](s) \text{ for all } s > a.^*$$

*I am indebted for this result to my former student David Gutierrez.

48.38. Use Exercise 48.37 to find the Laplace inverse transforms of the following functions:

a) $F(s) = \dfrac{1}{(s-2)(s^2+1)}$ **b)** $F(s) = \dfrac{2s+1}{(s-3)(s^2-2s+5)}$ **c)** $F(s) = \dfrac{e^{-3s}}{(s-1)(s^2-2s+5)}$

48.39. Find a function f such that

$$\mathcal{L}[f](s) = \frac{e^{-3s}(s-1)}{(s-1)^2+4}.$$

How would your answer change if this equation were replaced by

$$\mathcal{L}[f](s) = \frac{e^{-3(s-1)}(s-1)}{(s-1)^2+4}?$$

48.40. Find the Laplace transform of the piecewise continuous function f shown in Figure 48.6. Note: $f(t) = 0$ for all $t \in (4, \infty)$.

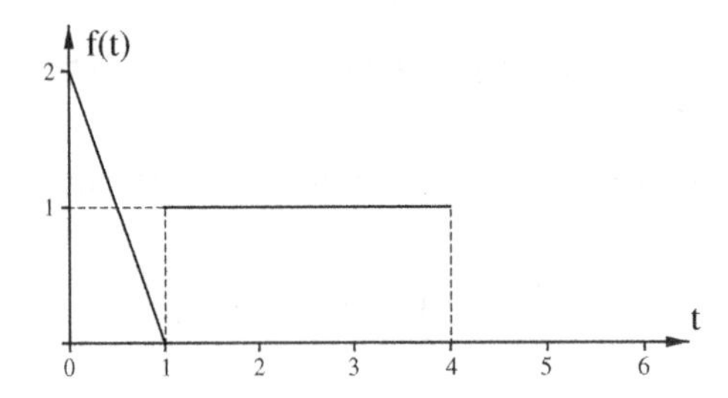

Figure 48.6: graph of f.

48.41. Explain why there cannot exist a differentiable function f of exponential order such that f' is of exponential order as well and $\mathcal{L}[f](s) = \sin(s)/s$.

48.42. Which of the following functions is/are periodic according to Definition 48.16?

a) $f : \mathbb{R} \to \mathbb{R}$, $f(t) := \sin(t)$,

b) $f : (-\infty, 0) \to \mathbb{R}$, $f(t) := \sin(t)$,

c) $f : (0, \infty) \to \mathbb{R}$, $f(t) := \cos(t)$,

d) $f : \mathbb{R} \to \mathbb{R}$, $f(t) := 1$.

Chapter 49

Applications of the Laplace Transform to Differential Equations

Solving Initial Value Problems

In Chapters 44 and 46 we saw how initial value problems corresponding to first- and second-order linear differential equations can be solved in a two step process by first finding a general solution and then determining the corresponding coefficients from the given initial conditions. In the present chapter we will study an alternative method that utilizes the computational properties of the Laplace transform and allows us to elegantly merge these two steps into one. To illustrate the central ideas, we consider a simple example:

49.1 Example. We wish to find the solution of the initial value problem

$$y'(t) - 2y(t) = e^t, \; y(0) = 2. \tag{49.1}$$

In applying the Laplace transform to the left-hand side of this equation and using Theorem 48.8a, f, we obtain

$$\mathcal{L}[y' - 2y](s) = \mathcal{L}[y'](s) - 2\mathcal{L}[y](s) = s\mathcal{L}[y](s) - y(0) - 2\mathcal{L}[y](s).$$

Substituting for $y(0)$ the value given in (49.1) yields

$$\mathcal{L}[y' - 2y](s) = (s-2)\mathcal{L}[y](s) - 2.$$

Furthermore, since $\mathcal{L}[e^t](s) = 1/(s-1)$ (see the table of transforms in Chapter 48), we may infer that

$$(s-2)\mathcal{L}[y](s) - 2 = \mathcal{L}[y' - 2y](s) = \mathcal{L}[e^t](s) = \frac{1}{s-1},$$

or equivalently,

$$\mathcal{L}[y](s) = \frac{1}{(s-1)(s-2)} + \frac{2}{s-2}.$$

This equation suggests that y can be determined as the inverse transform of the function $F(s) := 1/((s-1)(s-2)) + 2/(s-2)$, i.e.,

$$y(t) = \mathcal{L}^{-1}[F](t).$$

Using partial fractions decomposition as explained in Chapter 33 (and the next section), it is easy to see that

$$F(s) = \frac{3}{s-2} - \frac{1}{s-1},$$

and the inverse transform of this function is $3e^{2t} - e^t$. Thus, the solution of the initial value problem (49.1) is

$$y(t) = 3e^{2t} - e^t.$$

Remark. The crucial step in solving the initial value problem (47.1) was the evaluation of the Laplace transform of $y' - 2y$ via Theorem 48.8f. In particular, this step allowed us to make use of the given initial value, because according to Theorem 48.8f, we have $\mathcal{L}[y'](s) = s\mathcal{L}[y](s) - y(0) = s\mathcal{L}[y](s) - 2$. We notice, though, that this calculation requires the initial value to be given at $t = 0$! Consequently, *we will apply the Laplace transform to solve initial value problems only if the initial values are given at* $t = 0$ (which, however, is not a significant restriction).

It also needs to be mentioned that in Example 49.1 we implicitly assumed the Laplace transforms of y and y' to be well defined. It can be shown that this assumption is indeed justified as long as the nonhomogeneous term $f(t)$ (which in Example 49.1 was e^t) is continuous and of exponential order (see [Ra]). If it should ever happen that we encounter a nonhomogeneous term $f(t)$ that is not of exponential order, then the Laplace transform method cannot be applied, and we have to use instead the methods developed in Chapter 46.

49.2 Exercise. Solve the initial value problem $y'(t) + 3y(t) = 1$, $y(0) = 3$.

49.3 Example. Let us now consider an initial value problem involving a second-order linear differential equation:

$$\begin{aligned} y''(t) + 4y(t) &= 5e^{-t} \\ y(0) = 2,\ y'(0) &= 3. \end{aligned} \tag{49.2}$$

Using again Theorem 48.8a, f and the given initial conditions, we obtain

$$\begin{aligned} \mathcal{L}[y'' + 4y](s) &= \mathcal{L}[y''](s) + 4\mathcal{L}[y](s) = s^2\mathcal{L}[y](s) - sy(0) - y'(0) + 4\mathcal{L}[y](s) \\ &= (s^2+4)\mathcal{L}[y](s) - 2s - 3 \end{aligned}$$

and therefore,

$$(s^2+4)\mathcal{L}[y](s) - 2s - 3 = \mathcal{L}[5e^{-t}](s) = \frac{5}{s+1}.$$

Solving for $\mathcal{L}[y](s)$ yields

$$\mathcal{L}[y](s) = \frac{5}{(s+1)(s^2+4)} + \frac{2s+3}{s^2+4}. \tag{49.3}$$

To determine y from this equation, we need to find the inverse transform of the function on the right-hand side. This, however, requires that we find the partial fractions decomposition of a rational function whose denominator is a polynomial of degree greater than two, because the denominator $(s+1)(s^2+4)$ is of degree 3. Before we proceed to solve (49.3) for y, it is therefore appropriate to briefly discuss the general theory of partial fractions decomposition.

Partial Fractions Decomposition

Let us recall that, according to Theorem 45.9, every complex polynomial $p(z)$ can be written in the form

$$p(z) = a(z-z_1)^{m_1}\cdots(z-z_k)^{m_k}$$

where $z_1, \ldots, z_k$ are the distinct roots of $p(z)$. Furthermore, according to Theorem 45.12, every *real* polynomial $q(x)$ can be written in the form

$$\begin{aligned} q(x) &= b(x-d_1)^{p_1}\cdots(x-d_j)^{p_j}(x-z_1)^{q_1}(x-\bar{z}_1)^{q_1}\cdots(x-z_l)^{q_l}(x-\bar{z}_l)^{q_l} \\ &= b(x-d_1)^{p_1}\cdots(x-d_j)^{p_j}(x^2+b_1x+c_1)^{q_1}\cdots(x^2+b_lx+c_l)^{q_l}, \end{aligned}$$

where $d_1, \ldots, d_j$ are the distinct real roots of $q(x)$, $z_1, \bar{z}_1, \ldots, z_l, \bar{z}_l$ the distinct complex roots, and the coefficients b_k and c_k are respectively equal to $2\mathcal{R}e(z_k)$ and $|z_k|^2$ for all $k \in \{1, \ldots, l\}$. The above factorizations of real and complex polynomials are crucial for the formulation of the following theorem concerning the *partial fractions decomposition* of rational functions:

49.4 Theorem. a) *If $p(z)$ and $r(z)$ are complex polynomials such that the degree of $r(z)$ is less than the degree of $p(z)$ and*

$$p(z) = a(z - z_1)^{m_1} \cdots (z - z_k)^{m_k} \quad \textit{(as explained above)}$$

then there are complex numbers $A_{\alpha,\beta}$ such that

$$\begin{aligned} \frac{r(z)}{p(z)} &= \frac{A_{1,1}}{z - z_1} + \frac{A_{1,2}}{(z - z_1)^2} + \cdots + \frac{A_{1,m_1}}{(z - z_1)^{m_1}} \\ &+ \frac{A_{2,1}}{z - z_2} + \frac{A_{2,2}}{(z - z_2)^2} + \cdots + \frac{A_{2,m_2}}{(z - z_2)^{m_2}} \\ &\vdots \\ &+ \frac{A_{k,1}}{z - z_k} + \frac{A_{k,2}}{(z - z_k)^2} + \cdots + \frac{A_{k,m_k}}{(z - z_k)^{m_k}}. \end{aligned}$$

b) *If $q(x)$ and $r(x)$ are real polynomials such that the degree of $r(x)$ is less than the degree of $q(x)$ and*

$$\begin{aligned} q(x) &= b(x - d_1)^{p_1} \cdots (x - d_j)^{p_j} (x - z_1)^{q_1} (x - \bar{z}_1)^{q_1} \cdots (x - z_l)^{q_l} (x - \bar{z}_l)^{q_l} \\ &= b(x - d_1)^{p_1} \cdots (x - d_j)^{p_j} (x^2 + b_1 x + c_1)^{q_1} \cdots (x^2 + b_l x + c_l)^{q_l}, \end{aligned}$$

then there are real numbers $B_{\alpha,\beta}$, $C_{\alpha,\beta}$, and $D_{\alpha,\beta}$ such that

$$\begin{aligned} \frac{r(x)}{q(x)} &= \frac{D_{1,1}}{x - d_1} + \frac{D_{1,2}}{(x - d_1)^2} + \cdots + \frac{D_{1,p_1}}{(x - d_1)^{p_1}} \\ &+ \frac{D_{2,1}}{x - d_2} + \frac{D_{2,2}}{(x - d_2)^2} + \cdots + \frac{D_{2,p_2}}{(x - d_2)^{p_2}} \\ &\vdots \\ &+ \frac{D_{j,1}}{x - d_j} + \frac{D_{j,2}}{(x - d_j)^2} + \cdots + \frac{D_{j,p_j}}{(x - d_j)^{p_j}} \\ &+ \frac{B_{1,1}x + C_{1,1}}{x^2 + b_1 x + c_1} + \frac{B_{1,2}x + C_{1,2}}{(x^2 + b_1 x + c_1)^2} + \cdots + \frac{B_{1,q_1}x + C_{1,q_1}}{(x^2 + b_1 x + c_1)^{q_1}} \\ &+ \frac{B_{2,1}x + C_{2,1}}{x^2 + b_2 x + c_2} + \frac{B_{2,2}x + C_{2,2}}{(x^2 + b_2 x + c_2)^2} + \cdots + \frac{B_{2,q_2}x + C_{2,q_2}}{(x^2 + b_2 x + c_2)^{q_2}} \\ &\vdots \\ &+ \frac{B_{l,1}x + C_{l,1}}{x^2 + b_l x + c_l} + \frac{B_{l,2}x + C_{l,2}}{(x^2 + b_l x + c_l)^2} + \cdots + \frac{B_{l,q_l}x + C_{l,q_l}}{(x^2 + b_l x + c_l)^{q_l}}. \end{aligned}$$

Instead of attempting to give a general proof of this theorem (which is not difficult but tedious), we will consider some examples:

49.5 Example. We wish to find the real partial fractions decomposition (see Theorem 49.4b) of the rational function

$$f(x) := \frac{x^3 - x^2 + 1}{(x^2 - 2x + 2)^2 (x - 1)^2}.$$

Since the degree of the numerator is evidently less than the degree of the denominator and since the quadratic factor $x^2 - 2x + 2$ does not have any real roots, we may apply Theorem 49.4b to conclude that there are real numbers $B_{1,1}$, $B_{1,2}$, $C_{1,1}$, $C_{1,2}$, $D_{1,1}$, and $D_{1,2}$ such that

$$f(x) = \frac{B_{1,1}x + C_{1,1}}{x^2 - 2x + 2} + \frac{B_{1,2}x + C_{1,2}}{(x^2 - 2x + 2)^2} + \frac{D_{1,1}}{x - 1} + \frac{D_{1,2}}{(x - 1)^2}.$$

Multiplying both sides of this equation by $q(x) = (x^2 - 2x + 2)^2(x-1)^2$, we obtain

$$\begin{aligned} x^3 - x^2 + 1 = (B_{1,1}x + C_{1,1})(x^2 - 2x + 2)(x-1)^2 + (B_{1,2}x + C_{1,2})(x-1)^2 \\ + D_{1,1}(x^2 - 2x + 2)^2(x-1) + D_{1,2}(x^2 - 2x + 2)^2. \end{aligned} \tag{49.4}$$

To determine the coefficients $B_{\alpha,\beta}$, $C_{\alpha,\beta}$, and $D_{\alpha,\beta}$ we choose specific values for x (see also Chapter 33): setting x equal to 1, we see that $1 = D_{1,2}$. Filling this value for $D_{1,2}$ back into (49.4) yields

$$\begin{aligned} x^3 - x^2 + 1 &= (B_{1,1}x + C_{1,1})(x^2 - 2x + 2)(x-1)^2 + (B_{1,2}x + C_{1,2})(x-1)^2 \\ &\quad + D_{1,1}(x^2 - 2x + 2)^2(x-1) + (x^2 - 2x + 2)^2 \\ &= (B_{1,1}x + C_{1,1})(x^2 - 2x + 2)(x-1)^2 + (B_{1,2}x + C_{1,2})(x-1)^2 \\ &\quad + D_{1,1}(x^2 - 2x + 2)^2(x-1) + x^4 - 4x^3 + 8x^2 - 8x + 4, \end{aligned}$$

or equivalently,

$$\begin{aligned} -x^4 + 5x^3 - 9x^2 + 8x - 3 = (B_{1,1}x + C_{1,1})(x^2 - 2x + 2)(x-1)^2 + (B_{1,2}x + C_{1,2})(x-1)^2 \\ + D_{1,1}(x^2 - 2x + 2)^2(x-1). \end{aligned}$$

Since the right-hand side of this equation is divisible by $x - 1$, the left-hand side is necessarily divisible by $x - 1$ as well. Using polynomial division (i.e., long division), we find that

$$\frac{-x^4 + 5x^3 - 9x^2 + 8x - 3}{x - 1} = -x^3 + 4x^2 - 5x + 3$$

and therefore,

$$\begin{aligned} -x^3 + 4x^2 - 5x + 3 = (B_{1,1}x + C_{1,1})(x^2 - 2x + 2)(x-1) \\ + (B_{1,2}x + C_{1,2})(x-1) + D_{1,1}(x^2 - 2x + 2)^2. \end{aligned} \tag{49.5}$$

Setting again x equal to 1, we obtain $1 = D_{1,1}$. Filling this value back into (49.5), we may infer that

$$\begin{aligned} -x^3 + 4x^2 - 5x + 3 = (B_{1,1}x + C_{1,1})(x^2 - 2x + 2)(x-1) + (B_{1,2}x + C_{1,2})(x-1) \\ + x^4 - 4x^3 + 8x^2 - 8x + 4, \end{aligned}$$

or equivalently,

$$-x^4 + 3x^3 - 4x^2 + 3x - 1 = (B_{1,1}x + C_{1,1})(x^2 - 2x + 2)(x-1) + (B_{1,2}x + C_{1,2})(x-1).$$

Again, both sides of this equation are divisible by $x - 1$, and using polynomial division as above yields

$$-x^3 + 2x^2 - 2x + 1 = (B_{1,1}x + C_{1,1})(x^2 - 2x + 2) + (B_{1,2}x + C_{1,2}). \tag{49.6}$$

We continue to choose specific values for x, but this time complex values, because the roots of $x^2 - 2x + 2$ are the complex numbers $1 \pm i$. Substituting these values for x in (49.6), we obtain two equations:

$$\begin{aligned} x = 1 + i: &\quad 1 = B_{1,2}(1 + i) + C_{1,2}, \\ x = 1 - i: &\quad 1 = B_{1,2}(1 - i) + C_{1,2}. \end{aligned}$$

Solving for $B_{1,2}$ and $C_{1,2}$ (or comparing the real and imaginary parts in one of the equations) yields $B_{1,2} = 0$ and $C_{1,2} = 1$. It is of course no coincidence that $B_{1,2}$ and $C_{1,2}$ are real numbers, because this is part of the statement of Theorem 49.4b. Filling the values for $B_{1,2}$ and $C_{1,2}$ back into (49.6), it follows that

$$-x^3 + 2x^2 - 2x + 1 = (B_{1,1}x + C_{1,1})(x^2 - 2x + 2) + 1,$$

or equivalently,

$$-x^3 + 2x^2 - 2x = (B_{1,1}x + C_{1,1})(x^2 - 2x + 2).$$

Dividing both sides by $x^2 - 2x + 2$, we conclude that

$$-x = B_{1,1}x + C_{1,1},$$

and this equation trivially implies that $B_{1,1} = -1$ and $C_{1,1} = 0$. Thus, we arrive at the partial fractions decomposition

$$f(x) = \frac{-x}{x^2 - 2x + 2} + \frac{1}{(x^2 - 2x + 2)^2} + \frac{1}{x - 1} + \frac{1}{(x - 1)^2}.$$

Note: in Chapter 54 in Volume 2 we will rework this example using a different method that involves systems of linear equations. (The result, of course, will be the same.)

49.6 Example. Let us determine the real partial fractions decomposition of

$$f(x) := \frac{x^2}{(x^2 + 4)(x - 2)^3}.$$

According to Theorem 49.4b, there are real numbers $B_{1,1}$, $C_{1,1}$, $D_{1,1}$, $D_{1,2}$, and $D_{1,3}$ such that

$$f(x) = \frac{B_{1,1}x + C_{1,1}}{x^2 + 4} + \frac{D_{1,1}}{x - 2} + \frac{D_{1,2}}{(x - 2)^2} + \frac{D_{1,3}}{(x - 2)^3}.$$

Multiplying this equation by $(x^2 + 4)(x - 2)^3$, we obtain

$$x^2 = (B_{1,1}x + C_{1,1})(x - 2)^3 + D_{1,1}(x^2 + 4)(x - 2)^2 + D_{1,2}(x^2 + 4)(x - 2) + D_{1,3}(x^2 + 4),$$

and setting x equal to 2, we see that $D_{1,3} = 1/2$. Thus,

$$x^2 = (B_{1,1}x + C_{1,1})(x - 2)^3 + D_{1,1}(x^2 + 4)(x - 2)^2 + D_{1,2}(x^2 + 4)(x - 2) + \frac{1}{2}(x^2 + 4),$$

or equivalently,

$$\frac{1}{2}x^2 - 2 = (B_{1,1}x + C_{1,1})(x - 2)^3 + D_{1,1}(x^2 + 4)(x - 2)^2 + D_{1,2}(x^2 + 4)(x - 2).$$

Dividing both sides by $x - 2$ yields

$$\frac{1}{2}x + 1 = (B_{1,1}x + C_{1,1})(x - 2)^2 + D_{1,1}(x^2 + 4)(x - 2) + D_{1,2}(x^2 + 4).$$

Setting again x equal to 2 and repeating the same process, we find that $D_{1,2} = 1/4$ and

$$-\frac{1}{4}x^2 + \frac{1}{2}x = (B_{1,1}x + C_{1,1})(x - 2)^2 + D_{1,1}(x^2 + 4)(x - 2).$$

Hence

$$-\frac{1}{4}x = (B_{1,1}x + C_{1,1})(x - 2) + D_{1,1}(x^2 + 4).$$

Substituting yet again 2 for x yields $D_{1,1} = -1/16$, and therefore

$$\frac{1}{16}(x - 2) = B_{1,1}x + C_{1,1}.$$

Consequently, $B_{1,1} = 1/16$ and $C_{1,1} = -1/8$. Thus, the partial fractions decomposition of $f(x)$ is

$$f(x) = \frac{x/16 - 1/8}{(x^2 + 4)} + \frac{-1/16}{x - 2} + \frac{1/4}{(x - 2)^2} + \frac{1/2}{(x - 2)^3}.$$

49.7 Exercise. Find the real partial fractions decompositions of the following functions:

$$f(x) = \frac{x^2+1}{(x-1)(x-2)(x-3)} \quad \text{and} \quad g(x) = \frac{x^3+2x-1}{(x^2-4x+5)^2(x-1)}.$$

49.8 Example. Now we will work an example of a complex partial fractions decomposition for which the general form is given in Theorem 49.4a. This type of partial fractions decomposition will not be important for our discussion of Laplace transforms, but will be needed in Chapter 60 when we study exponential matrices. So let us consider the complex rational function

$$f(z) := \frac{z^2}{(z-i)^3(z+1)^2}.$$

According to Theorem 49.4a, there are complex numbers $A_{1,1}$, $A_{1,2}$, $A_{1,3}$, $A_{2,1}$, and $A_{2,2}$ such that

$$f(z) = \frac{A_{1,1}}{z-i} + \frac{A_{1,2}}{(z-i)^2} + \frac{A_{1,3}}{(z-i)^3} + \frac{A_{2,1}}{z+1} + \frac{A_{2,2}}{(z+1)^2}.$$

Multiplying both sides of this equation by $(z-i)^3(z+1)^2$, we obtain

$$z^2 = A_{1,1}(z-i)^2(z+1)^2 + A_{1,2}(z-i)(z+1)^2 + A_{1,3}(z+1)^2 + A_{2,1}(z-i)^3(z+1) + A_{2,2}(z-i)^3.$$

Substituting -1 for z, we see that $1 = -A_{2,2}(1+i)^3$, and therefore $A_{2,2} = -(1-i)^3/8 = 1/4 + i/4$. Thus,

$$\begin{aligned} z^2 = {} & A_{1,1}(z-i)^2(z+1)^2 + A_{1,2}(z-i)(z+1)^2 + A_{1,3}(z+1)^2 + A_{2,1}(z-i)^3(z+1) \\ & + \frac{1}{4}(z^3+3z^2-3z-1) + \frac{i}{4}(z^3-3z^2-3z+1), \end{aligned}$$

or equivalently,

$$\begin{aligned} -\frac{1}{4}(z^3-z^2-3z-1) - \frac{i}{4}(z^3-3z^2-3z+1) = {} & A_{1,1}(z-i)^2(z+1)^2 + A_{1,2}(z-i)(z+1)^2 \\ & + A_{1,3}(z+1)^2 + A_{2,1}(z-i)^3(z+1). \end{aligned}$$

Dividing both sides of this equation by $z+1$ yields

$$\begin{aligned} -\frac{1}{4}(z^2-2z-1) - \frac{i}{4}(z^2-4z+1) = {} & A_{1,1}(z-i)^2(z+1) + A_{1,2}(z-i)(z+1) \\ & + A_{1,3}(z+1) + A_{2,1}(z-i)^3. \end{aligned}$$

Again substituting -1 for z, we obtain $-1/2-3i/2 = -A_{2,1}(1+i)^3$, and by implication $A_{2,1} = 1/4 - i/2$. Hence

$$\begin{aligned} -\frac{1}{4}(z^2-2z-1) - \frac{i}{4}(z^2-4z+1) = {} & A_{1,1}(z-i)^2(z+1) + A_{1,2}(z-i)(z+1) + A_{1,3}(z+1) \\ & + \frac{1}{4}(z^3-6z^2-3z+2) - \frac{i}{2}(2z^3+3z^2-6z-1), \end{aligned}$$

or equivalently,

$$-\frac{1}{4}(z^3-5z^2-5z+1) + \frac{i}{2}(z^3+z^2-z-1) = A_{1,1}(z-i)^2(z+1) + A_{1,2}(z-i)(z+1) + A_{1,3}(z+1).$$

Dividing both sides by $z+1$ yields

$$-\frac{1}{4}(z^2-6z+1) + \frac{i}{2}(z^2-1) = A_{1,1}(z-i)^2 + A_{1,2}(z-i) + A_{1,3},$$

and setting z equal to i, we may infer that $i/2 = A_{1,3}$. Consequently,

$$-\frac{1}{4}(z^2 - 6z + 1) + \frac{i}{2}(z^2 - 2) = A_{1,1}(z-i)^2 + A_{1,2}(z-i).$$

In dividing both sides by $z - i$, we obtain

$$-\frac{1}{4}(z-4) + \frac{i}{4}(2z-1) = A_{1,1}(z-i) + A_{1,2}.$$

Substituting again i for z, it follows that $A_{1,2} = 1/2 - i/2$, and therefore,

$$-\frac{1}{4}(z-2) + \frac{i}{4}(2z+1) = A_{1,1}(z-i).$$

In performing another division by $z - i$, it is easy to see that $A_{1,1} = -1/4 + i/2$. Thus, the complex partial fractions decomposition of f is

$$f(z) = \frac{-1/4 + i/2}{z-i} + \frac{1/2 - i/2}{(z-i)^2} + \frac{i/2}{(z-i)^3} + \frac{1/4 - i/2}{z+1} + \frac{1/4 + i/4}{(z+1)^2}.$$

49.9 Exercise. Find the complex partial fractions decomposition of $f(z) = \dfrac{z^2+4}{(z-i)^2(z+i)}$.

Solving Initial Value Problems Continued

Our first objective in this section is to complete the solution of the initial value problem (49.2) in Example 49.3. Using partial fractions decomposition, it is easy to show that equation (49.3) can be written in the form

$$\mathcal{L}[y](s) = \frac{s}{s^2+4} + \frac{4}{s^2+4} + \frac{1}{s+1}. \tag{49.7}$$

49.10 Exercise. Verify this equation.

According to the table of transforms in Chapter 48, the inverse transform of the right-hand side in (49.7) is $\cos(2t) + 2\sin(2t) + e^{-t}$. Therefore, the solution of the initial value problem (49.2) is

$$y(t) = \cos(2t) + 2\sin(2t) + e^{-t}.$$

49.11 Example. Let us determine the solution of the initial value problem

$$\begin{aligned} y''(t) + 2y'(t) + 2y(t) &= \sin(t) \\ y(0) = 1,\ y'(0) &= -2. \end{aligned} \tag{49.8}$$

In applying the Laplace transform to both sides of the differential equation above, Theorem 48.8a, f allows us to infer that

$$s^2\mathcal{L}[y](s) - sy(0) - y'(0) + 2s\mathcal{L}[y](s) - 2y(0) + 2\mathcal{L}[y](s) = (s^2+2s+2)\mathcal{L}[y](s) - s = \frac{1}{s^2+1},$$

or equivalently,

$$\mathcal{L}[y](s) = \frac{s}{s^2+2s+2} + \frac{1}{(s^2+1)(s^2+2s+2)}.$$

Using partial fractions decomposition and completing the square in the denominator, it is easy to see that

$$\mathcal{L}[y](s) = \frac{7s/5 + 3/5}{s^2+2s+2} + \frac{-2s/5 + 1/5}{s^2+1} = \frac{7(s+1)/5}{(s+1)^2+1} - \frac{4/5}{(s+1)^2+1} - \frac{2s/5}{s^2+1} + \frac{1/5}{s^2+1}.$$

Given the table of transforms in Chapter 48, Theorem 48.8a, b implies that the inverse transform of the function on the right-hand side of this equation is $7e^{-t}\cos(t)/5 - 4e^{-t}\sin(t)/5 - 2\cos(t)/5 + \sin(t)/5$. Thus, the solution of the initial value problem (49.8) is

$$y(t) = \frac{7}{5}e^{-t}\cos(t) - \frac{4}{5}e^{-t}\sin(t) - \frac{2}{5}\cos(t) + \frac{1}{5}\sin(t).$$

49.12 Example. As a final example, we consider the initial value problem

$$y''(t) + y(t) = e^t + \sin(t)$$
$$y(0) = 0,\ y'(0) = 0.$$

Applying the Laplace transform yields

$$s^2\mathcal{L}[y](s) + \mathcal{L}[y](s) = \frac{1}{s-1} + \frac{1}{s^2+1},$$

and therefore,

$$\mathcal{L}[y](s) = \frac{1}{(s-1)(s^2+1)} + \frac{1}{(s^2+1)^2}. \tag{49.9}$$

Using partial fractions decomposition, it is easy to verify that

$$\frac{1}{(s-1)(s^2+1)} = \frac{1/2}{s-1} - \frac{s/2+1/2}{s^2+1},$$

and the inverse transform of this function is therefore

$$\frac{1}{2}e^t - \frac{1}{2}\cos(t) - \frac{1}{2}\sin(t).$$

To determine the inverse transform of $1/(s^2+1)^2$ (which is the second term on the right-hand side of (49.9)) is a little more complicated, but given the table of transforms in Chapter 48, the following algebraic manipulation suggests itself:

$$\frac{1}{(s^2+1)^2} = \frac{1}{2s}\cdot\frac{2s}{(s^2+1)^2}.$$

Since

$$\frac{2s}{(s^2+1)^2} = \mathcal{L}[t\sin(t)](s) \quad \text{(see the table of transforms)},$$

we may apply Theorem 48.8c to conclude that

$$\frac{1}{(s^2+1)^2} = \frac{1}{2}\mathcal{L}\left[\int_0^t \tau\sin(\tau)\,d\tau\right](s) = \mathcal{L}\left[\frac{1}{2}\sin(t) - \frac{1}{2}t\cos(t)\right](s).$$

Thus, the solution of the initial value problem is

$$y(t) = \frac{1}{2}e^t - \frac{1}{2}\cos(t) - \frac{1}{2}\sin(t) + \frac{1}{2}\sin(t) - \frac{1}{2}t\cos(t) = \frac{1}{2}e^t - \frac{1}{2}(1+t)\cos(t).$$

49.13 Exercise. Use the Laplace transform to solve the following initial value problems:

a) $y''(t) - y(t) = \cos(2t)$, $y(0) = 1$, $y'(0) = 1$

b) $y''(t) - 4y'(t) + 4y(t) = e^t$, $y(0) = 0$, $y'(0) = 0$

c) $y''(t) - 4y'(t) + 5y(t) = t$, $y(0) = 1$, $y'(0) = -1$

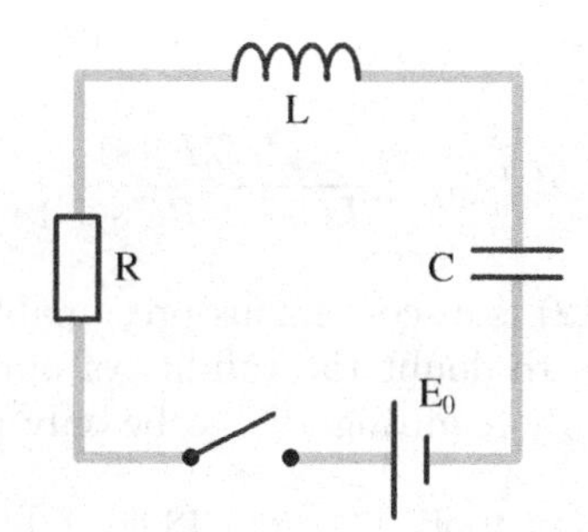

Figure 49.1: R-C-L circuit with a switch and a battery.

★ R-C-L Circuits with Discontinuous External Voltage Sources

49.14 Example. Let us consider an R-C-L circuit with a switch and a battery generating a constant difference in electric potential E_0 as shown in Figure 49.1. We assume that the switch is initially open and the circuit is completely passive (i.e., there is no charge on the capacitor and no current flowing through the inductor). Then at some point in time $t_0 \geq 0$ the switch is closed and a current begins to flow. The closing of the switch effectively models a *discontinuous* external voltage source generating a difference in electric potential $E(t)$ that satisfies the following conditions: $E(t) = 0$ for $t < t_0$ and $E(t) = E_0$ for $t \geq t_0$ (see Figure 49.2). In Chapter 46, p.371 we saw that in general an R-C-L circuit is

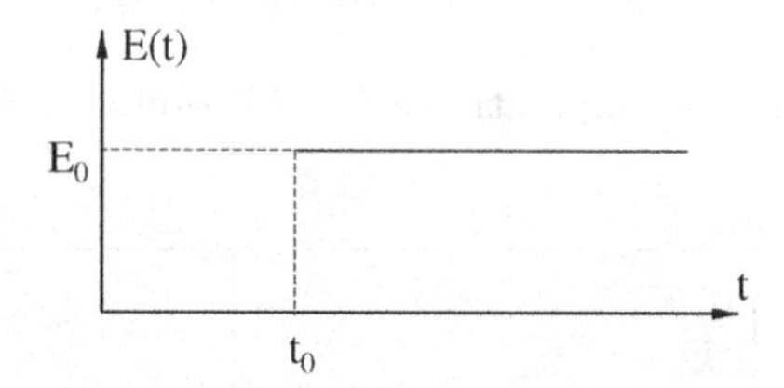

Figure 49.2: graph of the "switch on"-voltage $E(t)$.

described by the following equation:

$$RI(t) + \frac{Q(t)}{C} + LI'(t) = E(t). \tag{49.10}$$

Since the circuit is supposed to be passive for $t < t_0$, the initial conditions are

$$I(0) = 0 \quad \text{and} \quad Q(0) = 0. \tag{49.11}$$

To transform (49.10) into the equation $I''(t)+RI'(t)/L+I(t)/LC = E'(t)/L$ via differentiation, as we did in Chapter 46, is unfortunately not possible because the electric potential $E(t)$ is not differentiable. (The derivative of E is zero everywhere except at t_0, where it does not exist.) We will see, though, that the Laplace transform is ideally suited for handling problems of this sort.

Since $Q(0) = 0$ (by (49.11)) and $I(t) = Q'(t)$ (by (44.1)), we may infer that

$$Q(t) = Q(0) + \int_0^t I(\tau)\,d\tau = \int_0^t I(\tau)\,d\tau.$$

Therefore, (49.10) implies that

$$I'(t) + \frac{R}{L}I(t) + \frac{1}{LC}\int_0^t I(\tau)\,d\tau = \frac{1}{L}E(t). \tag{49.12}$$

An equation of this form is said to be an *integrodifferential equation.* In applying the Laplace transform to both sides of (49.12) and using Theorem 48.8a, c, f, it follows that

$$s\mathcal{L}[I](s) - I(0) + \frac{R}{L}\mathcal{L}[I](s) + \frac{1}{LCs}\mathcal{L}[I] = \frac{1}{L}\mathcal{L}[E](s).$$

Solving for $\mathcal{L}[I](s)$ and substituting 0 for $I(0) = 0$ (see (49.11)) yields

$$\mathcal{L}[I](s) = \frac{Cs\mathcal{L}[E](s)}{LCs^2 + RCs + 1}. \tag{49.13}$$

Remark. The fact that $E(t)$ in (49.12) is discontinuous presumably implies that $I'(t)$ is discontinuous as well. However, there is no reason to doubt the validity of our calculation, because Theorem 48.8f takes such a possibility into account by requiring $f^{(n)}$ to be only piecewise continuous.

Given the definition of $E(t)$, we may apply Theorem 48.8e and the result in Exercise 48.6 to conclude that

$$\mathcal{L}[E](s) = \frac{E_0 e^{-t_0 s}}{s}.$$

Therefore, (49.13) implies that

$$\mathcal{L}[I](s) = \frac{CE_0 e^{-t_0 s}}{LCs^2 + RCs + 1}.$$

To consider a concrete example, we assume that $R = 400\,\Omega$, $L = 0.2\,H$, and $C = 10^{-6}\,F$. Given these values, a simple algebraic manipulation shows that

$$\mathcal{L}[I](s) = \frac{5E_0 e^{-t_0 s}}{(s+1000)^2 + 2000^2},$$

and using the table of transforms in conjunction with Theorem 48.8a, b, e, the corresponding inverse transform is easily seen to be

$$\boxed{I(t) = \begin{cases} 0 & \text{if } t < t_0, \\ \dfrac{E_0}{400} e^{-1000(t-t_0)} \sin(2000(t-t_0)) & \text{if } t \geq t_0. \end{cases}}$$

Since $\lim_{t\to\infty} e^{-1000(t-t_0)} = 0$, the steady state solution is identically equal to zero. That, of course, is not surprising, because for $t \geq t_0$ the battery generates the constant difference in electric potential E_0, and what we should therefore expect to happen is simply that the current dies down once the capacitor is fully charged.

49.15 Example. Let us now consider an R-C-L circuit with an external voltage source generating a saw-tooth voltage $E(t)$ with amplitude E_0. More precisely, in denoting by g_α the saw-tooth function as defined in Chapter 48, p.394, we assume that

$$E(t) = E_0 g_\alpha(t)$$

for some $\alpha > 0$ and all $t \in [0, \infty)$. Furthermore, we assume that for $t < 0$ the circuit is completely passive, so that again the initial conditions in (49.11) apply. At time $t = 0$, the saw-tooth voltage source is switched on and an electric current starts to flow. As above, the circuit is described by the integrodifferential equation (49.12), and the Laplace transform of $I(t)$ is again given by equation (49.13). Since $E(t) = E_0 g_\alpha(t)$, we may use the result in (48.16) to infer that

$$\mathcal{L}[I](s) = \frac{CE_0(1+\alpha s)}{\alpha s(LCs^2 + RCs + 1)} - \frac{CE_0}{(LCs^2 + RCs + 1)(1 - e^{-\alpha s})}.$$

Substituting the same values for R, C, and L as in the preceding example yields

$$\mathcal{L}[I](s) = \frac{5E_0(1+\alpha s)}{\alpha s((s+1000)^2 + 2000^2)} - \frac{5E_0}{((s+1000)^2 + 2000^2)(1 - e^{-\alpha s})}. \tag{49.14}$$

In order to determine $I(t)$ we need to find the inverse transform of the expression of the right-hand side of this equation. Setting $b := 1000$ and $c := 2000$, and introducing the definitions

$$F(s) := \frac{5E_0(1+\alpha s)}{\alpha s((s+b)^2+c^2)},$$
$$G(s) := \frac{5E_0}{((s+b)^2+c^2)(1-e^{-\alpha s})},$$

we may rewrite (49.14) in the following equivalent form:

$$I(t) = \mathcal{L}^{-1}[F](t) - \mathcal{L}^{-1}[G](t). \tag{49.15}$$

Using partial fractions decomposition, it is easy to see that

$$F(s) = \frac{5E_0}{\alpha(b^2+c^2)}\left(\frac{1}{s} - \frac{s+b}{(s+b)^2+c^2} - \frac{b-\alpha(b^2+c^2)}{(s+b)^2+c^2}\right).$$

Consequently, Theorem 48.8a, b and the table of transforms on p.396 imply that

$$\mathcal{L}^{-1}[F](t) = \frac{5E_0}{\alpha(b^2+c^2)}\left(1 - e^{-bt}\cos(ct) - \frac{b-\alpha(b^2+c^2)}{c}e^{-bt}\sin(ct)\right). \tag{49.16}$$

To find the inverse transform of G from the defining equation for G above, we would certainly like to use the fact that the inverse transform of $c/((s+b)^2+c^2)$ is equal to $e^{-bt}\sin(ct)$, but the term $1-e^{-\alpha s}$ in the denominator needs to be accounted for as well, and for this reason we resort to a representation via infinite series: if s is strictly greater than zero, then $|e^{-\alpha s}| = e^{-\alpha s} < 1$, and therefore

$$\frac{1}{1-e^{-\alpha s}} = \sum_{k=0}^{\infty}(e^{-\alpha s})^k = \sum_{k=0}^{\infty} e^{-\alpha s k} \quad \text{(by Theorem 41.16).} \tag{49.17}$$

(Note: we will use an argument similar to the one in our derivation of Theorem 48.18, but in the reverse direction.) Given the result in (49.17), the defining equation for G implies that

$$G(s) = 5E_0 \sum_{k=0}^{\infty} \frac{e^{-\alpha s k}}{(s+b)^2+c^2}.$$

According to Theorem 48.8a, the transform of a sum is equal to the sum of the transforms and, consequently, the inverse transform of a sum is equal to the sum of the inverse transforms. Using more advanced analytical techniques than this text permits us to discuss, it can be shown (see [Kö1]) that under certain assumptions the same conclusion holds also for infinite sums. In other words, the inverse transform of an infinite series of functions is equal to the infinite series of the individual inverse transforms. Thus, in setting

$$G_k(s) := \frac{ce^{-\alpha s k}}{(s+b)^2+c^2}$$

we may infer that

$$\mathcal{L}^{-1}[G](t) = \frac{5E_0}{c}\sum_{k=0}^{\infty}\mathcal{L}^{-1}[G_k](t). \tag{49.18}$$

Using Theorem 48.8e and the table of transforms on p.396, it follows that

$$\mathcal{L}^{-1}[G_k](t) = \begin{cases} 0 & \text{if } 0 \le t < \alpha k, \\ e^{-b(t-\alpha k)}\sin(c(t-\alpha k)) & \text{if } t \ge \alpha k. \end{cases} \tag{49.19}$$

This result shows that $G_k(t) = 0$ whenever $k > t/\alpha$. Consequently, for any given t there are only finitely many values of k for which $\mathcal{L}^{-1}[G_k](t) \neq 0$, namely those that are less than or equal to t/α. In other words, the maximal value of k satisfying the condition $G_k(t) \neq 0$ is the greatest integer less than or equal to t/α. The standard notation for the greatest integer less than or equal to a given real number x is $[x]$ (e.g., $[4.1] = 4$, $[0.9] = 0$, $[-1.1] = -2$, etc.). Therefore, in combining (49.18) with (49.19), we obtain

$$\mathcal{L}^{-1}[G](t) = \frac{5E_0}{c} \sum_{k=0}^{[t/\alpha]} \mathcal{L}^{-1}[G_k](t) = \frac{5E_0}{c} \sum_{k=0}^{[t/\alpha]} e^{-b(t-\alpha k)} \sin(c(t-\alpha k)).$$

This is already a formula for the inverse transform of G, but it would be nice if we could actually evaluate the sum in order to obtain a simpler expression. For this purpose we first use Euler's formula to infer that

$$e^{-b(t-\alpha k)} \sin(c(t-\alpha k)) = \mathcal{Im}(e^{-b(t-\alpha k)} e^{ic(t-\alpha k)}) = \mathcal{Im}(e^{-(b-ic)t} e^{\alpha(b-ic)k}).$$

Now we recall that $\sum_{k=0}^{n} q^k = (1-q^{n+1})/(1-q)$ for all $q \neq 1$ (see (41.6)). Since this result is readily seen to be valid also for complex numbers different from one (by the same argument as in Example 41.15), we may set $q := e^{\alpha(b-ic)}$ to conclude that

$$\mathcal{L}^{-1}[G](t) = \frac{5E_0}{c} \mathcal{Im}\left(e^{-(b-ic)t} \sum_{k=0}^{[t/\alpha]} e^{\alpha(b-ic)k} \right) = \frac{5E_0}{c} \mathcal{Im}\left(e^{-(b-ic)t} \frac{1-e^{\alpha(b-ic)([t/\alpha]+1)}}{1-e^{\alpha(b-ic)}} \right).$$

(Note: q is different from 1 because $|q| = |e^{\alpha(b-ic)}| = e^{\alpha b} > 1$.) To determine the imaginary part on the right-hand side is a tedious but straightforward exercise in complex algebra. It turns out that

$$\mathcal{Im}\left(e^{-(b-ic)t} \frac{1-e^{\alpha(b-ic)([t/\alpha]+1)}}{1-e^{\alpha(b-ic)}} \right) = \Phi(t) - \Phi(t - \alpha([t/\alpha]+1))$$

where

$$\Phi(x) := \frac{e^{-bx} \sin(cx) - e^{b(\alpha - x)} \sin(c(\alpha + x))}{1 - 2e^{\alpha b} \cos(\alpha c) + e^{2\alpha b}}.$$

Hence

$$\mathcal{L}^{-1}[G](t) = \frac{5E_0}{c} \left(\Phi(t) - \Phi(t - \alpha([t/\alpha]+1)) \right).$$

Combining this equation with (49.15) and (49.16) finally yields

$$\boxed{I(t) = \frac{5E_0}{\alpha(b^2+c^2)} \left(1 - e^{-bt}\cos(ct) - \frac{b - \alpha(b^2+c^2)}{c} e^{-bt} \sin(ct) \right) - \frac{5E_0}{c} \left(\Phi(t) - \Phi(t - \alpha([t/\alpha]+1)) \right).}$$

Since $\lim_{t\to\infty} e^{-bt} = 0$, the steady state solution is

$$I_s(t) = \frac{5E_0}{\alpha(b^2+c^2)} + \frac{5E_0}{c} \Phi(t - \alpha([t/\alpha]+1)).$$

Figure 49.3 shows the graph of the steady state solution for $\alpha = 0.001\,s$ and $E_0 = 100\,V$.

49.16 Example. * As a final example, we consider an R-C-L circuit with an external voltage source generating a Morse-dot voltage with amplitude E_0, i.e., $E(t) = E_0 f_\alpha(t)$ for some $\alpha > 0$ where f_α is defined as in Chapter 48, p.393. As usual, we assume the circuit to be passive for $t < 0$ so that once again (49.11) and (49.13) are valid. Given these assumptions, equation (48.15) implies that

$$\mathcal{L}[I](s) = \frac{CE_0}{(LCs^2 + RCs + 1)(1 + e^{-\alpha s})}.$$

*Example 49.16 has been adopted with permission from the exposition in [WB], Example 6, p.653.

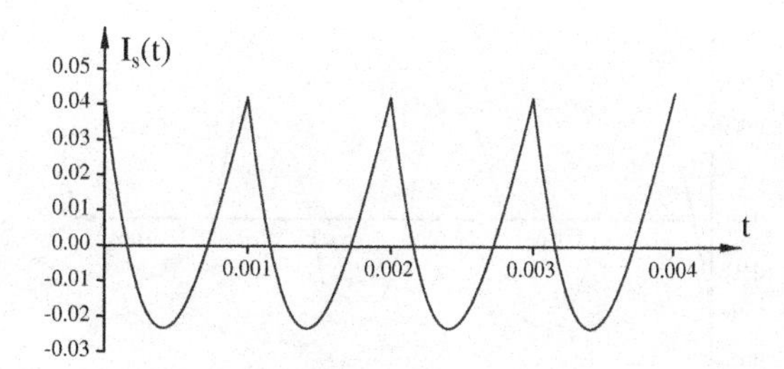

Figure 49.3: steady state solution for a saw-tooth voltage source.

Using the same values for R, L, C, b, and c as above yields

$$\mathcal{L}[I](s) = \frac{5E_0}{((s+b)^2+c^2)(1+e^{-\alpha s})}.$$

In order to find the inverse transform of the term on the right-hand side, we use again Theorem 41.16 to infer that

$$\mathcal{L}[I](s) = 5E_0\sum_{k=0}^{\infty}\frac{(-e^{-\alpha s})^k}{(s+b)^2+c^2} = 5E_0\sum_{k=0}^{\infty}\frac{(-1)^k e^{-\alpha ks}}{(s+b)^2+c^2}.$$

Setting now

$$H_k(s) := \frac{(-1)^k c e^{-\alpha ks}}{(s+b)^2+c^2},$$

it follows that

$$\mathcal{L}^{-1}[H_k](t) = \begin{cases} 0 & \text{if } 0 \le t < \alpha k, \\ e^{-b(t-\alpha k)}\sin(c(t-\alpha k)) & \text{if } t \ge \alpha k, \end{cases}$$

and relying on an argument similar to the one used in Example 49.15, we obtain

$$I(t) = \frac{5E_0}{c}\mathcal{Im}\left(e^{-(b-ic)t}\sum_{k=0}^{[t/\alpha]}(-1)^k e^{\alpha(b-ic)k}\right) = \frac{5E_0}{c}\mathcal{Im}\left(e^{-(b-ic)t}\frac{1-(-1)^{[t/\alpha]+1}e^{\alpha(b-ic)([t/\alpha]+1)}}{1+e^{\alpha(b-ic)}}\right).$$

Furthermore, it is not difficult to show that

$$\mathcal{Im}\left(e^{-(b-ic)t}\frac{1-(-1)^{[t/\alpha]+1}e^{\alpha(b-ic)([t/\alpha]+1)}}{1+e^{\alpha(b-ic)}}\right) = \Psi(t) + (-1)^{[t/\alpha]}\Psi(t-\alpha([t/\alpha]+1))$$

where

$$\Psi(x) := \frac{e^{-bx}\sin(cx) + e^{b(\alpha-x)}\sin(c(\alpha+x))}{1+2e^{\alpha b}\cos(\alpha c)+e^{2\alpha b}}.$$

Hence

$$\boxed{I(t) = \frac{5E_0}{c}\left(\Psi(t) + (-1)^{[t/\alpha]}\Psi(t-\alpha([t/\alpha]+1))\right).}$$

The steady state solution is, in this case,

$$I_s(t) = \frac{5E_0}{c}(-1)^{[t/\alpha]}\Psi(t-\alpha([t/\alpha]+1))$$

and the graph of $I_s(t)$ for $\alpha = 0.001\,s$ and $E_0 = 100\,V$ is shown in Figure 49.4.

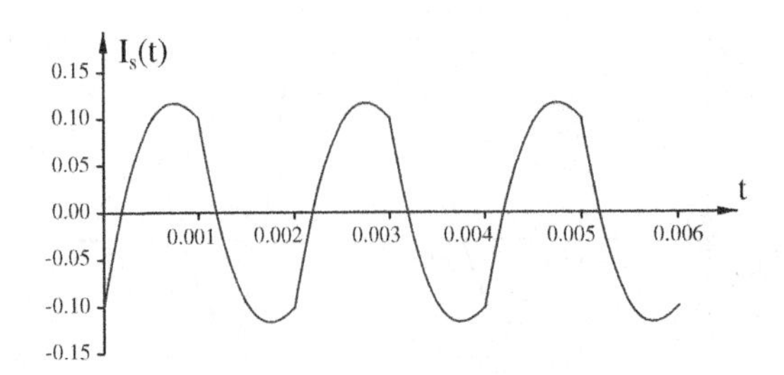

Figure 49.4: steady state solution for a Morse-dot voltage source.

Additional Exercises

49.17. Use the Laplace transform to solve the following initial value problems:

a) $y'(t) + 4y(t) = e^t + 2e^{-2t}$, $y(0) = -3$,

b) $y'(t) + y(t) = \sin(t)$, $y(0) = 1$,

c) $y'(t) - 2y(t) = \cosh(2t)$, $y(0) = 0$.

49.18. Find the complex partial fractions decomposition of $f(z) = \dfrac{z^2 + iz}{(z+2i)^2(z+1)}$.

49.19. Find the real partial fractions decomposition of the function $f(x) = \dfrac{x^2 - x + 1}{(x-1)^2(x^2+4x+5)^2(x+1)}$.

49.20. Use partial fractions decomposition to find the Laplace inverse transforms of the following functions:

a) $F(s) = \dfrac{e^{-3s}(s+2)}{(s+1)(s^2+4s+8)}$ **b)** $F(s) = \dfrac{1}{s(s-3)^4}$ **c)** $F(s) = \dfrac{se^{-s}}{(s^2+9)(s-1)^2}$

49.21. Use the Laplace transform to solve the following initial value problems:

a) $y''(t) + 2y'(t) + 2y(t) = e^{-t}\cos(t)$, $y(0) = 1$, $y'(0) = 3$,

b) $y''(t) + 2y'(t) + 2y(t) = e^{-t}\sin(t)$, $y(0) = 0$, $y'(0) = 2$,

c) $y''(t) + 2y'(t) - 3y(t) = f(t)$, $y(0) = 0$, $y'(0) = 0$, where $f(t) := \begin{cases} 2 & \text{if } 2 < t < 4, \\ 0 & \text{otherwise,} \end{cases}$

d) $y''(t) + 2y'(t) + y(t) = g(t)$, $y(0) = 0$, $y'(0) = 0$, where $g(t) := \begin{cases} 3(t-2) & \text{if } 2 < t < 3, \\ 0 & \text{otherwise.} \end{cases}$

49.22. Why would it not be feasible to use the Laplace transform to solve the initial value problem $y'(t) + 3y(t) = 1$, $y(1) = 3$?

49.23. Use the Laplace transform to determine the current $I(t)$ in an R-C-L circuit with external voltage $E(t) = e^{-t}$ under the assumption that the circuit is completely passive at time $t = 0$ and that $R = 400\,\Omega$, $L = 50\,H$, and $C = 10^{-3}\,F$.

49.24. If the switch in the circuit in Figure 49.1 is closed at some point in time $t_0 > 0$, then the resulting steady state solution...

a) has an increasing amplitude.

b) converges to zero.

c) is equal to zero.

49.25. ★ **a)** Given an R-C-L circuit with $R = 400\,\Omega$, $L = 0.2\,H$, and $C = 10^{-6}\,F$, determine the current $I(t)$ under the assumption that the circuit is completely passive at time $t = 0$ and that $E(t)$ is a periodic function of period α such that $E(t) = E_0 t^2/\alpha^2$ for all $t \in [0, \alpha)$. *Hint.* one of the first steps in your calculation will be to show that

$$\mathcal{L}[E](s) = \frac{E_0}{s}\left(1 + \frac{2}{\alpha s} + \frac{2}{\alpha^2 s^2} - \frac{1 + 2/(\alpha s)}{1 - e^{-\alpha s}}\right).$$

b) Determine the steady state solution $I_s(t)$ and plot its graph for $\alpha = 0.001\,s$ and $E_0 = 100\,V$.

Chapter 50

Numerical Solutions of Differential Equations and Falling Bodies

The Law of Falling Bodies Revisited

Central to our discussion of the law of falling bodies in Chapter 14 was the assumption that the gravitational acceleration g is constant. This assumption is well justified for falling bodies close to the surface of the earth, but fails to be accurate for objects dropped from a great height above the earth. In the latter case, we need to take into account that the gravitational force is inversely proportional to the square of the distance of the falling body from the center of the earth and proportional to the product of the mass of the earth and the mass of the falling body (see also our discussion in Chapter 9). The correct mathematical formulation of this law is as follows: if we denote by $y(t)$ the distance of the falling body from the center of the earth at time t, by M the mass of the earth, and by m the mass of the falling body (see Figure 50.1), then the gravitational force on m is equal to

$$-\frac{\gamma m M}{y(t)^2}, \tag{50.1}$$

where $\gamma \approx 6.672 \cdot 10^{-11} m^3 s^{-2} kg^{-1}$ is a constant of proportionality—the so called *gravitational constant.* For simplification, we will assume the mass of the falling body to be very small compared to the mass of the earth so that, in good approximation, the earth can be regarded to be at rest relative to the axis on which the movement of the falling body is measured; otherwise there would be a two-sided movement

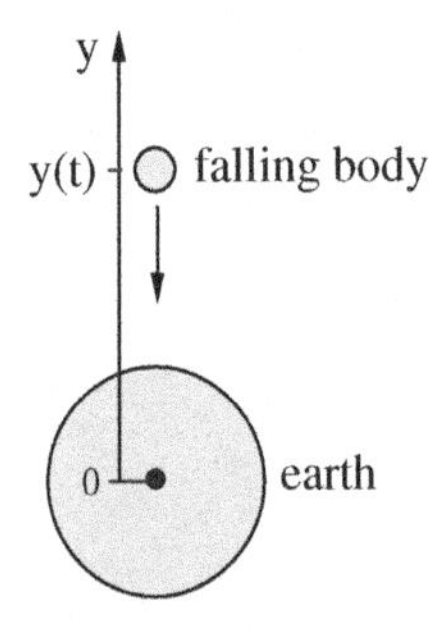

Figure 50.1: a falling body at a large distance from the earth.

of the falling body toward the earth and of the earth toward the falling body. According to Newton's fundamental law of mechanics (see Chapter 8), the force on the falling body is equal to its mass times its acceleration. Combining this observation with (50.1), we obtain

$$my''(t) = -\frac{\gamma m M}{y(t)^2},$$

or equivalently,

$$y''(t) = -\frac{\gamma M}{y(t)^2}. \tag{50.2}$$

Unfortunately, none of the methods that we have discussed so far applies to a second-order differential equation of this type—it is neither separable, nor homogeneous, nor linear. In such a case, the only possible option is often to find *numerical* (i.e., approximate) rather than exact solutions. In our age of high-speed electronic computing devices this is not a serious restriction, because very precise approximations can be generated very quickly. In fact, we will see at the end of this chapter that with a little ingenuity an implicit exact solution of equation (50.2) can still be found (and in a very special case even an explicit exact solution), but in general an explicit solution cannot be obtained without using some type of approximation technique.

Numerical Solutions of Differential Equations

Teacher: Our goal is to develop a general strategy for approximating the solution of a differential equation. As an example, we consider the initial value problem

$$y'(x) = x + \sin(y), \quad y(x_0) = y_0. \tag{50.3}$$

How can we find an estimate for the value of the solution y at some point $z > x_0$? (We could just as well choose a point $z < x_0$, but this is irrelevant.)

Sophie: Since we are looking for estimates we might try to use Taylor polynomials.

Teacher: I like your idea, but for simplicity's sake I suggest that we restrict ourselves—at least initially—to Taylor polynomials of degree one, that is, approximations by means of tangent lines.

Simplicio: I understand how tangent lines can be used to approximate the value of a function, but where is the function? If we do not know the solution of (50.3), how are we going to find the equation of its tangent line?

Teacher: From the differential equation itself! Let's begin with the point x_0 for which the value $y(x_0) = y_0$ is given. The first-order Taylor polynomial of y...

Simplicio: How can we talk about the "Taylor polynomial of y" if we don't have a clue what y actually looks like?

Teacher: Okay, let me phrase my statement slightly differently: *if* y is a solution of (50.3), *then* the first-order Taylor polynomial of y at x_0 is

$$P_{1,x_0}(x) = y(x_0) + y'(x_0)(x - x_0) = y_0 + y'(x_0)(x - x_0).$$

Given the differential equation in (50.3), we can now replace $y'(x_0)$ with $x_0 + \sin(y(x_0)) = x_0 + \sin(y_0)$, and thus express $P_{1,x_0}(x)$ only in terms of the given values x_0 and y_0:

$$P_{1,x_0}(x) = y_0 + (x_0 + \sin(y_0))(x - x_0).$$

So if x_1 is a point close to x_0, then

$$y(x_1) \approx P_{1,x_0}(x_1) = y_0 + (x_0 + \sin(y_0))(x_1 - x_0). \tag{50.4}$$

Simplicio: Why did you change your notation from z to x_1? Did we not set out to find an approximation for the value of y at z?

Teacher: The estimate in (50.4) is reasonably accurate only if x_1 is sufficiently close to x_0. Since we did not assume z to be close to x_0, it is not advisable to apply (50.4) to z directly because the resulting margin of error would likely be fairly large.

Simplicio: So we are stuck.

Teacher: No, we just need to repeat the process several times. To this end we subdivide the interval from x_0 to z into n subintervals of equal length $\Delta x = (z - x_0)/n$. The endpoints of the subintervals are $x_k = x_0 + k\Delta x$ for $k \in \{0, 1, \ldots, n\}$. Based on (50.4) an estimate for $y(x_1)$ is now given as follows:

$$y(x_1) \approx y_0 + (x_0 + \sin(y_0))(x_1 - x_0) = y_0 + (x_0 + \sin(y_0))\Delta x =: y_1.$$

To find an estimate y_2 for the value of y at x_2, we use the first-order Taylor polynomial of y at x_1 instead of x_0:

$$y(x_2) \approx P_{1,x_1}(x_2) = y(x_1) + y'(x_1)(x_2 - x_1) \approx y_1 + y'(x_1)\Delta x.$$

Since $y'(x_1) = x_1 + \sin(y(x_1)) \approx x_1 + \sin(y_1)$, we may infer that

$$y(x_2) \approx y_1 + (x_1 + \sin(y_1))\Delta x =: y_2.$$

In general, if the approximation $y(x_k) \approx y_k$ has been determined, then

$$y_{k+1} := y_k + (x_k + \sin(y_k))\Delta x \tag{50.5}$$

is an estimate for $y(x_{k+1})$. Thus, we have solved the problem of approximating the value of y at z, because $y(z) = y(x_n) \approx y_n$. To find y_n requires, of course, that we first calculate all the values from y_1 to y_{n-1}, and this is the reason why we usually have to use a computer to apply this method. The algorithm we have just described for finding successive approximations $y_1, y_2, y_3, \ldots$ using first-order Taylor polynomials (i.e., tangent lines) is known as *Euler's method.* A graphical illustration is given in Figure 50.2.

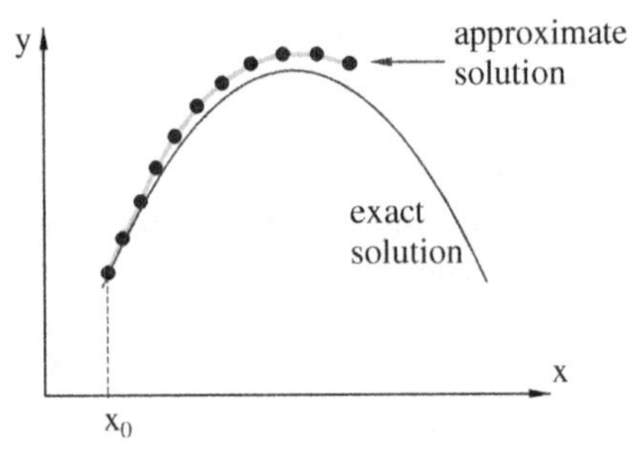

Figure 50.2: Euler's method.

50.1 Example. To give a numerical example, we consider the initial value problem (50.3) with the initial values $x_0 = 1$ and $y_0 = 2$. Let's say we wish to find an approximation for $y(2)$ (i.e., $z = 2$). Using five subdivisions along the interval $[1, 2]$ with $\Delta x = 0.2$, we obtain the following table of values:

y_0	2.0000	y_3	3.1142
y_1	2.3819	y_4	3.4396
y_2	2.7596	y_5	3.7409

The value $y_5 \approx 3.7409$ is the desired estimate for $y(2)$. A (very simple) PASCAL program that generates these values from the equation in (50.5) is shown in Figure 50.3.

Now let us repeat the experiment with ten subdivisions and $\Delta x = 0.1$. In this case, the table of values is

y_0	2.0000	y_4	2.7552	y_8	3.4248
y_1	2.1909	y_5	2.9328	y_9	3.5769
y_2	2.3823	y_6	3.1036	y_{10}	3.7247
y_3	2.5712	y_7	3.2674		

50.2 Exercise. Write a PASCAL program that generates the values in this table.

The new approximation $y_{10} \approx 3.7247$ for $y(2)$ is presumably a better estimate than the previous one, because we increased the number n of subintervals from five to ten.

```
program euler;
var  k: integer;
  y,dx: real;
begin
  y:=2;
  dx:=0.2;
  writeln(y);
  for k:=0 to 4 do
    begin
      y:=y+(1+k*dx+sin(y))*dx;
      writeln(y)
    end
end.
```

Figure 50.3: a PASCAL program simulating Euler's method.

Sophie: What does it mean to say that the approximation with ten subintervals is "presumably" better than the one with five? It certainly seems reasonable to think that with the length of the subintervals, the error in each step of Euler's method decreases, but since there are also more steps it is at least conceivable that the total error is in the end just as large.

Teacher: This is a valid concern, but it turns out that in "most" cases (and certainly in any case that we can reasonably expect to encounter in practical applications) the error can indeed be made arbitrarily small by increasing the number of subintervals. To illustrate this point we will now discuss a differential equation for which we actually know the exact solution so that we can examine how the accuracy of the approximation improves as n increases.

50.3 Example. Let us consider the initial value problem

$$y'(x) = \sqrt{1 - y(x)^2}, \quad y(x_0) = y_0. \tag{50.6}$$

It is easy to verify that for $x_0 = y_0 = 0$ the solution of this initial value problem on the interval $[-\pi/2, \pi/2]$ is $y(x) = \sin(x)$. Given this observation, we will now apply Euler's method to find an approximation for $y(1.2) = \sin(1.2) \approx 0.9320$. In analogy to (50.5), the successive approximations are described by the equation

$$y_{k+1} := y_k + \sqrt{1 - y_k^2}\Delta x.$$

Using six subdivisions with $\Delta x = 0.2$ yields:

y_0	0.0000	y_4	0.7426
y_1	0.2000	y_5	0.8765
y_2	0.3960	y_6	0.9728
y_3	0.5796		

50.4 Exercise. Use a calculator to verify the values in this table.

The error in the approximation is $|y_6 - y(1.2)| \approx |0.9728 - 0.9320| = 0.0408$. To reduce this error, we increase the number of subintervals from six to twelve with $\Delta x = 0.1$. The corresponding estimates are listed in the following table:

y_0	0.0000	y_5	0.4849	y_{10}	0.8586
y_1	0.1000	y_6	0.5724	y_{11}	0.9098
y_2	0.1995	y_7	0.6544	y_{12}	0.9513
y_3	0.2975	y_8	0.7300		
y_4	0.3930	y_9	0.7983		

50.5 Exercise. Write a PASCAL program that generates these values.

Now the error is $|y_{12} - y(1.2)| \approx |0.9513 - 0.9320| = 0.0193$. Thus we observe that at least in this example, the approximation does indeed appear to become more accurate as n increases.

50.6 Exercise. Use Euler's method to find an approximation for $y(1.2)$ with $\Delta x = 0.05$ (where, of course, y is again assumed to be a solution of (50.6) with $x_0 = y_0 = 0$).

To illustrate how the accuracy of our estimates can also be improved by increasing the degree of the approximating polynomials (rather than the number of subintervals) we will again consider the initial value problem (50.6). As before, we wish to approximate the value of the solution y at some point $z > x_0$, and again we use n subdivisions of the interval $[x_0, z]$ with $\Delta x = (z - x_0)/n$ and $x_k = x_0 + k\Delta x$. Using a Taylor polynomial of degree two, we obtain the following approximation for the value of y at x_1:

$$y(x_1) \approx P_{2,x_0}(x_1) = y(x_0) + y'(x_0)(x_1 - x_0) + \frac{y''(x_0)}{2}(x_1 - x_0)^2 = y_0 + y'(x_0)\Delta x + \frac{y''(x_0)}{2}\Delta x^2. \quad (50.7)$$

Since y is assumed to be a solution of the differential equation in (50.6), we may replace $y'(x_0)$ with $\sqrt{1 - y(x_0)^2} = \sqrt{1 - y_0^2}$. In order to determine $y''(x_0)$, we need to differentiate both sides of the differential equation in (50.6). This yields

$$y''(x) = -\frac{y(x)y'(x)}{\sqrt{1 - y(x)^2}},$$

and therefore

$$y''(x_0) = -\frac{y(x_0)y'(x_0)}{\sqrt{1 - y(x_0)^2}} = -\frac{y_0\sqrt{1 - y_0^2}}{\sqrt{1 - y_0^2}} = -y_0.$$

Thus, (50.7) can be written in the form

$$y(x_1) \approx y_0 + \sqrt{1 - y_0^2}\,\Delta x - \frac{y_0}{2}\Delta x^2 =: y_1.$$

To find an estimate for $y(x_2)$ we only need to substitute x_1 for x_0 and y_1 for y_0. Thus,

$$y(x_2) \approx y_1 + \sqrt{1 - y_1^2}\,\Delta x - \frac{y_1}{2}\Delta x^2 =: y_2.$$

In general, if y_k has been determined, then

$$y_{k+1} := y_k + \sqrt{1 - y_k^2}\,\Delta x - \frac{y_k}{2}\Delta x^2$$

is an approximation of $y(x_{k+1})$. For $x_0 = y_0 = 0$, $z = 1.2$, and $n = 12$ (i.e., $\Delta x = 0.1$) the corresponding sequence of values is shown in the following table:

y_0	0.0000	y_5	0.4802	y_{10}	0.8424
y_1	0.1000	y_6	0.5655	y_{11}	0.8921
y_2	0.1990	y_7	0.6451	y_{12}	0.9328
y_3	0.2960	y_8	0.7183		
y_4	0.3900	y_9	0.7843		

50.7 Exercise. Write a PASCAL program that generates these values.

As expected, the error $|y_{12} - y(1.2)| \approx |0.9328 - 0.9320| = 0.0008$ is significantly smaller than with Euler's method for the same number of subdivisions (see Example 50.3).

50.8 Exercise. Find an estimate for $y(1.2)$ by using a third-order Taylor approximation with $x_0 = y_0 = 0$ and $n = 12$.

In order to understand how the methods that we discussed so far can be adapted to the case of initial value problems involving second-order differential equations (such as (50.2)) we will now consider the initial value problem

$$\begin{aligned} y''(x) &= -y(x) \\ y(x_0) &= y_0,\ y'(x_0) = w_0. \end{aligned} \tag{50.8}$$

In dealing with second-order differential equations we do not only need to generate approximations for $y(x_k)$, but also for $y'(x_k)$. As before, we will work with n subdivisions of the interval $[x_0, z]$ for some $z > x_0$. Using first-order approximations for y and y', we obtain

$$\begin{aligned} y(x_1) &\approx y(x_0) + y'(x_0)(x_1 - x_0), \\ y'(x_1) &\approx y'(x_0) + y''(x_0)(x_1 - x_0). \end{aligned}$$

Given the initial values in (50.8) and the equation $y''(x_0) = -y(x_0) = -y_0$, it follows that

$$\begin{aligned} y(x_1) &\approx y_0 + w_0\Delta x =: y_1, \\ y'(x_1) &\approx w_0 - y_0\Delta x =: w_1. \end{aligned}$$

In general, if y_k and w_k have been determined, the approximations y_{k+1} and w_{k+1} for $y(x_{k+1})$ and $y'(x_{k+1})$ are given by the equations

$$\begin{aligned} y_{k+1} &:= y_k + w_k\Delta x, \\ w_{k+1} &:= w_k - y_k\Delta x. \end{aligned}$$

For $x_0 = y_0 = 0$, $w_0 = 1$, $z = \pi/2$, and $n = 15$ (i.e., $\Delta x = \pi/30$) we obtain the following tables of values:

y_0	0.0000	y_4	0.4143	y_8	0.7742	y_{12}	1.0139
y_1	0.1047	y_5	0.5121	y_9	0.8476	y_{13}	1.0489
y_2	0.2094	y_6	0.6054	y_{10}	0.9125	y_{14}	1.0728
y_3	0.3130	y_7	0.6931	y_{11}	0.9682	y_{15}	1.0852

w_0	1.0000	w_4	0.9343	w_8	0.7013	w_{12}	0.3345
w_1	1.0000	w_5	0.8909	w_9	0.6023	w_{13}	0.2284
w_2	0.9890	w_6	0.8373	w_{10}	0.5315	w_{14}	0.1185
w_3	0.9671	w_7	0.7739	w_{11}	0.4359	w_{15}	0.0062

50.9 Exercise. Write a PASCAL program that generates these tables.

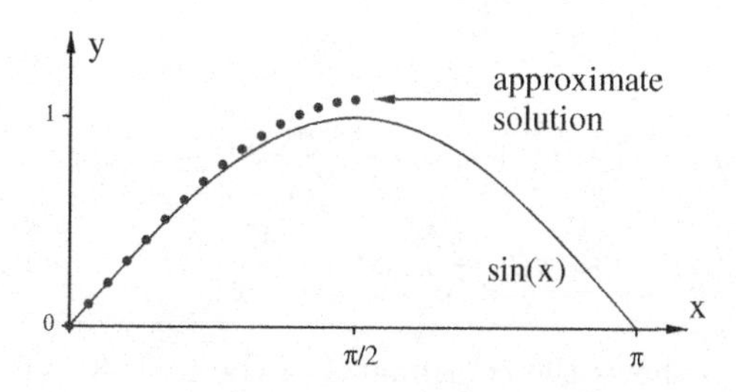

Figure 50.4: comparison of exact and approximate solution.

Since the *exact* solution of the initial value problem (50.8) with $x_0 = y_0 = 0$ and $w_0 = 1$ is obviously $y(x) = \sin(x)$, and since $\sin(\pi/2) = 1$, we may infer that the error in our approximation for $y(z) = y(\pi/2) = 1$ is $|1 - y_{15}| \approx |1 - 1.0852| = 0.0852$. Figure 50.4 shows the approximate values y_k in comparison to the exact solution $y(x) = \sin(x)$.

50.10 Exercise. Use second-order Taylor polynomials for y and y' to improve the approximation for the solution of (50.8) with $x_0 = y_0 = 0$, $w_0 = 1$, $z = \pi/2$, and $n = 15$.

An Approximate Solution of the Falling Body Problem

Our goal in this section is to find an approximate solution for equation (50.2) and to compare it graphically with the solution of the falling problem that we obtained in Chapter 14 under the assumption that the gravitational acceleration g is constant. To begin with, we determine an estimate for the constant γM which appears in (50.2). If $y(t)$ is equal to the radius R of the earth, then the gravitational acceleration $\gamma M/y(t)^2$ is equal to g, i.e., $\gamma M/R^2 = g$. Since the circumference of the earth is approximately 40000 km (at the equator), it follows that $R \approx 4 \cdot 10^7/2\pi\, m$. Thus, using the value $g \approx 9.81\, m/s^2$, we obtain

$$\gamma M = gR^2 \approx \frac{9.81 \cdot 16 \cdot 10^{14}}{4\pi^2} \approx 3.98 \cdot 10^{14}\, \frac{m^3}{s^2}.$$

Now suppose that at time $t = 0$ an object is dropped from a height d above the surface of the earth to descend on a straight line toward the center of the earth. Then $y(0) = R + d$, because, by definition, $y(t)$ measures the object's distance from the center of the earth (rather than the surface of the earth). Furthermore, the assumption that the object is at rest at time $t = 0$ implies that $y'(0) = 0$. Thus, we need to solve the following initial value problem:

$$\begin{aligned} y''(t) &= -\frac{\gamma M}{y(t)^2} \\ y(0) &= R + d,\ y'(0) = 0. \end{aligned} \tag{50.9}$$

As in the previous section, we denote by y_k and w_k the estimates for $y(t_k)$ and $y'(t_k)$. However, for better accuracy we will use second-order rather than first-order Taylor polynomials in our approximation algorithm. Assuming that the values of y_k and w_k have been determined, we find that

$$y(t_{k+1}) \approx y(t_k) + y'(t_k)(t_{k+1} - t_k) + \frac{y''(t_k)}{2}(t_{k+1} - t_k)^2 \approx y_k + w_k\Delta t + \frac{y''(t_k)}{2}\Delta t^2,$$
$$y'(t_{k+1}) \approx y'(t_k) + y''(t_k)(t_{k+1} - t_k) + \frac{y'''(t_k)}{2}(t_{k+1} - t_k)^2 \approx w_k + y''(t_k)\Delta t + \frac{y'''(t_k)}{2}\Delta t^2.$$

Differentiating both sides of the differential equation in (50.9) yields

$$y'''(t) = \frac{2\gamma M y'(t)}{y(t)^3},$$

and therefore,

$$y''(t_k) \approx -\frac{\gamma M}{y_k^2} \quad \text{and} \quad y'''(t_k) \approx \frac{2\gamma M w_k}{y_k^3}.$$

Hence

$$\begin{aligned} y(t_{k+1}) &\approx y_k + w_k\Delta t - \frac{\gamma M}{2y_k^2}\Delta t^2 =: y_{k+1}, \\ y'(t_{k+1}) &\approx w_k - \frac{\gamma M}{y_k^2}\Delta t + \frac{\gamma M w_k}{y_k^3}\Delta t^2 =: w_{k+1}. \end{aligned} \tag{50.10}$$

The dotted line in Figure 50.5 shows the coordinates of the first 40 points (t_k, y_k) generated by using the equations in (50.10) with $\Delta t = 100\, s$ and $d = 12000\, km = 1.2 \cdot 10^7\, m$ (d must be specified because the initial value for y is $d + R$). The curve ends at the line $y = R$, because if $y(t) = R$ the falling body has reached the surface of the earth. For comparison, Figure 50.5 also shows the graph that we would obtain if the gravitational acceleration were constant. In this case, the differential equation $y''(t) = -g$ (see Chapter 14) with $y(0) = d + R$ and $y'(0) = 0$ would yield the solution $y(t) = d + R - gt^2/2$. Thus, the graph of y would be a downward parabola with vertex at $(0, d + R)$.

50.11 Exercise. Write a PASCAL program that generates the coordinates for the points on the dotted graph in Figure 50.5.

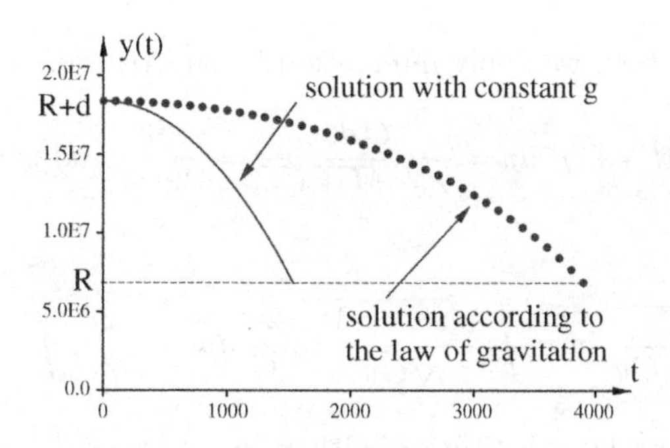

Figure 50.5: comparison of time-distance graphs for falling bodies.

Remark. Since the gravitational acceleration decreases with the square of the distance from the center of the earth, it is natural to expect that a body falling in the gravitational field of the earth will take longer to reach the surface of the earth than a body falling from the same distance with constant acceleration g. Figure 50.5 shows that this indeed is the case, because the t-coordinate of the point of intersection of the solution curve with the line $y = R$ is significantly smaller for the parabola $y(t) = d + R - gt^2/2$ than for the dotted graph representing the true solution with nonconstant g.

★ An Implicit Exact Solution of the Falling Body Problem

In order to find an implicit exact solution of the falling body problem, we multiply both sides of equation (50.2) by $y'(t)$:

$$y'(t)y''(t) = -\frac{\gamma M y'(t)}{y(t)^2}.$$

Using integration by substitution (with $u = y'(t)$ and $du = y''(t)dt$ on the left and $v = y(t)$ and $dv = y'(t)dt$ on the right), it is easy to see that

$$\frac{y'(t)^2}{2} = \frac{\gamma M}{y(t)} + C.$$

Thus,

$$y'(t) = \pm\sqrt{\frac{2(\gamma M + Cy(t))}{y(t)}},$$

or equivalently,

$$\sqrt{\frac{y(t)}{\gamma M + Cy(t)}}\,y'(t) = \pm\sqrt{2}.$$

Again using integration by substitution, it follows that

$$\int \sqrt{\frac{y}{\gamma M + Cy}}\,dy = \pm\sqrt{2}t + D. \tag{50.11}$$

If $C = 0$, a simple integration of $y^{1/2}$ shows that

$$\frac{2y^{3/2}}{3\sqrt{\gamma M}} = \pm\sqrt{2}t + D.$$

Therefore, in this very special case we find the *explicit* solution

$$\boxed{y(t) = \left(\frac{3}{2}\sqrt{\gamma M}(\pm\sqrt{2}t + D)\right)^{2/3}.}$$

If on the other hand $C \neq 0$, then we apply integration by substitution with

$$u = \sqrt{\gamma M + Cy},\ du = \frac{C\,dy}{2\sqrt{\gamma M + Cy}} = \frac{C\,dy}{2u},\ \text{and}\ y = \frac{u^2 - \gamma M}{C}$$

to infer that

$$\int \sqrt{\frac{y}{\gamma M + Cy}}\,dy = \int \frac{\sqrt{u^2 - \gamma M}}{\sqrt{C}u} \cdot \frac{2u}{C}\,du = \frac{2}{\sqrt{C}^3} \int \sqrt{u^2 - \gamma M}\,du.$$

Now using inverse integration by substitution with $u = \sqrt{\gamma M}\cosh(s)$ and $du = \sqrt{\gamma M}\sinh(s)ds$, we obtain

$$\begin{aligned}\int \sqrt{\frac{y}{\gamma M + Cy}}\,dy &= \frac{2\gamma M}{\sqrt{C}^3} \int |\sinh(s)|\sinh(s)\,ds = \pm\frac{2\gamma M}{\sqrt{C}^3} \int \sinh^2(s)\,ds \\ &= \pm\frac{2\gamma M}{\sqrt{C}^3} \int \frac{1}{4}(e^{2s} - 2 + e^{-2s})\,ds.\end{aligned}$$

Given the result in (50.11), we may thus conclude that

$$\frac{\gamma M}{2\sqrt{C}^3}\left(\frac{e^{2s}}{2} - 2s - \frac{e^{-2s}}{2}\right) = \pm\sqrt{2}t + D. \tag{50.12}$$

Using the definitions of the hyperbolic functions as introduced in Chapter 31, it is easy to see that $e^{2s}/2 - e^{-2s}/2 = 2\cosh(s)\sinh(s)$. Since $s = \operatorname{arcosh}(u/\sqrt{\gamma M}) = \operatorname{arcosh}(\sqrt{\gamma M + Cy}/\sqrt{\gamma M})$, it therefore follows that

$$\begin{aligned}\frac{e^{2s}}{2} - \frac{e^{-2s}}{2} &= 2\cosh\left(\operatorname{arcosh}\left(\frac{\sqrt{\gamma M + Cy}}{\sqrt{\gamma M}}\right)\right)\sinh\left(\operatorname{arcosh}\left(\frac{\sqrt{\gamma M + Cy}}{\sqrt{\gamma M}}\right)\right) \\ &= 2\cosh\left(\operatorname{arcosh}\left(\frac{\sqrt{\gamma M + Cy}}{\sqrt{\gamma M}}\right)\right)\sqrt{\cosh^2\left(\operatorname{arcosh}\left(\frac{\sqrt{\gamma M + Cy}}{\sqrt{\gamma M}}\right)\right) - 1} \\ &= \frac{2\sqrt{\gamma M + Cy}}{\sqrt{\gamma M}}\sqrt{\left(\frac{\sqrt{\gamma M + Cy}}{\sqrt{\gamma M}}\right)^2 - 1} \\ &= \frac{2\sqrt{Cy(\gamma M + Cy)}}{\gamma M}.\end{aligned}$$

Consequently, (50.12) can be written in the form

$$\boxed{\frac{\gamma M}{\sqrt{C}^3}\left(\frac{\sqrt{Cy(t)(\gamma M + Cy(t))}}{\gamma M} - \operatorname{arcosh}\left(\frac{\sqrt{\gamma M + Cy(t)}}{\sqrt{\gamma M}}\right)\right) = \pm\sqrt{2}t + D.}$$

This equation implicitly describes the solution $y(t)$ of (50.2) in the case where C is different from zero.

Additional Exercises

50.12. **a)** Find the exact solution of the initial value problem $y'(x) + 3y(x) = x$, $y(0) = 1$.

b) For the initial value problem in a), find the approximate value of $y(2)$ using Euler's method with 20 subintervals and compare your answer with the exact solution derived in a).

c) Rework b) using a second-order Taylor estimate in place of the tangent line approximation of Euler's method.

50.13. Repeat Exercise 50.12 for the initial value problem $y''(x) + 6y'(x) - 7y(x) = x$, $y(0) = 1$, $y'(0) = -1$.

50.14. Use Euler's method to find approximations for $y(2)$ for each of the initial value problems given below. You should work with 20 subintervals in each case.

a) $y'(x) = e^{y(x)^2/20}$, $y(0) = 1$,

b) $y'(x) = \sin(y(x)^2) + x$, $y(0) = 0$,

c) $y''(x) + \sin(y(x))y'(x) + y(x) = x$, $y(0) = 1$, $y'(0) = 0$.

50.15. Rework Exercise 50.14 using second-order Taylor approximations instead of Euler's method.

Chapter 51

Power Series Solutions

Given our discussion in Chapter 50, where we found approximate solutions of initial value problems by using Taylor *polynomials*, it is not surprising that in some cases we are able to find the *exact* solutions in the form of Taylor *series*. To illustrate this we will consider as a special case the homogeneous second-order linear differential equation

$$y''(x) + a(x)y'(x) + b(x)y(x) = 0 \quad \text{(see Chapter 46).} \tag{51.1}$$

The following theorem says that every solution of this differential equation can be represented by a power series if the coefficient functions a and b are equal to their Taylor series (for a proof of this theorem see [Cod]).

51.1 Theorem. *Assume that $a : I \to \mathbb{R}$ and $b : I \to \mathbb{R}$ are infinitely many times differentiable and equal to their Taylor series on the open interval I. If there exists an $x_0 \in I$ such that*

$$a(x) = \sum_{k=0}^{\infty} \frac{a^{(k)}(x_0)}{k!}(x - x_0)^k \quad \text{and} \quad b(x) = \sum_{k=0}^{\infty} \frac{b^{(k)}(x_0)}{k!}(x - x_0)^k$$

for all $x \in I$, then every solution y of (51.1) is infinitely many times differentiable on I and can be represented by a power series about x_0. More precisely, there are coefficients $c_k \in \mathbb{R}$ such that the power series $\sum_{k=0}^{\infty} c_k(x - x_0)^k$ is convergent for all $x \in I$ and

$$y(x) = \sum_{k=0}^{\infty} c_k(x - x_0)^k$$

is a solution of (51.1).

The uniqueness statement of Theorem 42.13 implies that the coefficients c_k are equal to $y^{(k)}(x_0)/k!$, but this observation is not helpful in finding the coefficients c_k because $y^{(k)}(x_0)$ can be evaluated only if the solution y is already known. Our objective is to determine y from the coefficients c_k—not the other way around.

51.2 Example. * Let us consider the initial value problem

$$\begin{aligned} &y''(x) + xy'(x) + y(x) = 0 \\ &y(0) = z_1,\ y'(0) = z_2. \end{aligned} \tag{51.2}$$

Since the coefficient functions $a(x) = x$ and $b(x) = 1$ are clearly equal to their MacLaurin series on $\mathbb{R}$, we may apply Theorem 51.1 to infer that the solution of (51.2) can be represented by a power series at

*Example 51.2 has been adopted with permission from the exposition in [WB], Example 7, p.135.

$x_0 = 0$. So there are coefficients $c_k \in \mathbb{R}$ such that

$$y(x) = \sum_{k=0}^{\infty} c_k x^k$$

for all $x \in \mathbb{R}$. According to Theorem 42.10, the first and second derivatives of y are

$$y'(x) = \sum_{k=1}^{\infty} k c_k x^{k-1},$$
$$y''(x) = \sum_{k=2}^{\infty} k(k-1) c_k x^{k-2}.$$

In order to determine the coefficients c_k we replace $y(x)$, $y'(x)$, and $y''(x)$ in the differential equation in (51.2) with the corresponding power series. This yields

$$0 = \sum_{k=2}^{\infty} (k-1) k c_k x^{k-2} + x \sum_{k=1}^{\infty} k c_k x^{k-1} + \sum_{k=0}^{\infty} c_k x^k. \tag{51.3}$$

Given the initial conditions in (51.2), we obtain

$$y(0) = c_0 = z_1 \quad \text{and} \quad y'(0) = c_1 = z_2. \tag{51.4}$$

To write the three series appearing on the right-hand side of equation (51.3) as a single series, the exponents of x need to be equal. For this reason, we change the index of summation in the first series in (51.3) and distribute the factor x in the second. This yields

$$\sum_{k=2}^{\infty} (k-1) k c_k x^{k-2} = \sum_{k=0}^{\infty} (k+1)(k+2) c_{k+2} x^k,$$
$$x \sum_{k=1}^{\infty} k c_k x^{k-1} = \sum_{k=1}^{\infty} k c_k x^k,$$

and by implication,

$$\begin{aligned} 0 &= \sum_{k=0}^{\infty} (k+1)(k+2) c_{k+2} x^k + \sum_{k=1}^{\infty} k c_k x^k + \sum_{k=0}^{\infty} c_k x^k \\ &= 2c_2 + \sum_{k=1}^{\infty} (k+1)(k+2) c_{k+2} x^k + \sum_{k=1}^{\infty} k c_k x^k + c_0 + \sum_{k=1}^{\infty} c_k x^k \\ &= c_0 + 2c_2 + \sum_{k=1}^{\infty} ((k+1)(k+2) c_{k+2} + k c_k + c_k) x^k \\ &= c_0 + 2c_2 + \sum_{k=1}^{\infty} (k+1)((k+2) c_{k+2} + c_k) x^k. \end{aligned} \tag{51.5}$$

Since this equation is valid for all $x \in \mathbb{R}$, the uniqueness statement of Theorem 42.13 allows us to infer that all the coefficients of the power series in (51.5) must be equal to zero, i.e.,

$$c_0 + 2c_2 = 0 \text{ and } (k+1)((k+2)c_{k+2} + c_k) = 0 \text{ for all integers } k \geq 1,$$

or equivalently,

$$c_2 = -\frac{c_0}{2} \text{ and } c_{k+2} = -\frac{c_k}{k+2} \text{ for all integers } k \geq 1. \tag{51.6}$$

Given these equations, it follows that

$$\begin{aligned}
c_4 &= -\frac{c_2}{4} = -\frac{1}{4}\cdot\frac{-c_0}{2} = (-1)^2\frac{c_0}{2^2 2!},\\
c_6 &= -\frac{c_4}{6} = -\frac{1}{6}\cdot\frac{(-1)^2 c_0}{2^2 2!} = (-1)^3\frac{c_0}{2^3 3!},\\
c_8 &= -\frac{c_6}{8} = -\frac{1}{8}\cdot\frac{(-1)^3 c_0}{2^3 3!} = (-1)^4\frac{c_0}{2^4 4!},\\
&\vdots
\end{aligned}$$

and, in general,

$$c_{2k} = (-1)^k\frac{c_0}{2^k k!} \text{ for all integers } k \geq 0. \tag{51.7}$$

To determine the coefficients with odd indices, we use again (51.6) to conclude that

$$\begin{aligned}
c_3 &= -\frac{c_1}{3},\\
c_5 &= -\frac{c_3}{5} = -\frac{1}{5}\cdot\frac{-c_1}{3} = (-1)^2\frac{2^2 2! c_1}{5!},\\
c_7 &= -\frac{c_5}{7} = -\frac{1}{7}\cdot\frac{(-1)^2 2^2 2! c_1}{5!} = (-1)^3\frac{2^3 3! c_1}{7!},\\
&\vdots
\end{aligned}$$

and, in general,

$$c_{2k+1} = (-1)^k\frac{2^k k! c_1}{(2k+1)!} \text{ for all } k \geq 0. \tag{51.8}$$

Combining (51.7) with (51.8) yields

$$\sum_{k=0}^{\infty} c_k x^k = c_0\sum_{k=0}^{\infty}\frac{(-1)^k}{2^k k!}x^{2k} + c_1\sum_{k=0}^{\infty}\frac{(-1)^k 2^k k!}{(2k+1)!}x^{2k+1}. \tag{51.9}$$

Since $e^x = \sum_{k=0}^{\infty} x^k/k!$, it follows that

$$\sum_{k=0}^{\infty}\frac{(-1)^k}{2^k k!}x^{2k} = e^{-x^2/2}.$$

Given this observation, the equations in (51.4) and (51.9) allow us to infer that the solution of the initial value problem (51.2) is

$$y(x) = z_1 e^{-x^2/2} + z_2\sum_{k=0}^{\infty}\frac{(-1)^k 2^k k!}{(2k+1)!}x^{2k+1}. \tag{51.10}$$

Since it can be shown (see the ⋆remark⋆ below) that

$$\sum_{k=0}^{\infty}\frac{(-1)^k 2^k k!}{(2k+1)!}x^{2k+1} = e^{-x^2/2}\int_0^x e^{t^2/2}\,dt, \tag{51.11}$$

the solution in (51.10) can be written in the form

$$\boxed{y(x) = z_1 e^{-x^2/2} + z_2 e^{-x^2/2}\int_0^x e^{t^2/2}\,dt.}$$

★ *Remark.* In order to prove (51.11) we will use the method of the reduction of the order which we discussed in Chapter 46, p.363. Using the same notation as in Chapter 46, we set $a(x) := x$ so that $A(x) := x^2/2$ is a corresponding antiderivative. Since $y_1(x) := e^{-x^2/2}$ is readily seen to be a solution of the differential equation in (51.2) (this also follows from the fact that the coefficients in the MacLaurin series of $e^{-x^2/2}$ satisfy (51.6) as demonstrated above), we may apply the formula derived in Chapter 46, p.363, to infer that

$$y_2(x) = y_1(x) \int_0^x \frac{e^{-A(t)}}{y_1(t)^2}\,dt = e^{-x^2/2} \int_0^x \frac{e^{-t^2/2}}{e^{-t^2}}\,dt = e^{-x^2/2} \int_0^x e^{t^2/2}\,dt$$

is a solution of the differential equation in (51.2) as well. To prove that

$$v(x) := \sum_{k=0}^{\infty} \frac{(-1)^k 2^k k!}{(2k+1)!} x^{2k+1}$$

is equal to $y_2(x)$, we first observe that $v(x)$ is a solution of the differential equation in (51.2). This is so because the coefficients of the series $\sum_{k=0}^{\infty}(-1)^k 2^k k! x^{2k+1}/(2k+1)!$ satisfy (51.6). (Alternatively, we can derive this conclusion from (51.10) because y_1 is a solution of the differential equation in (51.2), and therefore $v(x) = (y(x) - z_1 y_1(x))/z_2$ is a solution as well. Here we may assume z_2 to be different from zero because (51.10) holds for all $z_2 \in \mathbb{R}$.) Furthermore, it is easy to see that

$$\begin{aligned} y_2(0) &= v(0) = 0, \\ y_2'(0) &= v'(0) = 1. \end{aligned}$$

Consequently, v and y_2 are both solutions of the initial value problem

$$\begin{aligned} &y''(x) + xy'(x) + y(x) = 0 \\ &y(0) = 0,\ y'(0) = 1. \end{aligned}$$

Since solutions of initial value problems are always unique (see Theorem 46.2), it follows that $y_2 = v$, and the proof of (51.11) is thus complete. ★

51.3 Example. [†] We wish to solve the initial value problem

$$\begin{aligned} &y''(x) + xy(x) = 0 \\ &y(0) = z_1,\ y'(0) = z_2. \end{aligned} \tag{51.12}$$

According to Theorem 51.1, we can write the solution y of this initial value problem as a power series at $x_0 = 0$:

$$y(x) = \sum_{k=0}^{\infty} c_k x^x$$

for all $x \in \mathbb{R}$. Given this representation of y, the initial conditions in (51.12) imply that

$$y(0) = c_0 = z_1 \quad \text{and} \quad y'(0) = c_1 = z_2. \tag{51.13}$$

Since

$$y''(x) = \sum_{k=2}^{\infty} (k-1)k c_k x^{k-2} \quad \text{(by Theorem 42.10)},$$

the differential equation in (51.12) can be written in the form

$$\sum_{k=2}^{\infty} (k-1)k c_k x^{k-2} + x \sum_{k=0}^{\infty} c_k x^k = 0.$$

[†] Example 51.3 has been adopted with permission from the exposition in [WB], Example 8, p.138.

Changing the indices of summation yields

$$\begin{aligned}\sum_{k=2}^{\infty}(k-1)kc_k x^{k-2} &= 2c_2 + \sum_{k=1}^{\infty}(k+1)(k+2)c_{k+2}x^k, \\ x\sum_{k=0}^{\infty}c_k x^k &= \sum_{k=1}^{\infty}c_{k-1}x^k,\end{aligned} \tag{51.14}$$

and therefore

$$2c_2 + \sum_{k=1}^{\infty}((k+1)(k+2)c_{k+2} + c_{k-1})x^k = 0 \tag{51.15}$$

for all $x \in \mathbb{R}$.

51.4 Exercise. Verify the equations in (51.14).

As in Example 51.2, the uniqueness statement of Theorem 42.13 allows us to conclude that all the coefficients of the series in (51.15) are equal to zero. Hence

$$c_2 = 0 \text{ and } c_{k+2} = -\frac{c_{k-1}}{(k+1)(k+2)} \text{ for all integers } k \geq 1. \tag{51.16}$$

To find a general formula for the coefficients c_k, we first consider indices of the form $3k$ for $k \geq 1$:

$$\begin{aligned} c_3 &= -\frac{c_0}{2\cdot 3}, \\ c_6 &= -\frac{c_3}{5\cdot 6} = (-1)^2\frac{c_0}{2\cdot 3\cdot 5\cdot 6}, \\ c_9 &= -\frac{c_6}{8\cdot 9} = (-1)^3\frac{c_0}{2\cdot 3\cdot 5\cdot 6\cdot 8\cdot 9}, \\ &\vdots \end{aligned}$$

and in general

$$c_{3k} = (-1)^k\frac{c_0}{3^k k!(2\cdot 5\cdot 8\cdots(3k-1))}. \tag{51.17}$$

Next we examine coefficients with indices of the form $3k+1$ for $k \geq 1$:

$$\begin{aligned} c_4 &= -\frac{c_1}{3\cdot 4}, \\ c_7 &= -\frac{c_4}{6\cdot 7} = (-1)^2\frac{c_1}{3\cdot 4\cdot 6\cdot 7}, \\ c_{10} &= -\frac{c_7}{9\cdot 10} = (-1)^3\frac{c_1}{3\cdot 4\cdot 6\cdot 7\cdot 9\cdot 10}, \\ &\vdots \end{aligned}$$

and in general

$$c_{3k+1} = (-1)^k\frac{c_2}{3^k k!(4\cdot 7\cdot 10\cdots(3k+1))}. \tag{51.18}$$

Finally, given the equations in (51.16), it is easy to see that

$$c_{3k+2} = 0 \text{ for all } k \geq 0. \tag{51.19}$$

51.5 Exercise. Apply (51.16) to prove (51.19).

Using now (51.13) in conjunction with (51.17), (51.18), and (51.19), the solution of the initial value problem (51.12) turns out to be

$$\boxed{y(x) = z_1\left(1 + \sum_{k=1}^{\infty}\frac{(-1)^k x^{3k}}{3^k k!(2\cdot 5\cdot 8\cdots(3k-1))}\right) + z_2\left(x + \sum_{k=1}^{\infty}\frac{(-1)^k x^{3k+1}}{3^k k!(4\cdot 7\cdot 10\cdots(3k+1))}\right).}$$

Additional Exercises

51.6. Use a power series of the form $y(x) = \sum_{k=0}^{\infty} c_k (x - x_0)^k$ to solve the following initial value problems:

a) $y''(x) + x^2 y(x) = 0$, $y(0) = 0$, $y'(0) = 1$, $x_0 = 0$,
b) $(x-1)^2 y''(x) - 2y(x) = (x-1)^2/x^2 + (x-1)/x - 2$, $y(1) = 1$, $y'(1) = -1/2$, $x_0 = 0$,
c) $x^2 y''(x) + y(x) = 1/(1-x)$, $y(0) = 1$, $y'(0) = 1$, $x_0 = 0$.

Hint. To solve b) and c) you may want to apply Theorem 41.16.

51.7. Use a power series about the indicated point x_0 to determine the general solution of each of the differential equations given below.

a) $y''(x) - y'(x) = (4x^2 - 2x + 2)y(x)$, $x_0 = 0$,
b) $y''(x) - xy'(x) = (2x^2 + 5)y(x)$, $x_0 = 0$,
c) $x^2 y''(x) - xy'(x) + y(x) = 0$, $x_0 = 1$.

Hint. The best way to solve a) and b) is probably to find one solution using the power series approach and then a second solution from the formula on p.363 (as we did in the ⋆remark⋆ on p.427).

Epilogue

For the concerned reader we wish to give a brief account of how our three protagonists fared after completing this course.

Not surprisingly, *Sophie* easily passed her calculus exams with an A, at the top of her class. After graduating with the highest honors she embarked on a successful career in a major engineering firm.

The *Teacher* continued to pursue his calling and, for many years to come, dedicated himself to the mathematical education of aspiring young scientists and engineers. With his eloquent speech and firm command of the subject, he never failed to gain the admiration and respect of his listeners.

For *Simplicio*, however, things eventually took a turn for the worse. After receiving a straight F for his performance in the present course, he persevered for several years, but his dream of a passing grade in calculus never became reality. With his spirit broken, he soon found himself unable to take care of even his most basic needs and was forced to live out his days in a cardboard box at the back entrance of the math department. Here his only joy was to work on differentiation problems that the *Teacher*, in his compassion, occasionally dropped into the box. When death overtook him prematurely at the age of 42, *Simplicio's* worldly possessions consisted of nothing more than a stack of solutions to calculus problems, all of which, upon inspection, turned out to be wrong.

Appendix A

Trigonometric Identities and Polar Coordinates

Trigonometric Identities

Foundational to most trigonometric identities is the *theorem of Pythagoras* which asserts that for any right triangle with hypotenuse c and side lengths a and b (such as the shaded triangle in Figure A.1) we have

$$\boxed{c^2 = a^2 + b^2.} \tag{A.1}$$

In order to prove the theorem of Pythagoras, we take a look at Figure A.1. Here we see a square of

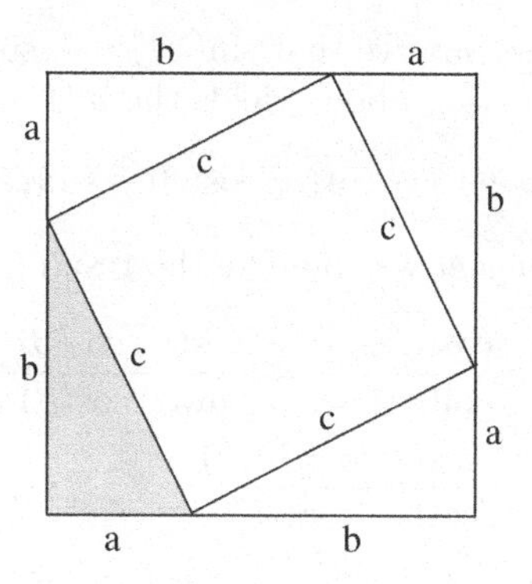

Figure A.1: proving the theorem of Pythagoras.

side length c inscribed in a larger square of side length $a + b$. Since the area of the larger square is on the one hand equal to $(a + b)^2$ and on the other hand equal to the area c^2 of the smaller square plus four times the area of the shaded right triangle, we obtain

$$a^2 + 2ab + b^2 = (a + b)^2 = c^2 + 4 \cdot \frac{1}{2}ab = c^2 + 2ab.$$

Subtracting $2ab$ from both sides yields the desired conclusion in (A.1).

The first and most important trigonometric identity that we are going to derive from the theorem of Pythagoras is the *addition law for sine.* For given angles α and β we wish to express $\sin(\alpha + \beta)$ in terms of $\sin(\alpha)$, $\cos(\alpha)$, $\sin(\beta)$, and $\cos(\beta)$. Considering the diagram in Figure A.2 and applying the definitions of the trigonometric functions as stated in Chapter 6, we observe that

$$\sin(\alpha+\beta) = b_1+c_2, \quad \sin(\alpha) = \frac{b_1}{c_1} = \frac{b_2}{c_2}, \quad \cos(\alpha) = \frac{a_1}{c_1} = \frac{a_2}{c_2}, \quad \sin(\beta) = a_2, \quad \cos(\beta) = c_1+b_2. \tag{A.2}$$

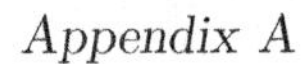

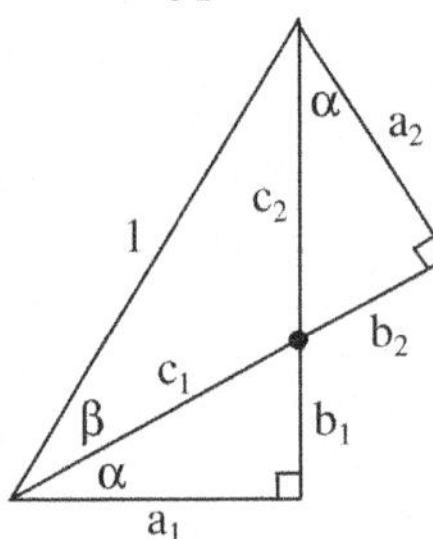

Figure A.2: derivation of the addition law.

Since $b_2^2 + a_2^2 = c_2^2$ (by the theorem of Pythagoras), the equations in (A.2) allow us to infer that

$$\begin{aligned}\sin(\alpha+\beta) &= b_1 + c_2 = b_1 + \frac{b_2^2 + a_2^2}{c_2} = b_1 + \frac{b_2 b_1}{c_1} + \frac{a_2^2}{c_2} = \frac{b_1}{c_1}(c_1 + b_2) + \frac{a_2}{c_2} \cdot a_2 \\ &= \sin(\alpha)\cos(\beta) + \cos(\alpha)\sin(\beta).\end{aligned}$$

Thus we have established the so-called *addition law for sine:*

$$\sin(\alpha+\beta) = \sin(\alpha)\cos(\beta) + \cos(\alpha)\sin(\beta). \tag{A.3}$$

Referring again to the definitions of sine and cosine in Chapter 6, it is easy to see that $\cos(\alpha) = \sin(\pi/2 - \alpha)$ and $\sin(\alpha) = \cos(\pi/2 - \alpha)$. Consequently, (A.3) implies that

$$\begin{aligned}\cos(\alpha+\beta) &= \sin\left(\left(\frac{\pi}{2} - \alpha\right) + (-\beta)\right) = \sin\left(\frac{\pi}{2} - \alpha\right)\cos(-\beta) + \cos\left(\frac{\pi}{2} - \alpha\right)\sin(-\beta) \\ &= \cos(\alpha)\cos(-\beta) + \sin(\alpha)\sin(-\beta).\end{aligned}$$

Taking a look at Figure 26.2, we also observe that $\sin(-t) = -\sin(t)$ and $\cos(-t) = \cos(t)$ for all $t \in \mathbb{R}$. Applying these identities to the equation above yields the *addition law for cosine:*

$$\cos(\alpha+\beta) = \cos(\alpha)\cos(\beta) - \sin(\alpha)\sin(\beta). \tag{A.4}$$

To derive the *addition law for cotangent* we combine the results in (A.3) and (A.4):

$$\begin{aligned}\cot(\alpha+\beta) &= \frac{\cos(\alpha+\beta)}{\sin(\alpha+\beta)} = \frac{\cos(\alpha)\cos(\beta) - \sin(\alpha)\sin(\beta)}{\sin(\alpha)\cos(\beta) + \cos(\alpha)\sin(\beta)} \\ &= \frac{\cot(\alpha)\cot(\beta) - 1}{\cot(\alpha) + \cot(\beta)}.\end{aligned} \tag{A.5}$$

A.1 Exercise. Use the addition law for cotangent and the fact that $\tan(t) = 1/\cot(t)$ to prove the following *addition law for tangent:*

$$\tan(\alpha+\beta) = \frac{\tan(\alpha) + \tan(\beta)}{1 - \tan(\alpha)\tan(\beta)}.$$

In order to find the *double angle formula for cotangent,* which we used in Chapter 6, we only need to set β equal to α in (A.5):

$$\cot(2\alpha) = \frac{\cot^2(\alpha) - 1}{2\cot(\alpha)}.$$

A.2 Exercise. Prove the following double angle formulae for sine, cosine, and tangent:

$$\begin{aligned}\sin(2\alpha) &= 2\sin(\alpha)\cos(\alpha), \\ \cos(2\alpha) &= \cos^2(\alpha) - \sin^2(\alpha) = 2\cos^2(\alpha) - 1 = 1 - 2\sin^2(\alpha), \\ \tan(2\alpha) &= \frac{2\tan(\alpha)}{1 - \tan^2(\alpha)}.\end{aligned}$$

For reference purposes we give here a list of some useful trigonometric formulae, most of which we have derived either in this appendix or in Chapter 26.

Pythagorean identities:

$$\sin^2(\alpha) + \cos^2(\alpha) = 1 \qquad 1 + \tan^2(\alpha) = \frac{1}{\cos^2(\alpha)} \qquad 1 + \cot^2(\alpha) = \frac{1}{\sin^2(\alpha)}$$

Addition laws:

$$\sin(\alpha + \beta) = \sin(\alpha)\cos(\beta) + \cos(\alpha)\sin(\beta) \qquad \tan(\alpha + \beta) = \frac{\tan(\alpha) + \tan(\beta)}{1 - \tan(\alpha)\tan(\beta)}$$

$$\cos(\alpha + \beta) = \cos(\alpha)\cos(\beta) - \sin(\alpha)\sin(\beta) \qquad \cot(\alpha + \beta) = \frac{\cot(\alpha)\cot(\beta) - 1}{\cot(\alpha) + \cot(\beta)}$$

Double angle formulae:

$$\begin{aligned} \sin(2\alpha) &= 2\sin(\alpha)\cos(\alpha) \\ \cos(2\alpha) &= \cos^2(\alpha) - \sin^2(\alpha) \\ &= 2\cos^2(\alpha) - 1 \\ &= 1 - 2\sin^2(\alpha) \end{aligned}$$

$$\tan(2\alpha) = \frac{2\tan(\alpha)}{1 - \tan^2(\alpha)} \qquad \cot(2\alpha) = \frac{\cot^2(\alpha) - 1}{2\cot(\alpha)}$$

Half angle formulae:

$$\sin\left(\frac{\alpha}{2}\right) = \pm\sqrt{\frac{1 - \cos(\alpha)}{2}} \qquad \cos\left(\frac{\alpha}{2}\right) = \pm\sqrt{\frac{1 + \cos(\alpha)}{2}} \qquad \tan\left(\frac{\alpha}{2}\right) = \frac{1 - \cos(\alpha)}{\sin(\alpha)} = \frac{\sin(\alpha)}{1 + \cos(\alpha)}$$

Cofunction indentities:

$$\sin\left(\frac{\pi}{2} - \alpha\right) = \cos(\alpha) \qquad \cos\left(\frac{\pi}{2} - \alpha\right) = \sin(\alpha) \qquad \tan\left(\frac{\pi}{2} - \alpha\right) = \cot(\alpha) \qquad \cot\left(\frac{\pi}{2} - \alpha\right) = \tan(\alpha)$$

Even-odd identities:

$$\sin(-\alpha) = -\sin(\alpha) \qquad \cos(-\alpha) = \cos(\alpha) \qquad \tan(-\alpha) = -\tan(\alpha) \qquad \cot(-\alpha) = -\cot(\alpha)$$

Two useful geometric facts that involve trigonometric functions are the laws of cosine and sine. For the

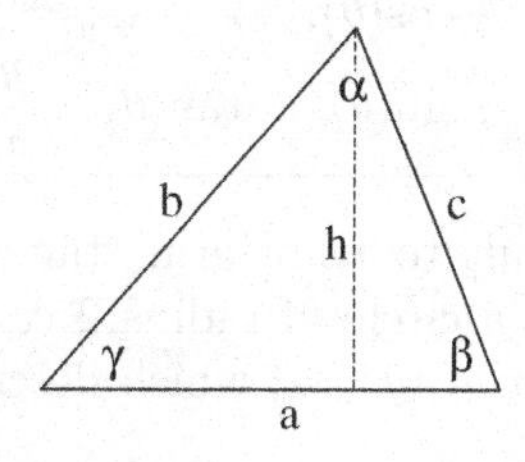

Figure A.3: a triangle.

triangle shown in Figure A.3 the *law of cosine* says that

$$c^2 = a^2 + b^2 - 2ab\cos(\gamma),$$

and according to the *law of sine*, we have

$$\boxed{\frac{\sin(\alpha)}{a} = \frac{\sin(\beta)}{b} = \frac{\sin(\gamma)}{c}.}$$

To prove the law of cosine, we denote by h the height of the triangle (see Figure A.3) and use the theorem of Pythagoras to infer that

$$\begin{aligned} c^2 &= h^2 + (a - b\cos(\gamma))^2 = (b\sin(\gamma))^2 + (a - b\cos(\gamma))^2 \\ &= a^2 + b^2(\sin^2(\gamma) + \cos^2(\gamma)) - 2ab\cos(\gamma) = a^2 + b^2 - 2ab\cos(\gamma) \end{aligned}$$

as desired. An even simpler proof establishes the law of sine:

$$\frac{\sin(\gamma)}{c} = \frac{h/b}{c} = \frac{h/c}{b} = \frac{\sin(\beta)}{b}.$$

The second identity $\sin(\gamma)/c = \sin(\alpha)/a$ can be derived in a completely analogous manner.

A.3 Exercise. Show that the laws of cosine and sine are also valid if one of the angles in the triangle is obtuse (i.e., greater than 90°).

Polar Coordinates

The *polar coordinates* of a point $P = (x, y)$ in an xy-coordinate system are the distance r from P to the origin $(0, 0)$ and the angle θ between the positive x-axis and the line that connects P with the origin (see Figure A.4). In applying the theorem of Pythagoras and the definitions of sine, cosine, and

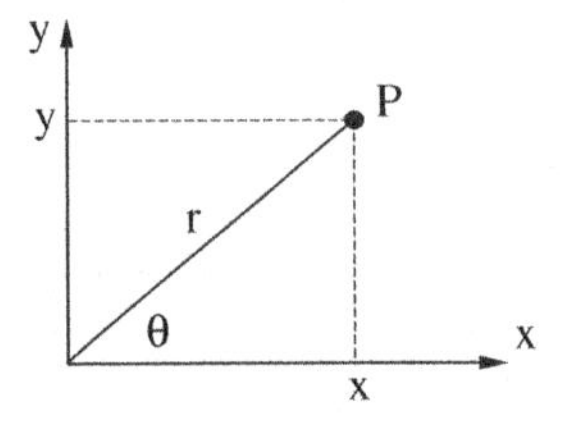

Figure A.4: polar coordinates.

tangent, as stated in Chapter 26, we easily see that the values of r and θ are related to x and y via the following transformation equations:

$$\boxed{\begin{aligned} x &= r\cos(\theta), & r &= \sqrt{x^2 + y^2}, \\ y &= r\sin(\theta), & \tan(\theta) &= \frac{y}{x}. \end{aligned}}$$

Given these equations, it is not difficult to transform standard xy-equations into polar form. For instance, the equation $x^2 + y^2 = 4$ for a circle of radius 2 can be written in polar form as $r^2 = 4$ or equivalently $r = 2$, and the equation $2x + 3y = 1$, which describes a straight line, is equivalent to the polar equation $2r\cos(\theta) + 3r\sin(\theta) = 1$.

A.4 Exercise. Transform the equation $x^2 + y^2 = 3xy$ into polar form.

In determining the polar coordinates of a point given in Cartesian xy-coordinates, we always need to be careful to identify correctly the quadrant in which θ is located. The following example provides an illustration:

A.5 Example. We wish to determine the polar coordinates r and θ of the point $(-2, 3)$. Since this point is located in the second quadrant (to the left of the y-axis and above the x-axis), we must determine θ as an angle between $\pi/2$ and π. According to the transformation equations above, it follows that

$$r = \sqrt{(-2)^2 + 3^2} = \sqrt{13},$$
$$\tan(\theta) = -\frac{3}{2}.$$

Using a calculator, we also find

$$\arctan\left(-\frac{3}{2}\right) \approx -0.983.$$

Consequently, the angle in the second quadrant at which tangent assumes the value $-3/2$ is $\theta = -0.983 + \pi$, and the polar coordinates of $(-2, 3)$ are therefore $(\theta, r) = (-0.983 + \pi, \sqrt{13})$.

A.6 Exercise. Find the polar coordinates of the following points: $(0, 2)$, $(1, 4)$, $(-2, 2)$, $(4, -3)$, and $(-2, -4)$.

Appendix B

Exponentials

For a positive integer n and a real number a, we define a^n as the n-fold product of a with itself, i.e.,

$$a^n = \underbrace{a \cdot \cdots \cdot a}_{n \text{ times}}.$$

Given this definition, it is easy to verify that for all positive integers n and m we have

$$a^n a^m = a^{n+m}, \tag{B.1}$$

$$(a^n)^m = a^{nm}. \tag{B.2}$$

To understand how the definition of exponentials can be extended to noninteger exponents, let us consider, for example, the expression $a^{1.5}$. It is obviously not meaningful to propose that we should multiply a by itself 1.5 times but, in light of (B.2), it appears reasonable to require that

$$a^{1.5} = a^{3/2} = (a^{1/2})^3.$$

The problem of defining $a^{1.5}$ is thus reduced to the problem of defining $a^{1/2}$. Arguing again from the assumed validity of (B.2), we further require that

$$(a^{1/2})^2 = a^{2 \cdot 1/2} = a. \tag{B.3}$$

Consequently, $a^{1/2}$ must be a number which, multiplied by itself, is equal to a. In other words, $a^{1/2}$ should be defined as the *square root* of a:

$$a^{1/2} := \sqrt{a}.$$

Note: at this point we need to assume a to be positive, because otherwise $\sqrt{a}$ is not a real number.

To give a definition for $a^{m/n}$ for arbitrary positive integers n and m we require, in analogy to (B.3), that

$$(a^{1/n})^n = a^{n \cdot 1/n} = a.$$

This equation shows that $a^{1/n}$ must be equal to the *nth root* of a, and $a^{m/n}$ is therefore properly defined as the n-th root of a raised to the m-th power:

$$\boxed{a^{m/n} := \sqrt[n]{a}^{\,m}.}$$

Having thus established a definition of a^r for all positive rational numbers $r = m/n$, we next consider the cases $r = 0$ and $r < 0$ (in this order). In assuming (B.1) to be valid, it follows that

$$a^1 = a^{1+0} = a^1 a^0,$$

and in canceling out a^1 we are thus led to define

$$a^0 := 1.$$

For $r < 0$ (and $-r > 0$) we argue again from the assumed validity of (B.1) to infer that

$$a^r a^{-r} = a^{r-r} = a^0 = 1.$$

Given this observation, we set

$$a^r := \frac{1}{a^{-r}}.$$

B.1 Exercise. Suppose r and s are rational numbers.

a) Show that $a^r a^s = a^{r+s}$ and $\dfrac{a^r}{a^s} = a^{r-s}$.

b) Compute $8^{2/3}$ and $16^{-3/4}$ without using a calculator.

The question we have left unanswered so far is how to define a^x in the case where x is not a rational number. In fact, it is not at all obvious that there is actually a question to be answered, because one might think that every real number can be expressed as a quotient of two integers. In other words, it is conceivable that every number is rational. Unfortunately, though, one of the disciples of the Greek mathematician Pythagoras refuted this assertion in the 6th century B.C. when he demonstrated $\sqrt{2}$ to be irrational: there are no two integers m and n such that $m/n = \sqrt{2}$!

★ *Remark.* The fact that $\sqrt{2}$ is not a rational number is readily established via a proof by contradiction. Suppose that m and n are positive integers such that $\sqrt{2} = m/n$. Without loss of generality we may assume m and n to be relatively prime (in the sense that m and n have no common divisors) because in the fraction m/n all common divisors can be canceled out. Since the equation $\sqrt{2} = m/n$ is obviously equivalent to the equation $2n^2 = m^2$, we may infer that m^2, and therefore also m itself, are even numbers. Thus, there must exist an integer k such that $m = 2k$. Substituting $2k$ for m in the equation $2n^2 = m^2$ yields $n^2 = 2k^2$, and by the same argument as above it follows that n is even as well. Having thus shown both m and n to be even, we have arrived at a contradiction to our assumption that m and n are relatively prime, because 2 is a factor common to both m and n. Consequently, $\sqrt{2}$ is indeed irrational. ★

Returning now to the problem of defining a^x for arbitrary real numbers x (including irrational numbers), we recall that one possible solution to this problem was provided in Chapter 23 where we defined a^x in terms of the natural logarithm and exponential functions. To outline a more intuitive approach, we make use of the fact that every real number x can be arbitrarily closely approximated by rational numbers r (for example, $\sqrt{2} \approx 1.41$ or $\sqrt{2} \approx 1.4142$). Without being completely precise, we define a^x as the limit of a^r as r approaches x in the rational numbers, i.e.,

$$\boxed{a^x := \lim_{\substack{r \to x \\ r \text{ rational}}} a^r.} \tag{B.4}$$

Based on this definition, it is possible to derive the same elementary rules for exponentials as in Theorem 23.2: for $a, b, x, y \in \mathbb{R}$ with $a, b > 0$ we have

a) $a^x a^y = a^{x+y}$,

b) $\dfrac{a^x}{a^y} = a^{x-y}$,

c) $(a^x)^y = a^{xy}$,

d) $a^0 = 1$,

e) $(ab)^x = a^x b^x$, and

f) $\left(\dfrac{a}{b}\right)^x = \dfrac{a^x}{b^x}$.

Appendix C

Conic Sections

The class of curves referred to as *conic sections* encompasses three different types of plane curves: *ellipses*, *parabolas*, and *hyperbolae*. The respective definitions of these three types are as follows:

a) Given two points F_1 and F_2 in the plane, we say that a curve $\mathcal{C}$ is an *ellipse* with *foci* at F_1 and F_2 if there is a constant $C > 0$ such that $\mathcal{C}$ consists of all points P in the plane for which the sum of the distances from P to F_1 and P to F_2 is equal to C.

b) Given a line L and a point F in the plane, we say that a curve $\mathcal{C}$ is a *parabola* with *directrix* L and *focus* F if $\mathcal{C}$ consists of all points P in the plane for which the distance from P to F is equal to the (minimal) distance from P to L.

c) Given two points F_1 and F_2 in the plane, we say that a curve $\mathcal{C}$ is a *hyperbola* with *foci* at F_1 and F_2 if there is a constant $C > 0$ such that $\mathcal{C}$ consists of all points P in the plane for which the absolute value of the difference of the distances from P to F_1 and P to F_2 is equal to C.

In order to show how these geometric definitions can be translated into algebraic equations, we consider first the case of an ellipse. Given two foci F_1 and F_2 in the plane as stated in a), we introduce an xy-coordinate system with the point of intersection of the ellipse with the positive x-axis at $(a, 0)$ and the foci at $(\pm c, 0)$ (see Figure C.1). Since the sum of the distances from $(a, 0)$ to F_1 and F_2 is easily seen

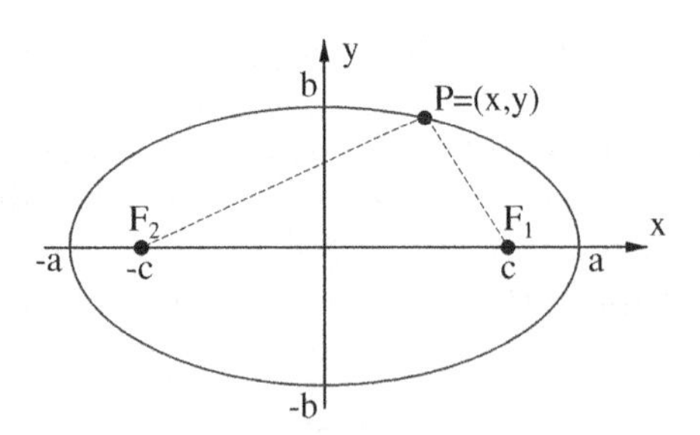

Figure C.1: an ellipse in standard position.

to be equal to $a+c+a-c = 2a$, the definition in a) implies that, in fact, the sum of the distances from any point $P = (x, y)$ on the ellipse to F_1 and F_2 is equal to $2a$ as well. (So the constant C, alluded to in the definition above, is equal to $2a$.) Using the theorem of Pythagoras, we may infer that the distances from P to F_1 and F_2 are equal to $\sqrt{(x+c)^2+y^2}$ and $\sqrt{(x-c)^2+y^2}$ respectively. Consequently, a point $P = (x, y)$ is on the ellipse if and only if

$$\sqrt{(x+c)^2+y^2} + \sqrt{(x-c)^2+y^2} = 2a.$$

To simplify this equation, we subtract $\sqrt{(x-c)^2+y^2}$ and then square both sides of the resulting equation. This yields

$$(x+c)^2+y^2 = \left(2a - \sqrt{(x-c)^2+y^2}\right)^2 = 4a^2 - 4a\sqrt{(x-c)^2+y^2} + (x-c)^2 + y^2,$$

or equivalently

$$a^2 - cx = a\sqrt{(x-c)^2 + y^2}.$$

Squaring both sides again, we obtain

$$(a^2 - cx)^2 = a^2((x-c)^2 + y^2),$$

and therefore

$$a^4 - 2a^2cx + c^2x^2 = a^2x^2 - 2a^2cx + a^2c^2 + a^2y^2.$$

With a few additional manipulations it is not difficult to show that this equation can be written in the form

$$\frac{x^2}{a^2} + \frac{y^2}{a^2 - c^2} = 1. \tag{C.1}$$

If $(0, b)$ is the point of intersection of the ellipse with the positive y-axis, then equation (C.1) must be satisfied for $x = 0$ and $y = b$. Hence

$$\frac{0^2}{a^2} + \frac{b^2}{a^2 - c^2} = 1,$$

or equivalently

$$\boxed{b^2 = a^2 - c^2.}$$

Substituting this result back into (C.1), we finally obtain the following defining equation for an ellipse in the standard position shown in Figure C.1:

$$\boxed{\frac{x^2}{a^2} + \frac{y^2}{b^2} = 1.}$$

The constants a and b are referred to as the major and minor radii of the ellipse.

C.1 Exercise. Find the defining equation for an ellipse in standard position that passes through the point $(2, 1)$ and has a focus at $(3, 0)$.

If the center of an ellipse is not at the origin, but at a point (x_0, y_0) (see Figure C.2), then the defining equation is of the form

$$\boxed{\frac{(x-x_0)^2}{a^2} + \frac{(y-y_0)^2}{b^2} = 1.}$$

C.2 Exercise. Explain why this equation describes the ellipse in Figure C.2.

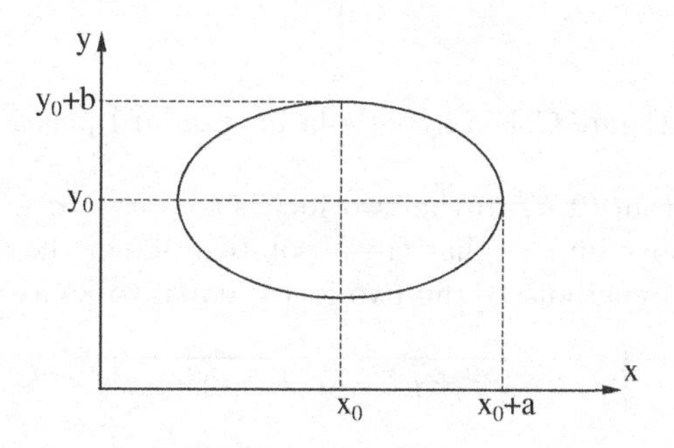

Figure C.2: an ellipse in shifted position.

Turning our attention now from the ellipse to the parabola, we introduce for a given line L and a point F (see the definition above) an xy-coordinate system such that $F = (p, 0)$ and L is described by the equation $x = -p$ for some $p > 0$ (see Figure C.3). Since, according to the definition in b) above,

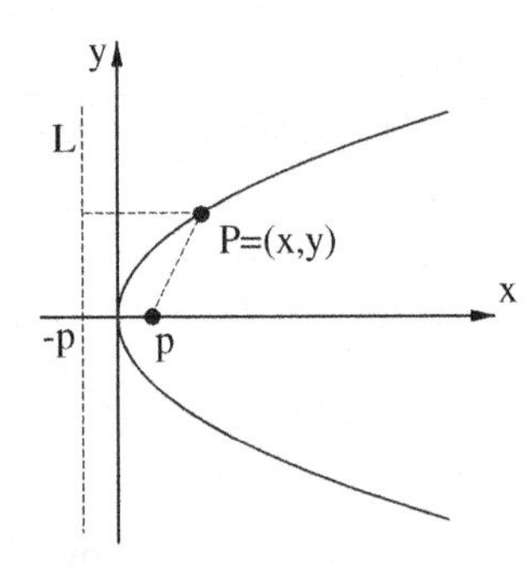

Figure C.3: a parabola in standard position.

the distance from any point $P = (x, y)$ on the parabola to F is equal to the distance from P to L, we obtain

$$\sqrt{(x-p)^2 + y^2} = x + p,$$

or equivalently

$$(x-p)^2 + y^2 = (x+p)^2.$$

Separating the variables in this equation, we arrive at the following defining equation for a parabola in the standard position shown in Figure C.3:

$$\boxed{y^2 = 4px.}$$

C.3 Exercise. Find the defining equation of a parabola in standard position that passes through the point $(5, 2)$.

To conclude our discussion, we will now discuss the case of a hyperbola in the standard position shown in Figure C.4 with foci at $(\pm c, 0)$ and x-intercepts at $(\pm a, 0)$. Since the absolute value of the

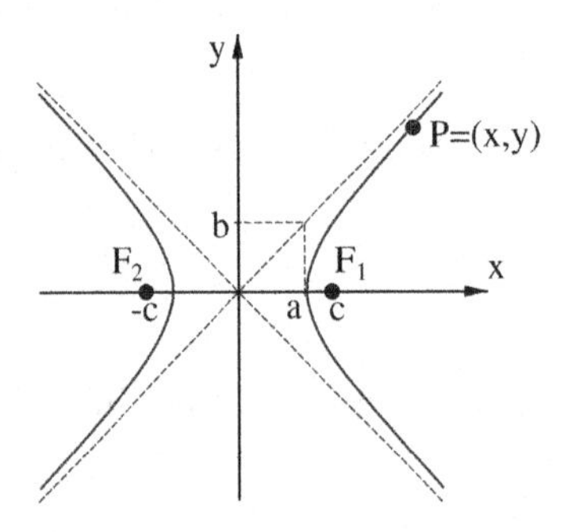

Figure C.4: a hyperbola in standard position.

difference of the distances from $(0, a)$ to the two foci is $|c - a - (c + a)| = 2a$, the definition of the hyperbola as stated in c) above implies that the absolute value of the difference of the distances from any point $P = (x, y)$ on the hyperbola to the two foci is equal to $2a$ as well. Consequently,

$$\left|\sqrt{(x-c)^2 + y^2} - \sqrt{(x+c)^2 + y^2}\right| = 2a, \tag{C.2}$$

and using elementary algebra, it is not difficult to show that this equation is equivalent to the equation

$$\frac{x^2}{a^2} - \frac{y^2}{c^2 - a^2} = 1. \tag{C.3}$$

C.4 Exercise. Verify that (C.2) and (C.3) are indeed equivalent.

In order to determine the relationship between the constants a, b, and c (see Figure C.4), we solve (C.3) for y:

$$y = \pm\frac{\sqrt{c^2 - a^2}}{a}\sqrt{x^2 - a^2}.$$

Dividing both sides by $x\sqrt{c^2 - a^2}/a$ and taking the limit as x tends to ∞ yields

$$\lim_{x\to\infty}\frac{y}{x\sqrt{c^2-a^2}/a} = \lim_{x\to\infty}\pm\frac{\sqrt{x^2-a^2}}{x} = \lim_{x\to\infty}\pm\sqrt{1-\frac{a^2}{x^2}} = \pm 1.$$

This calculation shows that the lines given by the equations $y = x\sqrt{c^2 - a^2}/a$ and $y = -x\sqrt{c^2 - a^2}/a$ are diagonal asymptotes to the hyperbola described by equation (C.3). Since the slope of these asymptotes in Figure C.4 is $\pm b/a$, it follows that

$$\frac{b}{a} = \frac{\sqrt{c^2 - a^2}}{a},$$

or equivalently

$$\boxed{b^2 = c^2 - a^2.}$$

Substituting this result back into (C.3), the defining equation for a hyperbola in the standard position shown in Figure C.4 can be written as follows:

$$\boxed{\frac{x^2}{a^2} - \frac{y^2}{b^2} = 1.} \tag{C.4}$$

C.5 Exercise. Find the defining equation for a hyperbola that passes through the point $(1, 3)$ and has a focus at $(1, 0)$.

If we interchange the positions of x and y in equation (C.4), then the resulting equation

$$\frac{y^2}{a^2} - \frac{x^2}{b^2} = 1$$

describes a *vertical* hyperbola with foci at positions $(0, c)$ and $(0, -c)$ on the y-axis (see Figure C.5). (Note: the hyperbola in Figure C.4 is naturally also referred to as a *horizontal* hyperbola.)

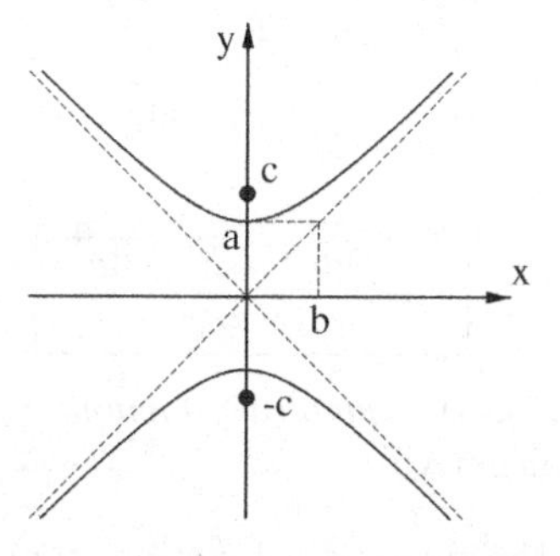

Figure C.5: a vertical hyperbola.

Remark. The name *conic section* is derived from the fact that any ellipse, parabola, or hyperbola can be represented by a curve of intersection of a double cone with a plane. The reader with an interest in this subject will find this and many other interesting facts about conic sections explained in standard texts on analytic geometry (see, for example, [Mu]).

Appendix D

Some Topics from Algebra

A frequently employed technique in algebraic manipulations is the factoring method known as *completing the square*. Given a quadratic function $f(x) = ax^2 + bx + c$ with $a \neq 0$, this technique works as follows:

$$\begin{aligned} ax^2 + bx + c &= a\left(x^2 + \frac{b}{a}x\right) + c = a\left(x^2 + 2\frac{b}{2a}x + \left(\frac{b}{2a}\right)^2 - \left(\frac{b}{2a}\right)^2\right) + c \\ &= a\left(\left(x + \frac{b}{2a}\right)^2 - \left(\frac{b}{2a}\right)^2\right) + c = a\left(x + \frac{b}{2a}\right)^2 - \frac{b^2}{4a} + c. \end{aligned} \tag{D.1}$$

This alternative representation of the quadratic term $ax^2 + bx + c$ allows us to infer that for $a > 0$ the function f assumes a global minimum at $x = -b/2a$ because for all $x \in \mathbb{R}$ we have $a(x + b/2a)^2 \geq 0$ with equality occurring at $x = -b/2a$. For $a < 0$, the function f also assumes a global extremum at $x = -b/2a$, but in this case the extremum is a maximum because for all $x \in \mathbb{R}$ we have $a(x + b/2a)^2 \leq 0$. Furthermore, the representation in (D.1) also allows us to determine the roots of f because in solving the equation

$$a\left(x + \frac{b}{2a}\right)^2 - \frac{b^2}{4a} + c = 0$$

for x, we find

$$\left(x + \frac{b}{2a}\right)^2 = \frac{b^2}{4a^2} - \frac{c}{a}$$

or, equivalently,

$$\boxed{x = \frac{-b \pm \sqrt{b^2 - 4ac}}{2a}.} \tag{D.2}$$

This result is commonly referred to as the *quadratic formula.* Note: the roots of f are complex (see Chapter 45) if the term $b^2 - 4ac$ is negative.

D.1 Exercise. Find the roots of the quadratic function $f(x) := 4x^2 - 3x - 5$, and determine the location of the global minimum of f.

Given the formula in (D.2), any quadratic function can be written as a product of two linear factors (see also Chapter 45) in the following way:

$$ax^2 + bx + c = a\left(x - \frac{-b + \sqrt{b^2 - 4ac}}{2a}\right)\left(x - \frac{-b - \sqrt{b^2 - 4ac}}{2a}\right).$$

D.2 Exercise. Write the quadratic function in Exercise D.1 as a product of two linear factors.

For reference purposes we conclude this appendix with a list of binomial factoring formulae.

Formulae for exponents 2 and 3:

$$\begin{array}{rcl} (a+b)^2 = a^2 + 2ab + b^2 & \quad & (a+b)^3 = a^3 + 3a^2b + 3ab^2 + b^3 \\ (a-b)^2 = a^2 - 2ab + b^2 & & (a-b)^3 = a^3 - 3a^2b + 3ab^2 - b^3 \\ a^2 - b^2 = (a-b)(a+b) & & a^3 - b^3 = (a-b)(a^2 + ab + b^2) \end{array}$$

Formulae for arbitrary integer exponents:

$$(a+b)^n = \sum_{k=0}^{n} \binom{n}{k} a^k b^{n-k} \text{ where } \binom{n}{k} = \frac{n!}{k!(n-k)!} \text{ and } a^n - b^n = (a-b) \sum_{k=0}^{n-1} a^k b^{n-1-k}.$$

Appendix E

★ Mathematical Induction

Given an infinite sequence of statements $A_1, A_2, \ldots, A_n, \ldots$ the *principle of mathematical induction* asserts that we may regard A_n to be valid for all positive integers n if

a) A_1 is true and

b) the validity of A_n for a given positive integer n implies the validity of A_{n+1}.

Intuitively, it is not at all difficult to understand that these two conditions do indeed guarantee the validity of A_n for all positive integers n. If, as assumed in a), the statement A_1 is true, then, according to b), the statement $A_{1+1} = A_2$ must be true as well. Therefore, another application of b) shows that also $A_{2+1} = A_3$ is true, and continuing in this way it follows that A_n is true for all positive integers n.

Remark. This argument makes the principle of mathematical induction appear plausible, but it does not amount to a proof. In fact, in rigorous treatments of the theory of numbers, the principle of mathematical induction is usually adopted as an axiom, and as such, in essence, accepted by faith.

E.1 Example. We wish to use the principle of mathematical induction to prove that for all positive integers n we have

$$\sum_{k=1}^{n} k = \frac{n(n+1)}{2}. \tag{E.1}$$

For a given n we denote by A_n the statement that equation (E.1) is valid for this particular value of n. Then A_1 is true because

$$\sum_{k=1}^{1} k = 1 = \frac{1 \cdot (1+1)}{2}.$$

Given this observation, the principle of mathematical induction will allow us to infer the validity of A_n for all positive integers n if we can show that the validity of A_n implies the validity of A_{n+1}. In other words, from the assumption that equation (E.1) is true for one particular n we need to be able to deduce that

$$\sum_{k=1}^{n+1} k = \frac{(n+1)((n+1)+1)}{2} = \frac{(n+1)(n+2)}{2}.$$

This, however, is rather simple because if (E.1) is true for a given n, then

$$\sum_{k=1}^{n+1} k = n+1+\sum_{k=1}^{n} k = n+1+\frac{n(n+1)}{2} = \frac{(n+1)(n+2)}{2}$$

as desired.

E.2 Exercise. Use mathematical induction to prove that $\sum_{k=1}^{n} k^2 = n(n+1)(2n+1)/6$ for all positive integers n.

E.3 Example. For $a, b \in \mathbb{R}$ we wish to use mathematical induction to prove the *binomial formula:*

$$(a+b)^n = \sum_{k=0}^{n} \binom{n}{k} a^k b^{n-k} \text{ for all positive integers } n. \tag{E.2}$$

Since $\binom{n}{k} = n!/(k!(n-k)!)$ (see Chapter 34), it follows that the binomial formula is valid for $n = 1$ because

$$\sum_{k=0}^{1} \binom{1}{k} a^k b^{1-k} = \binom{1}{0} a^0 b^1 + \binom{1}{1} a^1 b^0 = b + a = (a+b)^1.$$

Thus, according to the principle of mathematical induction, we only need to show that the binomial formula is valid for $n+1$ whenever it is valid for n. So let us assume that for a given positive integer n we have

$$(a+b)^n = \sum_{k=0}^{n} \binom{n}{k} a^k b^{n-k}.$$

Then

$$\begin{aligned}
(a+b)^{n+1} &= (a+b)(a+b)^n = (a+b) \sum_{k=0}^{n} \binom{n}{k} a^k b^{n-k} \\
&= \sum_{k=0}^{n} \binom{n}{k} a^{k+1} b^{n-k} + \sum_{k=0}^{n} \binom{n}{k} a^k b^{n+1-k} \\
&= \sum_{k=1}^{n+1} \binom{n}{k-1} a^k b^{n-(k-1)} + \sum_{k=0}^{n} \binom{n}{k} a^k b^{n+1-k} \qquad \text{(E.3)} \\
&= \binom{n}{n} a^{n+1} b^0 + \sum_{k=1}^{n} \left(\binom{n}{k-1} + \binom{n}{k} \right) a^k b^{n+1-k} + \binom{n}{0} a^0 b^{n+1} \\
&= \binom{n+1}{n+1} a^{n+1} b^0 + \sum_{k=1}^{n} \left(\binom{n}{k-1} + \binom{n}{k} \right) a^k b^{n+1-k} + \binom{n+1}{0} a^0 b^{n+1},
\end{aligned}$$

because $\binom{n}{n} = \binom{n}{0} = \binom{n+1}{n+1} = \binom{n+1}{0} = 1$. Furthermore,

$$\binom{n}{k-1} + \binom{n}{k} = \frac{n!}{(k-1)!(n+1-k)!} + \frac{n!}{k!(n-k)!} = \frac{n!(k+n+1-k)}{k!(n+1-k)!} = \frac{(n+1)!}{k!(n+1-k)!} = \binom{n+1}{k}.$$

Substituting for the corresponding expression in (E.3) yields

$$(a+b)^{n+1} = \binom{n+1}{n+1} a^{n+1} b^0 + \sum_{k=1}^{n} \binom{n+1}{k} a^k b^{n+1-k} + \binom{n+1}{0} a^0 b^{n+1} = \sum_{k=0}^{n+1} \binom{n+1}{k} a^k b^{n+1-k}$$

as desired.

Readers who are interested in a more detailed treatment of mathematical induction and its foundations in formal logic are encouraged to consult standard introductions to abstract mathematics such as [Jo].

Notes and Credits

Chapter 5. References for the exposition on the life and work of René Descartes are [Co2], [De1], [De2], [De3], [Ho], [KBW], [Kuh], and [Lav]. The information on Fermat is derived from [Ho] and [Si].

Quotations:

1. [Kuh], p.194
2. [De1], p.42
3. [De1], p.42
4. [De1], p.43
5. [De1], p.44
6. [Lav], p.87
7. [De1], p.42
8. [De3], p.46
9. [De2], p.78
10. [De1], p.295
11. [Si], p.33
12. [Ho], p.142
13. [Ho], p.142
14. The quote is attributed to Etienne Gilson and can be found in [Lav], p.89.

Chapter 9. The essential facts on Newton's life and work, except for the derivation of the universal law of gravitation, are based on the expositions in [Ant], [Ch], and [Ho].

Quotations:

1. [Ang], p.178
2. [Ch], p.1
3. [Ch], p.13
4. [Be], p.37
5. [Ant], p.23
6. [Ch], p.19
7. [Ch], p.54
8. [Ch], p.62
9. [Ch], p.62
10. [Ch], pp.73–74
11. [Ant], p.68

12. [Be], p.47
13. [Ho], p.179
14. [Ho], p.178
15. [Nei], p.28

Chapter 12. The technical information on CD technology is derived from the expositions in [Blo] and [AB]. The estimate for the storage capacity of a CLV disc can be found in [AB].

Chapter 15. Many key insights have been adopted from [Dr], [Kuh], and [Sob]. Furthermore, the footnote on p.131 is motivated by a comment in [Ta] on p.170.

Quotations:

1. [Ar1], p.377
2. [Ar1], p.387
3. [Sob], p.17
4. [Dr], pp.25–26
5. [Sob], p.22
6. [Ga], p.238
7. From Galileo's *Dialogue Concerning the Two Chief Systems of the World* as quoted in [Kö2], pp.137–138.
8. [Sob], p.66
9. Psalm 104:5, the *New International Version of the Bible*
10. [Sob], p.60
11. [Sob], p.63
12. [Stö], p.327
13. [Kuh], p.191
14. [Kuh], p.191
15. [Sob], p.78
16. [Sob], p.78
17. [Sob], p.78
18. [Sob], p.274
19. [Sob], p.275
20. [Sob], p.278
21. [Sob], p.278

Chapter 18. References for the discussion of Pythagorean philosophy are the expositions in [Co1], [Kl2], [Rob], and [Sti]. The remarks regarding the Neo-Pythagorean beliefs of Johannes Kepler are based on the discussion in [Kor]. A passage from the Copernican treatise *On the Revolutions of the Heavenly Spheres* that refers to Pythagorean cosmology can be found in [Kuh], p.149. Furthermore, the biographical notes on the life of Bernhard Riemann are largely based on the exposition in [Sti].

Quotations:

1. [Ar2], pp.503–504
2. [Sti], pp.216–217

3. [Ang], p.205

Chapter 20. The central ideas of the discussion on Nicolaus Cusanus are taken from [KBW] and [Stö]. The information on the life and philosophy of Leibniz is derived from [KBW] and [Stö] and, to a lesser extent, from [Co2] and [Rus]. Furthermore, the discussion on Leibniz's stepwise creation of the calculus is based on the exposition in [Kl1].

Chapter 25. The essential steps in the calculation of the volume of a honeycomb including Figures 25.3, 25.4, and 25.5 as well as part of Figure 25.1 have been adopted from the exposition in [Per]. The reference to the discovery by L. Fejes Toth can be found in [Hal], p447.

Quotation:

1. [Hal], p.447

Chapter 28. All the standard facts concerning the physics of a rainbow are derived from [Ja] and [HH].

Quotation:

1. Genesis 9:12-15, the *New International Version of the Bible.*

Chapter 33. The exposition in [BO] served as a reference for the discussion on sky diving.

Chapter 47. The images in Figures 47.1 (without the spirals indicated), 47.2 (without the spirals indicated), and 47.7 were provided by Dreamstime. Figure 47.1: © Raja Rc | Dreamstime.com; Figure 47.2: © Shaffandi | Dreamstime.com; Figure 47.7: © Christian Slanec | Dreamstime.com.

Chapter 49. Example 49.16 has been adopted with permission from the exposition in [WB], Example 6, p.653.

Chapter 51. Examples 51.2 and 51.3 can be found in [WB], Example 7, p.135 and Example 8, p.138. The material has been used with permission.

Bibliography

[AB] Ackerson, Bertholf, Choike, Stanley and Wolfe, *Red and Blue Laser CD's: How much Data can they Hold?*, COMAP, 1996.

[Ang] Anglin, W.S., *Mathematics: A Concise History and Philosophy*, Springer-Verlag, New York, 1994.

[Ant] Anthony, H. D., *Sir Isaac Newton*, Collier Books, New York, 1961.

[Ar1] Aristotle, *On the Heavens*, in *Great Books of the Western World*, Volume 8 by Robert Maynard Hutchins (Editor in Chief), Encyclopedia Britannica, Inc., Chicago, 1952.

[Ar2] Aristotle, *Metaphysics*, in *Great Books of the Western World*, Volume 8 by Robert Maynard Hutchins (Editor in Chief), Encyclopedia Britannica, Inc., Chicago, 1952.

[Be] Bell, Eric Temple, *On the Seashore*, in *Readings for Calculus* by Underwood Dudley (Editor), MAA Notes Number 31, The Mathematical Association of America, Washington, 1993.

[BL] Baxandall, P., Liebeck, H., *Vector Calculus*, Clarendon Press, Oxford, 1986.

[Bla] Blatter, Christian, *Analysis III*, Heidelberger Taschenbücher, Band 153, Springer-Verlag, Berlin, 1970.

[Blo] Bloomfield, Louis A., *How Things Work*, John Wiley and Sons, New York, 1996.

[Blu] Blume, Frank, *A Localized Probabilistic Approach to Circuit Analysis*, The American Mathematical Monthly, Volume 110, Number 10, pp.928–936.

[BO] Barger, V. and Olsson, M., *Classical Mechanics a Modern Perspective*, McGraw Hill Book Company, New York, 1973.

[Ca] Callahan, James J., *The Geometry of Spacetime*, Springer-Verlag, New York, 2000.

[Ch] Christianson, Gale E., *In the Presence of the Creator*, The Free Press, Macmillan, New York, 1984.

[Co1] Copleston, Frederick S.J., *A History of Philosophy*, Volume 1, Image Books, Doubleday, New York, 1993.

[Co2] Copleston, Frederick S.J., *A History of Philosophy*, Volume 4, Image Books, Doubleday, New York, 1993.

[Cod] Coddington, E. A., *Introduction to Ordinary Differential Equations*, Englewood Cliffs, Prentice-Hall, New Jersey, 1961.

[De1] Descartes, René, *Discourse on the Method of Rightly Conducting the Reason and Seeking for Truth in the Sciences*, in *Great Books of the Western World*, Volume 31 by Robert Maynard Hutchins (Editor in Chief), Encyclopedia Britannica, Inc., Chicago, 1952.

[De2] Descartes, René, *Meditations on the First Philosophy*, in *Great Books of the Western World*, Volume 31 by Robert Maynard Hutchins (Editor in Chief), Encyclopedia Britannica, Inc., Chicago, 1952.

[De3] Descartes, René, *Meditations*, in *The Library of Original Sources*, Volume VI by Oliver J. Thatcher (Editor in Chief), University Research Extension Company, Milwaukee, 1915.

[Dr] Drake, Stillman, *Galileo*, Hill and Wang, New York, 1980.

[DS] Dangello, Frank and Seyfried, Michael, *Introductory Real Analysis*, Houghton Mifflin Company, Boston, 2000.

[EI] Einstein, Albert and Infeld, Leopold, *The Evolution of Physics*, Simon and Schuster, New York, 1938.

[El] Elaydi, Saber N., *An Introduction to Difference Equations*, Undergraduate Texts in Mathematics, Springer-Verlag, New York, 1995.

[Er] Erwe, Friedhelm, *Differential- und Integralrechnung*, Zweiter Band, Bibliographisches Institut, Mannheim, 1962.

[Fe] Feynman, Richard P., *QED*, Princeton University Press, Princeton, New Jersey, 1985.

[Ga] Galileo Galilei, *The Two New Sciences*, in *Great Books of the Western World*, Volume 28 by Robert Maynard Hutchins (Editor in Chief), Encyclopedia Britannica, Inc., Chicago, 1952.

[Gol] Goldblatt, Robert, *Lectures on the Hyperreals*, Springer-Verlag, New York, 1998.

[Gou] Gould, Stephen Jay, *The Structure of Evolutionary Theory*, Harvard University Press, Cambridge, Massachusetts, 2002.

[Hal] Hales, Thomas C., *Cannonballs and Honeycombs*, Notices of the AMS, Volume 47, Number 4, April 2000.

[Ham] Hamming, Richard W., *The Art of Probability for Scientists and Engineers*, Perseus Books, Cambridge, Massachusetts, 1991.

[Har] Hartman, Philip, *Ordinary Differential Equations*, John Wiley & Sons, Inc., New York, 1964.

[HH] Hall, Rachel W., Higson, Nigel, *The Calculus of the Rainbow*, http://www.math.psu.edu/hall/newton/raindrop.pdf, 1998.

[Ho] Hollingdale, Stuart, *Makers of Mathematics*, Penguin Books, London, 1989.

[Ja] Janke, Steven, *Somewhere Within the Rainbow*, in *Applications of Calculus* by Phillip Straffin (Editor), MAA Notes Number 29, The Mathematical Association of America, Washington, 1993.

[Jo] Johnson, D.L., *Elements of Logic via Numbers and Sets*, Springer-Verlag, New York, 1998

[JP] Jacod, Jean, Protter, Philip, *Probability Essentials*, Universitext, Springer-Verlag, Berlin, 2000.

[KBW] Kunzmann, Peter, Burkard, Franz-Peter, Wiedmann, Franz, *dtv-Atlas Philosophie*, Deutscher Taschenbuch Verlag, München, 1991.

[Kl1] Kline, Morris, *The Creation of the Calculus*, in *Readings for Calculus* by Underwood Dudley (Editor), MAA Notes Number 31, The Mathematical Association of America, Washington, 1993.

[Kl2] Kline, Morris, *Mathematical Thought from Ancient to Modern Times*, Oxford University Press, New York, 1972.

[Kn] Kneller, Karl A., *Christianity and the Leaders of Modern Science*, Real-View-Books, Fraser, Michigan, 1995.

[Kö1] Körner, T. W., *Fourier Analysis*, Cambridge University Press, Cambridge, 1995.

[Kö2] Körner, T. W., *The Pleasures of Counting*, Cambridge University Press, Cambridge, 1998.

[Kor] Kors, Alan, *Galileo and the New Astronomy* in *Great Minds of the Western Intellectual Tradition*, Part III, Lecture 32 of *The Great Courses on Tape* series, The Teaching Company, 2000.

[Kuh] Kuhn, Thomas S., *The Copernican Revolution*, Harvard University Press, Cambridge, Massachusetts, 1999.

[Lan] Lancaster, P., *Theory of Matrices*, Academic Press, New York, 1969.

[Lav] Lavine, T.Z., *From Socrates to Sartre: the Philosophic Quest*, Bantam Books, New York, 1984.

[Lo] Lorenz, Konrad, *Behind the Mirror*, Harvest/HJB, New York, 1978

[Ma] Marsden, Jerrold E., *Elementary Classical Analysis*, W. H. Freeman and Company, New York, 1974.

[Mu] Murdoch, D.C., *Analytic Geometry with an Introduction to Vectors and Matrices*, John Wiley and Sons, New York, 1966.

[Nei] Neiman, Susan, *Evil in Modern Thought*, Princeton University Press, Princeton, New Jersey, 2002.

[New] Newman, James R., *The World of Mathematics*, Volume 1, Simon and Schuster, New York, 1956.

[Per] Peressini, Anthony L., *The Design of Honeycombs*, COMAP, 1980.

[Pet] Petersen, K.E., *Ergodic Theory*, Cambridge University Press, Cambridge, 1983.

[Ra] Rabenstein, Albert L., *Elementary Differential Equations with Linear Algebra*, Saunder, Harcourt Brace Jovanovich, Fort Worth, 1992.

[Rob] Robinson, Daniel N., *The Great Ideas of Philosophy*, Part I, Lecture 3, in *The Great Courses on Tape* series, The Teaching Company, 1997.

[Ros] Ross, Kenneth A., *Elementary Analysis: The Theory of Calculus*, Springer-Verlag, New York, 1980.

[Rud1] Rudin, Walter, *Principles of Mathematical Analysis*, 3rd edition, McGraw Hill Book Company, Auckland, 1976.

[Rud2] Rudin, Walter, *Real and Complex Analysis*, McGraw Hill Book Company, New York, 1987.

[Rus] Russell, Bertrand, *A History of Western Philosophy*, Simon and Schuster, New York, 1945.

[Si] Simmons, George F., *Fermat (1601-1665)*, in *Readings for Calculus* by Underwood Dudley (Editor), MAA Notes Number 31, The Mathematical Association of America, Washington, 1993.

[Sob] Sobel, Dava, *Galileo's Daughter*, Penguin Books, New York, 2000.

[Som1] Sommerfeld, Arnold, *Mechanics*, Lectures on Theoretical Physics, Volume 1, Academic Press, New York, 1964.

[Som2] Sommerfeld, Arnold, *Elektrodynamik*, Vorlesungen über Theoretische Physik, Band 3, Verlag Harri Deutsch, Frankfurt/M., 1977.

[Ste] Steiner, Mark, *The Applicability of Mathematics as a Philosophical Problem*, Harvard University Press, Cambridge, Massachusetts, 1998.

[Sti] Stillwell, John, *Mathematics and Its History*, Springer-Verlag, New York, 1989.

[Stö] Störig, Hans Joachim, *Kleine Weltgeschichte der Philosophie*, Fischer Taschenbuch Verlag, Stuttgart, 1999.

[Ta] Tannery, Paul, *Galileo and the Principle of Dynamics*, in *Galileo, Man of Science* by Ernan McMullin (Editor), Basic Books, Inc., New York, 1967.

[Un] Underwood, Dudley, (Editor), *Readings for Calculus*, MAA Notes Number 31, The Mathematical Association of America, Washington, 1993.

[Wa] Walter, W., *Gewöhnliche Differentialgleichungen*, Springer-Verlag, Berlin, 1976.

[WB] Wylie, C. Ray, Barrett, Louis C., *Advanced Engineering Mathematics*, McGraw-Hill, New York, 1995.

[Wr] Wright, Donald J., *Introduction to Linear Algebra*, McGraw-Hill, New York, 1999.

Index

A

B

C

D

E

F

G

H

I

J

K

L

M

N

O

P

Q

R

S

T

V

W